PALAEOHISTORIA

PALAEOHISTORIA

*ACTA ET COMMUNICATIONES
INSTITUTI BIO-ARCHAEOLOGICI
UNIVERSITATIS GRONINGANAE*

33/34

1991/1992

A.A.BALKEMA/ROTTERDAM/BROOKFIELD/1994

Editorial staff: Mette Bierma, J.W. Boersma, J.N. Lanting & Miriam Weijns
Address: Biologisch-Archaeologisch Instituut, Poststraat 6, 9712 ER Groningen, Netherlands

Published by

A.A. Balkema, P.O. Box 1675, 3000 BR Rotterdam, Netherlands

A.A. Balkema Publishers, Old Post Road, Brookfield, VT 05036, USA

ISSN 0552-9344

ISBN 90 5410 188 1
Printed in the Netherlands

CONTENTS

NOTES ON FOSSIL VERTEBRATES AND STONE TOOLS FROM SULAWESI, INDONESIA, AND THE STRATIGRAPHY OF THE NORTHERN WALANAE DEPRESSION

G.-J. BARTSTRA
Biologisch-Archaeologisch Instituut, Groningen, Netherlands

D.A. HOOIJER
Aert van Neslaan 101, Oegstgeest, Netherlands

B. KALLUPA
Suaka Peninggalan Sejarah dan Purbakala
Kompleks Benteng, Ujung Pandang, Sulawesi Selatan, Indonesia

M. ANWAR AKIB
Museum Prasejarah Calio, Beru, Cabenge, Sulawesi Selatan, Indonesia

ABSTRACT: These notes intend to clarify some details concerning the palaeontological and archaeological investigations in the northern Walanae depression in Sulawesi. The question of the age of the fossils and artifacts is elucidated, and it is argued that the fossils are older than the tools. Furthermore, information on the general geological framework is given, substantiating the view that the vertebrate remains erode from the upper part of the Walanae Formation. Also, four interesting fossil vertebrate finds are presented.

KEYWORDS: Southeast Asia, Sulawesi, Walanae depression, Walanae Formation, Walanae terraces, fossil vertebrate localities, *Archidiskodon-Celebochoerus* fauna, Palaeolithic Cabenge industry.

1. INTRODUCTION

The Plio-Pleistocene of the island of Sulawesi, formerly Celebes, has won a certain fame with its vertebrate fossils of the *Archidiskodon-Celebochoerus* fauna and its stone implements of the Cabenge industry.[1] In this paper we delve into the history of the research, and into the general Neogene stratigraphy of the area with the most promising fossil localities: the Walanae depression in the southwestern peninsula. This paper might be regarded as a sequel in a recent series on Sulawesi. The first one summarized our hypothesis that contrary to general belief the fossils and artifacts in the Walanae area are not equal or near-equal in age, and that the Cabenge industry could be associated with early *Homo sapiens* (Bartstra et al., 1991). A second paper dealt with newly acquired fossil remains, especially mentioning a stegodont molar from the northwestern part of Central Sulawesi (Bartstra & Hooijer, 1992). A third paper focuses on the artifacts of the Cabenge industry, and is also presented in this volume (Keates & Bartstra).

Figure 1 explains the geographic situation, with the area of interest. This prime region of the various finds lies directly south of a marked topographic depression (the so-called Singkang embayment (Beltz, 1944) or Tempe depression (van Bemmelen, 1949)) which separates the southwestern peninsula of Sulawesi from the central part of the island and which extends from the mouth of the River Sadang to the mouth of the River Cenrana (the shaded area in fig. 1). Until recently this depression was covered by the sea; the three present lakes are a vestige of this situation. The largest is Lake Tempe, and the River Walanae debouches herein. The drainage area of the Walanae, which stretches far to the south and is bordered by mountain ranges, is here referred to as the Walanae depression.

2. THE NOTION OF NON-CONTEMPORANEITY

One should approach the hazy shores of Sulawesi from the west, via the Makassar Strait. Sailing this sea, one can imagine being part of the first groups of *Homo sapiens* that travelled from Sundaland[2] eastwards and which saw Sulawesi appear as the new frontier. There is yet another reason to favour this western approach: one will be sailing renowned waters. In 1860 the great naturalist Wallace (the one who almost beat Darwin) expressed his idea of a zoogeographical boundary straight through the archipelago of Southeast Asia. To the west of Wallace's Line[3] there was an Asiatic animal world with highly developed mammals; to the east of it an Australian one with primitive types like the duck-billed platypus and marsupials. According to Wallace this boundary coincided with *inter alia* the deep Makassar Strait, which would mean that the fauna of Sulawesi should be totally different from that of Java or Borneo (Kalimantan).

However, when around the turn of the century the exploration of the interior of Sulawesi slowly got under way, it became clear that animals were living there of clearly Asiatic origin, like monkeys, buffaloes, and pigs. It was therefore assumed that land-bridges had existed, which had from time to time connected Sulawesi with mainland Asia, and across which animals had been able to migrate. The Sarasins (1901, 1905), for example, concluded that such early invasions had occurred, on the basis of the existence of archaic forms in the present-day fauna. The Sarasins were already undermining Wallace's Line by proposing that the fauna of Sulawesi is a mixed fauna, with a predominantly Asiatic character. In their opinion there never was a land-bridge to Borneo, but most probably there was one linking Sulawesi to Asia via the Philippines. This idea remained hypothetical, however, as during their expeditions (from 1893-1896 and from 1902-1903) the Sarasins did not find any fossils that could have provided them with supporting evidence. It was not until much later that such fossils came to light.

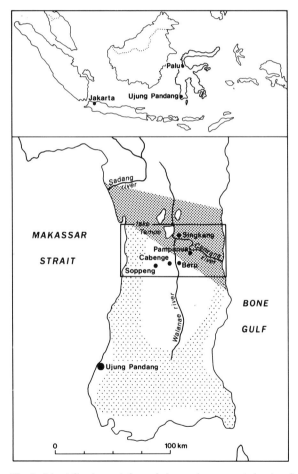

Fig. 1. Island Southeast Asia, and the southwestern peninsula of Sulawesi with geographic features and names referred to in the text. The shaded area (schematic) is the so-called Tempe depression. The dotted area (idem) indicates the Neogene island. The area within the rectangle is the main region of artifact sites and fossil localities.

In the summer of 1970 a dream came true for van Heekeren, a Dutch prehistorian and former employee with the Indonesian Archaeological Service. As organizer and co-leader of a scientific team, he was able to return to the area of his most cherished discoveries. In the hilly country east of the small town of Cabenge in South Sulawesi, in the area of the great Walanae river (figs 1 and 2), he had found in the years 1947-1950 fossil fragments of vertebrates together with heavily patinated flakes and cores of a presumably Palaeolithic stone industry (van Heekeren, 1949a; 1949b; 1949c).

Already in 1946 van Heekeren had started looking for fossils and implements on remnants of raised beaches along the Makassar Strait and Bone Gulf. He found them a year later in the interior of South Sulawesi (Bartstra, 1993). The fossil vertebrate fragments were brought to the Netherlands, where they were identified and described in a long succession of papers by one of us (D.A.H.), in those far-off days the newly appointed curator of the Dubois collection of fossil vertebrates at the Museum of Natural History at Leiden.[4] The first paper on the fossil fauna of Sulawesi appeared in 1948 and gave details of a giant suoid, *Celebochoerus heekereni*, with which species designation the finder of the new material was uniquely honoured. In the course of further publications, in which also elephantoid bones were described, the new fauna of southern Sulawesi became well-known under the name: *Archidiskodon-Celebochoerus* fauna (for summaries see: Hooijer, 1949b; 1960; 1975).

Meanwhile the political instability in Sulawesi and the worsening relations between the young republic of Indonesia and the Netherlands had made it impossible for van Heekeren to continue his research. Twenty long years elapsed, but at last in 1970 Uncle Bob (as van Heekeren was known to his friends) returned into the field for a prolonged period.[5]

The 1970 Joint Indonesian-Dutch Sulawesi Prehistoric Expedition was financed by WOTRO, the Netherlands Foundation for the Advancement of Tropical Research. In a final proposal to (presumably) WOTRO, written in January 1970, van Heekeren's ideas become quite clear.[6] He developed a working hypothesis for the field: the fossils and artifacts to the east of Cabenge are equally old and date from the Pleistocene. According to him, there is a good chance that *Homo erectus* (*Pithecanthropus erectus*) is to be found in the area. This latter possibility was van Heekeren's true dream, cherished from the days of the first finds, as is apparent from the small molars and molar fragments that he continually picked up, in the hope of finding *Pithecanthropus* at last, but which invariably turned out to come from suoids (Bartstra & Hooijer, 1992).

The members of the 1970 expedition were housed in the village of Beru, directly east of Cabenge (fig. 2), and for more than six weeks (in June, July and August) the surroundings were thoroughly explored, all on foot. We know, for three of us were members of the team: two (G.J.B. and B.K.) at that time still mere students of prehistory, but the third (D.A.H.) by then a well-known curator of fossil vertebrates at the Leiden museum.[7] The fourth author (M.A.A.) was a schoolboy then, often accompanying us on our trips. The geomorphology and stratigraphy of the Beru region became very familiar, and many artifacts and fossils were collected, from the surface as well as from excavations. In re-reading our fieldnotes from that period, we vividly remember the discussions on the high verandah of our home in the hot and lazy afternoons, van Heekeren clad in sarong and every now and then extinguishing his half-finished cigarettes. The discussions centred on the contemporaneity of artifacts and fossils, and, of course, on *Pithecanthropus erectus.*

In 1974 van Heekeren died unexpectedly. He was working on a monograph concerning the Sulawesi expedition, but only a few notes and the introductory pages of this manuscript seem to have survived. These became available to us some five years ago when van Heekeren's sons emptied a large cabin trunk in the attic of his former house. It is interesting to know what van Heekeren thought about his working hypothesis after the 1970 expedition, nurtured by six weeks of field research and discussions with fellow scientists.[8]

An indication might be found in van Heekeren's ideas on the chronology of the Indonesian prehistory, published posthumously in 1975. The stone implements from the surroundings of Beru, meanwhile officially termed the Palaeolithic Cabenge industry, are dated to the very beginning of the Upper Pleistocene (estimated age between 200 and 100 ka), while the same age is given to the fossil vertebrate remains of South Sulawesi, the *Archidiskodon-Celebochoerus* fauna. Van Heekeren was thus still convinced of a contemporaneity of artifacts and fossils. *Pithecanthropus* is still in the picture, albeit a late one, for in the chronology van Heekeren associates the tools from Sulawesi with *Homo soloensis*[9]; a thought-experiment, of course, for no hominid remains were unearthed during the expedition of 1970.

From the cabin trunk there also emerged an incomplete type-written report, possibly prepared in the fall of 1970 for WOTRO or the Indonesian Archaeological Service.[10] From this it becomes clear why van Heekeren's belief in the contemporaneity of fossils and artifacts could not be shattered. Although he mentions fossil localities around Beru where no artifacts could be traced (for instance Sompe and Celeko, fig. 2), he recollects the second excavation of the 1970 expedition where flake tools and vertebrate remains were found together 'in the very heart of the gravel'. Van Heekeren is referring here to the excavation at Marale (officially termed Beru or Baru II), immediately south of the main village of Beru, in which an abraded *Stegodon* molar had been found on top of a river-laid, gravelly, reddish sand. The superincumbent bed was a coarse, fluviatile gravel, with some worked flakes at the very top. In this report

van Heekeren does consider the possibility that the molar from Marale comes from older deposits, but in his heart he continues to believe in contemporaneity, to be demonstrated in forthcoming 'large-scale excavations'.

The research of 1970 had made it clear that the implementiferous stream gravels near Beru (including this superincumbent Marale gravel), apparently all remnants of a terrace system of the Walanae river, might actually be fairly recent from a geomorphological viewpoint, that is to say Upper Pleistocene at most. Thus in his 1970 fall report and in later publications van Heekeren was forced to take into account the possibility of this relatively recent age. A rather advanced technology exhibited by some of the flakes of the Cabenge industry could only confirm this dating, but van Heekeren's belief in contemporaneity would also imply that the vertebrate fossils of Beru were then of Upper Pleistocene age. Support for this viewpoint could be found in the peculiar status of the *Archidiskodon-Celebochoerus* fauna which had already become clear before 1970: impoverished, endemic, and insular, with full-sized species and dwarf descendants together. Nobody knows how long it takes for full-sized species to dwarf; therefore, an Upper Pleistocene age for some of the fossil remains did not seem improbable to van Heekeren.

The notes and introductory pages, including a table of contents, of the Sulawesi manuscript van Heekeren was working on, and which emerged from the trunk, indicate that the final volume might have been quite substantial. Reading the pages, it can nowhere be surmised that the author had had second thoughts as to the supposed contemporaneity of fossils and artifacts. On the contrary, van Heekeren writes about fossil vertebrates "recovered in association with Palaeolithic tools", and stone implements "always accompanied by remains of the *Archidiskodon-Celebochoerus* fauna". Thus it appears that van Heekeren had not essentially changed his views about the fossils and artifacts of South Sulawesi after the field campaign of 1970; at most he had become more cautious with regard to the age interpretation. As a final check one can consult van Heekeren's book about the Stone Age of Indonesia, published in 1972. This work is in fact a revised and extended version of an earlier publication on the same subject that appeared in 1957. The new text was undoubtedly concluded before the 1970 campaign in Sulawesi, but it should still have been possible to correct the galleys, if new results would have provided reason for doing so. But in the 1972 text it is still stated that the "Palaeolithic artifacts and the fossil vertebrates, or at least the larger part of the latter, are presumably of the same age".

In 1977 one of us (G.J.B.) voiced a different opinion in a paper on the stratigraphy of South Sulawesi: fossils and artifacts are not contemporaneous. The vertebrate remains of the *Archidiskodon-Celebochoerus* fauna occur in situ (autochthonous) in the top-sediment of the

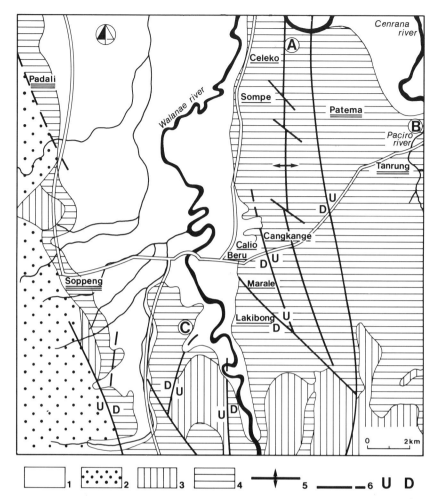

Fig. 2. Sketch map of the main area of interest in the northern Walanae depression, giving geomorphological and stratigraphical information, as well as names referred to in the text (partly after Sjahroel (1970), Sukamto (1975) and Sartono (1979)). 1. Alluvial; 2. Volcanic rock; 3. Limestone; 4. Walanae Formation; 5. Anticline; 6. Fault line; U. Up; D. Down. The volcanic rock, the limestone and the greater part of the sediments of the Walanae Formation are all Tertiary in age (see text). The alluvial deposits are mainly Holocene.

The amplitude of the anticlines to the right of the Walanae diminishes toward the east. Whereas the first anticline (A) is a distinct geomorphological feature, the second one (B) is difficult to recognise in the landscape. The crest of (B) is situated just outside the right edge of the above figure.

The various fossil localities are underlined: those clustered on the first anticline with a single line, those probably located on the western slope of the second anticline with a double line. Triple lines indicate the localities west of the Walanae. The town of Soppeng is not specifically mentioned in the text as a fossil locality. But there seem to exist some minor bone-bearing outcrops in the northern part of this town, probably belonging to the same sediment sequence which is exposed at Padali.

Not drawn on this map are the scattered remnants of riverterrace gravel, all Pleistocene in age, and overlying the Beru member of the Walanae Formation on the west-facing slope of the first anticline. This gravel is implementiferous (Palaeolithic Cabenge industry; Keates & Bartstra, this volume), and is matched by a gravelsheet on the left side of the Walanae, for example near Jampu (C).

The small town of Cabenge, which gave its name to the Palaeolithic Cabenge industry, but is not of particular importance as far as fossil finds are concerned, is also not located on this map (see fig. 1; and Keates & Bartstra, this volume).

bedrock of the region, around Beru consisting of partly consolidated sandstones and conglomerates. The Walanae river has cut into this bedrock, and therefore abraded (allochthonous) fragments of vertebrates are to be found in the Walanae terrace fills. Erosion of the bedrock also occurs where these sandstones and conglomerates outcrop, so that the fossil fragments become scattered over the fields. What Bartstra sees as

being most important, as it has been the source of all the confusion, is an unravelling of the various gravels that can be traced around Beru. These are either true river-terrace gravels, or the residue of conglomeratic bedrock sediment, or a local mixture of the two. Fossils of vertebrates occur in all three gravels. The Palaeolithic artifacts of the Cabenge industry can only be associated with the terrace gravel; it is even possible that they lie only on the surface in the fields around Beru, concentrated in true, prehistoric sites. Bartstra thus emphasizes a distinct difference in age between fossils and artifacts: the vertebrates are Lower Pleistocene or Upper Pliocene, but the implements are definitely Upper Pleistocene, and for a large part maybe even younger. *Pithecanthropus* or *Homo erectus* is not mentioned in the 1977 text, but it is clear that no finds of this particular hominid can be expected around Beru: the faunal remains are too old and the artifacts are too recent.

These ideas put forward in 1977 did not simply come out of the blue. As mentioned above, all four of us took part in the 1970 field trips, and thus we had the opportunity to make our own observations and to take notes. However, our initial approach to the questions of geomorphology and stratigraphy was biased by the postulate of contemporaneity, and it took quite some time before opposite views became established. In 1971 one of us (G.J.B.) published a popular report about the 1970 expedition, in which the contemporaneity of the fossils and artifacts was not yet contested. When another one of us (D.A.H.) published, in 1972, the fossil material that the expedition had yielded, the question of contemporaneity was not discussed. In an article published by Hooijer in 1973 it was still suggested that the extinction of the *Archidiskodon-Celebochoerus* fauna might have been caused by early man, in view of the presence of stone implements.

But the seed of doubt was sown definitively during the six weeks of fieldwork in 1970. With hindsight, it is curious to read in our notes several puzzling entries about the occurrence of fossils and artifacts. Why were artifacts never found in the fossiliferous consolidated sandstones and conglomerates? Why were they found only in loose gravels? Experience acquired in the years directly after 1970, during fieldwork on Java, Flores, and Timor, in which much attention was devoted to the genesis and morphology of river terraces, was very valuable for the ultimate verification of the view that the Walanae terrace gravels should be distinguished from the eroding cemented conglomerates, thus indicating that vertebrate localities and artifact sites around Beru are separate in terms of time and place. Many of our field observations, with which the emerging notion of non-contemporaneity could be tested, were shared with Sjahroel, the geologist of the 1970 expedition. All three of us have on many occasions accompanied Sjahroel on his trips in the field, and we have learnt a lot from him. In the end, our ideas on the local stratigraphic section differed from Sjahroel's (for instance, he saw Plio-Pleistocene beach gravel in many of the coarse clastics that we designated as Upper Pleistocene Walanae high terrace veneer), but on the basic stratigraphy of the area around Beru we all agreed. Thus, at the end of the 1970s the working hypothesis for palaeontological and archaeological fieldwork in the region was precisely the opposite of what it had been at the beginning of the decade.

In 1978 a new expedition was organized to the Beru area in cooperation with the Indonesian Archaeological Service and once again financed by WOTRO. During almost four months of fieldwork (May/June and September/October) with terrain surveys and small-scale excavations, it became possible to confirm the non-contemporaneity of artifacts and fossils, and new archaeological and geological data were acquired (Bartstra, 1978; Sartono, 1979). In 1980 the research around Beru was once more resumed, this time paid for by the University of Groningen. It included borings being made in the sediments of the three lakes in the Tempe depression, the remainders of the former sea connection, and palynological information was provided (Gremmen, 1990). During the eighties nearly every year the area around Beru was briefly visited and ideas about the manufacturers of the Cabenge stone tools became more and more established (Bartstra et al., 1991; Keates & Bartstra, this volume; compare also: Shutler, 1991).

The region to the north of the lakes was also investigated. In 1987 it became possible for the first time to travel by landrover across the thickly forested, central part of Sulawesi, that had previously only been accessible on foot, and a small expedition was organized to the area southeast of the small town of Palu (fig. 1). Our belief that vertebrate fossils were to be expected also in northern Central Sulawesi came true in 1992 with the find of a fragment of a mandibular portion with partial molar of *Stegodon* cf. *trigonocephalus* (Bartstra & Hooijer, 1992). We have recently been informed on more finds of vertebrate remains in this northern niche: evidently a portion of a molar of *Stegodon* sp.. Precise measurements are not yet available.

3. THE IDEA OF AN ANCIENT SHORE

Van Heekeren saw in the hilly landscape around Beru (fig. 2) only heavily dissected terraces, to be associated with the drainage pattern of the River Walanae. In one of his reports he refers to aggradation terraces, with at least five levels.[11] At the third and fourth level, 50 m and 75 m above sea-level, he found the fossils and artifacts, scattered on the surface. According to van Heekeren (1972) one of these levels was bordered 'on both sides' by cemented gravels and sandstones. Here he is probably referring to the scarps, the transitional slopes between the various terrace levels or treads, where consolidated clastics do indeed outcrop. In van Heekeren's view these gravels and sandstones form part of the terrace deposits.

In a paper published in 1977 one of us (G.J.B.) considerably diminished the importance of terrace sediments in the area around Beru. Such sediments do occur, but apart from accumulated clays in the low terrace they are only recognizable in rather thin, loose gravel veneers, laid down on the hill-slopes facing the river. Where a seemingly thicker fill occurs on the higher levels, slope wash must be reckoned with (local alluvium). The consolidated clastics that outcrop in this landscape have nothing to do with terrace sediments. These clastics form the bedrock of the region, pushed up in anticlinal ridges with a north-south orientation (fig. 2), consisting in the oldest parts (the cores of the anticlines) of fine-textured sediments such as clay shales, tuffaceous and calcareous sandstones, and marls, and in the younger parts (the outer layers of the anticlines) of coarser material such as gravelly sandstones and conglomerates. The whole sequence is graded and distinctly coarsening upward, and is therefore indicative of the shallowing of a former sedimentary basin.

A glance through the literature shows that the general geological situation of the area has been well known for some time. It was Wichmann (1890) who first mentioned a conspicuous anticlinal sandstone ridge with marine fossils, situated east of Lake Tempe and extending southwards. This observation is later confirmed by 't Hoen and Ziegler (1917), while they identify more anticlines. They attribute the layered anticlinal rock to what they define as the Walanae or Bone Formation. According to the two authors this is a typical basin fill, of Neogene age. Rutten (1927, 1932) estimates the thickness of this formation to be at least 3000 m. In his opinion folding took place around the transition from Pliocene to Pleistocene, after which 'severe denudation' occurred.

These subsiding basins, where large quantities of sediment could accumulate, are characteristic of the Neogene in many places in island Southeast Asia, when the greater part of the present archipelago was covered by the sea. In southern Sulawesi such a basin fill occurs not only in the Walanae depression (the study area of 't Hoen and Ziegler), but also to the east (Wichmann's anticline) and to the north of Lake Tempe, where for instance de Koning Knijff (1914) found extended deposits of sandstones and marly clays. One has to envisage the southern part of the southwest peninsula of Sulawesi at the beginning of the Neogene (Miocene) as a separate, U-shaped island (the dotted area in fig. 1), where western, southern and eastern mountain ranges enclosed an expanse of sea. This was the basin in which the sediments of the Walanae Formation developed: for a large part the erosional debris (the socalled extra-basinal deposits[12]) of the surrounding mountains and their foothills. In the south much tuffaceous material was also deposited; there is less of this in the north, but this latter region was more distant from the Neogene cores of volcanic activity.

The silting up of this basin began in the south, and as a result the regressive order of the sediments of the Walanae Formation is still best preserved in the north. When in the south deltaic, littoral, or even fluvial sediments were already laid down, in the north (around the present-day village of Beru) marine sedimentation was still taking place. The southern non-marine deposits have subsequently been eroded in the long eons of the Pliocene and Pleistocene, and it is only in the north (once again, around Beru), the last area of the former basin to become dry, that remnants of non-marine (deltaic and fluvial), or mixed marine/non-marine (estuarine and littoral) or shallow marine (nearshore, subtidal) deposit have been preserved, nowadays constituting the matrix of the *Archidiskodon-Celebochoerus* fauna.

The Sarasins (1901, 1905) invented the term Celebes molasse, encompassing various kinds of clay shales, sandstones, and conglomerates that they mapped during their expeditions, and that in their view were Neogene. In his overview of the geology of Indonesia, van Bemmelen (1949) does not explicitly use this term Celebes molasse with reference to the sedimentary rock of the Walanae Formation (although he does do this for similar gradational clastics in the eastern and south-eastern arm of Sulawesi; see also Marks, 1956). Celebes molasse and Walanae Formation are treated as one, however, on a later geologic map of Indonesia (Sheet Ujung Pandang; Sukamto, 1975). We are of the opinion that the outcropping consolidated clastics in the Beru area cannot simply be classified together with the poorly stratified, often loose masses of sediment that the Sarasins mapped as molasse in various parts of Sulawesi.

This problem of classification might be solved by setting apart, as Sartono (1979) has done, the coarser top part of the Walanae Formation and naming it as a separate unit. Sartono refers to this top part as the Beru Formation[13], reserving the old name Walanae Formation exclusively for the lower and finer clastic part of the total basin fill. In fact, Sartono makes this division into two formations on account of the stratigraphic position of a mass of limestone to the south of Beru (fig. 2), in his view intertonguing with the clastic sequence. This may be correct, but this limestone is a very local phenomenon, and cannot be decisive for a main division of the Walanae Formation. In the sections around Beru without limestone the placing of stratigraphic boundaries becomes very arbitrary. We are therefore more inclined to retain the original concept of the Walanae Formation, as envisaged by 't Hoen and Ziegler, and to continue applying it to the whole regressive sequence of clastic sediments which is the bedrock in the Walanae depression. As for Sartono's concept of a Beru Formation, we prefer to speak of a Beru member instead, situated at the very top of the Walanae Formation. For reasons explained above it must be emphasized that this Beru member only occurs in the northern part of the Walanae depression. The lower boundary of the Beru member does not necessarily coincide with the lower boundary of Sartono's Beru Formation; but an informed guess could be made concerning the total thickness of this member. Measured along the west flank of the first anticline, we give an estimate of between 300 and 500 metres.

It is not easy to give a lithostratigraphic definition of the Beru member. The upper boundary is clear: it is the unconformity (erosion surface) between the capped top sediment of the Walane Formation and the unconsolidated river-terrace gravels and clays. But the lower boundary presents problems. In the local sections of the Walanae Formation there is no definite level where coarse clastics become predominant and fines disappear. A coarsening upward is definitely present, as well as a thickening upward of the coarser strata, but the total picture is obscured in a bewildering variety of clays, sands, and gravels, consolidated and unconsolidated in layers, lenses, and tongues. The delineation of a shale unit, sandstone unit, and conglomerate unit in the local sections in the Beru area, with the aim of demonstrating this coarsening upward (Sjahroel, 1970; Sartono, 1979) may have theoretical value, but is of limited use in the field. One of the key characteristics of the Beru member is its fossil vertebrate content. Making use of faunal indices in placing boundaries in lithostratigraphic sequences is not to be recommended; but on the other hand it is clear that the Beru member constitutes a rather restricted local depositional environment within the former sedimentary basin with a deltaic, littoral, or estuarine facies, and that distinct lithological characteristics will reflect this situation. Much sedimentological research will be needed in the future to unravel the precise facies of the fossiliferous deposits at the various exposures. The clastics of the Beru member occur in all varieties, from fine-textured clayey and marly sandstones to rather coarse conglomerates with even cobble-sized components.

As for the identification of the Celebes molasse, we propose that the Walanae Formation be included in a related group of rock-stratigraphic sequences in Sulawesi, all of which show uniform conditions of graded sedimentation. The Walanae Formation would then no longer be simply equivalent to the Celebes molasse or, for example, to de Koning Knijff's sandstones and shales north of Lake Tempe, but would have an identity of its own, representative of the bedrock sediment in the Walanae depression alone.

Already in his first reports on the finds in Sulawesi van Heekeren mentions that the fossils found on the surface often show bits of cemented sediment, indicating that they have been washed out. This matrix belongs to the Beru member of the Walanae Formation, and not, as van Heekeren thought, to a terrace fill. Years ago already one of us (D.A.H., 1949b) was able to conclude from an analysis of fossil-adhering sediment that there exist around Beru at least two vertebrate horizons. At the moment it is better to state that no distinct fossil horizons exist, but that almost every stratum of the Beru member is fossiliferous. Vertebrate fossils are to be found everywhere in exposures of the top part of the Walanae Formation; the fossil localities referred to in this text and figures are just major exposures of the top layers. Taphonomic research could be very illuminating

here, and should also constitute a key element of future investigations.

The best preserved skeletal remains of the *Archidiskodon-Celebochoerus* fauna come from fine to medium-textured sandstone. Hooijer (1949b) has given an analysis of a sandstone matrix from Sompe, one of the most northerly fossil localities in the Walanae depression (fig. 2). It is lateritic, has detrital grains, the interstices are partly filled with amorphous limonitic silica and opaque components, there are a few pieces of quartz and veins of rhombohedral calcite, and the volcanic constituents are diopside and to a lesser degree alkaline feldspar. In all fossil localities on the river-facing slope of the first anticline east of the Walanae this sandstone is to be found, locally with intercalated thin gravel seams or lenses: in Celeko, Sompe, Calio, Cangkange, Marale and Lakibong (fig. 2).[14] Among ourselves, we often refer to this conspicuous, grey-yellowish, fine to medium-textured, vertebrate-bearing sandstone as the Sompe sandstone. There is also a more marly or argillaceous and very fine-textured variety, apparently intertonguing with this Sompe sandstone, which can be traced at many localities, but which is almost devoid of fossils. In those places where the Sompe sandstone becomes coarser or more gritty the profusion of complete fossils disappears, and one finds fragmented bone and isolated teeth or molars (and very often the rather distinct canines of *Celebochoerus*; see also the worn molar fragment in conglomeratic sandstone of fig. 3). These 'diminishing returns' are even more noticeable in the true conglomerates of the Beru member. These can still be called fossiliferous, but one finds only isolated, small, and often heavily abraded pieces of bone or molar.

The sub-tidal, estuarine, and deltaic environment of the Beru member is very clear at the two fossil localities of Patema and Padali, which have been discovered in the last few years (fig. 2).[15] The first is probably situated on the flank of a second anticline east of Beru; the other is situated on the left side of the Walanae. In these two localities fossils of land-based vertebrates are found, together with many remains of sea-dwelling species, like sharks and rays. Shells are also abundant. At Patema a vertebral body of *Celebochoerus* was picked up, still partly embedded in a matrix of calcareous sandstone with shell fragments. At Padali, a short distance away from a mandible of *Celebochoerus* still stuck in sandstone, a dental fragment of an eagle-ray has been discovered (Bartstra & Hooijer, 1992). Van Heekeren had also found remains of marine vertebrates, notably at the localities of Sompe and Celeko. Because he thought that the fossiliferous sediments were river terraces, complicated explanations were put forward to account for the occurrence of marine species in freshwater deposits (see also Hooijer, 1960).

It appears as if the presence of a sea-dwelling fauna in association with a land-based fauna is more pronounced at these most northerly localities than

Fig. 3. Portion of a large upper molar of *Stegodon* cf. *trigonocephalus* eroded from conglomeratic sandstone some two hundred metres east of Marale. Upper figure: crown view; lower figure: left side view (Museum of Prehistory, Calio; number: C 3.27.86). Scale 1:1.

Fig. 4. Portion of a lower molar of *Stegodon* cf. *trigonocephalus* from Tanrung. At this locality the small river of Telle, a tributary stream of the Paciro, cuts through the vertebrate-bearing sediments (see fig. 2). Left: crown view; right: left side view (Museum of Prehistory, Calio; number: TRG 12.01.91). Scale 1:1.

Fig. 5. Palatal view of a cranial portion of *Celebochoerus heekereni* embedded in sandstone, and found at Cangkange in 1978 (B.A.I.). Scale 1:2.

eisewhere. The feature might have to do with the Neogene regression of the sea having started in the south, and a migrating shoreline toward the north, which, however, in the end never quite reached the area. More details are needed concerning these matters, especially quantitative assessment, and we reiterate our plea for sedimentological and taphonomic investigation.

Worthy of mention is also the locality of Cangkange (find-spot of an almost complete skull of *Elephas celebensis* during the 1970 expedition, see Hooijer, 1972a; and a large portion of a cranium of *Celebochoerus*, fig. 5) where intercalated in the vertebrate-bearing Sompe sandstone a horizon of fossil wood occurs, which actually contains fragments of tree trunks (fig. 7).

Fig. 6. Fossil drift wood with traces of burrowing pelecypods. Scale 3:4.

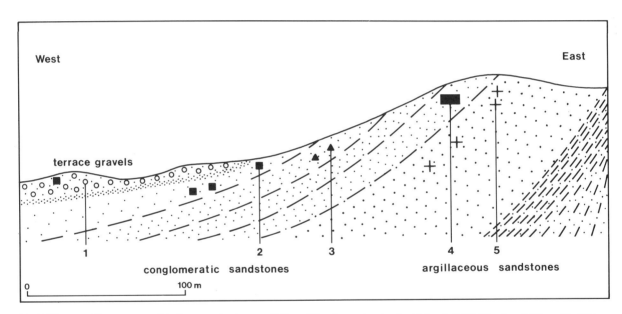

Fig. 7. Diagram of the upper part of the Beru member between Calio and Cangkange (partly after Sjahroel, 1970). The steep folding of the Walanae Formation is shown, as well as the considerable capping or truncation of the anticlines, outcome of a prolonged denudational history. Climbing the slope of this first anticline from west to east, the strata become progressively older and more fine-textured. Between terrace gravels and Beru member there is a distinct unconformity. The terrace gravels contain allochthonous fossil vertebrate remains and in situ Palaeolithic artifacts (1). In the Beru member are worn vertebrate fossils (2), mineralized wood remains (3), and the Cangkange cranial portions of *Elephas* and *Celebochoerus*, referred to in the text (4). Still higher up the slope marine invertebrates are to be found (5).

A few kilometres to the north, about halfway between Sompe and Beru (eastward into the field from kilometre post 23) a similar (or identical?) horizon is exposed, where tree-trunk fragments can be found with encrustation of shell and biogenic perforations (pelecypods; see fig. 6). These perforations are tunnels coated with calcareous shelly matter technically called 'the tube', which is white, thin and strong, and has no connection with the animal that makes it. If these tubes are found in fossil wood from marine or brackish water sediments they can safely be considered Teredinidae, family of Mollusca: Bivalvia, including the genus *Teredo*

Linnaeus or so-called ship-worms, very destructive pelecypods that first appear in the Jurassic and are well-known in the Cretaceous (Turner, 1966).

At the locality of Lakibong (find-spot of a very fine lower left first molar of *Elephas celebensis*; Bartstra & Hooijer, 1992)[16] the sediments with a marine fauna and a land-based fauna are clearly separated. Mollusc-bearing, argillaceous deposits occur distinctly higher on the west slope of the anticline than the vertebrate-bearing clastics. It is worth bearing in mind that the locality of Lakibong, stratigraphically seen, probably belongs to the oldest. Lakibong is situated farther south than any of the other localities, and thus must exhibit older strata. The various deposits at Lakibong are steeply folded.

Also at other exposures of the Beru member (e.g. in Cangkange) it can be seen that the layered sediments are folded almost vertically (fig. 8). Although Sulawesi is an island with a turbulent geological history, such steep folding indicates at least a Middle to Lower Pleistocene, if not Pliocene age for the fossiliferous deposits; another indication lies in the considerable capping of the anticlines. As mentioned above Rutten (1927) speaks of a Plio-Pleistocene folding of the Neogene Walanae Formation, and van Bemmelen (1949) supports this view. Hooijer (e.g. 1960, 1975) usually speaks of a Pleistocene age for the fossiliferous beds without any further specification, none of the fossil forms necessarily being Pliocene as to type. Palaeontological considerations, however, let Hooijer distinctly favour a Lower Pleistocene age; this in contrast to e.g. Sondaar (1981) who suggests the Middle to Upper Pleistocene.

Sartono (1979) gives an Upper Pliocene age for the vertebrates, on the basis of the foraminiferal content of vertebrate-bearing sandstone, deposited in a shallow marine and estuarine environment. The forams mentioned by Sartono are not sensitive to facies differences and are thus suitable for dating. In addition, an upper boundary of the Pliocene of 1.8 or 1.6 Ma has been assumed. However, it is not known where Sartono got his sandstone samples, whether from the immediate surroundings of Beru, or from farther south or north. As there has been a gradual silting-up of the former Neogene basin from the south, we wish to consider the possibility that part of the fossiliferous sandstones of the Beru member may be younger than 1.8 or 1.6 Ma, and is thus in fact Lower Pleistocene (Bartstra et al., 1991; we correct the date given by Allen, 1991: p. 247). This may be true especially for the vertebrate-bearing clastics of Celeko and Padali, and for conglomeratic sandstones, presumably still belonging to the Beru member, at a few minor but even more northerly outcrops towards the

Fig. 8. Exposure of the Beru member near Cangkange, along the road from Cabenge to Pampanua. The folded sediments consist of alternating claystones and sandstones. From this area comes the *Celebochoerus* skull of fig. 5, described in the text.

town of Singkang in the Tempe depression.

The best age determination would result from radiometric dating. However, preliminary attempts at U-series analysis of some vertebrate fragments from the surroundings of Beru were not very successful, ostensibly due to the low U-content of the bone. This in itself could be an indication of considerable age, but further attempts (including ESR) are under way.

4. THE EVIDENCE ON THE FOSSIL SCENE

The composition of the fossil fauna from the Walanae region was ascertained soon after its discovery. In 1947 the first vertebrate remains were collected by van Heekeren and sent to Holland; in 1948 three papers were published by Hooijer, on *Celebochoerus heekereni* nov. gen. nov. spec., *Testudo margae* nov. spec., and *Anoa depressicornis* (Smith) subsp. and *Babyrousa babyrussa beruensis* nov. subsp., respectively. A year later the identification of *Archidiskodon celebensis* nov. spec. was announced. The general term *Archidiskodon-Celebochoerus* fauna came into use at the beginning of the 1950s. It appeared to be of a peculiar, endemic, typical insular composition, characterized by dwarfing and giantism alike.

Archidiskodon celebensis is a distinct element of this fauna. The genus name *Archidiskodon* was abandoned in 1974, since when it has been customary to speak of

Elephas celebensis. This is a pygmy species; the shoulder height of adult individuals was approximately one and a half metres (e.g. Hooijer, 1949a; 1972a). At first sight it could be taken for a rather primitive species of *Elephas*, on the basis of the presence of functional premolars and the occurrence of tusks in the lower jaw, notably in male specimens. Initially Hooijer regarded *Elephas celebensis* as the only elephantid species with mandibular tusks, and he looked for its direct ancestor in a still unknown mastodontoid elephant. But it appears that this phenomenon of mandibular tusks also occurs elsewhere (*Primelephas gomphotheroides*, Upper Pliocene, Kenya), and, what is more important, that vestigial incisive chambers occur in mandibles of *Elephas planifrons*, a normal-sized species quite common in the Plio-Pleistocene of Southeast Asia (Maglio, 1973; Hooijer, 1972a; 1975). *Elephas celebensis* is now considered to be a direct descendant of *Elephas planifrons* (Groves, 1985), the reappearance of mandibular tusks being the result of dwarfing (paedomorphosis). The remains of *Elephas planifrons* have never been discovered in the fossil localities of the northern Walanae depression. This is significant, because it means that *Elephas celebensis* came to the area as an already dwarfed species. *Elephas planifrons* does not occur in the Philippines either, nor in Flores nor Timor.

In fact we are confronted here with a palaeontological enigma. This is because another proboscidean, of which the fossil remains have also been found in the Beru

Fig. 9. Browsing *Stegodon trigonocephalus*, showing in the male specimens the interesting feature of too closely implanted tusks, hampering the trunk. This drawing was made for Hooijer by the British artist Margaret Lambert. Hooijer published this picture for the first time in 1973.

member, occurs as both a normal and a dwarf-sized species: *Stegodon* cf. *trigonocephalus* and *Stegodon sompoensis*, respectively. Among the fossil material that van Heekeren had sent to Leiden, Hooijer (1953) initially could not identify anything more specific than *Stegodon* spec., although he believed that a dwarf species might have roamed the Walanae region. In 1964 this was confirmed by Hooijer by mentioning in a publication the new pygmy species of *Stegodon sompoensis*.[17] In 1972 it became possible to identify the large *Stegodon* cf. *trigonocephalus* among the extensive material collected during the 1970 expedition. Especially the male specimens of this browser could be rather large[18]; it is known that their tusks were so close together that the trunk could not reach between them (fig. 9).

A portion of a large molar of *Stegodon* cf. *trigonocephalus* has been found some 200 metres east of Marale (figs 2 and 3). It is the hinder end, heavily worn, with four ridges and the posterior talon. The foremost preserved ridge from behind is almost completely destroyed; the other three ridges are rather well preserved and, although worn, clearly were less high than wide in the unworn state. This is a feature of *Stegodon*. The last ridge is narrower than the second last ridge, 72 mm at base transversely against 88 mm in the penultimate ridge; it is only some 18 mm thick against some 26 mm in the second ridge. It is an upper molar, because the gingival line, the lower boundary of the crown enamel, falls off rootward at the penultimate ridge, and thus is convex toward the crown. There is no posterior pressure scar; this should certainly show on a molar in this advanced stage of wear caused by the molar following it behind. This is an indication that the present molar is the last molar, M^3. The three ridges preserved show transverse rows of subequal conelets and no dilatation in the middle, another *Stegodon* characteristic. The third ridge from behind is damaged buccally but certainly was some mm wider than the ridge behind it. The width of the fourth ridge from behind cannot be measured. This is all that can be observed in the present specimen of the posterior taper that characterizes last molars. Last upper molars of *Stegodon trigonocephalus* are widest near the middle, the fifth ridge from the front, as in two specimens from Flores (Hooijer, 1972b). The greatest width of the present molar, in the middle portion of the crown that is not preserved, may have been some 20 to 25 mm wider than is the second ridge from behind, or 108 to 113 mm. This is not outside the variation range in M^3 of the Java *Stegodon trigonocephalus*, however; we have a specimen in the Dubois collection that is 113 mm wide (Hooijer, 1955), and in Yogyakarta in 1972 we observed a specimen that is 119 mm wide. Last upper molars of *Stegodon trigonocephalus* have up to twelve ridges plus the posterior talon. In the present specimen the talon is a low series of conelets, some eleven in all, 61 mm wide but not very thick

anteroposteriorly (2-4 mm only); it rounds off the crown evenly behind. The molar portion is slightly curved anteroposteriorly, with one side flat and the other side convex. In upper molars the flat side is the lingual side; the present molar portion is from the right side. Two full ridges, the 2nd and the 3rd from behind, occupy an anteroposterior length of 55 mm, which means that the laminar frequency (number of ridges in 10 cm of anteroposterior length) is $100/55 \times 2 = 3.6$. These data indicate that this molar portion from 200 metres east of Marale agrees with the M^3 of the Java *Stegodon trigonocephalus*, which varies in crown width at base from 76 to 119 mm, and in laminar frequency from 3.25 to 5 (Hooijer, 1955); in the more anteriorly placed M^2 and M^1 the crown width becomes less and the laminar frequency higher.

A smaller molar portion of *Stegodon* cf. *trigonocephalus* has been obtained from Tanrung (figs 2 and 4). Both the last ridge and the talonid are unworn. The height of the ridge is 41 mm by a basal width of 67 mm, the *Stegodon* feature, the talonid 40 mm high by a width of 62 mm; such a large talonid may also be interpreted as the last ridge, in which case there would be an extremely small talonid or none at all. The rows of subequal conelets on the ridge and talonid or ridges confirm the identification as *Stegodon*. The forward inclination of the ridges shows that it is a lower molar, more probably from the right than from the left side; this cannot be ascertained from such a small crown portion. The laminar frequency, based upon the observation that the anteriormost preserved ridge is 19 mm anteroposteriorly, is $100/19 = 5.2$. These data indicate that the Sulawesi specimen is like the M_2 of *Stegodon trigonocephalus* from Java, which is 68-85 mm in basal width, 41-50 mm in crown height, and 3.8-6.5 in laminar frequency (Hooijer, 1955).

The above given two molar portions have to be recorded as *Stegodon* cf. *trigonocephalus*, because we cannot be quite certain that they do actually represent the species *Stegodon trigonocephalus*. Perhaps they merely agree with the Java *Stegodon* in molar morphology, but not in all trigonocephalous characters.

This brings us to the arrival of the Plio-Pleistocene proboscideans in southern Sulawesi. As explained, the vertebrate-bearing Beru member only occurs in the northern part of the Walanae depression; towards the south the suitable sediments have been eroded. This means that in the Beru area one is confronted with an already substantial, albeit impoverished and endemic fauna, without previous recorded history. How did this fauna come to the Neogene U-shaped island?

As yet there is no valid reason for placing the three proboscideans in the Beru area in a biostratigraphic sequence. The remains of *Elephas celebensis*, *Stegodon sompoensis*, and *Stegodon* cf. *trigonocephalus* all occur in the Beru member, and their worn allochthonous bones in the terrace sediments. It is supposed that

Stegodon sompoensis evolved from *Stegodon* cf. *trigonocephalus*, but we have no stratigraphic proof of this in the Beru area. All three proboscideans came together as fully developed species to southern Sulawesi, and lived there together. As one of us has explained concerning *Stegodon*, contemporaneity poses no problem: the larger one may have had five times the body size of the dwarfed species, a difference probably big enough to permit (peaceful) coexistence (Hooijer, 1970).

The idea of a biostratigraphic sequence in southern Sulawesi as far as *Stegodon* is concerned could possibly arise if one were to glance through the literature too fleetingly: after all, Hooijer (1972a) begins his exposé about *Stegodon* cf. *trigonocephalus* with the description of a few worn remains from terrace sediments (including the previously mentioned hinder end of an upper molar from the Marale excavation in 1970), while *Stegodon sompoensis* is initially mentioned from places where river terrace remnants do not exist (Sompe, Celeko). So one could gain the impression that the large stegodont appeared later on the scene than the small one, but this is not the case. The two above-described molar portions of the large *Stegodon* cf. *trigonocephalus*, for example, come from areas where there are no terraces, and they have clearly been derived from outcropping Beru member sediment.[19]

The situation observed thus far in southern Sulawesi is different from that in Timor or Flores, where also a normal-sized and a dwarf-sized *Stegodon* occur; but as far as Timor is concerned with only the pygmy one in the top strata (Astadiredja, 1972), and on Flores seemingly the reverse: only the normal-sized one in the top strata.[20]

For us, the idea of proboscideans that arrived swimming in southern Sulawesi (and if distances play a role, probably coming from the north) is now more appealing than the idea of legendary Stegoland (Hooijer, 1975; Sondaar, 1981; Braches & Shutler, 1984; Heany, 1985). The paucity of the Beru fauna is difficult to account for if landbridges existed. In the beginning of the 1970s the paucity could be explained by the small-scale nature of the research carried out. This is no longer possible; it appears as if the larger vertebrate genera and species of the fossiliferous Beru member are now all known.

Just as interesting as the arrival of the *Archidiskodon-Celebochoerus* fauna is its ultimate fate. As stated, in the region concerned the Neogene has been a period of prolonged subsidence and deposition. In the course of time the basin encompassed by the U-shaped island dried up, folding and faulting movements set in, and denudation severely capped the Walanae Formation. During this period, apparently the entire Pleistocene, conditions seemed not favourable for the preservation of bones of terrestrial or near-shore vertebrates. But the faunal composition must have changed. For instance, proboscidean remains are not found anymore in the

Holocene deposits east of Wallace's Line.

We have abandoned the idea, once toyed with, that the extinction of the *Archidiskodon-Celebochoerus* fauna was brought about by man. The implementiferous terrace fills in the Walanae depression do not contain autochthonous vertebrate fossils. In another paper we have mentioned Fooden (1969) and Musser (1987) and the notion that the *Archidiskodon-Celebochoerus* fauna could have disappeared as a result of competition with other faunas, when southern Sulawesi became attached to the central part (Bartstra & Hooijer, 1992). This must periodically have been the case during the Pleistocene, when the Tempe depression fell dry, due to eustatic sea level changes.

For the time being we have to view the occurrence of the *Archidiskodon-Celebochoerus* fauna in the northern Walanae region as a fleeting episode in the eons of geological time, a single still frame, as it were, in the faunal evolution of island Southeast Asia. It is not yet known how the *Archidiskodon-Celebochoerus* fauna came into being, or into what it evolved (Hooijer & Bartstra, in press).

Some light could be shed on this matter, however, by investigation into the parent sediment of a fossil skull of *Sus celebensis*, described by Hooijer in 1969. This fossil was found by a Dutch missionary in the 1930s, in the bed of the River Paciro, not very far east of Beru (fig. 2).[21] After changing hands a few times this skull finally ended up in the Zoölogical Museum in Amsterdam.

In his work on the subfossil fauna of the Toalian caves in southwestern Sulawesi, Hooijer (1950) had also devoted attention to *Sus celebensis*, an animal that nowadays still lives all over the island. From the comparison of subfossil and recent remains of this species, he had come to the conclusion that in contrast to a trend with other vertebrates, *Sus celebensis* had actually increased in size in the course of the Quaternary. The skull in the Amsterdam museum is marginally smaller than the subfossil remains from the caves and much smaller than recent specimens of *Sus celebensis*. Consequently Hooijer (1969) decided that it must be of Pleistocene age. The Amsterdam skull is also fossilized (mineralized) and thus not sub-fossil, as suggested by Hardjasasmita in 1987.

The locality and circumstances of this find are very interesting. The fossil has the appearance of a river cobble, but shows on closer inspection the palate and some incomplete molars (fig. 10). Every part of the fossil is abraded by fluvial transport, and it stands to reason that it came from farther upstream. The River Paciro has its source in the hills comprising the second anticline east of Beru (fig. 2). The rivulet of Tanrung, which cuts through the Beru member of the Walanae Formation at the locality of Tanrung (find-spot of one of the partial *Stegodon* molars described above), is in fact a tributary stream in the upper drainage area of the Paciro. However, as neither Tanrung (which apparently exhibits a normal *Archidiskodon-Celebochoerus* fauna)

Fig. 10. Cranial portion of *Sus celebensis*, found in the bed of the River Paciro. Upper figure: palatal view; lower figure: left side view. As described in the text, this fossil could provide a clue in unravelling the ultimate fate of the *Archidiskodon-Celebochoerus* fauna. Another picture of this skull portion has appeared in Hooijer 1969 (Instituut voor Taxonomische Zoölogie (Zoölogisch Museum), Amsterdam; number: ZMA 10910). Scale 1:1.

nor any other locality on the flanks of the two anticlines near Beru has so far yielded remains of *Sus celebensis*, the possibility exists that the River Paciro or one of its tributaries cuts through a Pleistocene fossiliferous sediment that stratigraphically overlies the Beru member, and that, as explained, must be rather rare. Caves are not present in the area concerned. Because more information on this *Sus celebensis* find would seem to be of great importance, research in the Paciro area is now under way (Hooijer & Bartstra, in press).

We wish to conclude these notes with a description of a cranial portion of *Celebochoerus heekereni*, the most prolific fossil vertebrate from the northern Walanae depression, the remains of which are found in nearly every Beru member sediment, thus constituting an indelible tribute to the prime hunter of this suid: Uncle Bob van Heekeren. This particular *Celebochoerus* fossil comes from Cangkange, and it stresses the importance of this locality that has already yielded more large skeletal elements (figs 2 and 5).

The palate, or what is left of it, agrees with that from Marale, 19th July 1970 (Hooijer, 1972a). Of the M^3 only that on the right side is preserved; both M^2 are present although incompletely so, and of the M^1 there is only the left, damaged laterally. The width of M^2 is 17.8 mm; the anteroposterior and transverse diameters of the crown of M^3 are 26.0 and 20.3 mm, respectively. The width of the M^2 and the length of the M^3 are perfectly intermediate between those of a male and a female palate from Marale, but the width of the M^3 is slightly in excess of that in the larger, presumably male, palate from Marale, 25th July, 1970. The maximum width found for M^3 in *Celebochoerus* is 21.5 mm, and thus exceeds that found for this Cangkange skull (Hooijer, 1972a). Therefore, as far as tooth size is concerned, this Cangkange specimen is neither very large nor very small; just average.

The surface of the skull is rather damaged except in the nasal region above. It is broken off anteriorly in front of the M^1 and thus the position of the canine is not shown. In a skull from Calio, the canine alveolus is seen to curve outward at the level of P^3 (Hooijer, 1972a). On the other hand, the nasal region, unlike that from Calio, is rather well preserved in this Cangkange skull, showing that there is no lateral angulation. The nasal bones are convex transversely but only preserved for some 4 cm (the naso-frontal suture is indistinct). The nasal width cannot be established, and neither can the frontal or the bizygomatic width. The latter is not far from some 20 cm in so far as the state of preservation permits judgment. This is considerably larger, by one-third to one-half, than that in recent Indonesian suid skulls like those of *Sus verrucosus* Müller & Schlegel or *Sus vittatus* Müller & Schlegel (Hardjasasmita, 1987). *Celebochoerus* is verrucose in the cross section of its lower canines, whereas *Sus vittatus* is *scrofa*-like in that respect (Hooijer, 1954). *Sus macrognathus* Dubois, a known fossil from Java is larger than these living forms. Only new and better preserved skulls of *Celebochoerus*, hopefully to be discovered in the future, will allow full craniometrical comparisons with other extinct or living suid species.

5. ACKNOWLEDGEMENTS

We wish to express our indebtedness to Prof.dr. R.P. Soejono, Prof.dr. Darmawan Mas'ud Rahman, Dr. R. Schutler Jr., and Dr. H.J. Veenstra, who all commented critically on this manuscript. We thank Prof.dr. E. Gittenberger, Dr. L. van der Valk, and Dr.ir. J. van der Plicht for their remarks on pelecypods and U-series dates. Dr. P.J.H. van Bree deserves our gratitude for the loan of the *Sus* skull from the Paciro river. Mr. T. van Heekeren provided us with enlightening details concerning his father's work. The maps in the above text were prepared by Mr. J.H. Zwier. The text itself was polished on several occasions by Mrs.drs. S. Ottway and Mrs. F.T. Cornelis. The lay-out of the final version is due to Mrs. M. Bierma.

6. NOTES

1. Formerly spelled: Tjabenge (van Heekeren, 1949a); a small town near the main sites (see Keates & Bartstra, this volume).
2. Due to Pleistocene eustatic sea level changes, the islands of Java, Sumatra, and Borneo often became attached to the Asiatic mainland. This expanse of land is called Sundaland (for map see e.g. Bellwood, 1985: p. 1).
3. The notion of a zoogeographic boundary is given in Wallace, 1860; the designation Wallace's Line is found in Huxley, 1868.
4. Dubois discovered *Pithecanthropus erectus* in Pleistocene sediments in Central Java at the end of the last century. During his excavations a mass of fossil vertebrate material became available, part of which was shipped to Holland and is now known as the Dubois collection.
5. In fact van Heekeren visited the Cabenge area for a couple of days already in 1968, when he saw his chance to slip away from a team investigating prehistoric cave sites in the Maros region, near Ujung Pandang.
6. H.R. van Heekeren & D.A. Hooijer, 1970. Petition for a joint Indonesian-Dutch expedition to Sulawesi, Flores and Timor in 1970. Unpublished report.
7. Hooijer actually participated in the team for only ten days: thereafter he spent several weeks exploring fossil localities in Flores and Timor.
8. The other leader of the 1970 expedition in the Beru area was R.P. Soejono, later director of the National Research Centre of Archaeology in Jakarta.
9. *Homo soloensis* was discovered in Central Java in the 1930s, and is regarded by several anthropologists as an advanced subspecies of *Homo erectus*.
10. H.R. van Heekeren, 1970 (?). The joint Indonesian-Dutch Sulawesi prehistoric expedition 1970. Unpublished report.
11. See report under note 10.
12. Extra-basinal (claystones or sandstones: brought into a basin from the outside) as opposed to intra-basinal (sediments that grew (bio)chemically in the waters of a basin); see e.g. G.M. Friedman & al., 1992.
13. Sartono mentions Berru Formation; Berru is the Buginese (local population) spelling of Beru.
14. In older literature or in Buginese spelling to be found as: Tjeleko, Sompo(h), Tjalio, Tjangkange and Lakiboong.
15. Patema also written as: Pattema.
16. In the caption of fig. 2 in this paper (Lutra), M^1 should be changed to M_1. Also, in fig. 4: M_3 should be M^3.
17. After the fossil locality of Sompo(h), now: Sompe.
18. According to Hooijer (1955) and Medway (1972), *Stegodon trigonocephalus* was a browser in shrubby or woodland vegetation.
19. The same holds true for a molar portion of a large stegodont from Padali, mentioned in Bartstra & Hooijer, 1992, picked up from eroding Beru member sediment.
20. Mentioned in: J. de Vos & R. van Zelst, Infusis (Intern informatieblad van het Nationaal Natuurhistorisch Museum te Leiden) 49, p. 5

21. Hooijer mentions: Salo Patjiro. The Buginese 'salo' means 'river', and Patjiro is nowadays written as: Paciro.

7. REFERENCES

ALLEN, H., 1991. Stegodonts and the dating of stone tool assemblages in island Southeast Asia. *Asian Perspectives* 30, pp. 243-265.

ASTADIREDJA, K.A.S., 1972. The *Stegodon*-bearing layers of Atambua, Timor. *Berita Dir. Geologi* 4, 30, p. 6.

BARTSTRA, G.-J., 1971. Op speurtocht naar de oermens. *Spiegel Historiael* 6, pp. 194-198.

BARTSTRA, G.-J., 1977. Walanae Formation and Walanae terraces in the stratigraphy of South Sulawesi (Celebes, Indonesia). *Quartär* 27/28, pp. 21-30.

BARTSTRA, G.-J., 1978. Note on new data concerning the fossil vertebrates and stone tools in the Walanae valley in South Sulawesi (Celebes). *Modern Quaternary Research in Southeast Asia* 4, pp. 71-72.

BARTSTRA, G.-J., 1993. Terugkeer naar Sulawesi. *Paleo-aktueel* 4, pp. 11-14.

BARTSTRA, G.-J., S.G. KEATES, BASOEKI & B. KALLUPA, 1991. On the dispersion of *Homo sapiens* in eastern Indonesia: the Palaeolithic of South Sulawesi. *Current Anthropology* 32, pp. 317-321.

BARTSTRA, G.-J. & D.A. HOOIJER, 1992. New finds of fossil vertebrates from Sulawesi, Indonesia. *Lutra* 35, pp. 113-122.

BELTZ, E.W., 1944. Principle sedimentary basins in the East Indies. *Bull. Am. Ass. Petr. Geol.* 28, pp. 1440-1454.

BELLWOOD, P., 1985. *Prehistory of the Indo-Malaysian Archipelago*. Sydney.

BEMMELEN, R.W. VAN, 1949. *The geology of Indonesia*. The Hague.

BRACHES, F. & R. SCHUTLER JR., 1984. The Philippines and Pleistocene dispersal of mammals in island Southeast Asia. *Philippine Quarterly of Culture and Society* 12, pp. 106-115.

FOODEN, J., 1969. *Taxonomy and evolution of the monkeys of Celebes (Primates: Cercopithecidae)* (= Bibliotheca Primatologia 10). Basel.

FRIEDMAN, G.M., J.E. SANDERS & D.C. KOPASKA-MERKEL, 1992. *Principles of sedimentary deposits*. New York.

GREMMEN, W.H.E., 1990. Palynological investigations in the Danau Tempe depression, southwest Sulawesi (Celebes), Indonesia. *Modern Quaternary Research in Southeast Asia* 11, pp. 123-134.

GROVES, C.P., 1985. Plio-Pleistocene mammals in island Southeast Asia. *Modern Quaternary Research in Southeast Asia* 9, pp. 43-54.

HARDJASASMITA, H.S., 1987. Taxonomy and phylogeny of the Suidae (Mammalia) in Indonesia. *Scripta Geologica* 85, pp. 1-68.

HEANEY, L.R., 1985. Zoogeographic evidence for Middle and Late Pleistocene land bridges to the Philippine islands. *Modern Quaternary Research in Southeast Asia* 9, pp. 127-143.

HEEKEREN, H.R. VAN, 1949a. Voorlopige mededeling over palaeolithische vondsten in Zuid-Celebes. *Oudheidk. Verslag* D (1941-1947), pp. 109-110.

HEEKEREN, H.R. VAN, 1949b. Early man and fossil vertebrates on the island of Celebes. *Nature* 163, 4143, p. 492.

HEEKEREN, H.R. VAN, 1949c. Preliminary note on Palaeolithic finds on the island of Celebes. *Chronica Naturae* 105, pp. 145-148.

HEEKEREN, H.R. VAN, 1972. *The Stone Age of Indonesia*, 2nd rev.ed. (= Verhand.Kon.Inst. Taal-, Land- en Volkenkunde 61). The Hague.

HEEKEREN, H.R. VAN, 1975. Chronology of the Indonesian prehistory. *Modern Quaternary Research in Southeast Asia* 1, pp. 47-51.

HOEN, C.W.A.P.'T & K.G.J. ZIEGLER, 1917(1915). Verslag over de resultaten van geologisch-mijnbouwkundige verkenningen en opsporingen in Zuidwest-Celebes. *Jaarboek van het Mijnwezen in Ned. Oost-Indië* 44, pp. 235-335.

HOOIJER, D.A., 1948a. Pleistocene vertebrates from Celebes, I. *Celebochoerus heekereni* nov.gen.nov.spec. *Prov. Kon. Ned. Akad. v. Wetenschappen* 51, pp. 1024-1032.

HOOIJER, D.A., 1948b. Pleistocene vertebrates from Celebes, II. *Testudo margae* nov.spec. *Prov. Kon. Ned. Akad. v. Wetenschappen* 51, pp. 1169-1182.

HOOIJER, D.A., 1948c. Pleistocene vertebrates from Celebes, III. *Anoa depressicornis* (Smith) subsp., and *Babyrousa babyrussa beruensis* nov.subsp. *Prov. Kon. Ned. Akad. v. Wetenschappen* 51, 10, pp. 1322-1330.

HOOIJER, D.A., 1949a. Pleistocene vertebrates from Celebes, IV. *Archidiskodon celebensis* nov.spec. *Zool. Mededelingen* (Rijksmus. Nat. Hist. Leiden) 30, pp. 205-226.

HOOIJER, D.A., 1949b. The Pleistocene vertebrates of southern Celebes. *Chronica Naturae* 105, pp. 148-150.

HOOIJER, D.A., 1950. Man and other mammals from Toalian sites in south-western Celebes. *Verh. Kon. Ned. Akad. v. Wetensch.* 46, pp. 1-164.

HOOIJER, D.A., 1953. Pleistocene vertebrates from Celebes, VI. *Stegodon* spec. *Zool. Mededelingen* (Rijksmus. Nat. Hist. Leiden) 32, pp. 107-112.

HOOIJER, D.A., 1954. Pleistocene vertebrates from Celebes, VIII. Dentition and skeleton of *Celebochoerus heekereni* Hooijer. *Zool. Verhandelingen* (Rijksmus. Nat. Hist. Leiden) 24, pp. 1-46.

HOOIJER, D.A., 1955. Fossil Proboscidea from the Malay Archipelago and the Punjab. *Zool. Verhandelingen* (Rijksmus. Nat. Hist. Leiden) 28, pp. 1-146.

HOOIJER, D.A., 1960. The Pleistocene vertebrate fauna of Celebes. *Asian Perspectives* 11, pp. 71-76.

HOOIJER, D.A., 1964. Pleistocene vertebrates from Celebes, XII. Notes on pygmy stegodonts. *Zool. Mededelingen* (Rijksmus. Nat. Hist. Leiden) 40, pp. 38-44.

HOOIJER, D.A., 1969. Pleistocene vertebrates from Celebes, XIII. *Sus celebensis* Müller & Schlegel, 1845. *Beaufortia* (Zoöl. Museum Amsterdam) 16, pp. 215-218.

HOOIJER, D.A., 1970. Pleistocene South-east Asiatic pygmy stegodonts. *Nature* 225, pp. 474-475.

HOOIJER, D.A., 1972a. Pleistocene vertebrates from Celebes, XIV. Additions to the *Archidiskodon-Celebochoerus* fauna. *Zool. Mededelingen* (Rijksmus. Nat. Hist. Leiden) 46, pp. 1-16.

HOOIJER, D.A., 1972b. *Stegodon trigonocephalus florensis* Hooijer and *Stegodon timorensis* Sartono from the Pleistocene of Flores and Timor. *Proc. Kon. Ned. Akad. van Wetensch.* B, 75, pp. 12-33.

HOOIJER, D.A., 1973. Reuzenschildpadden en dwergolifanten. *Museologia* 1, 10, pp. 9-14.

HOOIJER, D.A., 1975. Quaternary mammals west and east of Wallace's Line. *Modern Quaternary Research in Southeast Asia* 1, pp. 37-46.

HOOIJER, D.A. & G.-J. BARTSTRA, in press. Fossils and artifacts from southwestern Sulawesi (Celebes) *Proceedings Pithecathropus Centennial*, Leiden 1993, Session B.

HUXLEY, T.H., 1868. On the classification and distribution of the Alectoromorphae and Heteromorphae. *Proc. Zool. Soc. London* 1868, pp. 294-319.

KEATES, S.G. & G.-J. BARTSTRA, this volume. Island migration of early modern *Homo sapiens* in Southeast Asia: the artifacts from the Walanae depression, Sulawesi, Indonesia.

KONING KNIJFF, J. DE, 1914(1912). Geologische gegevens omtrent gedeelten der afdelingen Loewoe, Paré Paré en Boni van het gouvernement Celebes en onderhoorigheden. *Jaarboek van het Mijnwezen in Ned. Oost-Indië* 41, pp. 277-312.

MAGLIO, V.J., 1973. *Origin and evolution of the Elephantidae* (= Trans. Amer. Philos. Soc., new series 63, 3). Philadelphia.

MARKS, P., 1956. *Stratigraphic lexicon of Indonesia* (= Publikasi keilmuan 31, seri geologi). Bandung.

MEDWAY, LORD, 1972. The Quaternary mammals of Malesia: a review. *Univ. Hull Dept. Geogr. misc. Series* 13, pp. 63-83.

MUSSER, G.G., 1987. The mammals of Sulawesi. In: T.C. Whitmore (ed.), *Biogeographical evolution of the Malay Archipelago*. Oxford, pp. 73-93.

RUTTEN, L.M.R., 1927. *Voordrachten over de geologie van Nederlandsch Oost-Indië*. Groningen.

RUTTEN, L.M.R., 1932. *De geologie van Ned. Indië*. Den Haag.

SARASIN, P. & F. SARASIN, 1901. *Entwurf einer geografisch geologischen Beschreibung der Insel Celebes*. Wiesbaden.

SARASIN, P. & F. SARASIN, 1905. *Reisen in Celebes, ausgeführt in den Jahren 1893-1896 und 1902-1903*. Wiesbaden.

SARTONO, S., 1979. The age of the vertebrate fossils and artefacts from Cabenge in South Sulawesi, Indonesia. *Modern Quaternary Research in Southeast Asia* 5, pp. 65-81.

SHUTLER, JR., R., 1991. Colonization, expansion, and successful adaptation in Southeast Asia, New Guinea and Australia 40,000-10,000 B.P. *Asian Profile* 19, pp. 151-157.

SJAHROEL, 1970. Penjelidikan geologi pendahuluan daerah bahagian tengah Kabupaten Soppeng di Sulawesi Selatan. Bandung, unpubl., 31 pp.

SONDAAR, P.Y., 1981. The *Geochelone* faunas of the Indonesian archipelago and their paleogeographical and biostratigraphical significance. *Modern Quaternary Research in Southeast Asia* 6, pp. 111-119.

SUKAMTO, R., 1975. *Peta geologi Indonesia, lembar Ujung Pandang*. Bandung.

TURNER, R.D., 1966. *A survey and illustrated catalogue of the Teredinidae (Mollusca: Bivalvia)*. Cambridge (Mass.).

WALLACE, A.R., 1860. On the zoological geography of the Malay Archipelago. *Journ. Linn. Soc. London* 4, pp. 172-184.

WICHMANN, A., 1890. Bericht über eine im Jahre 1888-1889 ausgeführte Reise nach dem Indischen Archipel. *Tijdschr. Kon. Ned. Aardr. Gen.*, pp. 921-994.

ISLAND MIGRATION OF EARLY MODERN *HOMO SAPIENS* IN SOUTHEAST ASIA: THE ARTIFACTS FROM THE WALANAE DEPRESSION, SULAWESI, INDONESIA.

S.G. KEATES
School of Anthropology and Museum Ethnography, Oxford, Great Britain

G.-J. BARTSTRA
Biologisch-Archaeologisch Instituut, Groningen, Netherlands

ABSTRACT: Surveys and excavations conducted in the northern Walanae depression[1] in South Sulawesi have resulted in the recovery of lithic artifacts from Late Pleistocene river terraces. A number of these artifacts may represent evidence of the earliest known occupation of the island, and this may be of relevance to the first occupation of Australia. The uncomplex technology of the artifacts is of special interest regarding discussions on the presumed cultural complexity of early modern *Homo sapiens*.

KEYWORDS: Southeast Asia, Sulawesi, Walanae depression, Walanae terraces, Palaeolithic Cabenge industry, lithic artifacts, cores, flakes.

1. INTRODUCTION

Java is the only island in Southeast Asia which preserves evidence of *Homo erectus* occupation with a date of about 1 million years, coincident with the earliest arrival of hominids in the Far East from the African continent (Pope, 1983; Bartstra, 1983; Bartstra & Basoeki, 1989).[2] The antiquity for the occupation of other islands in the region by (presumably) anatomically modern humans extends back to about 30 000 to 40 000 years ago (Fox, 1970; Harrisson, 1970; 1978; Glover, 1981; for an overview see Bellwood, 1985). These dates lie within the time-frame of the hitherto earliest known occupation of the Australian continent (Pearce & Barbetti, 1981), but seem at odds with the most recently determined age of about 50 000 years ago from northern Australia (Roberts et al., 1990). According to Shutler (1984) *Homo sapiens sapiens* might have entered Southeast Asia as long as 70 000 years ago. The earliest radiometrically dated occupation site in Sulawesi is Leang Burung 2, a cave located in the Maros region in the southwest peninsula (c. 30 000 BP; Glover, 1981).

The 2000 m deep Makassar Strait which separates Sulawesi from the islands of the Sunda shelf appears to have formed an effective geographical barrier to hominid migration until the Late Pleistocene. The Sunda shelf was periodically dry during the Pleistocene (Berggren & van Couvering, 1979). Thiel (1987) suggests that the Southeast Asia islands were first colonized during the Late Pleistocene as a response to high sea-levels which reduced food resources caused by a reduction in available land. She assumes that a period of rising sea-level between 53 000 and 45 000 BP (50 000 and 40 000 BP, see Chappell & Thom, 1977) based on sea-level curve calculations of Chappell and Thom provided the stimulus and opportunity to initiate sea-travel. The updated sea-level curve of Chappell & Shackleton (1986), however, suggests that a higher sea-level rise occurred between 64 000 and 59 000 years ago. Clark (1991) has recently criticized Thiel's hypothesis as too simplistic.

In this article we discuss what may represent the earliest evidence for human colonization of Sulawesi. Among artifacts collected from river terraces in the northern part of the Walanae depression in Southwest Sulawesi, many specimens probably predate the assemblages from Leang Burung 2. The Walanae sites have not yet yielded datable materials; the vertebrate fossils discovered in this region (Hooijer, 1949, 1960, 1975) bear no stratigraphical relationship to the artifacts (Sjahroel, 1970; Bartstra, 1977, 1978; Sartono, 1979; Bartstra & Hooijer, 1992; Hooijer & Bartstra, in press; Bartstra et al., this volume). Recently, we have proposed a relative chronological framework for the Walanae artifacts based on terrace morphology and comparison to radiometrically dated assemblages known from a number of caves in the Maros region (Bartstra et al., 1991).

2. ARCHAEOLOGICAL RESEARCH IN THE WALANAE DEPRESSION

Artifacts and vertebrate fossils in the northern Walanae depression were first recognized by van Heekeren (1949, 1958) mainly in the vicinity of the village of Beru (Bartstra et al., 1991; Bartstra et al., this volume; fig. 1). Subsequent, more extensive fieldwork has shown that these artifacts (which have entered the textbooks as the Palaeolithic Cabenge industry, formerly Tjabenge industry, van Heekeren, 1972) are associated with three or four river terraces (Bartstra, 1977; Sartono, 1979).

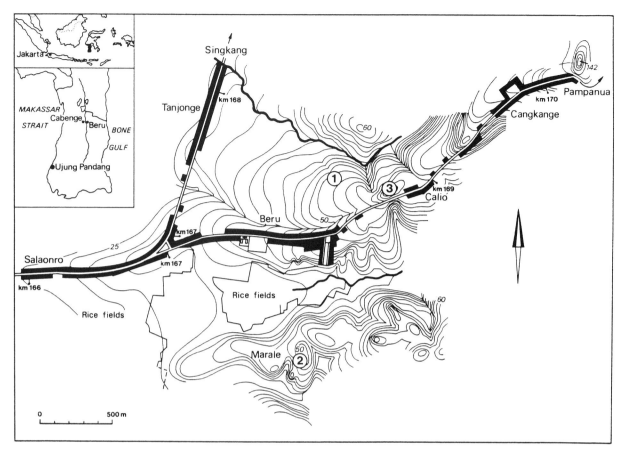

Fig. 1. Island Southeast Asia, the southwestern peninsula of Sulawesi, and the surroundings of the village of Beru in the northern Walanae depression. The ciphers 1, 2, 3 refer to the former (1970) excavations of Beru I (also known as Calio I), Beru II (Calio II or Marale), and Beru III (Calio III). These small excavations, intended to collect possible in situ artifacts, were in streamterrace deposits between 50 and 75 metres above the mean waterlevel of the Walanae river. On the above map, this river is located about one kilometre west of Beru. Lithic artifacts were collected in the very top (gravel) part of the pit profiles. They also could and still can be found on the surface in the vicinity of these excavations.

Exposed through erosion and excavation artifacts have over the years been collected from gravel and shallow unconsolidated sand-clay deposits from the right bank of the northern Walanae river (Bartstra, 1978; Bartstra et al., 1991). It is from these artifacts that the sample described in this paper derives. Surveys have shown that many lithic specimens are distributed at a height of 50 to 75 m above the river, e.g. at Beru, Calio, and Marale (fig. 1). Investigation on the left bank of the Walanae has shown similar implementiferous terrace deposits.

Very rolled and patinated core and flake artifacts occur in situ in high terrace gravel, while cores and flakes with less pronounced fluvial wear are found on the surface of the high terrace, and on lower levels. Small lithic specimens with no evidence of redeposition and no traces of abrasion are confined to elevations (terrace and non-terrace) in the proximity of the Walanae river, not only in the northern part of the drainage area but also in the southern part (Bartstra, 1978).

Analyses of the geomorphology of the Walanae terraces indicate a Late Pleistocene age for the artifacts (Sjahroel, 1970; Bartstra, 1977, 1978; Bartstra et al., 1991; Bartstra et al. this volume). On the basis of the stratigraphical context, lithic technology, and to some extent the degree of mechanical and chemical weathering, the sample of 28 lithic specimens discussed in this paper has provisionally been classified into the three indicated units. This random sample of surface and excavated specimens gives a fair insight into the variety of the lithic technology in the Beru area and for that matter in the whole northern Walanae depression. The main concern of this paper is to provide typological and technological details.

3. THE ARTIFACTS FROM THE WALANAE DEPRESSION

The sample comprises four bifacially flaked pebbles and three bifacially flaked cobbles (so-called 'chopping-tools'); two pointed partial bifaces ('proto-handaxes')

Table 1. Frequenies of lithic categories.

Category	n	%
Bifacial pebbles	4	14.3
Bifacial cobbles	3	10.7
Partial bifaces	2	7.1
Uniface	1	3.6
Cores	7	25.0
Flakes	11	39.3
Total	28	100.0

Table 2. Frequencies of raw materials.

Category	Sil. tuffaceous mat. n	Silic.tuff n	Silic.limestone n
Bifacial pebbles	3	0	1
Bifacial cobbles	1	2	0
Partial bifaces	1	1	0
Uniface	0	1	0
Cores	0	4	3
Flakes	0	11	0
Total	5	19	4

Table 3. Length, width, and thickness of the Walanae lithic sample (in cm).

Category	Specimen No.	Length	Width	Thickness
Core	M 70/2	3.65	3.5	2.65
Flake	M 70/4	3.8	2.7	0.85
Core	M 70/1	4.2	3.45	3.4
Flake	M 70/U	4.3	2.9	1.2
Core	M 70/4	4.4	4.1	3.6
Bif. pebble	M 70/3	4.65	3.9	2.6
Bif. pebble	K 87/6	4.8	4.6	3.0
Bif. pebble	B 70/1	5.2	4.7	3.35
Flake	P 87/B	5.25	3.8	1.6
Core	K 87/7	5.3	5.2	4.0
Core	M 70/A	5.9	4.25	4.2
Bif. pebble	M 70/B	6.1	5.1	4.3
Flake	P 87/A	6.6	3.1	1.6
Flake	M 70/S	7.0	4.6	2.45
Flake	M 70/Q	7.2	5.1	1.6
Core	K 87/3	7.3	7.25	5.0
Flake	K 87/5	7.5	7.0	2.9
Flake	M 70/R	7.7	4.5	2.15
Flake	M 70/T	7.85	4.4	1.9
Flake	M 70/P	8.85	6.0	2.35
Flake	K 87/4	8.9	8.0	2.75
Bif. cobble	K 70/1	9.0	8.0	6.5
Bif. cobble	K 87/2	9.5	7.7	5.1
Bif. cobble	K 70/2	9.6	8.2	6.5
Core	K 70/3	9.6	8.15	5.5
Biface	P 78/15	10.4	8.8	3.7
Biface	K 87/1	12.3	9.15	4.8
Uniface	B I 79/100	12.9	11.9	7.75

and one pointed uniface; seven cores; and eleven flakes (table 1). The specimens were mainly manufactured by direct hard-hammer percussion on locally available river-bed nodules of silicified limestone and silicified tuff varieties (table 2). Since most of these specimens exhibit fluvial wear, it is not possible to positively determine the cause of edge damage, i.e. natural abrasion or use-wear.

The following two lithic categories, A and B, conform typologically to artifacts first described by Movius (1943: p. 351) as 'chopping-tools'. As pointed out in an earlier paper (Bartstra et al., 1991), these artifacts might deserve an alternative terminology devoid of functional connotation. Nodule size based on the Wentworth (1922) classification of clast size, where the distinction be—tween a pebble and a cobble is a size limit of smaller or larger than 64 mm in diameter, and the extent of modification (unifacial or bifacial flake removal), are the defining attributes of pebble and cobble artifacts. The use of either pebbles or cobbles may indicate selection for size and weight by the tool makers. As both types of nodules were readily available for manufacture, this could be of functional significance.

A. The four bifacial pebble artifacts derive from Kecce[3] (n=1), Marale (n=2), and the very topsoil of the former excavation of Beru I (n=1; fig. 1). The specimens are all but one made of silicified tuffaceous materials; the exception is of silicified limestone (B 70/1).[4] These artifacts range in size from 4.65×3.9×2.6 to 6.1×5.1×4.3 cm (table 3). Dorsal modification is more extensive than on the ventral aspect and the worked edges are sinuous in shape. The one specimen from Kecce (K 87/6) is the least cortical specimen and ventral modification is limited to the upper half of this face. The shaped edge is acute and largely unworn, and exhibits very localized, minute wear in contrast to the cortical edges where this is more extensive. Modification on an almost wholly cortical specimen (M 70/3) covers one third of the dorsal face with two relatively shallow flake scars on either end of the obverse of the slightly sinuous worked edge. The small protrusion in the central part of the modified edge is only very slightly worn (also evident on K 70/1), while the neighbouring margins are comparable in degree of wear to the bifacial cobble artifacts. This pattern may indicate utilization of these edges. The edge of the most extensively flaked specimen (M 70/B) is worked to a round point with limited step-fracturing on its dorsal side and limited dorsal and ventral cortex. The dorsal aspect exhibits natural abrasion. The Marale specimen M 70/3 exhibits also slight fluvial abrasion along it's edge.

B. The three bifacially flaked cobble artifacts were all collected at Kecce and are manufactured of silicified tuff (K 70/1, K 70/2, K 87/2). The size range is 9.0×8.0×6.5 cm to 9.6×8.2×6.5 cm. The specimens K 70/1 and K 70/2 (fig. 2) were modified by the detachment of a few flakes through alternate retouch resulting in sinuous edges occupying about half of each specimens'

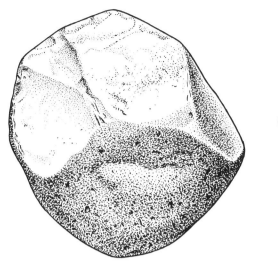

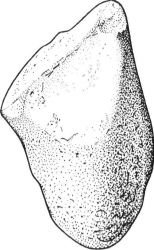

Fig. 2. Dorsal and right side views of the bifacially flaked cobble artifact K 70/2. Scale 2:3.

Fig. 3. Dorsal view of the bifacially flaked cobble artifact K 87/2. Scale 1:1.

periphery, and pronounced ridges at the intersection of flake scars. Size and depth of flake scars are variable with modification most extensive on the dorsal and largely limited to this face. In one case (K 70/1) two flake scars cover almost the entire dorsal surface. In contrast, the worked edge of K 87/2 (fig. 3) is relatively straight. Its non-cortical dorsal aspect shows more extensive modification than the obverse where one large flake was removed parallel to the dorsally worked edge associated with a small and shallow scar. The pattern of modification suggests that the knapper intended to produce an artifact with a straight edge by detaching one large ventral flake. The tuff variety of this specimen may have allowed more controlled flaking than the tuff of the other two artifacts.

C. The two pointed partial bifaces (K 87/1, P 78/15) and the pointed uniface (B I 79/100) were found at Kecce, Paroto[5], and in the immediate surroundings of the Beru I excavation (fig. 1). These specimens measure from 10.4×8.8×3.7 cm to 12.9×11.9×7.75 cm and are thus the largest of the sample (table 3). Flaking on these cobble nodules was carried out vertical to the long axis, although the Kecce specimen also shows limited horizontal modification. On the bifaces the dorsal face is more extensively modified, and on all specimens shaping of the points is limited to the dorsal aspect and flaking is concentrated on the left edge. The edges are usually sinuous, and the size and depth of scars is variable and frequently associated with step-flakes. One of the bifaces (K 87/1; figs 4 and 5) exhibits more

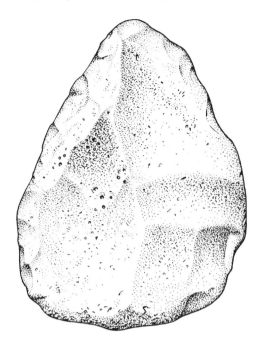
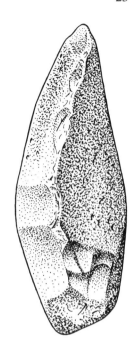

Fig. 4. Dorsal and right side views of biface K 87/1. Scale 2:3. Compare fig. 5.

regular flake scars, a smaller point, and is also more extensively modified compared to the other two specimens. Thinning of the lateral edges was conducted by invasive secondary retouch along the left dorsal edge and detachment of a single large flake from the left ventral edge placed parallel to the long axis. These features are possibly of functional significance, and the chipping along both edges may derive from utilization. Most of the ventral face is cortical. Modification of the Paroto specimen is less extensive with cortex covering the lower half of the dorsal face (P 78/15; figs 6 and 7)[6], and with a broader point than on the former specimen. The largest specimen is the uniface (B I 79/100; figs 8 and 9). The broader point compared to the bifaces is possible related to the size of this artifact. Small flake scars on the dorsal periphery of the point may be attributable to utilization. The relatively strong fluvial abrasion on the bifaces and the uniface hampers positive identification of their use as tools, but it seems that the aim of the knappers was to produce pointed artifacts with sinuous edges, attributes which are possibly of functional importance.

D. The cores were found at Kecce (n=3) and Marale (n=4) and usually exhibit slight natural abrasion. The specimens are all in silicified tuff or limestone, and include both small and large cores with a size range of 3.65×3.5×2.65 cm to 9.6×8.15×5.5 cm. The three cores from Kecce include the largest core in the sample. This specimen is a flat-based, thick, and relatively steep-edged piece (K 70/3), similar to the so-called 'horsehoof' cores. About one half of the specimen was modified. The irregular pattern of flake scars may be attributable to the vesicular limestone which also has enamel and radiolaria inclusions. Edge fracturing distributed along

the base of the cortical part is possibly natural damage. The other cores (K 87/7, K 87/3) are very similar in morphology. These double-platform cores exhibit removal of mainly large flakes around the entire periphery at an acute angle. The larger core preserves more cortex compared to the smaller specimen, and on both flake scars are frequently step-flaked. The smaller striking platforms on both specimens are almost flat, in contrast to the larger irregular platforms. The larger four-sided core shows fluvial abrasion, while the smaller five-sided core is in fresh condition. The Marale cores (M 70/2, M 70/1, M 70/4) are small double-platform cores. One of these is a six-sided specimen and very similar in shape to the double-platform cores from Kecce. Step fracturing occurs on two of these cores, especially on the core which also exhibits stronger edge blunting. The third specimen (M 70/1) was more extensively worked with a limited cortical area. Flake morphology and size of this specimen are less regular compared to the two other cores from this locality. The only bipolar core (M 70/A), of heavily patinated silicified limestone, is quadrilateral in shape with small to large sized flake scars and an islet of cortex on one striking platform. Edge blunting is unevenly distributed on most of the flake scar margins.

E. The Walanae flakes finally were found at Marale (n=7), Kecce (n=2), and Paroto (n=2). All are unifacially worked specimens of usually irregular shape. The sample includes one specimen of flake-blade proportions. The specimens are mainly in silicified tuff, usually of a fine-textured variety. Cortex on these specimens is limited. The flakes were detached from transverse (n=7) and end (n=4) angles by direct hard-hammer percussion. Striking platforms are plain (n=10), cortical (n=1), and

Fig. 5. Upper figure: ventral view of biface K 87/1. Compare fig. 4.
Lower figure: dorsal view. Scale 1:1.

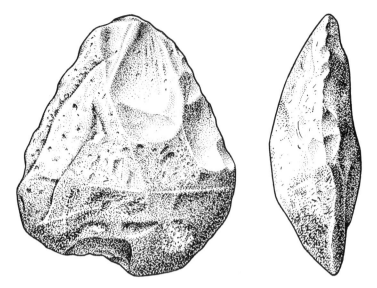

Fig. 6. Dorsal and right side views of biface P 78/15. Scale 2:3. Compare fig. 7.

Fig. 7. Ventral view of biface P 78/15. Compare fig. 6. Scale 1:1.

faceted (n=1). The size range is 3.8×2.7×0.85 cm to 8.9×8.0×2.75 cm (table 3).

The 'waisted' shape of the largest flake (K 87/4; 8.9×8.0×2.75) was formed by multi-directional modification of it's lateral edges. The central part of this face is cortical and exhibits limited stepped flaking on

the right dorsal. One flake scar is on the cortical and heavily patinated ventral face. The sinuous proximal end and lateral edges may be use-worn (less marked on the right edge). However, this is the most abraded flake and the abrasion is too pronounced to classify this specimen with any certainty as a tool.

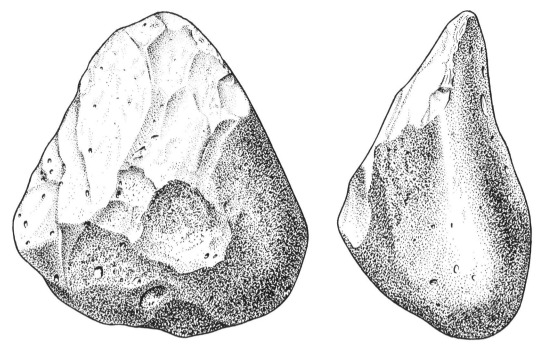

Fig. 8. Dorsal and right side views of uniface B I 79/100. Scale 2:3. Compare fig. 9.

Fig. 9. Left side view of uniface B I 79/100. Compare fig. 8. Scale 1:1.

Among the flakes with the most pronounced fluvial wear is a pointed flake with cortex extending almost around the whole periphery (K 87/5). Secondary retouch and possible use-wear is limited to the oblique margin of the point. Similar in wear and patination are three flakes from Marale. On one of these (M 70/Q) the only non-cortical edge exhibits three small and continuous flake scars which may be attributable to use-damage.

Two flakes of roughly triangular and similar shape (M 70/P; M 70/T) both retain cortical left lateral edges which are thick and positioned at a relatively steep-angle. On the cortical edge of the keeled flake (M 70/P) the invasive notch just below the retouched and blunted point may have formed through utilization including part of the associated oblique margin. The edge below the point of the sinuous right edge was modified by the removal of two long, narrow and obliquely angled flakes. Both lateral edges of the smaller specimen (M 70/T) exhibit what appears to represent invasive use-wear. This is less pronounced on the notch of the cortical edge which is located half-way below the blunt point of which the apex seems to have been removed. Prior to utilization of the non-cortical edge one narrow flake was detached from below the point, extending 3/4 down the edge. The slight denticulation on this thin edge may have formed through use-damage.

Two flakes in vesicular tuff indicate selection for cortical edge utilization. Of two roughly quadrilateral flakes one (M 70/R) may show use-wear on both denticulated lateral edges. The right margin is thin compared to the left cortical edge. The second specimen is a keeled flake (M 70/S). The concavities on the left edge and on the right cortical edge may be use-damaged. The larger of two keeled flakes from Paroto is a non-cortical flake-blade with slight distal end retouch due either to utilization or natural wear (P 87/A). The denticulated lateral edges may have formed through use-wear which is more invasive on the right edge. The second Paroto artifact is pointed and one of three small sized flakes (P 87/B). The distal point is distinctly acute and positioned at a left angle. Invasive secondary retouch on the cortical, slightly oblique left edge extends from the distal point half-way down this edge. This area and the notch on the lower right edge may be use-worn.

A small roughly quadrilateral and bilaterally notched flake (M 70/V) of limestone is in its shape and flaking pattern very similar to the large 'waisted' flake described above (K 87/4). It's left edge was formed by the removal of one flake (exposing what appear to be fossilized scales), while the right step-flaked edge exhibits two flake scars. Cortex is limited to the central part of this specimen. Both edges may be utilized. This artifact is less thick compared to a mainly non-cortical flake (M 70/U) on which both lateral oblique margins exhibit secondary retouch and possible use-wear, more pronounced on the right edge. The apex of the blunt point may have been removed.

4. CHRONOLOGICAL AND TYPOLOGICAL CONTEXT OF THE WALANAE ARTIFACTS

The Walanae artifacts have on considerations of terrace association, findspot, typology, and observations of fluvial wear and patination, confirmed in the above described sample, been divided into three temporal groups (Bartstra, 1978; Bartstra et al., 1991). Comparison to assemblages from three Late Pleistocene and Holocene cave sites in the Maros region near the southwestern coast of Southwest Sulawesi provides a provisional chronological framework for the artifacts (Bartstra et al., 1991). These limestone caves document a temporal, though possibly interrupted record of human occupation from c. 30 000 years BP to c. 2000 years BP (Mulvaney & Soejono, 1970, 1971; Glover, 1981; Bellwood, 1985).

The described bifacial cobble artifacts, partial pointed bifaces, the pointed uniface, large cores, and the large waisted flake, are the most abraded and patinated artifacts from the Walanae area. These artifacts derive from high terrace gravel, which has a sheet-like distribution on the river-facing slopes of the hills bordering the Walanae. The artifacts lie possibly not too far from the place(s) where they were manufactured and utilized (Bartstra, 1978; Bartstra et al., 1991). The described categories have not been found anywhere else in Sulawesi, including the oldest, radiometrically dated site in Sulawesi, the partially excavated Leang Burung 2 cave (Glover, 1981, and pers. comm.; Bartstra et al., 1991). The described artifacts may therefore represent the earliest evidence

of human occupation in Sulawesi and may indicate that the initial adaptation to this island environment was based on a flake and core tool technology.

Cores and flakes with better preservation of flake scars are found on the surface (tread) of the high terrace of the Walanae river and in those deposits which may constitute lower terrace levels (Bartstra, 1978). Some of the above described bifacial pebble artifacts and flakes can be included in this second group (e.g. K 87/6, M 70/B, and M 70/U). The small sized flakes with their slight fluvial wear and patination might also be included in this category. According to Glover (1981, and pers. comm.) these artifacts are comparable to the Late Pleistocene Leang Burung 2 industry.

Only two specimens in the sample might belong to the third recognized group of small cores and flakes with no evidence at all of fluvial transport and found on hilltops and other high territory throughout the entire drainage system of the Walanae river (B 70/1 and M 70/1). We consider this third group as definitely Holocene and synchroneous with the later cave artifacts of the Maros region ('Toalian').

The archaeological sequence of the Ulu Leang cave site in the Maros region is of particular importance here. The Ulu Leang assemblage includes the denticulated variety of the so-called Maros projectile-points which appear for the first time at Ulu Leang from about 6200 BP until the site was abandoned at about 3000 BP (Glover, 1976). These points have also been excavated at the later site of Leang Burung 1 in association with pottery (Mulvaney & Soejono, 1970; 1971). No substantial change in stone artifact technology of the Maros assemblages has been recognized until about 7000 years ago with the appearance of an increase in flake tools and a trend toward smaller sized tools (Presland, 1980). The diminution in relative artifact size in the Maros sites may indicate that the unrolled bifacial pebble artifacts, cores, and small flakes from the northern Walanae depression are younger than the abraded specimens, reflecting an evolution towards smaller size in that area too.

The Walanae artifacts were originally attributed to the Cabenge 'flake' industry of South Sulawesi by van Heekeren (1958). The smaller flake specimens bear out van Heekeren's (1958; 1972) observation of the technological and morphological similarities he recognized between the artifacts from Cabenge and from the site of Sangiran in Java, in particular the keeled flake-blades. However, although commonalities be-tween these flakes to a number of the Walanae artifacts are apparent, this cannot in any way indicate a cultural or proximate temporal relationship. This is an important point, since the frequent similarity of temporally separate assemblages constitutes a major feature of Far Eastern Palaeolithic technology (e.g. see Ikawa-Smith, 1978; Aigner, 1981; Pope, 1988). If one has to look for similarities or resemblances it is more adequate to compare the Walanae artifacts with the so-called

Pacitanian (formerly Patjitanian) of the south coast of Java. Especially the core artifacts make a comparison on typological and technological grounds worthwhile, setting aside for the moment any notions of cultural affinity.

The Walanae technology is characterized by the generally consistent use of hard-hammer percussion, preferential selection of fine-textured raw materials, and uncomplex modification. The faceted striking platform of one flake exhibits evidence of rough core preparation (but not Levallois technology). Faceted striking platforms have also been noted at Leang Burung 2 (Glover, 1981). The mainly vertical flaking on the bifaces and the uniface is comparable to that observed on the pebble and cobble artifacts. However, it is the total morphological pattern of the bifaces and the uniface that makes these artifacts characteristic in the selection of larger nodules, more extensive flaking of one face compared to the obverse, their pointed shape with modification of the points limited to the dorsal aspect and mainly to the left side, and the 'arched' form of the dorsal. The Kecce biface (K 87/1; figs 4 and 5) is the most extensively worked specimen and also shows more secondary retouch along it's left dorsal edge. The morphology and size of these artifacts is very similar to an artifact collected by H.G.A. van Panhuys in southwestern Halmaheira, Indonesia.[7] This specimen is in silicified limestone and described as an 'hand-axe on a river pebble' (cobble, 13×11 cm). In their shape and in the technique of mainly vertical flake removal, the pointed partial bifaces and the pointed uniface from Sulawesi and the specimen from Halmaheira are very similar to the Pacitanian artifacts classified as 'hand-axes' or 'proto-hand-axes' (von Koenigswald, 1939; Movius, 1948: pp. 358, 361; van Heekeren, 1972: pp. 37, 41). However, these so-called handaxes are not to be confused with those known from the Acheulean of Africa and Europe, since they are in most cases not of Acheulean technique. We prefer to classify these proto-handaxes and handaxes as cores, unifacial or bifacial pebble and cobble artifacts, picks, or pointed unifaces and bifaces.

The similarities between the pebble and cobble artifacts and the bifaces and the uniface suggests that the latter technique may have developed from the former one. Movius (1944: p. 101; 1948: p. 361) thought that the proto-handaxes and handaxes of the Far East were a local development. The probably earliest dated artifacts in Southeast Asia are unifacial cobble artifacts ('choppers') from early Middle Pleistocene contexts in northern Thailand (Pope et al., 1981; 1986; see also Jacob et al., 1978). This may indicate that bifacially modified artifacts are a later temporal phenomenon, but at present we have too few sites in Southeast Asia to support this interpretation.

It is of interest that the two pointed partial bifaces and the pointed uniface were not more extensively modified (e.g. into handaxes) despite the tractable raw materials

in which they were made. This indicates that the material was not a limiting factor, and that other explanations must be sought to understand the technology (see below). These artifacts were manufactured by modern *H. sapiens*, but lack the complexity which researchers familiar with European and Near Eastern assemblages cite as a characteristic component of the cultural character of this species (e.g. Binford, 1985; Mellars, 1989; Klein, 1989).

5. FUNCTIONAL ASPECTS

At this stage of research it is indeterminable whether the specimens reflect different activity facies. Secondary retouch is usually not extensive and occurs on the partial bifaces, the uniface, and on four flakes. The pattern of flake removal and the usually sinuous edges on the bifacial pebble and cobble artifacts suggests that these specimens were made for use as tools and may indicate thinning of edges for functional purposes. The variable degrees of edge damage on some or all of these artifacts may represent use-wear, but this cannot be ascertained with specimens found in a fluvial depositionary context, as is the case with the Walanae implements.

Some of these artifacts, including the not recovered flakes detached from the pebble and cobble artifacts may have been used for the manufacture of non-lithic tools. The notched flakes and indications for selection of cortical edge utilization may be reflective of wood-working processes. One notched flake from Leang Burung 2 on which the cortical edge was utilized is illustrated by Glover (1981: fig. 7c). A number of authors familiar with the Pleistocene record of Far Eastern lithic technology have argued that the use of non-lithic technology could explain the conservative and generally amorphous character of the local industries (Gorman, 1970; van Heekeren, 1972; Hutterer, 1977 contra 1985; Harrisson, 1978; Ikawa-Smith, 1978; Pope, 1983; 1989). It has also been suggested (Pope, 1988; 1989) that the distribution of bamboo may coincide with the distribution of chopper-chopping tools as defined by Movius (1944; 1948). The ubiquitous availability of bamboo and other suitable plant material for tool production in the forested environments of the Far East may have presented the triggering mechanism in the evolution of an evolved non-lithic technology of which the Walanae stone artifacts may form a part. Thus, modelling hominid behaviour in the Far East solely by means of lithic technology, disregards consideration of the palaeoenvironmental context. The informality of the Walanae artifacts, although manu–factured by modern *H. sapiens* in tractable materials may be more suitably explained in terms of an emphasis on non-lithic tool production. The use of a limited variety of raw materials, i.e. silicified tuff and silicified limestone may be referable to the ubiquitous availability of these rocks compared to other materials. However, it cannot be excluded that they were preferentially selected, because of their effectiveness in processing such plants as bamboo and wood. While these considerations may contribute to a better understanding of hominid behaviour in this region, it should also be noted that some standardization of manufacture is indicated with regard to the pointed bifaces and the uniface. These artifacts may represent a lithic facies restricted to the Walanae, and recent discoveries of further specimens in the area seems to promise the possibility of identifying a regionally characteristic technology for Sulawesi.

6. CONCLUSION

The geomorphological, typological, and sea-level change evidence may indicate that some of the Walanae artifacts (the large flake, large cores, bifacially worked cobbles, and the partial bifaces and uniface) may be representative of the earliest phase of hominid occupation of Sulawesi. These artifacts and the earliest dates for hominid occupation of the Australian continent (Roberts et al., 1990), suggest that sea-travel in island Southeast Asia was initiated before 50 000 years ago. We therefore suggest that the chronometric evidences from the Niah and Leang Burung 2 localities should be interpreted as the minimum age for sea migration within island Southeast Asia. Although further research is necessary to detail the stratigraphic context of the Walanae artifacts, we believe that they nevertheless represent important evidence for Late Pleistocene and Early Holocene behaviour of early modern *H. sapiens*. The Walanae artifacts as a whole reinforce the conservative and uncomplex quality of Southeast Asian stone tool technology which does not seem to have changed significantly with the emergence of modern humans.

Further surveys and possible excavations with the aim to recover secure primary context evidence of human occupation is planned in the Walanae depression. This may lead to a better understanding not only of the process and pattern of early human colonization of island Southeast Asia, Australia and New Guinea, but also of the regional behavioural characteristics where early modern human occupation in the world has been documented.

7. ACKNOWLEDGEMENTS

It was not possible to depict all the artifacts of the described sample. We have therefore concentrated on a few distinct core implements. Other artifacts will be illustrated in forthcoming papers. They are all available for study at the Biologisch-Archaeologisch Instituut in Groningen.

We wish to thank Dr. D.A. Hooijer and Dr. R. Shutler Jr. for critical reading of the manuscript. The drawings were prepared by Mr. J.M. Smit and J.H. Zwier. Typing and lay-out were done by Mrs. I. Cornelis and Mrs. M. Bierma.

8. NOTES

1. The drainage area of the Walanae river is here referred to as the Walanae depression (see Bartstra et al. this volume).
2. Recently it has been suggested that *Homo erectus* could have used boats or rafts to reach the islands east of Java (J. de Vos & R. van Zelst, Infusis (Intern informatieblad van het Nationaal Natuurhistorisch Museum te Leiden) 49, p. 5).
3. The site of Kecce is not shown in figure 1. It is situated about 3 km southeast of Marale and consists of some low hills (local outcrop of the Walanae Formation; see Bartstra et al., this volume) with a gravel sheet on the river facing slope. This streamgravel is implementiferous, and is part of the Walanae riverterrace system.
4. The codes refer to the numbers on the artifacts: the letter gives the site, the first two numbers the year of the find.
5. The site of Paroto also lies outside the area depicted in figure 1. Paroto lies approximately one kilometre south of Kecce and consists of gravel-strewn hills. For more locational details of these sites on the right bank of the River Walanae, see caption of figure 2 in Bartstra et al., this volume.
6. Figure 7 depicts the number BI/15/78 on the ventral side of the artifact. This number is obsolete and even wrong, and should be read as: P 78/15. The biface is definitely from Paroto.
7. See frontispiece of volume 8 of the series *Modern Quaternary Research in Southeast Asia*, Balkema, Rotterdam.

9. REFERENCES

AIGNER, J., 1981. *Archaeological remains in Pleistocene China*. München.

BARTSTRA, G.-J., 1977. Walanae Formation and Walanae terraces in the stratigraphy of South Sulawesi (Celebes, Indonesia). *Quartär* 27/28, pp. 21-30.

BARTSTRA, G.-J., 1978. Note on new data concerning the fossil vertebrates and stone tools in the Walanae valley in South Sulawesi (Celebes). *Modern Quaternary Research in Southeast Asia* 4, pp. 71-72.

BARTSTRA, G.-J., 1983. Some remarks upon: fossil man from Java, his age, and his tools. *Bijdragen Taal-, Land- en Volkenkunde* 139, pp. 421-434.

BARTSTRA, G.-J. & BASOEKI, 1989. Recent work on the Pleistocene and the Palaeolithic of Java. *Current Anthropology* 30, pp. 241-244.

BARTSTRA, G.-J. & D.A. HOOIJER, 1992. New finds of fossil vertebrates from Sulawesi, Indonesia. *Lutra* 35, pp. 113-122.

BARTSTRA, G.-J., S.G. KEATES, BASOEKI & KALLUPA, 1991. On the dispersion of *Homo sapiens* in Eastern Indonesia: The Palaeolithic of South Sulawesi. *Current Anthropology* 32, pp. 317-321.

BARTSTRA, G.-J., D.A. HOOIJER, B. KALLUPA & M. ANWAR AKIB, this volume. Notes on fossil vertebrates and stone tools from Sulawesi, Indonesia, and the stratigraphy of the northern Walanae depression.

BELLWOOD, P., 1985. *Prehistory of the Indo-Malaysian Archipelago*. Sydney.

BERGGREN, W.A. & J.A. VAN COUVERING, 1979. The Quaternary. In: R.A. Robinson & C. Teichart (eds), *Treatise on invertebrate paleonotlogy*. Lawrence, Kansas, pp. 505-543.

BINFORD, L.R., 1985. Human ancestors: Changing views of their behavior. *Journal of Anthropological Archaeology* 4, pp. 492-327.

CHAPPELL, J. & B.G. THOM, 1977. Sea levels and coasts. In: J. Allen, J. Golson & R. Jones (eds), *Sunda and Sahul*. New York, pp. 275-292.

CHAPPELL, J. & N.J. SHACKLETON, 1986. Oxygen isotopes and sea level. *Nature* 324, pp. 137-140.

CLARK, J.T., 1991. Early settlement in the Indo-Pacific. *Journal of Anthropological Archaeology* 10, pp. 27-53.

FOX, R.B., 1970. *The Tabon Caves: archaeological explorations and excavations on Palawan Island, Philippines* (= National Museum of the Philippines, Monograph 1).

GLOVER, I.C., 1976. The Late Stone Age in eastern Indonesia. *World Archaeology* 9, pp. 42-61.

GLOVER, I.C., 1981. Leang Burung 2: An Upper Palaeolithic rock shelter in South Sulawesi, Indonesia. *Modern Quaternary Research in Southeast Asia* 6, pp. 1-38.

GORMAN, C.F., 1970. Excavations at Spirit Cave, North Thailand: Some interim interpretations. *Asian Perspectives* 13, pp. 79-107.

HARRISSON, T., 1970. The prehistory of Borneo. *Asian Perspectives* 13, pp. 17-45.

HARRISSON, T., 1978. Present status and problems for Palaeolithic studies in Borneo and elsewhere. In: F. Ikawa-Smith (ed.), *Early Paleolithic in South and East Asia*. The Hague, pp. 37-57.

HEEKEREN, H.R. VAN, 1949. Early man and fossil vertebrates on the island of Celebes. *Nature* 163, pp. 492.

HEEKEREN, H.R. VAN, 1958 (1960). The Tjabenge flake industry from South Celebes. *Asian Perspectives* 2, pp. 77-81.

HEEKEREN, H.R. VAN, 1972. *The Stone Age of Indonesia*, 2nd rev.ed. (=Verhand. Kon. Inst. Taal-, Land- en Volkenkunde 61). The Hague.

HOOIJER, D.A., 1949. The Pleistocene vertebrates of southern Celebes. *Chronica Naturae* 105, pp. 148-150.

HOOIJER, D.A., 1958 (1960). The Pleistocene vertebrate fauna of Celebes. *Asian Perspectives* 11, pp. 71-76.

HOOIJER, D.A., 1975. Quaternary mammals west and east of Wallace's Line. *Modern Quaternary Research in Southeast Asia* 1, pp. 37-46.

HOOIJER, D.A. & G.-J. BARTSTRA, in press. Fossils and artifacts from southwestern Sulawesi. *Proceedings Pithe–canthropus Centennial, Leiden 1993, Session B*.

HUTTERER, K.L., 1977. Reinterpreting the Southeast Asian Palaeolithic. In: J. Allen, J. Golson & R. Jones (eds), *Sunda and Sahul*. London: Academic Press, pp. 31-71.

HUTTERER, K.L., 1985. The Pleistocene archaeology of Southeast Asia in regional context. *Modern Quaternary Research in Southeast Asia* 9, pp. 1-23.

IKAWA-SMITH, F., 1978. Introduction: The Early Paleolithic tradition of East Asia. In: F. Ikawa Smith (ed.), *Early Paleolithic in South and East Asia*. The Hague, pp. 1-10.

JACOB, T., R.P. SOEJONO, L. FREEMAN & F.H. BROWN, 1978. Stone tools from Mid-Pleistocene sediments in Java. *Science* 202, pp. 885-887.

KLEIN, R.G., 1989. *The Human Career*. Chicago.

KOENIGSWALD, G.H.R. VON, 1939. Das Pleistocän Javas. *Quartär* 2, pp. 28-53.

MELLARS, P., 1989. Major issues in the emergence of modern humans. *Current Anthropology* 30, pp. 349-385.

MOVIUS, H.L., 1943. The Stone Age of Burma. In: H. de Terra & H.L. Movius (eds), Research on Early man in Burma. *Transactions of the American Philosophical Society* 32, pp. 341-393.

MOVIUS, H.L., 1944. *Early man and Pleistocene stratigraphy in southern and eastern Asia* (= Papers of the Peabody Museum of American Archaeology and Ethnology 19, 3). Cambridge (Mass.).

MOVIUS, H.L., 1948 (1949). The Lower Palaeolithic cultures of southern and eastern Asia. *Transactions of the American Philosophical Society* 38, pp. 329-420.

MULVANEY, D.J. & R.P. SOEJONO, 1970. The Australian-Indonesian archaeological expedition to Sulawesi. *Asian Perspectives* 13, pp. 163-178.

MULVANEY, D.J. & R.P. SOEJONO, 1971. Archaeology in Sulawesi, Indonesia. *Antiquity* 45, pp. 26-33.

PEARCE, R. & M. BARBETTI, 1981. A 38 000-year-old archaeological site at Upper Swan, Western Australia. *Archaeology in Oceania* 16, pp. 173-178.

POPE, G.G., 1983. Evidence on the age of the Asian hominidae. *Proceedings, National Academy of Sciences U.S.A.* 80, pp. 4988-4992.

POPE, G.G., 1988. Recent advances in Far Eastern Paleoanthropology. *Annual Review of Anthropology* 17, pp. 43-77.

POPE, G.G., 1989. Bamboo and human evolution. *Natural History*, October, pp. 48-57.

POPE, G.G., D.W. FRAYER, M. LIANGCHAREON, P. KULASING & S. NAKABANLANG, 1981. Paleoanthropological investigations of the Thai-American Expedition in Northern Thailand (1978-1980): An interim report. *Asian Perspectives* 21, pp. 148-163.

POPE, G.G., S. BARR, A. MACDONALD & S. NAKABANLANG, 1986. Earliest radiometrically dated artifacts from mainland Southeast Asia. *Current Anthropology* 27, 3, pp. 275-279.

PRESLAND, G., 1980. Continuity in Indonesian lithic traditions. *The Artefact* 5 (1-2), pp. 19-46.

ROBERTS, R.G., R. JONES & M.A. SMITH, 1990. Thermoluminescence dating of a 50000-year-old-human occupation site in northern Australia. *Nature* 345, pp. 153-156.

SARTONO, S., 1979. The age of the vertebrate fossils and artefacts from Cabenge in South Sulawesi, Indonesia. *Modern Quaternary Research in Southeast Asia* 5, pp. 65-81.

SHUTLER, R., Jr., 1984. The emergence of *Homo sapiens* in Southeast Asia, and other aspects of hominid evolution in East Asia. In: R.O. Whyte (ed.), *The evolution of the East Asian environment* 2. Hong Kong, pp. 818-821.

SJAHROEL, A.A., 1970. Penjelidikan geologi pendahuluan daerah bahagian tengah Kabupaten Soppeng di Sulawesi Selatan. Bandung, unpubl., 31 pp.

THIEL, B., 1987. Early settlement of the Philippines, Eastern Indonesia, and Australia-New Guinea: a new hypothesis. *Current Anthropology* 28, pp. 236-241.

WENTWORTH, C.K., 1922. A scale of grade and class terms for clastic sediments. *Journal of Geology* 30, pp. 377-392.

INTRASITE SPATIAL ANALYSIS
AND THE MAGLEMOSIAN SITE OF BARMOSE I

DICK STAPERT

Biologisch-Archaeologisch Instituut, Groningen, Netherlands

ABSTRACT: The Maglemosian site of Barmose I in Denmark (Blankholm, 1991) is analysed by the ring and sector method. This is a simple method for within-site spatial analysis, based on the use of rings and sectors around hearths. The results of the analysis are contrasted with the ideas put forward by Blankholm. The ring and sector method makes it possible to demonstrate whether a hearth was inside a dwelling or in the open. Contrary to the assumption by Blankholm, it was found that the hearth of Barmose I must have been located in the open air, not inside a hut, despite the presence of bark flooring. A general conclusion is that complex computerized procedures, such as several of the clustering techniques applied by Blankholm, are not well suited to analyse open-air sites of this type, consisting of a central hearth with an artefact concentration around it. Their level of resolution is set too high for such situations, where many different activities were performed in a small area near the hearth, overlapping each other in space. Indications are that Barmose I was a hunting camp, occupied by a small group of men; there are no good arguments for the presence of women.

KEYWORDS: Intrasite spatial analysis, ring and sector method, Mesolithic, Maglemosian, dwelling structures.

1. INTRODUCTION

Intrasite spatial analysis, the study of spatial patterns within a site, has developed into a somewhat schizophrenic field. On the one hand, there are archaeologists who try to analyse distribution maps by visual inspection and simple descriptive means. On the other hand, a whole series of complicated mathematical and statistical techniques, which generally require the help of computers, have been developed by several archaeologists since about 1970. In general, these two groups of archaeologists hardly communicate with each other in any fruitful way. The first group mostly consists of field archaeologists, who themselves excavated the sites under discussion. The second group is composed of statistically oriented archaeologists, who play computer games with sites that more often than not were excavated by colleagues not well-versed in mathematics.

The 'mathematicians' attempt to quantitatively describe spatial patterns, by contouring, clustering, establishing patterns of covariation between artefact types, etcetera. They expect to recognize spatial patterns that, so it is claimed, are difficult to ascertain in any other way, and to do so 'objectively'. In most cases, the outcomes of the mathematical procedures are eventually rephrased in a descriptive way, and interpreted in terms of 'activity areas'. For more details about the techniques currently in use, the reader is referred to several recent publications (Hietala, 1984; Carr, 1984; 1985; Whallon, 1984; Kent, 1987; Blankholm, 1991).

In many cases the mathematical/statistical techniques are quite complicated, which has discouraged many archaeologists from applying them, or even from trying to understand them. Moreover, quite a few of these techniques do not seem to work well, producing results

TERMS AND ATTRIBUTES

D: Distance to the centre of the hearth.

R: The 'richest site-half'. The site-half with the highest number of tools. At Barmose I: sectors 5, 6, 7 and 8.

P: The 'poorest site-half'. The site-half with the lowest number of tools. At Barmose I: sectors 1, 2, 3 and 4.

A: The quarter within R with the highest proportion (with respect to the total number of tools per quarter) of 'projectiles' (in the case of Barmose I: microliths). At Barmose I: sectors 7 and 8.

B: The quarter within R with the lowest proportion of 'projectiles'. At Barmose I: sectors 5 and 6.

Asymmetry index: The proportion of tools present in R to the total amount of tools. At Barmose I: 63.0%.

Centrifugal index: Mean D of the cores/mean D of the tools. At Barmose I: 1.07.

Scraper/projectile D index: Mean D of the scrapers/mean D of the 'projectiles'. At Barmose I: 1.14.

Tools/cores in R index: % of N tools in R/% of N cores in R. At Barmose I: 1.19.

Core index: N cores/N tools. At Barmose I: 0.20.

Projectile/burin index: N 'projectiles'/N burins. In the case of Barmose I 'projectiles' are microliths. At Barmose I: 1.83.

A/B Fisher p: The p according to the Fisher Exact Probability Test (Siegel, 1956) for the difference in proportions of 'projectiles' to 'other tools' between the quarters A and B. In the case of Barmose I 'other tools' are: scrapers, burins, core/flake axes, denticulated/notched pieces, and blade/flake knives. At Barmose I: 0.32.

31

that either are hardly interpretable or could easily have been obtained by simpler means. It is also true to say that many mathematical techniques have underlying models or assumptions that are not really adapted to the analysis of archaeological residues. Therefore, complex techniques will at best produce a mixture of potentially valuable information and meaningless 'artefacts' created by the mathematical procedures, and it may be impossible to disentangle these components. Uncritical application of these techniques thus may easily lead to serious cases of over- or misinterpretation. It is unrealistic to believe that a mathematical or statistical procedure can be developed that brings out all spatial patterns existing at a given site. These are of many kinds, because many different site-formation processes have played a part (e.g. Schiffer, 1976; Binford, 1983). We should be pleased whenever a technique demonstrates at least some interpretable patterns in a satisfactory way.

Given the situation described above, there is no reason to abandon the use of simple approaches to intrasite spatial analysis alongside those involving complex computerized procedures. In this article one such method is introduced, which is based on the use of rings and sectors around 'domestic hearths' (Stapert, 1989; 1990a; 1990b; Stapert & Terberger, 1989). The idea behind this method is that the domestic hearth was a focal point, attracting many activities – irrespective of whether it was inside or outside a dwelling (e.g. Binford, 1983; Olive & Taborin, 1989; Yellen, 1977). The ring and sector method is therefore feature-oriented. It should be clear that this method does not claim to detect all possible spatial patterns in sites. It is directed at describing and interpreting global spatial patterns that relate to the domestic hearth. It is essentially a way of partitioning space (in two related ways: rings and sectors), which seems more suited than any regular grid structure to analyse sites where the global spatial structure is determined by the presence of a central hearth.

So far, the method has been applied to twelve concentrations of Pincevent (Late Magdalenian), four concentrations of Gönnersdorf (Late Magdalenian), and to several other Late Palaeolithic and Mesolithic sites in northwestern Europe. In this article I will use the ring and sector method to analyse the Maglemosian site of Barmose I in Denmark, and the outcomes will be contrasted with the results of Blankholm (1991), who analysed the same site using four different computerized procedures. It should be stressed that the main goal of this article is not so much to contribute to the knowledge about the Maglemosian, though I hope it will, but to explore the potential of the ring and sector method, compared to other techniques of spatial analysis.

2. THE SITE OF BARMOSE I

A book by Blankholm (1991) on intrasite spatial analysis appeared recently. Blankholm's text is especially useful as a technical compendium: no fewer than ten different mathematical/statistical techniques are described and illustrated in much detail. Four of these techniques, considered by Blankholm to be the most effective (k-means analysis, unconstrained clustering, correspondence analysis, and his own 'presab'), are applied by him to the early Maglemosian site of Barmose I. Most of the other techniques described in his book seem to have disappointed Blankholm: "A perusal (...) rules out Index of Segregation/Aggregation, DANOVA, Morisita's Index, Hodder and Okell's A-index and Carr's Coefficient of Polythetic Association from further consideration. None of these methods have, in fact, proven capable of revealing anything of significance at all." (Blankholm, 1991: p. 167).

The site of Barmose I was discovered in 1966, and excavated by A.D. Johansson in 1967-1971. It is dated to 9170 BP by five accelerator dates (Fischer, 1991). In the middle of the find concentration was a large hearth, measuring about 2.5×1.7 m (see fig. 1), with sand and clay and quite a lot of charcoal. Around the hearth, remnants of sheets of bark were found. To the NNE of the hearth, about 1.6 m outside its periphery, a large stone was encountered (a sitting stone? see also section 11).

A first test pit is not indicated on the drawing in figure 1, because its position is not exactly known. It was probably located in the northwestern part of the site (Blankholm, 1991: pp. 185, 204), but elsewhere it is stated to have been in the northeastern part (Blankholm, 1991: p. 186). It only partly disturbed the culture layer, and the amount of artefacts from this test pit is not fully

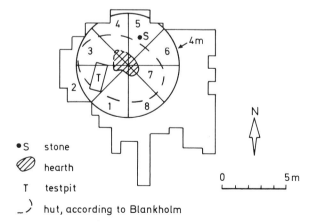

●S   stone
▨   hearth
T   testpit
⌐⌐)   hut, according to Blankholm

Fig. 1. Barmose I, general site plan. The area within 4 m from the hearth centre is analysed by the ring and sector method; it is divided into 8 sectors. Note the test pit in sector 2, and the large stone in sector 5. Also indicated is the outline of the hut, as assumed by Blankholm (1991: map overlay 2).

known (Blankholm, 1991: p. 186). A second, more regular test pit was located to the SW of the hearth. It measured 2×1 m (see fig. 1), according to Blankholm's drawing (Blankholm, 1991: map overlay 2), not 3×1 m as stated in his text (Blankholm, 1991: p. 186). The artefact content of this test pit was not considered by Blankholm, and is not plotted on his distribution maps; the number of artefacts is merely said to be low (Blankholm, 1991: p. 204). In my opinion, the disturbing effect of these two test pits on any kind of spatial analysis is taken too lightly in Blankholm's discussion of the results (Blankholm, 1991, p. 204), especially as he fails to inform us of their artefact contents.

Blankholm's 'presab' technique produces clusters, and the best solution according to Blankholm consists of a configuration of 19 clusters (which are not homogeneous in space: see fig. 2). The contents of these clusters are given in a table in the form of presence/absence data (Blankholm, 1991: Table 70). For example, cluster 2 is characterized only by burins, cluster 17 by notch remnants ('microburins'), splintered pieces ('square knives'), denticulated/notched pieces, and cores. Subsequently, these clusters, which are widely scattered in space, are grouped into 15 'activity areas'. This is essentially an intuitive procedure; it is not at all clear how this grouping is achieved. It can be seen in

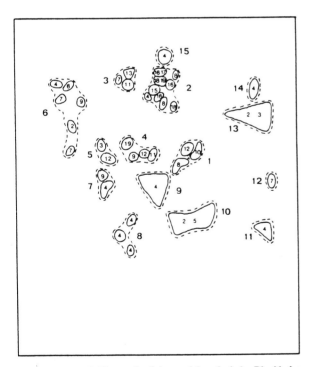

Fig. 2. Barmose I. The result of the spatial analysis by Blankholm (1991), on the basis of his 'presab' technique. Nineteen 'clusters', which are not homogeneous in space, are grouped into fifteen 'activity areas'. The represented area of 12×14 m was only partially excavated (see fig. 1). After Blankholm, 1991: fig. 125.

Blankholm's picture (fig. 2) that very different clusters (in terms of content) are grouped together. One wonders why Blankholm should first do a lot of calculating if in the end, to make some sense of the results, he resorts to intuitive grouping in a way that is not at all different from what an 'old-fashioned' archaeologist would do on the basis of the distribution maps. The resulting 15 'activity areas' are then loosely described and interpreted, for example as 'general work areas'. To my mind the results of such procedures are rather disappointing, and I find output such as figure 2 very hard to interpret; such pictures seem to obfuscate rather than clarify.

Even more disturbing is the fact that Blankholm's analysis proceeds on the basis of several unproven assumptions, which are not critically tested. The most important of these is the idea that a hut was present at the site, with the hearth located at the centre of its interior. The demonstration that a dwelling was present should be one of the goals of intrasite spatial analysis, not an assumption to start with! It will be realized that the interpretation of any 'patterns', established with the help of whatever mathematical technique, will be very different, depending on whether or not the presence of a dwelling is assumed. Blankholm's arguments for the presence of a hut are: "... the sharp inflection in the debitage distribution, the remnants of horizontal bark flooring, and a hearth of sand and clay with conspicuous amounts of charcoal and burnt flint ..." (Blankholm, 1991: pp. 184, 185). The hut is indicated on a drawing (Blankholm, 1991: map overlay 2) as an oval outline, with a diameter of 6.9×4.7 m (see fig. 1); the entrance is supposed to have been to the east (Blankholm, 1991: p. 204).

Blankholm's arguments are not conclusive, however. The presence of bark flooring is not necessarily indicative of a dwelling (see also Bokelmann, 1986; and section 6). An abrupt change in local artefact density, as indicated by Blankholm in the northeastern part of the site, is also not a conclusive argument: there are many other mechanisms that could have caused such a pattern. Moreover, this phenomenon does not show up clearly either on the artefact density map (fig. 3) or in the ring diagrams (to be discussed in later sections). It is also completely unclear to me why the presence of a large hearth, with sand and clay and a lot of charcoal, should be regarded as evidence for a dwelling around it. In fact, one would expect such large and dirty hearth areas to be located outside, in the open air (e.g. Binford, 1983). Another argument of Blankholm's is the supposed 'marginal distribution' of the cores (Blankholm, 1991: p. 185). It is true that in many Late Palaeolithic and Mesolithic sites the cores are located more peripherally, with respect to the central hearth, than tools. This tendency, called the 'centrifugal effect', however, is no proof of the existence of a dwelling. At many sites with open-air hearths, the centrifugal effect can be shown to have been operative (Stapert, 1989). Moreover, at

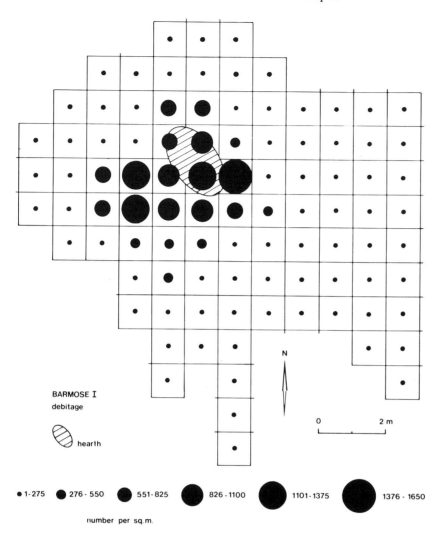

BARMOSE I
debitage

hearth

● 1-275   ● 276 - 550   ● 551-825   ● 826 -1100   ● 1101-1375   ● 1376 - 1650

number per sq. m.

Fig. 3. Barmose I. Density of flint 'debitage'. This map is organized according to the principles outlined by Cziesla (1990). Based on data in Blankholm, 1991: fig. 100.

Barmose I the centrifugal effect is rather weakly developed (see section 4): the cores cannot be said to be located significantly farther away from the hearth than the tools.

In his book Blankholm mentions his 'standard method for delineation of Maglemosian hut floors' (Blankholm, 1991: p. 185; see also Blankholm, 1984; 1987); this method seems to consist of simply equating a selected density contour line with the outline of a supposed hut.[1] Such procedures are meaningless. We need solid arguments for assuming the presence of a dwelling structure around a hearth, not conjectures without any foundation.

It is my opinion that the hearth of Barmose I was in the open air, and I will present arguments for this hypothesis in following sections of this article.

Not surprisingly, Blankholm's summary of his analyses of Barmose I clearly is determined by his idea that the hearth was inside a dwelling: "Basically what we can see is first a distinction between use of inside and

outside space. As to the inside, there is generally indication of at least three general multipurpose work areas around the hearth in the central and eastern part (where the entrance presumably has been) of the dwelling, whereas there are several indications suggesting that the western end of the floor was an area of low activity of different kinds, storage or sleeping. As to the outside, the content of the activity is more varied and thus indicates more differentiated uses." (Blankholm, 1991: p. 204; note the accumulation of vague terms in a single sentence).

For me it is hard to find such results very interesting. Most of Blankholm's picture simply reflects his unfounded assumption of a dwelling structure around the hearth. The 'area of low activity' in the west is no more than another way of telling us that the density of artefacts is low there. All in all, my impression is that the four computerized procedures applied by Blankholm did not perform very well; in my opinion at least we have not learned very much about Barmose I. A general problem with this kind of approach seems to be that there are no guidelines for interpreting the results of such rather mechanical mathematical operations. These do not seem to be directed at answering specific questions, and we are essentially left in the dark as to what the outcomes might mean. See section 11 for a further discussion.

3. RINGS AND SECTORS

If the domestic hearth is taken as the focal point, two ways of partitioning space are appropriate: using rings and sectors around the centre of the hearth, as depicted schematically in figure 4. The ring method is extremely simple: frequencies of artefacts are counted in rings of 0.5 m width around the hearth centre. It is advisable to count the ring frequencies per sector, because it may be fruitful to combine the sector and ring approaches, as we will see below. The distribution of artefact frequencies in the rings can be illustrated in the form of histograms, in which 0 on the X-axis is the centre of the hearth. It is important to note that we are not discussing densities here, in terms of numbers of artefacts per square metre. The rings only serve as a graphical illustration of the method, and in fact it would be more precise to speak of distance classes. The distance between an artefact location and the hearth centre is called 'D'.

The sector method investigates frequencies in sectors around the hearth centre. The choice of the number of sectors employed is arbitrary; in my experience a number of eight in most cases is adequate. The sectors should of course be equally large. With the sectors we are dealing with data that are much weaker than the distance data used in the ring method. Distance data can be considered as measurements in the ratio scale (Siegel, 1956), allowing many statistical manipulations (though in general nonparametric statistics are preferable). Frequencies in sectors around the hearth, on the other hand, constitute measurements in the nominal scale, despite the fact that the frequencies themselves are counted in the ratio scale. The same is true for frequencies in cells of a grid structure of whatever kind. For more details about the ring and sector method, and its problems and applications, the reader is referred to previous publications (Stapert, 1989; 1990b; Stapert & Terberger, 1989).[2]

In the case of Barmose I, the rings up to 4 m from the hearth centre are approximately complete (fig. 1). Therefore, it was decided to limit the analysis to that area. Artefacts found farther away are omitted, but they are relatively few. In total, 322 tools of 7 types were present within 4 m from the hearth centre, and only 41 were found beyond the 4 m limit (11.3% of total). Within 4 m from the hearth centre the frequencies and percentages of the 7 tool types are as follows:

	N	%	% of N 6 types (N = 149)
Splintered pieces ('square knives')	173	53.7	–
Scrapers	35	10.9	23.5
Microliths	33	10.2	22.1
Core axes (4) and flake axes (23)	27	8.4	18.1
Denticulated/notched pieces	20	6.2	13.4
Burins	18	5.6	12.1
Blade/flake knives	16	5.0	10.7
Total	322	100.0	99.9

Splintered pieces (which Blankholm calls 'square knives') are very numerous, and in fact do not constitute a formal tool class, as the splintering is no intentional retouch but probably the result of some heavy use, the nature of which is unclear to us (Blankholm suggests: 'light duty and precision work on bone/antler, wood and hides': Blankholm, 1991: p. 189; see also Eickhoff, 1988). Therefore, the percentages of the 6 formal tool types, based on their total of 149, are also given in the above list.

Apart from the tools, the ring and sector distributions of cores and 'microburins' (notch remnants) too are studied. There were 66 cores within 4 m from the hearth centre. Beyond the 4 m limit, 15 cores were present (18.5% of total). This is a higher proportion than in the case of the tools, indicating that, on average, cores indeed tend to be located somewhat further from the

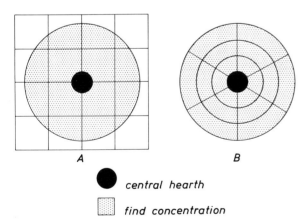

A   B

● *central hearth*

▨ *find concentration*

Fig. 4. Schematic representation of two different approaches to intrasite spatial analysis of flint concentrations around a central hearth. Many conventional analyses are based on a regular grid, as in A. In this article B is advocated, the ring and sector method. The latter system of subdividing space seems much more appropriate for the analysis of more-or-less circular artefact concentrations around a central hearth. The analysis of system B takes place in two phases. In the first, distances are measured between the locations of artefacts and the centre of the hearth. This is usually done in classes (rings) of 0.5 m width. In the second phase the numbers of artefacts per sector are counted. In most cases a number of eight sectors works best. Combining the ring and sector data may be fruitful; therefore, it is advisable to measure the distances per sector.

hearth than the tools. The difference is not significant, however. According to the chi-square two-sample test (Siegel, 1956), 0.1 < p (two-tailed) < 0.2.

None of the small 'microburins' (N = 16) were located more than 4 m from the hearth centre. They were probably left on the spot at the place where microliths were manufactured. This work was done close to the hearth, to the east of it (see section 8 and table 4).

The location of the first test pit and its contents are unknown to us. Therefore, it is impossible to estimate to what degree this pit affects our analysis. The contents of the regular 2x1 m test pit are also unknown, but its location is known. I have positioned the ring and sector system in such a way that only one sector is affected by this test pit: sector 2 (fig. 1). All ring diagrams in this article are based exclusively on the seven other sectors (the data from sector 2 are omitted). Concerning the sector data, it should be remembered throughout this article that the frequencies in sector 2 must be considered as minimum estimates.

4. THE CENTRIFUGAL EFFECT

Binford (1983) provided useful descriptions about people's spatial behaviour in relation to outdoor hearths, which can be summarized in his 'hearth model' (fig. 5). He distinguishes 'drop' and 'toss zones'. Drop zones are found close to the hearth in the form of a semicircle, where small debris fall to the ground during all sorts of activities, and generally are left lying.[3] Larger pieces of refuse end up in the toss zones. Two toss zones are distinguished: a 'backward toss zone' which lies in the form of an arc around the drop zone (on the same side of the hearth), and a 'forward toss zone' on the opposite, unoccupied side of the hearth. Near an open-air hearth, the drop zone and the backward toss zone are located on the side where the people sat and worked, to windward of the hearth. An important point to note is that pieces of refuse arrive in the toss zones individually, by being tossed or kicked away, gradually accumulating there in the course of the occupation period. This is in contrast to dumps, where waste is discarded collectively. Dumps are mostly found at some distance from the hearth, and it seems that at Barmose I dumps were absent (or, alternatively, located outside the excavated area).

There are two important differences between the drop zone and the toss zones. The first is that toss zones are clearly more peripheral with respect to the hearth, at any rate in an overall sense. There is a certain overlap, however, in terms of distance to the hearth, between the drop zone and the forward toss zone (indicated by means of broken lines in fig. 5). The second is the size of the items that end up in them: small objects in the drop zone, larger ones in the toss zones. Hence we are dealing with a size-sorting process: a tendency towards spatial segregation of finer and coarser refuse. On the

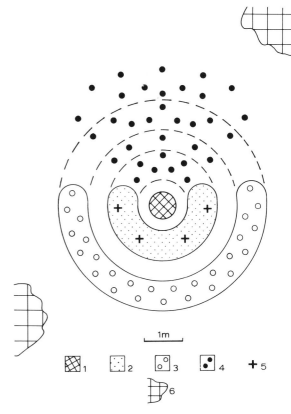

Fig. 5. Sketch of Binford's hearth model (after Binford, 1983: p. 153, with minor alterations). 1. Hearth; 2. Drop zone; 3. Backward toss zone; 4. Forward toss zone; 5. Seating positions of four people; 6. Dumps.

whole the coarser items have a greater chance than the small ones of ending up in the periphery of the site. This pattern has since long been known to archaeologists: many distribution plans show that cores (mostly the largest flint artefacts) mainly occur in the periphery of sites. The tendency for larger objects to end up farther away from the hearth is called the 'centrifugal effect'. A clear centrifugal effect would be expected especially if a backward toss zone existed.

The strength of the centrifugal effect can easily be quantified by means of the 'centrifugal index': mean D of the cores/mean D of the tools (D is distance to the hearth centre). In a sample of 18 Upper/Late Palaeolithic sites, all of them supposed to have had open-air hearths, this index was found to range from 0.61 to 2.35 (Stapert, 1989). An important finding was that not all sites show a clear centrifugal effect; at some sites cores were on average even somewhat closer to the hearth than tools. In the case of Barmose I the centrifugal effect is only weakly developed. The centrifugal index in this case is 1.07, which is too close to 1 to be significant.

It is easy to see why the ring method is well suited for studying the centrifugal effect. It should show up in ring distributions if we divide the artefacts into size-classes.

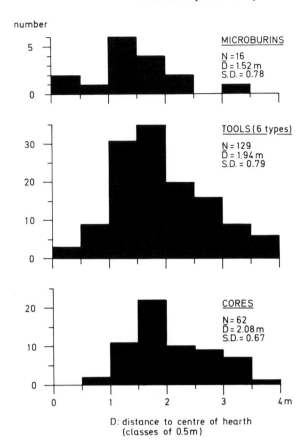

number

Fig. 6. Barmose I. Ring diagrams of 'microburins' (notch remnants), tools of six formal types (microliths + burins + denticulated/notched pieces + blade/flake knives + scrapers + core/flake axes) and cores. On the X-axis, 0 is the centre of the hearth. Note that the data from sector 2 are omitted, because of the test pit located in that sector (see fig. 1). It can be seen that the three distributions are clearly unimodal.

Table 1. Barmose I. Mean distances to the centre of the hearth. Only locations within 4 m from the hearth centre, excluding the locations in sector 2 (see main text, section 2). D: Distance to the hearth centre.

	N	Mean D	Stand. dev.
Microliths	33	1.81	0.78
Burins	16	1.77	0.67
Denticulated/notched pieces	20	1.91	0.70
Blade/flake knives	14	1.91	0.63
Scrapers	30	2.06	0.84
Core (3) and flake axes (13)	16	2.19	1.00
Total 6 types	129	1.94	0.79
Splintered pieces	166	1.94	0.79
Total 7 types	295	1.94	0.79
'Microburins'	16	1.52	0.78
Cores	62	2.08	0.67

Barmose I seems to belong to a group of sites where the centrifugal effect was largely absent. Other sites showing no clear centrifugal effect are Marsangy N19 (Schmider, 1979; 1984), Bro I (Andersen, 1973) and the three units at Pincevent Habitation 1 (Leroi-Gourhan & Brézillon, 1966). In fact, these sites show no clear evidence for the existence of either forward or backward toss zones. It has been hypothesized that such sites were occupied by men only; they may, for example, have been hunting camps (see section 10). In all cases the hearths at these sites appear to have been located in the open air, not within dwellings. At sites where a dwelling structure can be observed archaeologically, as at two concentrations of Gönnersdorf (see section 5), there is always a marked centrifugal effect (Stapert, 1989; 1990a). Thus, the absence of a clear centrifugal effect in Barmose I can be considered to be an indication for the absence of a hut around the hearth. Apart from establishing the presence or absence of the centrifugal effect, the ring method provides us with yet another way of approaching this important question, which will be discussed in the next section.

5. UNIMODAL AND BIMODAL RING DISTRIBUTIONS

Within a dwelling with a central hearth, the centrifugal movements are of course restricted by the walls. Therefore, one may expect much of the refuse to be carried outside and dumped en masse. One type of dump is characteristic of dwellings: the door dump (Binford, 1983). People simply throw their larger pieces of rubbish out through the entrance, to the left or to the right. This phenomenon in itself can be considered as contributing to the strength of the centrifugal effect. However, inside the dwelling the centrifugal effect will also be operative, though generally not in all directions.

For this purpose I have divided the artefacts of Barmose I into three groups, from small to large: 'microburins', tools (6 types, i.e. without splintered pieces) and cores. Artefact frequencies of these three groups in rings 0.5 m wide (excluding the locations in sector 2) are presented in figure 6. It can be seen that no clear centrifugal is present within 4 m from the hearth centre: the distribution of the cores has the same mode as that of the tools (in the 1.5-2 m class). There is a small difference, however, in the mean distances of tools and cores, as we have noted above. After combining the data into frequencies per 1 m rings, it is possible to test the difference between the cores and the tools by means of the chi-square two-sample test in a valid way (Siegel, 1956). The conclusion must be that there is no clear difference between the tools and the cores in this respect: $0.5 < p$ (two-tailed) < 0.7. Therefore, Blankholm's argument regarding the more peripheral location of cores with respect to the hearth, compared to that of tools, cannot be upheld. There is only a weak tendency, which cannot be shown to be significant in a statistical sense.[4]

The walls of the dwelling then serve as a barrier. The refuse gradually accumulates against them in the course of the occupation, again with a relatively high proportion of coarse material. This I have termed the 'barrier effect'.

When we consider the ring distribution of all the tools taken together, the sites investigated so far with the help of the ring and sector method seem to fall into two groups: those with unimodal and those with bimodal ring distributions (Stapert, 1989; 1990a; 1990b). Most analysed sites show unimodal ring distributions; this applies for example to all 12 analysed concentrations at Pincevent (Leroi-Gourhan & Brézillon, 1966; 1972; Julien et al., 1988), Oldeholtwolde (Stapert et al., 1986), Bro I, Marsangy N19, Olbrachcice 8 East (Burdukiewicz, 1986) and concentrations I and IV of Niederbieber (Bolus et al., 1988; Winter, 1986; 1987). As an example, the unimodal distribution of Niederbieber I is illustrated in figure 7. At none of the sites for which I have obtained unimodal ring distributions were any archaeological traces of tents of huts observed.

At the site of Gönnersdorf two concentrations occur with clear traces of the existence of tents. At Gönnersdorf I the presence of a tent is evident from a circular arrangement of postholes (Bosinski, 1979), at Gönnersdorf IV from the presence of a ring of large stones around the hearth, which can be interpreted as a tent ring (Bosinski, 1981; Terberger, 1988). At both concentrations the centrifugal effect is very strong: it not only resulted in the cores being far away from the hearth – in fact, most are located outside the dwellings –, but also affected the tools. When we consider the ring distributions of all tools combined in Gönnersdorf I and IV, their bimodal character is immediately apparent (see fig. 7 for the distribution of Gönnersdorf IV). The first peak lies at c. 1 m from the hearth centre; a second, higher one at c. 2.5 m. This second peak is generated mainly by the larger tools (such as blade scrapers and burins), and it coincides with the tent ring. The first peak can be interpreted as the drop zone near the hearth. It is made up especially of the small backed bladelets, with hardly any larger tools. In other words: only small objects are left near the hearth, while the larger ones, including tools, are removed from the central part of the tent.

In my opinion, the second peak results from the combined centrifugal and barrier effects. Two important points emerge from investigating the tents of Gönnersdorf: a) in a dwelling the centrifugal effect is stronger than it is around a hearth in the open air; b) the tent wall is made visible through the barrier effect, which results in a bimodal ring distribution. In other words: my interpretation of the second peak is that the centrifugal movements occurring inside a dwelling with a central hearth are stopped by the walls, in due time resulting in a second peak in the ring distribution that roughly coincides with the walls of the dwelling. More than 4 m away from the hearths, we often see a third peak at Gönnersdorf (not illustrated in fig. 7),

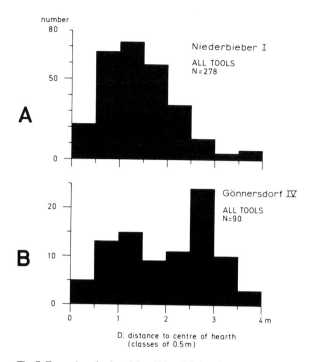

Fig. 7. Examples of unimodal and bimodal ring distributions of tools within 4 m from the hearth centres. Unimodal distributions, such as A, are considered to be characteristic of open air hearths, while bimodal ones, such as B, are associated with hearths inside dwellings (see main text, section 5). Compare with fig. 6.

which can be interpreted as resulting largely from the door dumps. For a more detailed discussion of Gönnersdorf, the reader is referred to other texts (Stapert, 1989; 1990a; Stapert & Terberger, 1989). It should be noted here, however, that at Concentrations II and III, where dwelling structures are not visible archaeologically, the same type of bimodal ring distributions have been obtained, suggesting that these sites too had tents (at least during one of their occupation phases: Stapert & Terberger, 1989). Other archaeologically 'invisible' tents have been demonstrated at Etiolles P15 (Olive, 1988) and at Verberie D1 (Audouze et al., 1981; Symens, 1986).

The analysis of the tents at Gönnersdorf I and IV has provided us with a method of demonstrating the presence of a dwelling with the help of the ring method. We can now classify archaeological residues with a central hearth into two types: those with unimodal and those with bimodal (or trimodal) frequency distributions of distances between tool locations and the hearth centres.

In the case of bimodal distributions we are dealing with hearths inside dwellings. Unimodal ring distributions will in general be characteristic of hearths in the open air. Of course, there are various complications. For example, if the hearth was located

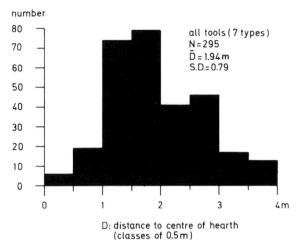

Fig. 8. Barmose I (7 sectors). Ring distributions of all the tools, of seven types: the six formal types represented in figure 6 and the splintered pieces. It can be seen that this distribution is slightly bimodal, which is caused by the splintered pieces (see fig. 15).

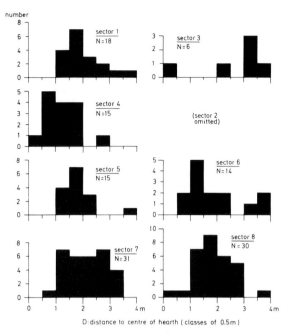

Fig. 9. Barmose I (7 sectors). Ring distributions per sector of the tools of the six formal types. Note the almost empty area within 3 m from the hearth centre in sector 3. Compare with figures 10 and 11, where the same phenomenon can be observed.

eccentrically inside a dwelling, we would need ring distributions per sector to demonstrate the presence of walls, and it will usually be profitable to study such distributions. For this, however, the numbers of artefacts per sector should not be too low. In many cases such a detailed approach is ruled out because of insufficient frequencies. In the case of Barmose I, fortunately, numbers of artefacts are sufficiently high to allow a sectorwise study of ring distributions.

In the following section I will investigate what the ring and sector method can contribute in this case, as regards the presence or absence of a hut around the hearth.

6. BARMOSE I: THE DWELLING HYPOTHESIS

The ring distributions of 'microburins', tools of the six formal types taken together, and cores, are illustrated in figure 6. The data from the area within 4 m from the hearth centre are used, excluding sector 2. It can be seen that all three distributions are unimodal. This suggests that the hearth at Barmose I was in the open air. However, if we include the splintered pieces in the analysis, the resulting diagram of all the tools (fig. 8) becomes slightly bimodal. Still, this distribution is not really comparable to the diagram of Gönnersdorf IV (fig. 7), as the second peak is not very conspicuous. As noted, the small second peak in figure 8 is mainly caused by the splintered pieces, which is immediately apparent from the ring distribution of that artefact group, illustrated in figure 15.

Given this situation, we need to investigate this matter more fully in this case, by studying the ring distributions per sector. In figures 9-11, I have presented the ring distributions for seven sectors (not sector 2) of

the splintered pieces, the tools of the 6 formal types taken together, and the cores. Most distributions are clearly unimodal. The second peak noted above is caused by a phenomenon that shows up in the distributions of all three artefact groups: a distant peak in sector 3, while the space within 2.5 to 3 m from the hearth centre in that sector is almost empty. It is possible that this phenomenon is the result of the first test pit, of which we do not know the exact location (see section 2). This seems unlikely, however, because the other sectors do not show high peaks between 2.5 and 3.5 m from the hearth centre. If the empty space in sector 3 was caused by testpitting, and if sector 3 originally possessed a unimodal distribution similar to those of sectors 1 and 4, the number of artefacts in sector 3 must have been extremely high to account for the frequencies in the rings between 2.5 and 3.5 m. This seems unlikely, because the western half of the site of Barmose I as a whole is characterized by low tool frequencies (see also section 7). We get the impression, therefore, that an area near the hearth in sector 3 was avoided during occupation. Possibly this area remained largely devoid of artefacts because it was covered by organic material that left no archaeological trace (wood?). I shall come back to this phenomenon in later sections of this article.

The conclusion of the analysis of the ring distributions at Barmose I can be no other than that the hearth was located in the open air. Of course, this conclusion does not exclude the possibility that a hut or other type of dwelling was present at Barmose I. There could have

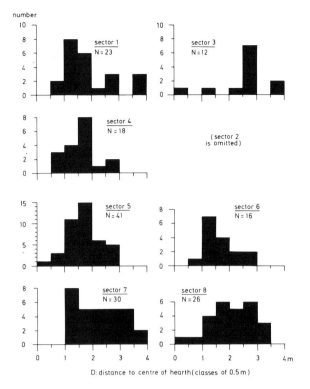

Fig. 10. Barmose I. Ring distributions per sector of the splintered pieces.

been a dwelling at some distance from the hearth, possibly outside the excavated terrain. We have no way of investigating this possibility, however.

In the above, two arguments were presented for the hypothesis that the hearth of Barmose I was not located inside a dwelling: 1) the absence of a strong centrifugal effect; 2) the fact that the ring distributions of tools (and cores) are essentially unimodal. Furthermore, it was noted earlier that, quite apart from these arguments, we would in any case expect such a large and dirty hearth area to be located outside. Evidently, the presence of bark floors does not seem to be associated exclusively with the interior of a dwelling, as supposed by Blankholm.

At the site of Duvensee several concentrations around large hearths were excavated by Bokelmann, and here too bark floors were present near the hearths (e.g. Bokelmann, 1986; 1989; Bokelmann et al., 1981; 1985). Bokelmann is of the opinion that the hearths of Duvensee were open-air ones (e.g. Bokelmann, 1989: p. 17); according to him the bark floors functioned to insulate the occupied area against groundwater. Grøn (e.g. 1987a: p. 304), however, has proposed that these were the sites of 'single-family dwellings'. The sites of Duvensee 8 and 13 were analysed with the ring and sector method. In both cases unimodal ring distributions were obtained, suggesting that Bokelmann is quite right: the bark floors were not inside dwellings. As an example the

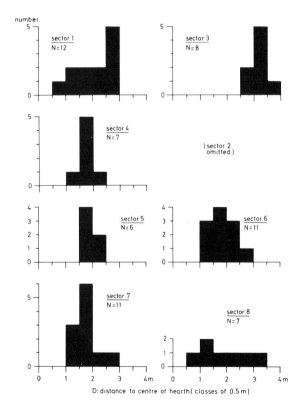

Fig. 11. Barmose I. Ring distributions per sector of the cores.

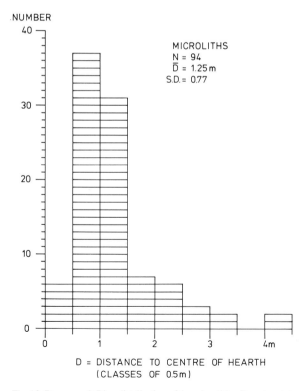

Fig. 12. Duvensee 8. Ring distribution of the microliths. Based on data

diagram for the microliths of Duvensee 8 is reproduced (fig. 12).

7. DROP AND TOSS ZONES

If we look at distribution maps of tools, many sites of the type discussed here, artefact concentrations around central hearths, show a marked asymmetry, in the sense that many more tools are found on one side of the hearth than on the opposite side. If artefact concentrations around hearths were created in the open air, as at Barmose I, the existence of a prevailing wind direction is a likely explanation.

First, however, I want to quantify this asymmetry and to establish that it is real. In order to investigate this, the concentration is divided into two halves so as to maximize the difference between the numbers of tools in the two halves. In other words, we seek four adjacent sectors that have a higher total of tools than all other combinations of four adjacent sectors. Although the number of tools in sector 2 must be considered as a minimum estimate, I do not expect this to affect our analysis very much, as in the western half of the site tool frequencies are relatively low everywhere. Throughout this article the site-half with the highest total number of tools will be called the 'richest site-half' or 'R', and the other half the 'poorest site-half' or 'P'. In the case of Barmose I, the richest site-half is composed of sectors 5, 6, 7 and 8, i.e. the eastern half of the concentration (fig. 13). The asymmetry can be quantified easily by calculating what percentage of the total number of tools is present in R. In Barmose I this is 63%: almost two-thirds of all the tools are in the eastern half. We then want to investigate whether this difference could have

arisen by chance. It is usual to apply the chi-square one-sample test in such cases (Siegel, 1956). It was found that the asymmetry is significant: p (two-tailed) < 0.001.

We have seen that the residue of Barmose I was most probably created in the open air. This means that people would have sat mainly on one side of the hearth: to windward, in order to avoid the smoke. The next question therefore is: was the occupied side of the hearth located in the richest site-half or in the poorest? In other words: is the drop zone in the site-half with high tool density, or in the opposite half? This is not a trivial question, because we cannot know a priori where most of the tools were eventually discarded: in the forward toss zone, or in the drop zone and the backward toss zone (see fig. 5). I have discussed this question *in extenso* in another article (Stapert, 1989), and do not want to repeat all the arguments here. The answer is unambiguous: the drop zone was (mostly) located in the richest site-half.

One of the problems with ethnoarchaeological observations such as Binford's hearth model, if we want to use them for archaeological interpretations, can be elucidated by the concept of time depth. The model depicted in figure 5 in fact illustrates the situation at a given moment. With archaeological sites, however, we are mostly dealing with a residue of an occupation of some duration, perhaps in the order of weeks or even months. Even if at any given moment during occupation the spatial 'organization' of the site at Barmose I resembled the model of figure 5, its lay-out did not necessarily remain unchanged. For example, if during occupation the wind direction changed several times, the whole system would have rotated around the hearth repeatedly. If the wind mostly came from the same direction, the resulting residue would still roughly resemble the model.

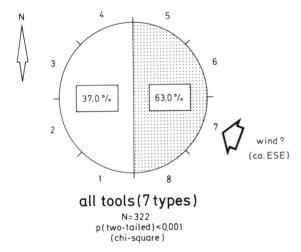

all tools (7 types)

N=322
p(two-tailed)<0.001
(chi-square)

Fig. 13. Barmose I. Reconstruction of the prevailing wind direction during habitation. The site is divided into two halves so as to maximize the difference between the numbers of tools (7 types) in the two halves. The percentages of N tools in the two halves are indicated. See main text, section 7.

Table 2. Barmose I. Frequencies in rings of 0.5 m width around the centre of the hearth, excluding the locations in sector 2. Groups of artefacts: 1. Microliths; 2. Burins; 3. Denticulated/notched pieces; 4. Blade/flake knives; 5. Splintered pieces ('square knives'); 6. Scrapers; 7. Core/flake axes; 8. Total tools (7 types); 9. 'Microburins'; 10. Cores.

Rings	Groups of artefacts									
	1	2	3	4	5	6	7	8	9	10
0-0.5 m	3	0	0	0	3	0	0	6	2	0
0.5-1	1	1	2	1	10	2	2	19	1	2
1-1.5	11	4	3	4	43	7	2	74	6	11
1.5-2	5	8	8	3	44	7	4	79	4	22
2-2.5	6	1	4	2	21	5	2	41	2	10
2.5-3	5	1	1	4	30	4	1	46	0	9
3-3.5	2	0	1	0	8	3	3	17	1	7
3.5-4	0	1	1	0	7	2	2	13	0	1
Total	33	16	20	14	166	30	16	295	16	62

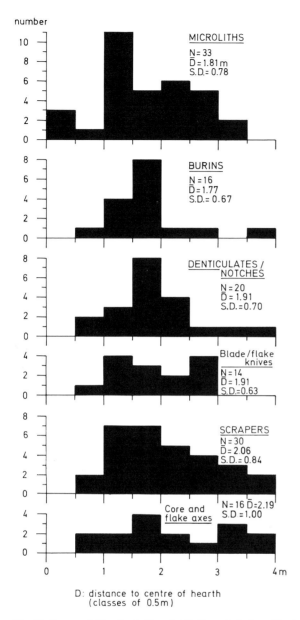

Fig. 14. Barmose I (7 sectors). Ring distributions of six formal tool types.

is that in the poorest site-half some special activities, which were not very time- and flint-consuming, were performed while the drop zone was located in the richest site-half.

The sector method provides us with a way of investigating whether or not the poorest site-half was the forward toss zone. We have noted that especially larger pieces of refuse tend to end up in the toss zones. Thus, if a forward toss zone existed in the poorest site-half, we would expect the proportion of cores, with respect to that of tools, to be higher in that site-half than in the richest site-half. In the case of Barmose I we would expect this all the more since a backward toss zone seems to have been only weakly developed (no clear centrifugal effect can be demonstrated: section 4).

The numbers of tools (7 types) and cores in the two site-halves will be found in table 4. It can be seen that the proportion of cores in the poorest site-half indeed is somewhat higher than that of tools. Of all the cores, 47% are present in P, of all the tools only 37%. However, the difference is not very large, and cannot be shown to be significant according to the chi-square two-sample test (Siegel, 1956): $0.1 < p$ (two-tailed) < 0.2. Though the poorest site-half probably served as the forward toss zone during occupation, this tendency towards spatial segregation of tools and cores is only weakly developed, just as in the case of the backward toss zone. We should reckon, therefore, that part of the tools present in the poorest site-half were left there after playing a functional role in that area. Thus, either the drop zone was in that half during a part of the occupation period with a deviating wind direction, or some special activities went on there while the drop zone was in the eastern half of the concentration. For the last possibility at any rate

However, if there was no prevailing wind direction during the period of occupation, the end product would definitely be a palimpsest residue, even if at any given moment the site's structure was similar to Binford's model.

All this leaves us with at least three possible processes to account for the presence of artefacts in the poorest site-half, which may all have been operative in the course of occupation. The first possibility is that also the poorest site-half contained the drop zone for some time, but for a shorter timespan than the richest site-half did. The second is that the poorest site-half was the forward toss zone for most or all of the time. The third possibility

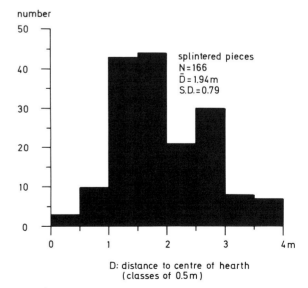

Fig. 15. Barmose I (7 sectors). Ring distribution of the splintered pieces ('square knives').

there are some arguments, which I will discuss in section 8.

It is of interest to note here that the presence of toss zones cannot be demonstrated very clearly in the case of Barmose I. Both tendencies towards spatial segregation of cores and tools which would indicate toss zones – a strong centrifugal effect, and a clearly higher proportion of cores in P compared to that of tools – are only weakly developed at Barmose I, and cannot be proven to be significant in a statistical sense. The occupants clearly did not bother very much about clearing up during occupation; this conclusion at any rate seems to apply to flint artefacts.

As concluded above, the drop zone must have been located in the eastern half of the concentration (R) most of the time. Since it is probable that the hearth of Barmose I was in the open air, it seems possible to reconstruct the prevailing wind direction during occupation. As the southern quarter within R contains more tools than the northern quarter (see table 5), and also because sector 1 has more tools than sector 4 (see note 3), the wind arrow in figure 13 is not placed in the middle of R, but is shifted somewhat towards the south, suggesting that the prevailing wind direction was roughly ESE.

8. TOOL TYPES AND RING DISTRIBUTIONS

It is of interest to study the ring distributions for the various tool types separately. The ring diagrams of the six formal tool types are presented in figure 14; the diagram of the splintered pieces is given in figure 15. Most diagrams are reasonably unimodal. However, especially the distribution of the splintered pieces seems to be bimodal. It was noted above that this bimodality is largely due to the fact that in sector 3 the area within 2.5 m from the hearth centre is almost devoid of artefacts, while the only peak in that sector occurs between 2.5 and 3.5 m; this applies especially to the splintered pieces (see figs 9-11). At this point it was decided to prepare separate ring diagrams of the richest and the poorest site-halves, for all the tools (7 types) taken together (fig. 16). It can clearly be seen that the diagram of the richest site-half, where the people would mostly have sat and worked, is a regular and unimodal dis-tribution; most of the tools are located between 1 and 2 m from the hearth centre. This diagram supports the conclusion reached above, viz that the hearth of Barmose I was in the open air. The diagram of the poorest site-half (sector 2 omitted) shows bimodality, which, as we have seen, is largely caused by the distant peak in sector 3.

In view of this situation, it would perhaps have been better to prepare ring diagrams for the separate tool types, based only on the locations in the richest site-half. However, numbers per tool type would then become

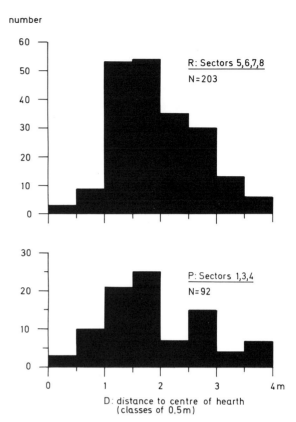

number

Fig. 16. Barmose I (7 sectors). Ring distributions of all tools, for the richest and poorest site-halves separately (see fig. 13). Note that the diagram for the richest site-half (R) is regularly unimodal. The diagram of the poorest site-half (P) is bimodal. In this case we are not dealing with a real bimodality (see the sectorwise ring diagrams in figs 9-11); this pattern is the result of the deviating situation in sector 3 (see main text, section 8).

so low that these diagrams would no longer be very informative. Moreover, more than two-thirds (69%) of all the tools represented in figures 9-11 occurred in the richest site-half, so any patterns present in that half would dominate the picture (see also Stapert, 1989).

The diagram of the microliths among all the tool types has its mode closest to the hearth centre: in the 1-1.5 m class. Most other tool types have the mode in the 1.5-2 m class. The core and flake axes are on average located farthest from the hearth. Also scrapers are on average located relatively far from the hearth. This pattern, with 'projectiles' (microliths) located close to the hearth and scrapers away from it, seems to be very common at sites where scrapers were made on blades. For example, it applies to 11 of the 12 analysed con-centrations of Pincevent (Stapert, 1989), and to many other Upper or Late Palaeolithic sites. I have explained this pattern as due to 'retooling' (Keeley, 1982). It is probable that heat was needed when new flint insets were fixed into their shafts with the help of, for example, birch tar (Moss & Newcomer, 1981; Moss, 1983), and this could be the reason why the repairing of weapons (and other tools with flint insets) took place close to the hearth. During the Upper and Late Palaeolithic, scrapers were mostly used to work hides (see e.g. Juel Jensen, 1988). Because of the fact that many types of hide-working required quite a lot of working space, scrapers would have ended up farther away from the hearth. During the Mesolithic, however, many scrapers were used to work wood (Juel Jensen, 1988). Most tasks carried out by means of tools such as borers and burins possibly required neither fire nor a large amount of space, so that these tools tended to be used and discarded at intermediate distances from the hearth.

It is possible to quantify the above-mentioned pattern by a simple index, analogous to the centrifugal index: the ratio of mean D of the scrapers to mean D of the microliths. In the case of Barmose I this index has a value of 1.14. This is a relatively low figure. In the case of Pincevent, this index is mostly above 1.5 (Stapert, 1989). Moreover, the difference between microliths and scrapers cannot be proved to be significant in a statistical sense (after combining the frequencies in two rings of 2 m width, the Fisher Test results in a p of 0.37). Of course, it is true that 'a behaviourally meaningful or relevant relationship is not necessarily statistically significant' (Blankholm, 1991: p. 43). However, in such cases we must have good arguments for believing that any patterns are indeed meaningful. The pattern of microliths (or other insets of projectiles) lying close to the hearth with scrapers farther away, shows up in many sites, and in several of them can be proved to be significant. Therefore, I believe that the same types of formation processes could have been at work at Barmose I. It is also possible, however, that this pattern is weakly developed in Barmose I because wood-working was dominant over hide-working in the case of the scrapers.

I shall now return to the deviating picture in the poorest site-half (fig. 16). In sector 3, the area up to 2.5-3 m from the hearth centre is relatively empty. This is most probably a real phenomenon, and not one caused by testpitting. One way to approach this phenomenon is to investigate in what proportions the various tool types are present in the richest and poorest site-halves. The data can be found in table 4. We noted above that about 63% of all the tools are located in the richest site-half. If we look at the individual tool types, however, some variation is apparent. The most conspicuous deviation from the general picture is exhibited by the axes. Whereas all the other types show proportions above 50%, and mostly above 60%, in the richest site-half, of the axes only 33% are present in this area. Moreover, there is a marked concentration in sector 2 (see table 3). This is all the more surprising since this sector is heavily disturbed by the 2x1 m testpit, of which the contents are unknown to us. Therefore, we may expect that the number of axes in sector 2 originally was even higher. Core and flake axes make up 8.4% of all the tools (7 types) within 4 m

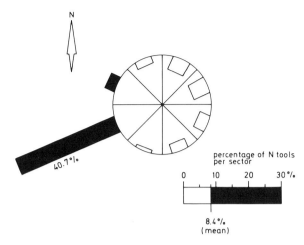

Fig. 17. Barmose I. Sector percentages of axes (N = 27), based on the total numbers of tools (7 types) per sector. Note the high percentage in sector 2.

from the hearth centre (see section 3). In figure 17, I have indicated the percentages per sector (based on the total number of tools per sector). Axes make up more than 40% of all the tools in sector 2.

This is a remarkable phenomenon. Clearly, the spatial distribution of axes is completely different from that of all the other tool types. This can also be shown statistically (see table 4: remarks 2 and 3). Moreover, we may be fairly sure that the concentration of axes in sector 2 is behaviourally relevant. For example, it is unlikely that most axes ended up in sector 2 because they were tossed out there. If the concentration of axes in the poorest site-half were due to their having been discarded in the forward toss zone, we would expect the same or a higher proportion of cores also to be located in that site-half, which is not the case (of all the axes 67% are in P, of the cores only 47%).

The concentration of axes is found immediately to the south of the relatively empty area in sector 3. If axes played a functional role in the working of wood (this seems to be the case, according to unpublished research by N. Symens; see Grøn, 1987a: p. 314), we can now offer the following explanation. Sector 3 might have contained a wood pile, for example as a fuel supply. This would make sense in several ways. It would explain the empty area in sector 3, and the concentration of axes in sector 2. Moreover, if a wood pile was indeed present, sector 3 would have been the most logical choice for its location. As we have seen, it is probable that the hearth of Barmose I was in the open air, and we have reconstructed the prevailing wind direction during occupation as ESE (fig. 13). Sector 3 is located opposite the tool-richest part of the site, which is composed of sectors 7 and 8 (see table 3). Moreover, the presence of a wood pile relatively close to the hearth does seem to be a reasonable proposition, given the fact that the large hearth of Barmose I contained quite a lot of charcoal.

Table 3. Barmose I. Artefact frequencies in 8 sectors around the centre of the hearth (for sector boundaries see fig. 1). Only locations within 4 m from the centre of the hearth. Note: frequencies in sector 2 should be considered as minimum estimates (see main text, section 2). Artefact groups: 1. Microliths; 2. Burins; 3. Denticulated/notched pieces; 4. Blade/flake knives; 5. Splintered pieces ('square knives'); 6. Scrapers; 7. Core and flake axes; 8. Total tools (7 types); 9. 'Microburins'; 10. Cores.

Sectors	Groups of artefacts									
	1	2	3	4	5	6	7	8	9	10
1	5	2	1	2	23	5	3	41	1	12
2	0	2	0	2	7	5	11	27	0	4
3	2	0	0	1	12	1	2	18	0	8
4	3	3	1	1	18	5	2	33	1	7
5	3	1	4	2	41	3	2	56	5	6
6	3	3	3	2	16	2	1	30	2	11
7	10	3	5	3	30	7	3	61	5	11
8	7	4	6	3	26	7	3	56	2	7
Total	33	18	20	16	173	35	27	322	16	66

The use of fire must have been of considerable importance to the inhabitants.

9. THE TWO QUARTERS WITHIN THE RICHEST SITE-HALF

Although women of several hunter/gatherer groups participate in some forms of hunting, this is usually the work of men. The sexual division of labour with 185 ethnographically studied peoples is discussed by Murdock & Provost (1973). Hunting large land fauna is done exclusively by men with 96.5% of the 144 peoples for which there are relevant data in the tables of Murdock & Provost (1973). With the remaining 3.5%, hunting is done predominantly by men. A very interesting aspect of this matter is that even in cases where women participate in hunting, there is a world-wide taboo on their handling weapons that cut or penetrate the animals, thus drawing blood (Testart, 1986). Although we shall never know for sure, this pattern may well have been in existence in Late Palaeolithic and Mesolithic times. This assumption leads to the conclusion that 'projectiles', such as microliths, most probably were made and left behind by men. Therefore, microliths would be the only tool type to be associated with one of the sexes. If used microliths are found, located relatively close to the hearth, we may be fairly sure that at least one man was present at the site, who among other things repaired his hunting equipment.

What about the women? Is it possible to find evidence relating to their presence or absence at a given site? The topic of gender in archaeology has been discussed recently in several publications (e.g. Conkey & Spector, 1984; Gero & Conkey, 1991). It has to be admitted, however, that sound empirical evidence regarding such questions is often lacking at Stone Age sites. We have

no a priori indications to postulate sex-specificity for tool types such as burins and scrapers, though there seems to be a tendency among subrecent hunter/gatherers for most hide-working to be done by women. Among the 185 peoples studied by Murdock & Provost (1973), there are 40 which can be classified as hunter/gatherers, i.e. peoples whose livelihood is provided for more than 90% by hunting, fishing and gathering. For 27 peoples among these 40, data concerning the sex-specificity of hide-working are available. With c. 59% of these 27 groups hide-working was done exclusively by women, and with 11% predominantly by women. With 22%, hide-working was done exclusively or predominantly by men. This would mean that scrapers, most of which played a functional role in hide-working (e.g. Juel Jensen, 1988), were used more frequently by women than by men. Even if this were true, however, it would not help us very much in the interpretation of individual sites, because it is probable that men also engaged in hide-working, for example at hunting camps.

In preceding sections I concluded that the richest site-half, in terms of tool numbers, is the area where people would have sat and worked most of the time. Let us assume that a nuclear family lived here. In that case we may postulate that of the two quarters constituting the richest site-half, one was occupied by a man and the other by a woman. We know that a sexual division of domestic space within dwellings is a common phenomenon with hunter/gatherers (e.g. Faegre, 1979; Grøn, 1989). Fixed areas for men and women may also

Table 4. Barmose I. Frequencies and proportions in 'R': the richest site-half (sectors 5, 6, 7, and 8) and 'P': the poorest site-half (sectors 1, 2, 3 and 4). Only locations within 4 m from the centre of the hearth. Artefact groups: 1. Microliths; 2. Burins; 3. Denticulated/notched pieces; 4. Blade/flake knives; 5. Splintered pieces ('square knives'); 6. Scrapers; 7. Core/flake axes; 8. Total tools (7 types); 9. 'Microburins'; 10. Cores.

Artefact groups	R		P		N
	Number	%	Number	%	
1	23	69.7	10	30.3	33
2	11	61.1	7	38.9	18
3	18	90.0	2	10.0	20
4	10	62.5	6	37.5	16
5	113	65.3	60	34.7	173
6	19	54.3	16	45.7	35
7	9	33.3	18	66.7	27
8	203	63.0	119	37.0	322
9	14	87.5	2	12.5	16
10	35	53.0	31	47.0	66

Remarks: 1. The difference between cores and all tools (7 types) is not significant: $0.1 < p$ (two-tailed) < 0.2 (chi-square two-sample test: Siegel, 1956).
2. The difference between core/flake axes and splintered pieces is significant: $0.001 < p$ (two-tailed) < 0.01.
3. The difference between core/flake axes and all the other tools (except splintered pieces) is significant: $0.001 < p$ (two-tailed) < 0.01.

46   D. Stapert

be expected around hearths in the open air (Marshall,
1959; 1973; Tindale, 1972). With the Bushmen, the
building of shelters mostly is the work of women. 'It
takes the women only three-quarters of an hour to an
hour to build their shelters, but half the time at least the
women's whim is not to build shelters at all. In this case
they sometimes put up two sticks to symbolize the
entrance of the shelter so that the family may orient
itself as to which side is the man's side and which the
woman's side of the fire." (Marshall, 1973: p. 97),

We would then expect the proportions of microliths
and 'other tools' to be different in the two quarters. This
would be because hunting gear was repaired only by
men. Even if the other tool types were used by both men
and women, this would lead to differences in the
proportion of microliths with respect to the other tools.
Of course, there are many problems to consider. Since
we are dealing with a small and intensively used area,
we have to anticipate smearing processes. If the wind
direction changed several times, mixing would occur as
a result of rotation around the hearth, blurring such
patterns.

Moreover, if a larger group was occupying the site,
consisting of several men and women, it would be much
harder to demonstrate sexual division of space. In the
case of Barmose I, however, we have reasons to believe
that the group of occupants was relatively small. The
drop zone is located quite close to the hearth (see fig.
20). Since only a semicircle is available for sitting near
to an open-air hearth, the distance between the drop
zone and the hearth will become larger when a greater
number of people are present (Binford, 1983; Stapert,
1989). In the case of Barmose I, the presence of only two
or three adults seems to be a reasonable proposition.

Despite such problems, we should, when dealing with
open-air hearths occupied by families, expect a
difference between the two quarters to be demonstrable
in many cases, which we would not expect if several
persons of the same sex were present. This implication
can be investigated statistically. The richest site-half is
divided into two quarters. The quarter with the highest
proportion of 'projectiles' is called 'A', the other 'B'.
The frequencies of 'projectiles' and of 'other tools' in
A and B are counted (table 5). In the case of Barmose I
the 'projectiles' are microliths, and the 'other tools' are
scrapers + burins + denticulated/notched pieces + blade/
flake knives + core/flake axes. (The splintered pieces
will be considered separately.) We then want to test the
null hypothesis, which states that there are no differences
between A and B, regarding the proportions of
'projectiles' to 'other tools' present in them. The
alternative hypothesis is that the proportion of
'projectiles' is significantly higher in A than in B. This
can be investigated by the Fisher Exact Probability Test
(Siegel, 1956). In the case of Barmose I, there is no
significant difference between the two quarters: p =
0.32 (see fig. 18). Moreover, I have also tested the

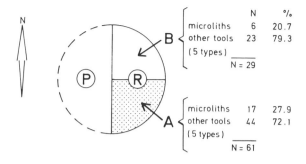

Difference in proportions of microliths and other tools between
A and B is not significant.

Fisher Test: p = 0.32.

Fig. 18. Barmose I. The richest site-half (R) is divided into two
quarters, A and B. These two quarters are not demonstrably different as regards the proportions of microliths and other tools (see also
table 5).

difference between A and B for all possible combinations
of two tool types (among the six formal types). All 15
combinations produced Fisher p's above 0.30. Therefore,
as far as the six formal tool types are concerned, no
differences between A and B can be demonstrated.
Since microliths were found to be located relatively
close to the hearth, the conclusion of this exercise
should be that one or several men were present, but
probably no women.

Another argument in this respect is the spatial
distribution of the small 'microburins'. These are waste
from the production of microliths. Of their total of 16,
14 are in R: the eastern site-half (87.5%). Because these
small objects will have been tossed away less frequently
than larger artefacts, this again indicates that the drop
zone was in R. Quarters A and B in R contained equal
numbers (7) of microburins, suggesting that microliths
were produced in both quarters during occupation.

It was concluded that the two quarters within R do not
differ as regards the proportions of the six formal tool
types. However, they are different in two other respects.
The first is that A is about twice as rich in tools of the six
formal tool types as B (totals are 61 and 29, respectively).
The numbers of splintered pieces in the two quarters,
however, are about the same (56 and 57, respectively).
The difference in proportion of the splintered pieces,
with respect to the total of the other tools, can be shown
to be significant (see table 5: remark 3). Thus, the
situation can be summarized as follows: in both quarters
the amount of splintered pieces is the same, but in
quarter A there are twice as many other tools as in B,
though their proportions are similar in both quarters.
Though in no way conclusively, this seems to suggest
the presence of at least two men, who performed similar
types of activity, but with a different intensity as regards
the tools other than splintered pieces.

Table 5. Barmose I. Comparison between the two quarters A and B within 'R', the richest site-half (sectors 5, 6, 7 and 8).

	B Sectors 5+6 Number	%	A Sectors 7+8 Number	%	N
Scrapers	5	17.2	14	23.0	19
Burins	4	13.8	7	11.5	11
Microliths	6	20.7	17	27.9	23
Core/flake axes	3	10.3	6	9.8	9
Dentic./notched pieces	7	24.1	11	18.0	18
Blade/flake knives	4	13.8	6	9.8	10
Total 6 types	29	99.9	61	100.0	90
Splintered pieces	57	66.3	56	47.9	113
Total 6 other types	29	33.7	61	52.1	90
Total 7 types	86	100.0	117	100.0	203

Remarks: 1. Differences between all pairs among the 6 formal tool types, regarding their proportions in the two quarters, are not significant: all 15 p's resulting from application of the Fisher Exact Probability Test (Siegel, 1956) are above 0.30.
2. The difference between microliths and total tools of 5 other types (i.e., without splintered pieces) is not significant: p = 0.32 (Fisher Test).
3. The difference between splintered pieces and total other tools (6 types) is significant: p = 0.01 (Fisher Test).

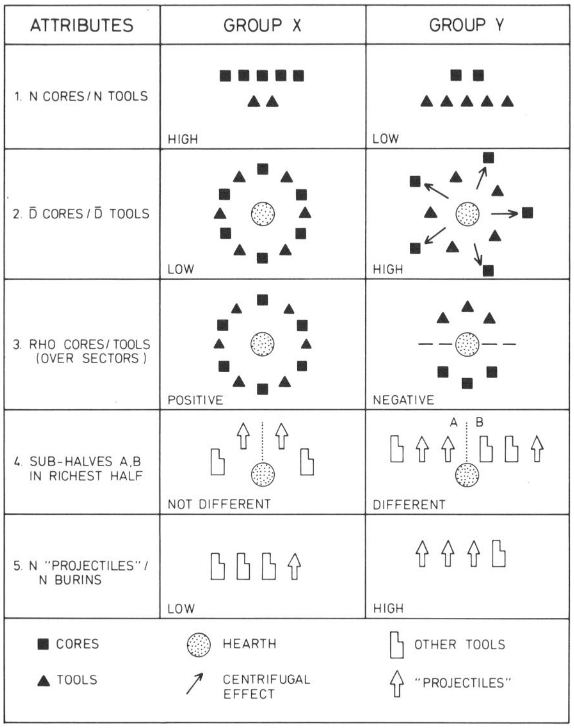

10. COMPARING BARMOSE I WITH SEVERAL OTHER ANALYSED SITES

In a previous article (Stapert, 1989), I have analysed eighteen other sites (Upper/Late Paleolithic) for which it is probable that the hearths were in the open air (on the basis of the tools' showing a unimodal ring distribution). It was found that some sites show a significant difference between quarters A and B within the richest site-halves, and others do not. It was also found that the sites with different A's and B's are furthermore characterized by the presence of a clear centrifugal effect, and of spatial segregation of tools and cores in their sector distributions. Thus, sites showing clear indications of the presence of both forward and backward toss zones also tend to have different quarters within R, whereas sites that appear not to have had clear toss zones also show no differences between the two quarters in R. On the basis of these attributes, therefore, it seems as if two site-types can be defined, which I have called 'Group X' and 'Group Y', and it was found that these two groups also differ in several other aspects (fig. 19).

Group X includes sites such as Habitation 1 at Pincevent (Leroi-Gourhan & Brézillon, 1966), Marsangy N19 (Schmider, 1979; 1984; 1988), and Bro I (Andersen, 1973). These sites do not show a clear centrifugal effect ('centrifugal index' smaller than 1.15), there is no tendency towards spatial segregation of cores and tools in sector distributions (% of N cores in R not significantly smaller than that of tools), there is a relatively high proportion of cores to tools, the ratio of

Fig. 19. Classification of 18 Upper/Late Palaeolithic sites with open-air hearths into two groups, X and Y (Stapert, 1989). The sites of Group X show no centrifugal effect, nor any spatial segregation of tools and cores in their sector distributions; moreover, these sites show a relatively large number of cores in proportion to tools, and a large number of burins in proportion to 'projectiles'. For the sites of Group Y the reverse is true. Sites in Group Y show a tendency for quarters within the richest site-halves to differ in terms of proportions of 'projectiles' to other tools, while this is not the case at sites assigned to Group X. One explanation could be that most sites of Group Y were occupied by families, and those of Group X by small groups of men (e.g. hunting camps). Barmose I can be placed in Group X.

'projectiles' to burins is low ('projectile/burin index' mostly smaller than 1.25), and there is no clear difference between the two quarters within the richest site-half.

Group Y includes sites such as Niveau IV-20 at Pincevent (Leroi-Gourhan & Brézillon, 1972; Julien et al., 1988), Oldeholtwolde (Stapert et al., 1986), Olbrachcice 8 east (Burdukiewicz, 1986) and Niederbieber (Bolus et al., 1988; Winter, 1986; 1987). These sites show a clear centrifugal effect ('centrifugal index' above 1.20), a tendency towards spatial segregation of cores and tools in sector distributions (% of N cores in R clearly smaller than that of tools), a relatively low proportion of cores to tools, a high ratio of 'projectiles' to burins ('projectile/burin index' mostly above 1.25), and clear differences between the two quarters within the richest site-half.

It was hypothesized that most sites of Group X were occupied by men only, while at most sites of Group Y women were also present. In other words: Group X might represent hunting camps or male 'special purpose camps', and Group Y family camps. For more details concerning this site grouping, the reader is referred to a previous publication (Stapert, 1989).

Barmose I clearly belongs to Group X, as defined above. Toss zones seem to have been only weakly developed (spatial segregation of cores and tools in their ring and sector distributions cannot be shown to be significant in a statistical sense), there is a relatively large proportion of cores compared to tools, and the two quarters within R cannot be shown to be different. Compared to the Upper/Late Palaeolithic sites placed in Group X, however, there are fewer burins in proportion to 'projectiles'.

All in all, indications are that Barmose I was occupied by a few men only; we have no sound indications for the presence of women.

11. DISCUSSION AND SOME CONCLUSIONS

This section consists of two parts. First I will summarize my results for the site of Barmose I. In the second part I will evaluate the performance of the ring and sector method.

The hypothesis of a dwelling structure around the hearth of Barmose I has to be rejected on the basis of the ring distributions: occupation took place in the open air.

People must have been sitting and working to the east of the hearth most of the time, which suggests a prevailing wind from the east. Apart from some variability in local tool density, the whole eastern half, which is the richest site-half in terms of tool numbers, seems to have been a single 'general activity area'. Probably many different activities went on here, including flint-working, and I cannot see much reason for functional differentiation within this area. Most artefacts in this area seem to be located in a drop situation, as described by Binford (1983). Indications for the existence of distinct toss zones are weak, and not significant in a statistical sense. In other words, continual clearing up during occupation hardly took place. This suggests that the occupants of the site anticipated only a short stay at this locality.

A second 'activity area' is located on the opposite side of the hearth, to the west of it. Here a more specific activity is indicated, involving the use of axes; it can be suggested that especially wood-working took place here from time to time. This activity seems to be associated with a relatively empty space near the hearth where possibly a wood pile was present. Though not significantly so, microliths are found relatively close to the hearth, and scrapers farther away. This pattern, which is also common at Upper and Late Palaeolithic sites, can be explained by assuming that in the retooling of 'projectiles' (microliths) heat was needed, while hide-working (scrapers) required quite a lot of space. The two quarters within the drop zone in the eastern half of the site cannot be shown to be different, in terms of

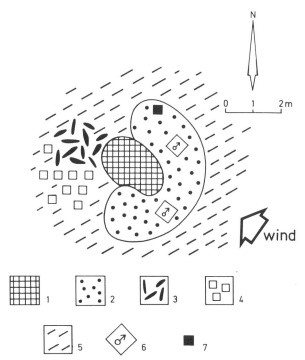

Fig. 20. Barmose I. Sketch in which the interpretations resulting from the analysis by the ring and sector method have been summarized. It is concluded that the hearth of Barmose I was in the open air. The prevailing wind during habitation is thought to have been roughly from the east. Perhaps two men were present; there are no good reasons for assuming the presence of women. Key: 1. Hearth; 2. Drop zone; 3. Wood pile?; 4. Concentration of axes; 5. Toss zones; 6. Suggested seating positions; 7. Large stone (seat?).

the proportions in which six formal tool types are represented. This suggests that a few persons of the same sex were present, which, in view of the presence of microliths relatively close to the hearth, must have been men. Perhaps the most likely interpretation of the site would be that it was a hunting camp. However, it should be remembered that we do not really know what tasks were carried out with the help of the numerous splintered pieces. The hypotheses resulting from my analysis have been summarized in a general sketch of the site (fig. 20). The large stone (diameter about 30 cm, see fig. 1) is now seen to be located in the drop zone to the east of the hearth. A reasonable suggestion therefore is that it was a seat.

In my opinion the ring and sector method has performed well in this case. The effectiveness of the method seems to be due to several factors. First of all, it is adapted to the global structure of sites such as Barmose I, where a central hearth, which clearly was the focus of all sorts of activities, defines the spatial 'organization'. It links up with ethnoarchaeological models, such as Binford's hearth model. For example, the presence or absence of toss zones can be investigated satisfactorily with the help of ring and sector distributions. Moreover, the method is simple and, above all, transparent. It contains no inherent assumptions of a statistical nature, which encumber many of the more complex approaches to intrasite spatial analysis.

The method seems to make it possible to demonstrate whether hearths lie in the open or inside a tent, which in my opinion is a prerequisite for any meaningful spatial analysis. Another important aspect is that the ring and sector method makes it possible to compare different sites as to their global spatial layout. Several attributes which are investigated with this method, can be summarized in the form of simple indexes. For example, it is possible to describe quantitatively two different tendencies towards spatial segregation of cores and tools – both suggesting the presence of toss zones – by the 'centrifugal index' and the 'tools/cores in R index'. This has resulted in a grouping of the analysed sites into two types: sites with and without clear indications for the existence of toss zones (Groups Y and X, respectively: fig. 19). Interpretations attached to this finding may be arbitrary, but the statistical patterns seem to be quite convincing, and hence are interesting.

One general result of my analysis is, once again, that archaeological residues around 'domestic hearths' in the open air present us with a somewhat frustrating situation, as far as intrasite spatial analysis is concerned. In such cases we should not attempt to demonstrate discrete 'activity areas' by complex procedures. Since the central hearth attracted many different activities, it can hardly be expected that the separate activity areas should still be recognizable, as these would have become blurred in this small but intensively used area. Of various types of activity many episodes must have occurred around the hearth, and these will have had different results in terms of the numbers of tools that were discarded, and the size and shape that waste scatters took, and it is to be expected that the residues of many episodes of different activities will overlap in space. As Carr has put it: "Co-occurrences between different artefact types in this situation reflect the common social context in which they were used, rather than use in a common activity." (Carr, 1984: p. 115). In other words, the level of resolution of most of the complex techniques, as used by Blankholm, is set much too high to be appropriate in such cases. It seems unrealistic to produce 15, 16 or 19 'clusters' by mathematical means (unconstrained clustering, correspondence analysis and 'presab', respectively) in the case of Barmose I and similar sites. Such outcomes are hardly interpretable, because these techniques expect too much 'expressiveness' from archaeological sites of this type. In such cases we are dealing with a kind of spatial 'autocorrelation' within the richest site-halves (e.g. Simek, 1984; Yellen, 1977). Therefore, we should concentrate first of all on global patterns, and try to make sense of these in the light of, for example, ethnoarchaeological observations, instead of expecting

miracles from very detailed mathematical procedures. This problem was recognized by Blankholm, as appears from the following citation: "... Simek (1984) has convincingly demonstrated the need to take the effect of spatial autocorrelation of behaviourally independent tasks around features, fixtures and centres of social activity into account." (Blankholm, 1991: p. 48). Evidently, given the detailed analyses in the rest of his book, Blankholm did not consider Barmose I to be an example of this problem. In my opinion, however, Barmose I is a typical example, and the same is true for many other sites.

Complex techniques are often applied rather mechanically, without an adequate theoretical framework to guide interpretation of the results. One only needs to imagine a site which had several occupations, and where during each occupation a great deal of rotation around the hearth occurred because of changing wind directions. The resulting residue will be a palimpsest. Cluster techniques will still produce clusters, however, because this is what they are designed to do, and archaeologists will then try to interpret these clusters in terms of 'activity areas'. This is because even in such situations the spatial distributions will not have become totally random – there will always be local irregularities. Therefore, what we need are ways to bridge the gap between the static data, including patterns produced by computer procedures, and realistic interpretations. To do this, interpretive models are needed. One of the best ways to obtain these is by making ethnoarchaeological observations operational for archaeology; this is what Binford (1983) has called 'decoding the archaeological record'.[5]

12. ACKNOWLEDGEMENTS

Thanks are due to Miriam Weijns and J.M. Smit (B.A.I.) for drawing work, and to Xandra Bardet for expertly improving my English text. I thank Tineke Looijenga (Groningen) for translating parts of Blankholm (1984). A first draft of this article was critically read by Bjarne Grønnow (Copenhagen), whose comments I have very much appreciated. Thanks are due to Jeannette van der Post (Groningen), for removing several typing errors.

13. NOTES

1. In Blankholm (1984: p. 63; 1987: pp. 109, 110), it is stated that the best approximation of the hut outline at Maglemosian sites is the contour line of the mean number of artefacts per square metre, calculated over the site as a whole.
2. Somewhat similar methods were employed by Dekin (1976) and Hull (1987). What distinguishes these methods from the ring and sector method is that they are not hearth-oriented (though they could be adapted in this sense).
3. In reality, the drop zone probably was mostly somewhat larger than a semicircle (see e.g. Binford, 1983: figs 89, 90). It may cover 5 or even 6 sectors (of a total of 8). Gallay (1989) describes groups

of 4 to 7 Touareg men, drinking tea around a fire. In his figure (Gallay, 1989: fig. 3), a group of 6 men is indicated, sitting inside a 'drop zone' 1 to 2 m from the hearth centre, to windward of it. Of the total circumference of 360°, they occupy about 260°.

4. B. Grønnow (Copenhagen) kindly informed me of the following. 'The core- and in particular the flake-axes of the Barmose group are the largest and most heavy flint objects, so I guess that they have to be taken into consideration in the centrifugal analysis.' (letter of 16-VIII-1991). Therefore, in this case it would be appropriate to compare the axes to all the other tools taken together, as regards their distance to the hearth centre. As can be seen in table 1, the core- and flake-axes indeed are, on average, located farthest from the hearth among the tool classes. After combining the data in two rings of 2 m width, it is possible to apply the chi-square two-sample test (Siegel, 1956) in a valid way. Again, however, the difference cannot be shown to be significant in a statistical sense ($0.5 < p$ (two-tailed) < 0.7), though there is a tendency towards it.

5. In 1993, Blankholm critically reacted to this article; his comments are published in this volume of *Palaeohistoria*, as is the case with my reply.

14. REFERENCES

ANDERSEN, S.H., 1973. Bro. En senglacial bobplads på Fyn. *Kuml* 1972, pp. 7-60.

AUDOUZE, F., D. CAHEN, L.H. KEELEY & B. SCHMIDER, 1981. Le site magdalénien du Buisson Campin à Verberie (Oise). *Gallia Préhistoire* 24, pp. 99-143.

BINFORD, L.R., 1983. *In pursuit of the past. Decoding the archaeological record*. London.

BLANKHOLM, H.P., 1984. Maglemosekulturens hyttegrundrids. En undersøgelse af bebyggelse og adfaerdsmønstre i tidlig Mesolitisk tid. *Aarbøger for Nordisk Oldkyndighed og Historie* 1984, pp. 61-77.

BLANKHOLM, H.P., 1987. Maglemosian hutfloors: an analysis of the dwelling unit, social unit and intra-site behavioural patterns in early Mesolithic Southern Scandinavia. In: P. Rowley-Conwy, M. Zvelebil & H.P. Blankholm (eds), *Mesolithic Northwest Europe: recent trends*. Sheffield, pp. 109-120.

BLANKHOLM, H.P., 1991. *Intrasite spatial analysis in theory and practice*. Aarhus.

BOKELMANN, K., 1986. Rast unter Bäumen; ein ephemerer mesolithischer Lagerplatz aus dem Duvenseer Moor. *Offa* 43, pp. 149-163.

BOKELMANN, K., 1989. Eine mesolithische Kiefernrindenmatte aus dem Duvenseer Moor. *Offa* 46, pp. 17-22.

BOKELMANN, K., F.-R. AVERDIECK & H. WILLKOMM, 1981. Duvensee, Wohnplatz 8; neue Aspekte zur Sammelwirtschaft im frühen Mesolithikum. *Offa* 38, pp. 21-40.

BOKELMANN, K., F.-R. AVERDIECK & H. WILLKOMM, 1985. Duvensee, Wohnplatz 13. *Offa* 42, pp. 13-33.

BOLUS, M., G. BOSINSKI, H. FLOSS, H. HUSMANN, U. STODIEK, M. STREET, T. TERBERGER & D. WINTER, 1988. Le sequence Bølling – Dryas III en Rhenanie. In: M. Otte (ed.), *De la Loire à l'Oder; les civilisations du Paléolithique final dans le nord-ouest européen* (= BAR Intern. Series 444 (ii)). Oxford, pp. 475-509.

BOSINSKI, G., 1979. *Die Ausgrabungen in Gönnersdorf 1968-1976 und die Siedlungsbefunde der Grabung 1968* (= Gönnersdorf 3). Wiesbaden.

BOSINSKI, G, 1981. *Gönnersdorf. Eiszeitjäger am Mittelrhein*. Koblenz.

BURDUKIEWICZ, J.M., 1986. *The late Pleistocene shouldered point assemblages in Western Europe*. Leiden.

CARR, C., 1984. The nature of organization of intrasite archaeological records and spatial analytic approaches to their investigation. *Advances in archaeological method and theory* 7, pp. 103-222.

CARR, C. (ed.), 1985. *For concordance in archaeological analysis. Bridging data structure, quantitative technique, and theory*. Kansas City.

CONKEY, M. & J.D. SPECTOR, 1984. Archaeology and the study of gender. *Advances in archaeological method and theory* 7, pp. 1-38.

CZIESLA, E., 1990. *Siedlungsdynamik auf steinzeitlichen Fundplätzen; methodische Aspekte zur Analyse latenter Strukturen* (= Studies in Modern Archaeology 2). Bonn.

DEKIN Jr., A.A., 1976. Elliptical analysis: an heuristic technique for the analysis of artefact clusters. In: M.S. Maxwell (ed.), *Eastern arctic prehistory: Paleoeskimo problems* (= Memoirs of the Society for American Archaeology 31). pp. 79-88.

EICKHOFF, S., 1988. Ausgesplitterte Stücke, Kostenki-Enden und 'retuschierte Bruchkanten'; einige Aspekte zur Untersuchung der Artefakte aus westeuropäischer Feuerstein auf dem Magdalénien-Fundplatz Gönnersdorf. *Archäologische Informationen* 11, pp. 136-144.

FAEGRE, T., 1979. *Tents. Architecture of the nomads*. London.

FISCHER, A., 1991. Pioneers in deglaciated landscapes: the expansion and adaptation of Late Palaeolithic societies in southern Scandinavia. In: N. Barton, A.J. Roberts & D.A. Roe (eds), *The Late Glacial in North-West Europe* (= CBA report 77). London, pp. 100-121.

GALLAY, A., 1989. Vivre autour d'un feu; recherche d'une problématique d'analyse archéologique. In: M. Olive & Y. Taborin (eds), *Nature et fonction des foyers préhistoriques. Actes du Colloque International de Nemours, 12-13-14 mai 1987* (= Mémoires du Musée de Préhistoire d'Ile de France 2). Nemours, pp. 101-112.

GERO, J.M. & M.W. CONKEY (eds), 1991. *Engendering archaeology; women and prehistory*. Oxford.

GRØN, O., 1987a. Seasonal variation in Maglemosian group size and structure; a new model. *Current Anthropology* 28, pp. 303-327.

GRØN, O., 1987b. Dwelling organization – a key to the understanding of social structure in Old Stone Age societies? An example from the Maglemose Culture. In: J.K. Kozlowski & S.K. Kozlowski (eds), *New in Stone Age archaeology* (= Archaeologia Interregionalis, Varia 230). Warsaw, pp. 63-83.

GRØN, O., 1989. General spatial behaviour in small dwellings: a preliminary study in ethnoarchaeology and social psychology. In: C. Bonsall (ed.), *The Mesolithic in Europe. Papers presented at the third symposium*. Edinburgh, pp. 99-105.

HIETALA, H.J. (ed.), 1984. *Intrasite spatial analysis in archaeology*. Cambridge.

HULL, K.L., 1987. Identification of cultural site formation processes through microdebitage analysis. *American Antiquity* 52, pp. 772-783.

JUEL JENSEN, H., 1988. Functional analysis of prehistoric flint tools by high-power microscopy: a review of Westeuropean research. *Journal of World Prehistory* 2, pp. 53-88.

JULIEN, M., F. AUDOUZE, D. BAFFIER, P. BODU, P. COUDRET, F. DAVID, G. GAUCHER, C. KARLIN, M. LARRIÈRE, P. MASSON, M. OLIVE, M. ORLIAC, N. PIGEOT, J.L. RIEU, B. SCHMIDER & Y. TABORIN, 1988. Organisation de l'espace et fonction des habitats magdaléniens du bassin parisien. In: M. Otte (ed.), *De la Loire à l'Oder. Les civilisations du Paléolithique final dans le nord-ouest européen* (= BAR Intern. Series 444 (i)). Oxford, pp. 85-123.

KEELEY, L.H. 1982. Hafting and retooling: effects on the archaeological record. *American Antiquity* 47, pp. 798-809.

KENT, S. (ed.), 1987. *Method and theory for activity area research. An ethnoarchaeological approach*. New York.

LEROI-GOURHAN, A. & M. BRÉZILLON, 1966. L'Habitation Magdalénienne no. 1 de Pincevent près Montereau (Seine-et-Marne). *Gallia Préhistoire* 9, pp. 263-385.

LEROI-GOURHAN, A. & M. BRÉZILLON, 1972. *Fouilles de Pincevent. Essai d'analyse ethnographique d'un habitat Magdalénien (la section 36)* (= VIIe Suppl. Gallia Préhistoire). Paris.

MARSHALL, L., 1959. Marriage among !Kung Bushmen. *Africa* 29 (4). London.

MARSHALL, L. 1973. Each side of the fire. In: M. Douglas (ed.), *Rules and meanings. The anthropology of everyday knowledge.* Harmondsworth, pp. 95-97.

MOSS, E.H., 1983. *The functional analysis of flint implements. Pincevent and Pont d'Ambon: two case-studies from the French Final Palaeolithic* (= BAR Intern. Series 177). Oxford.

MOSS, E.H. & M.H. NEWCOMER, 1981. Reconstruction of tool use at Pincevent: microwear and experiments. *Studia Praehistoria Belgica* 2, pp. 289-312.

MURDOCK, G.P. & C. PROVOST, 1973. Factors in the division of labor by sex: a cross-cultural analysis. *Ethnology* 12, pp. 203-225.

OLIVE, M., 1988. *Une habitation magdalénienne d'Etiolles. L'unité P15* (= Mémoire Soc. Préh. Franç. 20). Paris.

OLIVE, M. & Y. TABORIN (eds), 1989. *Nature et fonction des foyers préhistoriques. Actes du Colloque International de Nemours, 12-13-14 mai 1987* (= Mémoires du Musée d'Ile de France 2). Nemours.

SCHIFFER, M., 1976. *Behavioral archaeology.* New York.

SCHMIDER, B., 1979. Un nouveau faciès du Magdalénien du Bassin Parisien: l'industrie du gisement du Pré-des-Forges à Marsangy (Yonne). In: *La fin des temps glaciaires en Europe.* Bordeax, pp. 763-771.

SCHMIDER, B., 1984. Les habitations Magdaléniennes de Marsangy (Vallée de Yonne). In: H. Berke, J. Hahn & C.-J. Kind (eds), *Jungpaläolithische Siedlungsstrukturen in Europa.* Tübingen, pp. 169-180.

SCHMIDER, B., 1988. Un outil spécialisé dans le Magdalénien du Bassin Parisien: le bec; sa place dans l'habitat. *Revue archéologique de Picardie* 1/2, pp. 195-200.

SIEGEL, S., 1956. *Nonparametric statistics for the behavioural sciences.* Tokyo.

SIMEK, J., 1984. Integrating pattern and context in spatial archaeology. *Journal of archaeological science* 11, pp. 405-420.

STAPERT, D., 1989. The ring and sector method: intrasite spatial analysis of Stone Age sites, with special reference to Pincevent. *Palaeohistoria* 31, pp. 1-57.

STAPERT, D., 1990a. Within the tent or outside? Spatial patterns in Late Palaeolithic sites. *Helinium* 30, pp. 14-35.

STAPERT, D., 1990b. Middle Palaeolithic dwellings: fact or fiction? Some applications of the ring and sector method. *Palaeohistoria* 32, pp. 1-19.

STAPERT, D., J.S. KRIST & A.L. ZANDBERGEN, 1986. Oldeholtwolde, a Late Hamburgian site in The Netherlands. In: D. Roe (ed.), *Studies in the Upper Palaeolithic of Britain and NW Europe* (= BAR Intern. Series 296). Oxford, pp. 187-226.

STAPERT, D. & T. TERBERGER, 1989. Gönnersdorf Concentration III: investigating the possibility of multiple occupations. *Palaeohistoria* 31, pp. 59-95.

SYMENS, N., 1986. A functional analysis of selected stone artefacts from the Magdalenian site at Verberie, France. *Journal of field archaeology* 13, pp. 214-222.

TERBERGER, T., 1988. Ein Zeltringbefund des Magdalenien-Fundplatzes Gönnersdorf. *Jahrbuch des Römisch-Germanischen Zentralmuseums Mainz* 35, pp. 137-159.

TESTART, A., 1986. *Essai sur les fondements de la division sexuelle du travail chez les chasseurs-cueilleurs.* Paris.

TINDALE, N.B., 1972. The Pitjandara. In: M.G. Bicchieri (ed.), *Hunters and gatherers today.* New York.

WHALLON, R., 1984. Unconstrained clustering for the analysis of spatial distributions in archaeology. In: H.J. Hietala (ed.), *Intrasite spatial analysis in archaeology.* Cambridge, pp. 242-277.

WINTER, D., 1986. Der spätpaläolithische Fundplatz Niederbieber; Fläche 50/14-56-20. Magisterarbeit, Köln.

WINTER, D., 1987. Retoucheure des spätpaläolithischen Fundplatzes Niederbieber/Neuwieder Becken (Fläche 50/14-56/20). *Archäologisches Korrespondenzblatt* 17, pp. 295-309.

YELLEN, J.E., 1977. *Archaeological approaches to the present. Models for reconstructing the past.* New York.

RINGS, SECTORS AND BARMOSE I: A REPLY TO STAPERT

H.P. BLANKHOLM

Byagervej 204, Beder, Denmark

ABSTRACT: This paper comments critically on Stapert's (1992 and this volume) treatment of the Danish early Maglemosian site Barmose I. Basically, he displays an unfortunate mixture of the uninformed use of Maglemosian material on the one hand and methodological inconsistencies on the other.

He fails to consult the excavator's publications and also largely fails to refer to other primary site publications or sources on the Maglemosian, thus ignoring much relevant information. Consequently, his treatment of the site, classification, and hut floors suffers from the lack of knowledge on the nature and specifics of the data.

His ring and sector method has severe theoretical, statistical and operational constraints and has neither been tested on independent ethno-archaeological material for which the behavioural parameters are known, nor matched consistently with other methodologies. It is not proven that the ring and sector method is capable of delivering information of relevance for behavioural interpretation. Thoroughly, his discussions lack balance and consistency.

His conclusion, that Barmose I was an open air hunting camp with a central hearth and only used by two men, remains unfounded and contradicted by the primary data.

KEYWORDS: Intrasite spatial analysis, Barmose I, hut floors, classification, ring and sector method, multivariate methods, ethno-archaeology, Mask Site.

1. INTRODUCTION

In his recently published thesis 'Rings and sectors: Intrasite spatial analysis of Stone Age sites', Stapert (1992) launches a view of the Danish early Maglemosian site Barmose I very much at variance with previous interpretations of the site (Johansson, 1971; 1990; Blankholm, 1991). As will become apparent, this discrepancy seems to be rooted in Stapert's unfortunate mixture of the uninformed use of Maglemosian material on the one hand and his methodological inconsistencies on the other.

In the first instance Stapert completely fails to refer to or consult the excavator's original and latest publications of the site (Johansson, 1971; 1990). Except for Bokelmann (1986; 1989), Bokelmann et al. (1981; 1985), Grøn (1987a; 1987b; 1989) and Blankholm (1984; 1987; 1991) he also fails to refer to any other primary site publication or sources on the Maglemosian. By doing so, much important information is ignored. For instance, it is clearly stated in Johansson (1990: p. 43) that two lumps of resin show the toothmarks of a child of approximately 7-8 years of age, and of a young person not less than 11 years of age, respectively. This effectively refutes Stapert's notion of a men-only hunting camp (1992: p. 157).

In the following I will briefly deal in turn with some pertinent aspects of the Barmose I data; the site, classification, hut floors, and the ring and sector method.

2. THE SITE

From the outset it must be stressed that my 1991 application of the best of spatial analytical methods (i.e. k-means Analysis, Unconstrained Clustering, Correspondence Analysis, and Presab) to Barmose I was not a test, but rather a demonstration (a tutorial if one wishes). Stapert seems to forget this completely. In fact, I (Blankholm, 1991: p. 183, see also the Danish summary p. 233) stated:

> The theoretical potential, practical limitations, and relative power and efficiency of the selected methods should now be readily apparent from the preceding chapters. In fact, all students and professionals with a basic knowledge of quantitative methods and spatial analysis should be able to proceed from here and perform their own analysis. However, in conclusion I will demonstrate how the best of methods may be applied to a purely archaeological example.

The real test, in fact, was on ethno-archaeological material (The Mask Site (Binford, 1978)) for which the behavioural parameters are known. This test clearly proved that intrasite spatial analyses are capable of delivering relevant and important information on spatial data structures for behavioural interpretation (Blankholm, 1991: p. 211).

Since the Barmose I site was already published (Johansson, 1971) and was about to be re-published in extenso (Johansson, 1990), I refrained from going into details in my general summary of the site (1991: pp.

184-186); simply assuming that any analyst naturally would consult the site publications, or, if necessary, the excavator for more detailed information, as I did myself.

In his short rendition of the site based on my own summary (Blankholm, 1991: pp. 184-186), Stapert (1992: p. 145) seems particularly concerned about the effect of the test pit and test excavation on his analysis and interpretation. He could have spared himself those worries (not least concerning his sector 2 (Stapert, 1992: fig. 1, p. 170)) had he read the excavator's latest publication (Johansson, 1990). There it can be seen from the distribution plans that the large 2x1 m test excavation contained 19 tools, which, to put things straight, is less than half of the 41 tools (or 11.3% of Stapert's total) Stapert nonchalantly excludes from his own analysis and thus consequently ignores the behaviourally interesting areas in the eastern periphery of the site (Johansson, 1990: p. 47; Blankholm, 1991: p. 192 ff.); apparently because he cannot get his rings to fit (Stapert, 1992: p. 147)! I guess this puts his argument (Stapert, 1992: p. 145):

> ..., the disturbing effect of these two test pits on any kind of spatial analysis is taken too lightly in Blankholm's discussion of the results...

into the right perspective. If anyone has taken Barmose I and its data too lightly it manifestly is Stapert. On the contrary, my own analysis and interpretations did fully consider the contents and effect of the regular test pit (Blankholm, 1991: pp. 185-186, 204).

3. CLASSIFICATION

It is of some interest that Stapert (1992: p. 147), without having read the excavator's publications (Johansson, 1971, 1990) and without having seen the material first-hand, relegates the tool class 'square knives' to mere 'splintered pieces' and then goes on to treat them seperately from the rest of the tool classes in his analysis. I wonder how he can justify the following assertion (Stapert, 1992: p. 147):

> Splintered pieces (which Blankholm calls 'square knives') are very numerous, and in fact do not constitute a formal tool class, as the splintering is no intentional retouch but probably the result of some heavy use...

If Stapert really wishes to redefine the material he should do so based on first-hand knowledge of the data, not on sheer speculation. Certainly square knives is a formal tool class in its own right and as such should have been treated consistently with the other artifact classes. He also seemingly is dissatisfied with the term 'micro-burin', which he rather wants to call 'notched remnants'. Although the term 'microburin' may be unfortunate, it nevertheless is a standard and well-recognized term in European Mesolithic typology. At the time of writing, I saw, and see now, no compelling reason to change it.

4. HUT FLOORS

After a discussion of the Presab method for spatial analysis (Blankholm, 1991: pp. 151-65, 199-202), Stapert (1992: p. 145) writes:

> Even more disturbing is the fact that Blankholm's analysis proceeds on the basis of several unproven assumptions, which are not critically tested. The most important of these is the idea that a hut was present at the site, with the hearth located at the centre of its interior. The demonstration that a dwelling was present should be one of the goals of intrasite spatial analysis, not an assumption to start with! It will be realized that the interpretation of any 'patterns', established with the help of whatever mathematical technique, will be very different, depending on whether or not the presence of a dwelling is assumed.

It is self-evident that the demonstration that a dwelling was present should be one of the goals of intrasite spatial analysis, and this also was exactly what was done. The problem is that Stapert confuses matters. In my analysis I used a well-established spatial analytical method, based on intimate knowledge of and experience with preserved Maglemosian hut floors and associated material distributions (Blankholm, 1981; 1984; 1987), for evaluating the Barmose I case. The outline then was modified according to other contextual knowledge (Blankholm, 1991: p. 185). I should kindly recommend Stapert to read Blankholm 1984 and 1987 in detail. None of the multivariate methods (apparently complex methods to Stapert) tested by myself in 1991 were ever designed or geared to demonstrate dwellings per se. Consequently they were not used to this end!

Certain aspects of the distribution of debitage (waste) is essential for delineating hut floors on Maglemosian settlements. However, Stapert excludes waste/flint-knapping debris from his Barmose I analysis altogether apparently for several, in my view, highly dubious presumptions not mentioned in his text on Barmose I proper. Instead they are enumerated in a methodological section of his book called 'Some choices' (Stapert, 1992: pp. 28-29). Also, in his discussion of the behavioural meaning of the distribution of cores (Stapert, 1992: p. 146), he does not recognize their conspicuous oval distribution. In my view this distribution is not consistent with the Binfordian toss model neither in static, nor in rotary seating mode. Again, I see no reason to change my (and the excavator's) original view on its significance (Blankholm, 1991: p. 185).

In fact Stapert's treatment seems to have more than its share of unwarranted presumptions, biases and prejudices against prehistoric behaviours, their multi-variate nature and firmly tested methods to their resolution. Unfortunately Stapert does not report the coordinate of the origin for his rings and sectors, so I am unable to redo his analysis including debitage.

Anyway, the partly analytically demonstrated floor on Barmose I is based on much firmer and solid knowledge and facts than the presumption that Stapert wishes to replace it with and then depart from: i.e. an open air, central domestic hearth. In the first instance he bases his research on two unwarranted assumptions or postulates: a) the spatial organization of a site is defined by a central hearth, and b) that the central hearth (itself a presumption to begin with) was the focal point of the daily life of the inhabitants, regardless of whether it was inside or outside a dwelling. While this sometimes may have been the case, I could enumerate a great number of other things that could have structured or partly defined the spatial layout of a site, including dwellings. Moreover, Stapert's analysis carries with it, or imposes, a series of notions from his previous investigations of Late/Upper Palaeolithic and Mesolithic sites, widely scattered in time and space across northwestern Europe, which may not be applicable to the Maglemosian. In my opinion his methodology clearly would have gained had he tested it on independent, ethno-archaeological material for which the behavioural parameters and internal arrangements are known and then presented a positive result (see below).

Also Stapert, in his assessment of his own method, forgets that some dwellings, such as tents or light structures, might not necessarily leave any archaeologically visible traces, which of course has a bearing on his grouping of hearths into open air and inside hearths.

5. THE RING AND SECTOR METHOD

Perhaps not surprisingly, Stapert starts off his theoretical considerations with a very insufficient and unbalanced critique of mathematical/statistical approaches and methods of others, largely emphasizing their drawbacks and completely forgetting their advantages (Stapert, 1992: p. 144). In fact, one gets the impression that Stapert has felt it necessary to play down the capabilities of other methods in order to advance his own. In this perspective a statement on his own method is interesting (1992: p. 144):

> It should be clear that this method does not claim to detect all possible spatial patterns in sites.

While this certainly is true for the ring and sector method, he fails to report that neither have any such claims been made for any other method (Blankholm, 1991).

Stapert's seeming preponderance for grabbing things out of context and apparent unwillingness to see things in a broader perspective is also reflected in the opening paragraph of his Section 2 (Stapert, 1992: p. 145). For instance, in his rendition of some of the results from my 1991 publication he fails to report that the various methods and techniques were, in fact, rigorously and scientifically tested on ethno-archaeological material (see above). Clearly my aim was objectively to evaluate and find those methods that were capable of delivering

information of relevance for behavioural interpretation (Blankholm, 1991: p. 53 ff.).

Back to rings and sectors. Stapert (1992: p. 144) claims that the main goal of his paper is:

> ..., to explore the potential of the ring and sector method, compared to other techniques of spatial analysis.

Basically Stapert falls short of his aim. Firstly, he opted to try it out on a site (Barmose I) for which he never sought relevant information in the original publications. Secondly, it is surprising that since so much of his argumentation and modelling is linked to Binford's 'hearth model' (Binford, 1983), he does not take the only right and straightforward consequence and test his techniques on the Mask Site data, not least since those data are equally available in my 1991 publication (Blankholm, 1991, Appendix A)! His application of the ring and sector method to Gönnersdorf (Bosinski, 1979; 1981) was not a real, independent, test as the material and structural remains were interpreted prior to analysis. Thirdly, nowhere does he embark on any rigorous procedure for the testing and comparison of the different methods although such procedures were readily available in my book (Blankholm, 1991: p. 55 ff.).

With both data sets (The Mask Site and Barmose I) ready at hand, it should have been no problem for Stapert to match his method in a more appropriate fashion. Instead he defers to what appears to me an unbalanced and inconsistent discussion.

As to the ring and sector method proper it suffers from a number of constraints. Being a feature oriented approach (which is fine and commendable (Blankholm, 1991: p. 25)), Stapert (1992: p. 144) states:

> It is directed at describing and interpreting global spatial patterns that relate to the domestic hearth.

This requires that a domestic hearth be demonstrated and not simply assumed in the first place, and the method then applied subsequently. He (Stapert, 1992: p. 144) also states:

> It is essentially a way of partitioning space (in two related ways: rings and sectors), which seems more suited than any regular grid structure to analyse sites where the global spatial structure is determined by the presence of a central hearth.

This again would require the demonstration prior to analysis that the global structure is determined by a central hearth.

In contrast, none of the multivariate methods make such extremely narrow assumptions. In fact, the latter are developed and operate under a minimum of constraints according to modern concepts of past human behaviour, in which activity areas are expected to be of widely differing size, shape, density, composition and internal patterns of covariation and association, and in which the above characteristics are treated as variables (Whallon, 1984, Blankholm, 1991). This, indeed, is far from Stapert's simplistic notion that they essentially are cluster procedures.

Operationally and regarding interpretation, the ring and sector method is linked to the work of Dekin (1976) and Yellen (1977) and in particular to Binford's (1983) 'hearth model' and the associated concepts of toss and drop zones in a very narrow fashion, ignoring the fact that people need not always arrange themselves or do things in semi-circles around hearths.

The use of the ring and sector method constrains the association and covariation of piece-plotted artifact classes and consequently much information is lost for interpretation.

The position of the rings and sectors clearly is arbitrary, as is the selection of radii for the rings and size of the sectors. This is a draw-back. For instance, the selection of radii can be made so as to reveal, respectively mask, important variability, just as in the case of selecting intervals for histograms (also frequently used by Stapert) and the selection of contours for distribution plans (Blankholm, 1991: p. 79). Moreover Stapert's (1992: p. 158) statement:

> It contains no inherent assumptions of a statistical nature, which encumber many of the more complex approaches to intrasite spatial analysis.

is not correct. In fact, the statistical analysis and significance tests Stapert applies to the ring and sector method minimally require that comparisons are made over units of equal size. Where this is not possible, areal extent of the pertinent units must be accounted for in the calculation of the expected values (e.g. Dalton et al., 1972). Merely consider that the area of a circle grows exponentially relative to r, while areal increase between circles drawn with equal increments of r grows linearly. For example, some of Stapert's calculations are based on 2 rings of 2 m width (Stapert, 1992: pp. 153, 159). In these cases the area of the inner and outer rings are 12.56 and 37.68 sq.m, respectively, which of course must be considered in significance tests.

Talking statistics, it is also of interest to note that no mention is made of the very large standard deviations and their possible analytical consequences for his distance measures per tool class (Stapert, 1992: table 1).

Some general problems of Stapert's treatment of Barmose I was dealt with above. A number of more specific aspects will be dealt with in turn below. On p. 145 and again on p. 146 Stapert comments critically on the use of my own method, Presab, and related interpretation of Barmose I. His critique merely reflects his uninformed use of the data (already commented upon) and seeming unwillingness to understand modern multivariate approaches to spatial analysis generally and specifically. In the first place, the Presab procedures are clearly described in Blankholm, 1991: pp. 151-164). As to his statement (Stapert, 1992: p. 146):

> A general problem with this kind of approach seems to be that there are no guidelines for interpreting the results of such rather mechanical mathematical operations. These do not seem to be directed at answering specific questions, and we are essentially left in the dark as to what the outcomes might mean.

the answer is that multivariate methods, such as k-means Analysis, Unconstrained Clustering, Correspondence Analysis, and Presab, based as they are on modern concepts of past human behaviour, in fact are so versatile that they can be applied to answer or elucidate a wide range of questions, and not simply one single specific one. To me this is a great advantage. As to the results and as a comment to Stapert's appraisal of 'old-fashioned' ways (Stapert, 1992: p. 145), I can only refer to my own conclusions (Blankholm, 1991: p. 203):

> The Barmose I example has minimally shown that all the methods have proved useful for gaining a quicker and much clearer overview of the pattern of variation and covariation between and among the involved item categories than would have been possible from visual inspection.
>
> With all due respect, a very generalized interpretation such as Johansson's (1990), for instance, which suggests only two activity areas, would be quite common among professionals these days as well.
>
> My own preliminary and visual inspection indicated that there would be more areas of immediate interest, and possibly also different uses, both inside and outside the dwelling.
>
> However, the spatial analytical methods instantly revealed an even greater number of interesting areas and patterns of variation and covariation, and also through the aggregate statistics allowed for quick and precise descriptions and assessments of their contents, centres, sizes, etc., such that different areas could be readily compared. That this is all desirable need not be emphasized.

As to the application of the ring and sector method, it is based on the partitioning of space around the 'central hearth'. In this case one is left wondering what constitutes the centre of the roughly 2.4x1.5 m large hearth area. Stapert does not report, either that it is uncertain that the whole area ever was in use at one and the same time (Johansson, 1990: p. 14), or why he selects a fairly asymmetrical position within the hearth area as his centre for calculations. His results might well have been different had he chosen a more standardized and replicable way of defining a centre. In fact, one gets the impression that it was not the centre of the hearth that governed the imposition of rings and sectors, but rather the desire to fit in as many circles as possible within the constraints of the excavation outline, plus the desire to have the most of the test excavation within one single sector (sector 2, Stapert, 1992: fig. 1, p. 170) that determined Stapert's 'centre' of the hearth!

Moreover Stapert is on very shaky ground as to his interpretations of sexual division of labour and prevailing wind direction. In the first place we have no micro-wear determinations and consequently no firm functional determinations of tools or tool classes for Barmose I. Also the strict caution by many a micro-wear analyst that (Dumont, 1987: p. 88):

..., no single tool type can be confidently correlated to either a single manner-of-use or worked material on a scale greater than that of the individual site.

is ignored, quite apart from the fact that we have no way of relating specific tasks or tools to specific sexes nor can we assess prehistoric wind directions. Simply, the answer to such questions is blowing in the wind!

The conclusion to his analysis, that Barmose I was an open air hunting camp with a central hearth and only used by two men, remains unfounded and contradicted by the primary data, i.e. the two lumps of resin show the tooth-marks of a child of approximately 7-8 years of age, and of a young person not less 11 years of age, respectively (Johansson, 1990: p. 43).

6. CONCLUSION

Some years ago I wrote in a comment to the ring and sector techniques (Blankholm, 1991: pp. 25-26):

Stapert is to be commended for his concern with a feature-oriented approach, yet being global in nature, and limited in scope as to their applicability, they are neither geared to handle local variability, nor versatile.

I see no reason to change my position. Also, I see no reason to change my opinion of Barmose I. The analytical result and interpretation of activity areas would have been almost exactly the same, even if a hut floor had not been defined. Stapert has chosen the wrong context for presenting his case and has been largely uninformed about's the nature and specifics of his database. It remains Stapert's task to demonstrate on independent, ethno-archaeological material, for which the behavioural parameters and structural setting are known, that the ring and sector method is capable of delivering information of relevance for behavioural interpretation.

As to his treatment of Barmose I, I find it eminently unsatisfactory and unconvincing.

7. REFERENCES

BINFORD, L.R., 1978, Dimensional analysis of behavior and site structure: Learning from an Eskimo hunting stand. *American Antiquity* 43, pp. 330-361.

BINFORD, L.R., 1983 *In pursuit of the past. Decoding the archaeological record.* London.

BLANKHOLM, H.P., 1981. Aspects of the Maglemose settlement in Denmark. *Veröffentlichungen des Museums für Ur- und Frühgeschichte Potsdam 14/15* (= B. Gramsch (ed.), Mesolitikum in Europa; 12ts Internationales Symposium, Potsdam 3 bis 8 April 1978), pp. 410-414.

BLANKHOLM, H.P., 1984. Maglemosekulturens Hyttegrundrids. En undersøgelse af bebyggelse og adfærdsmønstre i tidlig Mesolitisk tid. *Årbøger for Nordisk Oldkyndighed og Historie*, pp. 61-77.

BLANKHOLM, H.P., 1987. Maglemosian hutfloors: an analysis of the dwelling unit, social unit and intra-site behavioural patterns in early Mesolithic southern Scandinavia. In: P. Rowley-Conwy, M. Zvelebil & H.P. Blankholm (eds), *Mesolithic Northwest Europe: Recent trends*, Sheffield. pp. 109-120.

BLANKHOLM, H.P., 1991. *Intrasite spatial analysis in theory and practice.* Århus.

BOKELMANN, K., 1986. Rast unter Bäumen; ein ephemerer mesolithischer Lagerplatz aus dem Duvenseer Moor. *Offa* 43, pp. 149-163.

BOKELMANN, K., 1989. Eine mesolithische Kiefernrindenmatte aus dem Duvenseer Moor. *Offa* 46, pp. 17-22.

BOKELMANN, K., F.-R. AVERDIECK & H. WILLKOMM, 1981. Duvensee, Wohnplatz 8, neue Aspekte zur Sammelwirtschaft im frühen Mesolithikum. *Offa* 38, pp. 21-40.

BOKELMANN, K., F.-R. AVERDIECK & H. WILLKOMM, 1985. Duvensee, Wohnplatz 13". *Offa* 42, pp. 13-33.

BOSINSKI, G., 1979. *Die Ausgrabungen in Gönnersdorf 1968-1976 und die Siedlungsbefunde der Grabung 1968* (= Gönnersdorf Band 3). Wiesbaden.

Bosinski, G., 1981. *Gönnersdorf. Eiszeitjäger am Mittelrhein.* Koblenz.

DALTON, R., J. GARLICK, R. MINSHULL & A. ROBINSON, 1972. *Correlation techniques in geography.* London.

DEKIN, A.A., 1976. Elliptical analysis: an heuristic technique for the analysis of artefact clusters. In: M.S. Maxwell (ed.), Eastern arctic prehistory: Paleoeskimo problems, *Memoirs of the Society for American Archaeology* 31. pp. 79-88.

DUMONT, J.V., 1987. Mesolithic microwear research in Northwest Europe. In: P. Rowley-Conwy, M. Zvelebil & H.P. Blankholm (eds), *Mesolithic Northwest Europe: Recent trends.* Sheffield, pp. 82-92.

GRØN, O., 1987a. Seasonal variation in Maglemosian group size and structure: a new model. *Current Anthropology* 28, pp. 303-327.

GRØN, O., 1987b. Dwelling organization – a key to the understanding of social structure in Old Stone Age societies? An example from the Maglemose culture. In: J.K. Kozlowski & S.K. Kozlowski (eds), *New in Stone Age archaeology* (= Archaeologia Interregionalis, Varia vol. 230). Warsaw, pp. 63-83.

GRØN, O., 1989. General spatial behaviour in small dwellings: a preliminary study in ethno-archaeology and social psycology. In: C. Bonsall (ed.), *The Mesolithic in Europe. Papers presented at the third symposium.* Edinburgh, pp. 99-105.

JOHANSSON, A.D., 1971. Barmose-gruppen. Præboreale bopladsfund med skiveøkser i Sydsjælland. *Historisk samfund for Præstø Amt* 1968.

JOHANSSON, A.D., 1990. *Barmosegruppen. Præboreale bopladsfund i Sydsjælland.* Århus.

STAPERT, D., 1992. Rings and sectors: Intrasite spatial analysis of Stone Age sites. Groningen.

WHALLON, R., 1984. Unconstrained clustering for the analysis of spatial distributions in archaeology. In: H.J. Hietala (ed.), *Intrasite spatial analysis in archaeology.* Cambridge, pp. 242-277.

YELLEN, J.E., 1977. *Archaeological approaches to the present. Models for reconstructing the past.* New York.

INSIDE OR OUTSIDE: THAT IS THE QUESTION;
SOME COMMENTS ON THE ARTICLE BY H.P. BLANKHOLM

DICK STAPERT

Biologisch-Archaeologisch Instituut, Groningen, Netherlands

Blankholm's critical remarks on the interpretation of the Barmose I site, originally published in my thesis (Stapert, 1992) but reprinted in this volume of *Palaeohistoria* at the request of the editorial board, ask for a reply.

The book by Johansson (1990), unfortunately, was not in our library when I wrote my paper (first half of 1991). Johansson described "two lumps of resin that show the tooth-marks of a child of approximately 7-8 years of age, and of a young person not less than 11 years of age". According to Blankholm, "this effectively refutes Stapert's notion of a men-only hunting camp". What kind of argument is this? Can children not be male? One can easily imagine that some boys accompanied their fathers during a hunting trip (e.g. Spencer, 1976 (1959): p. 241), and this possibility was mentioned in my book, though not with respect to Barmose (Stapert, 1992: p. 77). It may be of interest that the tooth-marks were found in resin, a substance often used for hafting microliths – not an unlikely activity at a hunting camp. (see also Bang-Andersen, 1988: p. 348)

One of the main problems addressed in my article is the question whether the hearth of Barmose was inside a hut or in the open air. The ring-diagrams obtained for the site convinced me that we are dealing with a hearth in the open air. Both test pits were located in the western half of the site. The eastern site-half, which was the richer in terms of tool numbers, was excavated completely. The ring-diagrams for the eastern half should therefore be given more weight than the diagrams for the western half. As can be seen in my article (Stapert, 1992: fig. 16), this half shows a classical unimodal distribution. There is no indication at all of any 'wall effects', and the mode falls between 1 and 2 m from the hearth centre – perfectly plausible in the case of an open-air hearth.

I am still somewhat baffled by Blankholm's idea that selecting a density contour line is a meaningful 'analytical method' to establish the presence of a hut outline. Of course, one can always draw a contour line, for example the one based on the mean number of artefacts per square metre (Blankholm's 'standard method for delineation of Maglemosian hut floors'). But this will never prove anything, least of all the presence of a hut wall. Quite apart of any 'statistical' arguments, one would in any case expect such a large and messy hearth area to be located outside. Blankholm states that he also used 'other contextual knowledge' in this connection. I suppose he refers to phenomena such as the presence of bark flooring and the relatively peripheral location of the cores. In my article I discussed these, and found them inadequate as indicators for the existence of a hut (sections 2, 4, 6).

Contrary to what Blankholm states, the spatial distribution of flint waste was discussed in my article. In figure 3 a density map of flint waste is given, based on the principles outlined by Cziesla (1990). It can be seen that the material is quite tightly clustered near the hearth, and that the density gradually decreases outwards. This pattern can also be observed at many other sites, and seems to be characteristic of outdoor hearths. In the case of a hearth located inside a dwelling, one would expect a zone of richer squares accentuating the walls, as a result of the 'barrier effect' (see Stapert, 1989). One example of this phenomenon is provided by the Middle Palaeolithic site of Buhlen, where a distinct tent ring consisting of large stones was excavated (see Stapert, 1990: fig. 3). Regarding the cores, I agree that they occurred somewhat more peripherally than tools (mean distances are 2.08 m and 1.94 m, respectively; see my table 1). However, the difference is slight, and cannot be shown to be significant in a statistical sense. Moreover, in the ring-diagrams it can be seen that the two distance distributions have the same mode, in the 1.5-2 m class (see my fig. 6).

Blankholm states that "... Stapert ... forgets that some dwellings, such as tents or light structures, might not necessarily leave any archaeologically visible traces, which of course has a bearing on his grouping of hearths into open air and inside hearths". Blankholm here completely misses the point of my work. The ring and sector method produces indications for the presence or absence of dwellings independently of archaeologically visible features, and this is exactly what makes it potentially very useful. That is why I stated in my book: "An important reason why it is desirable to have an independent method for establishing the presence or absence of dwellings, based on the *structures latentes*, is the circumstance that Palaeolithic dwellings might easily leave no archaeologically visible traces, even in sites with perfect in situ preservation" (Stapert, 1992: p. 193).

So far, archaeologically invisible tents or huts have been demonstrated at Gönnersdorf II and III, Verberie E1 and Etiolles P15, by means of the ring and sector method. In other cases, it has been shown that tents or huts envisaged by archaeologists probably did not exist: Rheindahlen, Pincevent, Olbrachcice 8, Orp East, Barmose. This is not simply a 'grouping', but a result of analysis.

Gönnersdorf I was carefully interpreted by Bosinski (1979) as a large tent, on the basis of archaeologically visible features. That is exactly why this site was appropriate for testing the ring and sector method. Again the same crucial point has escaped Blankholm: this method offers an independent way of demonstrating the presence or absence of tent walls.

If one can be reasonably sure whether a dwelling structure was present or not, a whole series of other questions can be addressed, for example the possible existence of gender patterns. Therefore, as stated in my book, "... this should be the first step in any meaningful intrasite spatial analysis" (Stapert, 1992: p. 34). For this step to be taken, sound methodology is needed, not conjectures without any foundation, such as equating a density contour line with a hut outline.

The ring and sector method was designed for "... sites characterised by the presence of a hearth closely associated in space with an artefact scatter" (Stapert, 1992: p. 11). I speak of 'central hearths' in such cases, but I do not, of course, mean that the hearths were in the exact geometrical centre of the artefact scatters. The idea behind the method is that these hearths were a focal point in the daily life of a small group of people. The hearth attracted many activities, and also played an important role in social life. Blankholm seems to doubt this, but to my mind this is obvious. Blankholm asks where the centre of the ring and sector system was located in the case of Barmose I; he can find it in figure 1. I can assure him that shifting it a few decimetres would not alter the results in any significant way.

I did not attempt to 'play down' other methods of spatial analysis, although I did not see any point in abandoning simple methods in favour of complex computerized procedures. What I did criticize was Blankholm's interpretation of the patterns created by his computer programmes. In the final section of my article, I expressed my doubts about his interpretation by stating that Blankholm expects too much 'expressiveness' from archaeological sites of this type (Stapert, 1992: p. 158). Clear-cut 'activity areas', sharply delineated in space, should not be expected to show up at such sites. This is why I emphasized that we should start by looking for global patterns, instead of expecting miracles from detailed mathematical manipulations. Blankholm informs us that Johansson 'suggests only two activity areas', which Blankholm

clearly thinks to be a very poor interpretation. As for me, I can only agree with Johansson, because this is a realistic assessment of what we can observe, without falling into the trap of over-interpretation, which is what Blankholm did.

Blankholm notes that the "position of the rings and sectors clearly is arbitrary, as is the selection of radii for the rings and size of the sectors". Of course it is, as I have clearly stated. It would even be an enhancement if we could vary the ring width, the number of sectors, and even the 'centre' of the whole system. In this way we would be able to find the optimal parameters of the method in every case. Note that this would not imply any manipulation of the data; it would optimalize the visibility of any patterns present in a site. It is impracticable to do this by hand; therefore, dedicated software would be very useful. For the past year and a half, G.R. Boekschoten (Groningen) has been engaged in developing a computer programme for applying the ring and sector method, which is now operational (Boekschoten & Stapert, in press).

Blankholm (and other colleagues) have wondered why I did not calculate densities per ring, in terms of numbers of artefacts per square metre, since the rings grow in surface area going outwards from the centre. I believe it is not advisable to do so. The ring approach is meant to reveal patterns in the distribution of artefacts in terms of distance to the hearth. Therefore, it would be more precise to speak of distance classes. The rings only serve as a graphical illustration of the method. "Calculating densities per ring would only transform the data, and moreover give the false impression that the artefacts are scattered evenly in the rings ..." (Stapert, 1992: p. 31). In other words, one would obtain averaged densities per ring, and these do not have much value. Imagine that a woman cleaned a hide, using five scrapers. The distance to the hearth of these five scrapers is the issue of interest here, not their averaged density over whichever ring in which they were located.

Blankholm thinks the ring and sector method suffers from 'extremely narrow assumptions'. By contrast, he believes "... that multivariate methods, such as k-means Analysis, Unconstrained Clustering, Correspondence Analysis, and Presab, based as they are on modern concepts of past human behaviour, in fact are so versatile that they can be applied to answer or elucidate a wide range of questions, and not simply one specific one". Maybe so, but we should be modest enough to admit that it can be very difficult even to answer one simple question, as for example whether the hearth of Barmose lay in a hut or in the open. This is also a very basic question, and Blankholm's multivariate methods have not provided any answer, nor were they designed to do so, as he kindly informs us. In fact, they do not give answers to specific questions at all, which to my mind makes them rather unattractive. What we need are realistic interpretive models, coupled to transparent analytical techniques.

Nothing is lost by using simple methods; much can become hopelessly entangled by the use of very complex procedures.

REFERENCES

BANG-ANDERSEN, S., 1988. Mesolithic adaptations in the southern Norwegian highlands. In: C. Bonsall (ed) *The Mesolithic in Europe*, Edinburgh, pp. 338-350.

BLANKHOLM, H.P., 1991. *Intrasite spatial analysis in theory and practice*. Aarhus.

BOEKSCHOTEN, G.R. & D. STAPERT. in press. 'Rings & Sectors': een computerprogramma voor ruimtelijke analyse. *Paleoaktueel 5*.

BOSINSKI, G., 1979. *Die Ausgrabungen in Gönnersdorf 1968-1976 und die Siedlungsbefunde der Grabung 1968* (= Gönnersdorf 3). Wiesbaden.

CZIESLA, E., 1990. *Siedlungsdynamik auf steinzeitlichen Fundplätzen; methodische Aspekte zur Analyse latenter Strukturen* (= Studies in Modern Archaeology 2). Bonn.

JOHANSSON, A.D., 1990. *Barmosegruppen. Praeboreale bopladsfund i Sydsjaelland*. Aarhus.

SPENCER, R.F., 1976 (1959). *The North Alaskan Eskimo; a study in ecology and society*. New York.

STAPERT, D., 1989. The ring and sector method: intrasite spatial analysis of Stone Age sites, with special reference to Pincevent. *Palaeohistoria* 31, pp. 1-57.

STAPERT, D., 1990. Middle Palaeolithic dwellings: fact or fiction? Some applications of the ring and sector method. *Palaeohistoria* 32, pp. 1-19.

STAPERT, D., 1992. Rings and sectors: intrasite spatial analysis of Stone Age sites. Thesis, University of Groningen.

STAPERT, D. & T. TERBERGER, 1989. Gönnersdorf Concentration III: investigating the possibility of multiple occupations. *Palaeohistoria* 31, pp. 59-95.

SOME FINAL WORDS ON STAPERT'S TREATMENT OF BARMOSE I

H.P. BLANKHOLM
Byagervej 204, Beder, Denmark

Following my reply to Stapert (this volume) I had expected him to re-analyse the site and begin arguing from solid knowledge of the data. I still see no references to particulars or contextual information in Johansson's (1971; 1990) publications, nor do I see any test of the ring and sector method on independent ethno-archaeological data.

As should be evident from my review, and to use his own phrasing, it manifestly is the ring and sector method and its application to Barmose I that amounts to conjectures without foundation.

His 'test' on two sections of Gönnersdorf (Bosinski, 1979), and his apparent claim that it is applicable to the vast range of site structures, features and sizes and shapes of dwellings from the Late Palaeolithic and Mesolithic of Central, western and northern Europe is unconvincing. Negative evidence plays an important role in support of his hypothesis of open air hearths or sites. We need solid information on the distributional trends across a considerably larger and broader selection of dwellings from different cultural and behavioural contexts before we may adequately assess the possible significance of variability in 'modalities' of associated artifact distributions. Even if we for discussion accept Stapert's 'test' on Gönnersdorf and accept that strong 'centrifugal effect' and bimodal distribution may be indicative of Gönnersdorf type dwellings, it does not necessarily follow that lack of 'centrifugal effect' and unimodal distribution indicate open-air hearths without dwellings! This also is one of the reasons why I wish to see such an allegedly universal method tested on ethno-archaeological sites both with and without dwellings and open-air hearths.

In contrast, the method for delineating Maglemosian hut floors (Blankholm, 1981; 1984; 1987) never has been claimed applicable beyond Maglemosian contexts proper. In other words, it is culturally and behaviourally specific. Secondly, I have never claimed in Stapert's fashion that lack of indication of a typical Maglemosian hut floor would be indicative of an open-air Maglemosian hearth or site. That simply would be carrying things too far.

Stapert needs not be baffled about the use of the mean number of artifacts per square metre contour for delineating Maglemosian hut floors. Simply, given the nature of the distributional trends of flint artifacts on and around the preserved floors, it remains a demonstrated fact that this contour is the best approximation to the wall lines! Also, contrary to what Stapert believes, the presence of a hearth does not in itself demonstrate the presence of a hut floor. Instead, it is contextual and corroborative evidence given other, primary, indications. In fact, several locations also show indications of outside open-air hearths. Stapert's notion of messiness of hearth areas, as an indication for their open-air location, once again only reveals his prejudices against past and present human behaviour.

On to Barmose I. Stapert writes: "There is no indication at all of any `wall effects'".

In fact there is (Johansson, 1971; 1990; Blankholm, 1991). The case is that Stapert's method is incapable of revealing them. Stapert's method is so 'hearth-and-circle-fixed' that it seems unable to detect structures or dwellings of other shapes (particularly if hearths are off-set from the dwelling's centre), for example Maglemosian hut floors that generally are rectangular or subrectangular in shape, measuring 5-9 m in length and 3-5 m in width and often with the hearth off-set towards one end.

Stapert (1992: fig. 3) provided a density map of flint waste. Apart from a few reflections, however, he made no scientific analysis. Clearly, a density map based on Cziesla's (1990) principles, with a set of contour lines at 275 piece intervals, is inadequate for the context. Simply, it is not scaled to reveal inflections (wall lines) or boundary effects. I strongly recommend the reader to see the sharp inflection in the distribution of debitage (Blankholm, 1991: fig. 100). In my view, this very clearly shows the boundary effect of the wall line. Irrespective of whichever method is used, the study of such phenomena also requires the detailed study of distributional trends across the site and the careful investigation of sections and artificial profiles. Moreover, the investigation of average weight per piece of flint per unit (Johansson, 1990) may be useful. All this also is contextual information, but again ignored by Stapert.

I need an exact coordinate (a set of numbers) for the origin of the ring and sector system in Stapert's investigation; not a position on a greatly reduced and off-scaled map, which need recalculation. As to his comment: "I can assure him that shifting it a few decimetres would not alter the results in any significant way".

I need no assurance or assertions from someone, who, for instance, investigates sites without consulting the excavator's publications and then unfoundedly proceeds to critisize other peoples methodologies on the basis of the outcome! In fact, we are not discussing a few decimetres, but a distance in the range of c. 0.5 m. If, for example, the centre is moved to a more appropriate coordinate, for instance 5,10.33, a brief inspection of the distribution plans for scrapers and microliths (Blankholm, 1991: figs 89 and 91) and using 0.33 m intervals would seem to indicate trimodal and bimodal distributions, respectively.

Stapert's ring and sector approach and notion of 'distance classes' still runs counter to basic statistics (Blankholm, this volume). Even if he abandons significance tests, for which area must be accounted for in the calculation of the expected values, counts are still obtained from increasingly larger (sampling) areas. At least counts must be weighted relative to the sizes of the (sampling) areas.

Stapert's notion of 'expressiveness' reveals a limited scope of our discipline. If we wish to contribute to archaeology, or more modestly to activity area research, we certainly need to be realistic, begin to understand that prehistoric behaviour is multivariate in nature, penetrate deeper into the variability, and consequently apply multivariate approaches to its resolution. Contrarywise, much effort may be wasted in developing simplistic and highly constrained spatial analytical methods and models, like the ring and sector method, that basically seem more aimed at confirming the obvious or trivial.

I also wonder how Stapert can argue: "... they do not give answers to specific questions at all ...".

There is a large body of good examples in the literature where multivariate methods have been used successfully to answer both specific and general questions, in all

modesty including my own analyses of the Mask Site (Binford, 1978) and Barmose I (Blankholm, 1991 with references).

In archaeological examples one should first ask questions appropriate and relevant for the given cultural, behavioural and site specific context, then select methodologies accordingly (there is no single the method or panacea approach to spatial analysis (Blankholm, 1991)), and then proceed to apply those methods based on solid knowledge of the data. Stapert failed to do so with Barmose I.

REFERENCES

BINFORD, L.R., 1978, Dimensional analysis of behavior and site structure: Learning from an Eskimo hunting stand. *American Antiquity* 43, pp. 330-361.
BLANKHOLM, H.P., 1981. Aspects of the Maglemose settlement in Denmark. *Veröffentlichungen des Museums für Ur- und Frühgeschichte Potsdam 14/15* (= B. Gramsch (ed.), Mesolitikum in Europa; 12tes Internationales Symposium, Potsdam 3 bis 8 April 1978), pp. 410-414.
BLANKHOLM, H.P., 1984. Maglemosekulturens Hyttegrundrids. En undersøgelse af bebyggelse og adfærdsmønstre i tidlig Mesolitisk tid. *Årbøger for Nordisk Oldkyndighed og Historie*, pp. 61-77.
BLANKHOLM, H.P., 1987. Maglemosian hutfloors: an analysis of the dwelling unit, social unit and intra-site behavioural patterns in early Mesolithic southern Scandinavia. In: P. Rowley-Conwy, M. Zvelebil & H.P. Blankholm (eds), *Mesolithic Northwest Europe: Recent trends*, Sheffield. pp. 109-120.
BLANKHOLM, H.P., 1991. *Intrasite spatial analysis in theory and practice*. Århus.
BOSINSKI, G., 1979. *Die Ausgrabungen in Gönnersdorf 1968-1976 und die Siedlungsbefunde der Grabung 1968* (= Gönnersdorf Band 3). Wiesbaden.
CZEISLA, E., 1990. *Siedlungsdynamik auf steinzeitlichen Fundplätzen; methodische Aspekte zur Analyse latenter Strukturen* (= Studies in Modern Archaeology 2). Bonn.
JOHANSSON, A.D., 1971. Barmose-gruppen. Præboreale bopladsfund med skiveøkser i Sydsjælland. *Historisk samfund for Præstø Amt* 1968, pp. ???.
JOHANSSON, A.D., 1990. *Barmosegruppen. Præboreale bopladsfund i Sydsjælland*. Århus.
STAPERT, D., 1992. Rings and sectors: Intrasite spatial analysis of Stone Age sites. Groningen.

THE PLANT HUSBANDRY OF ACERAMIC ÇAYÖNÜ, SE TURKEY

WILLEM VAN ZEIST & GERRIT JAN DE ROLLER
Biologisch-Archaeologisch Instituut, Groningen, Netherlands

ABSTRACT: The present paper deals with the results of the examination of plant remains recovered from aceramic Neolithic levels at Çayönü in southeastern Anatolia. The cultivated plants included einkorn and emmer wheat, field pea, lentil and bitter vetch. It is not clear whether grass pea and chickpea were crop plants of the aceramic farmers. Barley was neither cultivated nor gathered intentionally. Pulses predominate over cereals in the archaeobotanical record. A characteristic feature of Çayönü is the wild type emmer wheat grains in the lower occupation levels. Wild pistachio (*Pistacia atlantica/khinjuk*) fruits were collected intensively. The large numbers of wild vetch (*Vicia* sp.) seeds suggest that these were gathered purposely.

KEYWORDS: aceramic Neolithic, SE Anatolia, plant domestication, hulled wheats, pulse crops, bitter vetch, wild fruit collecting.

1. INTRODUCTION

1.1. Location and environmental conditions

The prehistoric mound of Çayönü Tepesi is located c. 7 km southwest of Ergani, in the Diyarbakir Province of southeastern Anatolia, at 38°16'N, 39°43'E (fig. 1). The site is situated on a small tributary of the upper Tigris (Dicle) River, in a rather broad valley at the foot of the East Taurus Mountains, at an elevation of c. 830 metres.

1.1.1. *Physical environment*

No long-term weather records are available for the Çayönü area (Ergani). In table 1, data for three stations which are closest to Çayönü are presented. Climatic (hydrothermic) diagrams for the stations concerned, Diyarbakir, Elazığ and Malatya, are shown in Zohary (1973: fig. 8). For Ergani, which is about halfway between Diyarbakir and Elazığ, mean annual precipitation may be estimated at about 450 mm. Winters are relatively mild (estimated mean January temperature somewhat above 0°C) and summers are fairly hot (estimated mean July temperature 28°C). The dry period lasts from June through September. The Çayönü area does not experience the continental climatic conditions of the interior of eastern Anatolia. Precipitation is brought in by prevailing southwesterly winds from the Eastern Mediterranean. The climate of the area is suitable for rain-fed agriculture.

To the authors' knowledge there are no specialist reports available on the soil conditions in the Çayönü area. As mentioned by Çambel & Braidwood (1983) and Stewart (1976) the valley bottom consists of calcareous, reddish, silty clay which provides good arable soil. To the south of the stream on which the site is located the landscape is dominated by weathered, bare limestone outcrops.

1.1.2. *The vegetation*

Under the present climatic conditions a considerable part of southeastern Turkey, including the Çayönü area, would naturally, that is to say without the interference of man, be covered by woodland: open forest with a tree cover of at most 50% (fig. 2). The woodland in this part of Turkey belongs to Walter's (1956) East-Anatolian oak-juniper region. Arboreal components include *Quercus brantii, Q. infectoria, Q. boissieri, Acer cinerascens (A. monspessulanum* ssp. *cinerascens), Pyrus syriaca, Crataegus azarolus, Pistacia atlantica, P. khinjuk* and *Juniperus oxycedrus.* Summer drought prevents a more dense tree growth. As a result of the destructive activities of man most of the woodland has degraded into shrub vegetation or has disappeared altogether. Holy places are often the last refuges of trees. The Çayönü area is an example of a landscape where only a few poor tree stands testify to the original woodland vegetation.

To the south, with decreasing precipitation, the woodland gives way to steppe, usually via a transitional (almond-pistachio) forest-steppe zone. At present most of the steppe vegetation is dominated by *Artemisia (herba-alba)*, but this may be due to intensive grazing which results in the replacement of the herb-rich grass steppe (with many palatable species) by *Artemisia* steppe rich in thorn-cushion species. The 'steppe island' in the Diyarbakir basin is somewhat puzzling,

W. van Zeist & G.J. de Roller

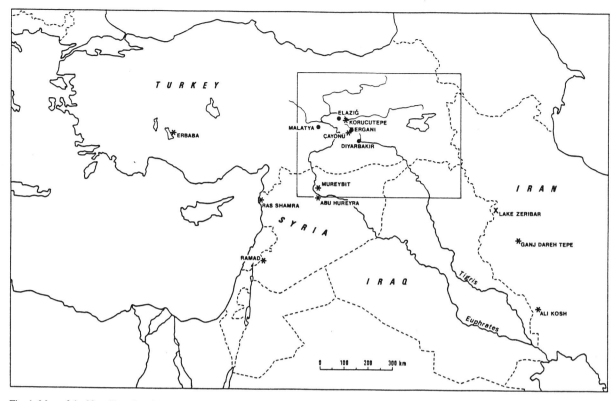

Fig. 1. Map of the Near East showing the location of sites mentioned in the present paper. For the framed area a vegetation map is shown in fig. 2.

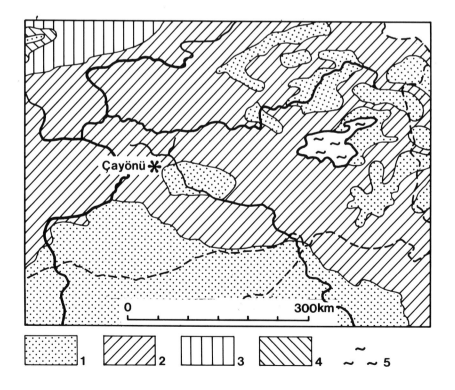

Fig. 2. Map of (assumed) natural vegetation. Section of map published by van Zeist & Bottema (1991: fig. 4), slightly simplified. 1. Steppe vegetation; 2. SE Anatolian mixed-oak woodland; 3. Montane Euxinian forest; 4. Xero-Euxinian woodland; 5. Lake Van.

Table 1. Climatic data for three stations in SE Turkey (after Alex, 1985).

	Diyarbakir	Elaziǧ	Malatya
Elevation (in m)	660	1105	998
Mean annual temperature (in °C)	16.0	13.2	13.6
Mean January temperature	2.5	–1.0	–0.1
Mean July temperature	30.2	26.8	26.7
Mean annual precipitation (in mm)	495	415	395.8
Month with maximum precipitation	Dec.	May	April
Mean precipitation in wettest month	79.0	62.2	61.8
Mean precipitation in driest month (July)	0.4	0.7	0.9
Dry summer period	June-Sept.	June-Sept.	June-Sept.

because climatically it cannot easily be understood. Mean annual precipitation at Diyarbakir (495 mm) is even higher than at Elaziǧ (415 mm) and Malatya (396 mm), which are both situated in naturally wooded areas. Mean summer temperatures are higher at Diyarbakir (table 1), implying a more intensive drought period, which could possibly account for the steppe vegetation. On the other hand, one wonders to what extent the predominantly volcanic soils there create unfavourable conditions for woodland vegetation. The dry limestone outcrops near Çayönü may originally have borne steppe vegetation.

1.2. The site

The information below on Çayönü Tepesi has been taken mainly from Braidwood et al. (1981), Schirmer (1990) and Özdoǧan & Özdoǧan (1989), supplemented by personal information from Professor Robert J. Braidwood and Dr. Mehmet Özdoǧan. The site appears to measure about 150 by 250 m, with an overall inhabited area not yet well determined; it may have reached over three hectares. At its highest point it stands about 3 m above the level of the neighbouring fields. The core of the site is formed by the pre-pottery (aceramic) Neolithic mound (phase I), the surface area of which is not exactly known because it is partly covered by younger occupation deposits. Occupation phase II almost completely surrounds the pre-pottery mound and comprises mainly pre-Halafian (pottery Neolithic) layers. Minor and discontinuous deposits in various parts of the mound date from the Early Bronze Age (EBAII and III) to the Early Iron Age (phase III).

Excavations at Çayönü, which through 1988 were largely confined to the pre-pottery mound, started in 1964 as a joint project of the Prehistory Section of Istanbul University and the Oriental Institute of the University of Chicago, under the direction of Professors Halet Çambel and Robert J. Braidwood. Later, in 1978, the expedition was joined by a team from the Institut für Baugeschichte, University of Karlsruhe, headed by Professor Wolf Schirmer. In 1989 Dr. Mehmet Özdoǧan from the Prehistory Section of Istanbul University succeeded Çambel and Braidwood as field director. Since 1964, thirteen campaigns of archaeological excavation have been undertaken, plus three more under Özdoǧan.

Radiocarbon determinations date the pre-pottery occupation between 9200 and 8700 BP, which after calibration should approximately correspond with 8250-7750 BC in calendar years. In this main prehistoric phase (phase I), a number of so-called 'sub-phases' are

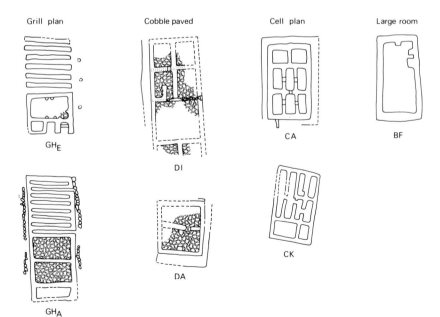

Fig. 3. Schematized plans of main domestic building types at aceramic Çayönü. Redrawn from Özdoǧan & Özdoǧan (1989: fig. 2).

Fig. 4. To the left, at a higher stratigraphic level, three cell plan buildings; at a lower stratigraphic level, grill plan buildings (photo Çayönü archive, Istanbul).

Table 2. Periodization in sub-phases of the aceramic occupation (phase I) at Çayönü. Various sub-phases have been further subdivided, e.g. c1, c2, c3.

lr	Large room	
c/lr	Cell/Large room transition	
c	Cell House (Cell plan)	
cp	Cobble-paved house	Former
ch	Channelled house	Intermediate
cp	Basal pits	Sub-phase
g	Grill house (Grill plan)	
r	Rounded huts (Round plan)	

distinguished. The sub-phases are mainly characterized by houses of different plan types (see figs 3-5). In this paper, the botanical samples are attributed to eight sub-phases (cf. tables 3 and 4). Below there follows a brief description of the house plans, from deepest to uppermost in stratigraphic order, characteristic of the various sub-phases. In parentheses are the sub-phase letter indications (see also table 2).

Round or oval huts, with a sunken floor, were recovered from the bottom levels: the Rounded huts or Round plan (r) sub-phase. The walls were constructed in wattle and daub technique, sometimes on stone-laid foundations.

The second sub-phase is characterized by the so-called grill-plan buildings: the Grill plan or Grill house (g) sub-phase. The rectangular buildings of this sub-phase, measuring c. 5.5 by 11 m, are tripartite in ground plan and approximately north-south oriented. The northern part of the buildings has a substructure of parallel narrow wall strips built of small stones, the grills. The superstructure of these houses is believed to have been made of light organic material.

The succeeding Basal pits (bp) sub-phase is somewhat puzzling in that it suggests that the inhabitants of the site changed their fairly roomy houses for sunken huts of about 3-5 m in diameter[1].

In the Channelled house (ch) sub-phase, the grill substructure was replaced by a solid stone foundation, interrupted by parallel drainage channels.

In the Cobble-paved (cp) sub-phase houses, a paved floor covers the entire space of the room.

The ground floor plan of the Cell plan or Cell house (c) sub-phase is somewhat different from those of the preceding sub-phases. The term 'cell plan' derives from the very small dimensions of the individual chambers, whose walls support the overlying living floor. In addition, the stone foundations are known to have supported mud-brick walls.

The final two sub-phases, Cell/Large room (c/lr),

Fig. 5. The structure in front represents the remains of two buildings: the right part is made up of the front courtyard and cellular section of a grill plan building (the grill structure itself is still covered by the clayey floor); the left part is a cobble paved building, partly running over the grill plan house. Behind this composite structure, a grill plan building. At the right end of the latter building, the remains of a rounded hut (Round plan sub-phase) (photo Çayönü archive, Istanbul).

and Large room (lr), (for which the terms Early Large room and Late Large room are preferable) have houses of the Large room plan type. They are distinguished from one another on the basis of stratigraphy and other recognizable differences.

Within most sub-phases, a number of main architectural layers can be observed, e.g. Cell plan c1, c2 and c3.

In addition to the domestic buildings, a few buildings of special function, different in plan and construction techniques, have been uncovered (e.g. Schirmer, 1983). One of them is the so-called 'skull building', so named when some 70 human skulls were uncovered in the first season of its excavation. Continued excavations revealed that in earlier aspects of this building (at least five major phases suggesting rebuilding are distinguished), skeletons and parts of skeletons had also been stored here. The skull building, in its various rebuilding aspects, largely corresponds with the Cobble-paved sub-phase.

The house plans and those of other non-domestic buildings are indicated by a two-letter code, e.g. GD, DI and BM (skull building).

One may wonder whether the different building types characterizing the sub-phases point to discontinuity in the habitation. However, the differences in house plan between sub-phases are less abrupt than they seem at first sight. Except for the earliest round houses and the so-called basal pits, a gradual development of architecture can be observed.

Two sectors of the aceramic mound have been excavated (eastern and western excavation areas) separated by a c. 30 m wide, largely unexcavated zone. The approximate extent of the excavated areas appears from figure 6 in which the distribution of botanical samples included in the present publication is shown. Of the Cell plan sub-phase, which has been excavated most extensively, an estimated 20-30% had been unearthed by 1987 (Özdoğan & Özdoğan, 1989).

1.3. Archaeobotanical research

In 1968, the first attempts were made to retrieve floral remains from the Çayönü site, with the aim of obtaining information on the exploitation of cultivated and wild plants by the aceramic inhabitants. Professor Robert B. Stewart (Houston State University, Huntsville, Texas) initiated the flotation of soil samples from the site. The results were disappointing in that Stewart did not succeed in recovering identifiable plant remains. In the next field campaign in 1970, the first author was more successful. Appreciable numbers of seeds, fruits and other plant remains could be brought to light (van Zeist, 1972; this paper is referred to below as '1972 publication'). Stewart's negative results can probably be ascribed to the fact that the material he had access to was only available from layers near the surface of the mound. As a consequence of alternate drying and moistening of the soil and of root action, the carbonized plant remains had disintegrated in the upper layers.

In 1972, Stewart returned to Çayönü (fourth season) and examined a large number of samples secured from 109 provenances (Stewart, 1976). This time the results were quite satisfactory; fair quantities of plant material were recovered, including a cache of more than 1500 emmer wheat grains. Stewart (1976) noted that the 1972 campaign had yielded only a few additions to the inventory of plant taxa demonstrated in 1970. Whether or not this made the excavators believe that little new information was to be expected from a continuation of archaeobotanical research on the site, the fact is that in following field seasons no botanist was scheduled as a member of the excavation team. It was not until the 1985 campaign that archaeobotanical fieldwork was taken up again, this time by the second author of this report. Sampling for archaeological plant remains was continued in 1986 and 1987.

2. THE SAMPLES

2.1. Field and laboratory procedures

The conditions at Çayönü are such that only carbonized vegetable material and possible mineralized (organic material replaced by calcium phosphate) plant remains will have been preserved. The plant remains were recovered in the field by means of manual water flotation of samples of occupational soil. The volume of soil floated varies considerably, but the majority of the samples measured between 15 and 40 litres.

The samples were taken partly by members of the excavation team and partly by the palaeobotanist during the periods the latter took part in the field work (WvZ in 1970, GJdR in 1985, 1986, 1987). The flotation was carried out by the botanist. Samples were secured from places which looked promising because of the presence of charcoal and ashes or because seeds were observed with the naked eye (so-called judgement samples). The presence of particular features was also occasion for botanical sampling. No systematic sampling in the sense of some kind of random sampling has been considered, irrespective of the question whether or not this may have been practicable.

The samples taken in 1970 during the stay of the palaeobotanist in the field were examined in the Çayönü excavation house. Of this material only part was taken to the laboratory in Groningen for checking identifications and for measuring and drawing seed and fruit types. In preparing the present report it was regretted that some of the 1970 material was not available for re-examination. Some of the samples floated in 1985, 1986 and 1987 were provisionally examined in the field, but all the material was taken to Groningen. In the laboratory the samples were examined according to standard procedures.

A considerable number of samples turned out to be disappointingly poor in seeds, fruits and chaff remains. Wood charcoal was examined whenever the pieces

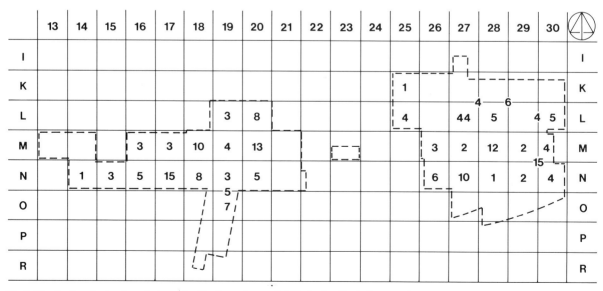

Fig. 6. Distribution of palaeobotanical samples. For discussion, see 2.1. The broken line is the approximate limit of the eastern and western excavation areas (end 1987 field season).

were large enough to be handled. Frequently the wood structure had suffered much from heat, which prevented identification.

The samples included in the present publication are mainly from the 1970, 1985, 1986 and 1987 field seasons. A few samples were secured by the excavators in 1979, 1981 and 1984. Where appropriate, the results of the 1972 field season (Stewart, 1976) will be included in the discussion, but the data of the latter campaign have not been published in such a way that they could be incorporated in those of the other years. The fact that only a few field seasons are reasonably well covered archaeobotanically inevitably leads to a very uneven distribution of the botanical samples over the areas excavated. This is clear from figure 6, in which the numbers of samples which yielded floral remains other than wood charcoal are given per square. For some of the 1970 samples the exact location within the present grid system, which was laid out in 1978, could not be established, hence the numbers between squares. It is true that the number of samples examined is greater than that shown in figure 6 because some of them yielded no seeds or chaff. However, the latter samples would not have changed the overall picture of the sample distribution. It goes without saying that this incomplete coverage has provided less detailed information on the Çayönü plant husbandry than one could obtain from a more thorough study of the deposits excavated.

A fairly large number of accidentally burned mudbrick samples had been secured by the excavators for botanical examination. Inspection of these samples showed that the imprints of vegetable matter could generally not be identified ('grass stems'and such like) and therefore the mudbricks would not essentially supplement the charred seed record. For that reason it was decided to discard the mudbrick samples.

Mr. Michael K. Davis, New York, a Çayönü expedition member of many years'standing, has expertly identified the sub-phase attributions of the botanical samples.

2.2. Presentation of the results

With a few exceptions no data of individual samples are presented, but in tables 3 and 4 (for the eastern and western areas respectively) the total numbers of seeds, fruits and other plant remains other than wood charcoal are shown per occupation sub-phase. In addition, the sample frequencies, expressed as percentages, are given. Seed types for which not even the family could be identified are not listed. As for the numbers of *Triticum* spikelet forks, two glume bases are counted as one spikelet fork. It was decided not to publish the full data as this would have led to long lists with predominantly zero scores[2]. A small number of individual samples are presented as examples of the botanical contents of certain types of contexts (tables 12, 21, 23-25).

As usual measurements are presented. In addition to descriptions and illustrations, measurements should document the archaeological plant remains. So far, little use has been made of measurements in evaluating the archaeobotanical data, particularly in comparing between sites. A complicating factor is the fact that we hardly know to what extent the dimensions have been affected by the carbonization.

In this paper the term 'seeds'has been used somewhat loosely and may also denote anatomically-defined fruits.

3. COMMENTS ON PLANT REMAINS

The comments on the seed and fruit types presented in this section are in part similar to those in the 1972 publication. As this final report aims at presenting all relevant information on the Çayönü floral record, some repetition of what has been written in an earlier paper is unavoidable. Moreover, the 1972 publication may not always readily be available to the reader. The illustrations in the 1972 publication are also shown again, in addition to new drawings of plant remains. Drawings have been prepared for only some of the seed and fruit types established for Çayönü. For various other types reference is made to illustrations in archaeobotanical literature: 'Korucutepe': van Zeist & Bakker-Heeres, 1975b; 'Ramad': van Zeist & Bakker-Heeres, 1982(1985); 'Ras Shamra': van Zeist & Bakker-Heeres, 1984(1986)a. 'Cf.'indicates that the Çayönü seed does not exactly conform to that illustrated.

Except for some crop-plant species, seed and fruit types recovered only from pottery Neolithic samples (table 3) are left out of consideration.

3.1. Cereals

By far the majority of the cereal remains are of hulled wheats. Particularly chaff, in the form of spikelet forks and glume bases (see figs 8:5,6 and 9:2,3), is well represented. As mentioned above, in the tables two glume bases are counted as one spikelet fork. No free-threshing wheat has been identified for aceramic levels, but some naked wheat grains (*Triticum durum/aestivum*) were recovered from pottery Neolithic samples.

3.1.1. *Emmer wheat*

In the 1972 publication attention is focussed on the occurrence of wild emmer wheat (*Triticum dicoccoides*) grains in addition to those of the domestic type, *Triticum dicoccum*. Wild emmer wheat kernels are characterized by an oblong shape and a longitudinally straight or only slightly curved dorsal side. The flat ventral side has a narrow furrow. *Triticum dicoccum*-type grains are spindle-shaped in outline, showing the greatest breadth in the middle of the kernel. The ventral side is longitudinally straight or somewhat concave. Examples

Table 3. Eastern excavation area. Total numbers of seeds, etc. (Σ) and sample frequencies expressed as percentages (% fr) per sub-phase. x Present; xx Frequent; xxx. Very frequent.

Sub-phase Number of samples	r 16		g 17		bp 23		ch 21		cp 24	
	Σ	%fr	Σ	%fr	Σ	%fr	Σ	%fr	Σ	%fr
Triticum boeoticum	-	-	-	-	4.3	17	0.6	5	1.5	8
Triticum boeoticum/monococcum	-	-	-	-	2.3	9	1	5	1	4
Triticum monococcum	-	-	1.1	12	2.6	9	1	5	1	4
Triticum dicoccoides	-	-	-	-	6.5	17	5	14	-	-
Triticum dicoccoides/dicoccum	-	-	0.9	12	5.5	13	5.3	19	-	-
Triticum monococcum/dicoccum	-	-	1	6	-	-	-	-	-	-
Triticum dicoccum	-	-	-	-	1.5	4	-	-	1	4
Triticum aestivum/durum	-	-	-	-	-	-	-	-	-	-
Triticum sp.	-	-	-	-	1	4	-	-	-	-
Triticum spikelet forks	9	56	116	59	249	56	128	71	254	71
Triticum rachis internodes	-	-	-	-	-	-	-	-	-	-
Hordeum (cf.) *spontaneum*	-	-	0.8	12	1.8	13	4	5	-	-
Hordeum rachis internodes	-	-	3	6	2	4	2	10	1	4
Unident. rachis internodes	-	-	-	-	-	-	-	-	-	-
Cereal grain fragments	xx	38	xx	24	x	27	x	19	x	42
Pisum	-	-	1	6	31	35	20.5	38	32.5	25
Lens	1	6	-	-	1.5	9	1.5	10	6	13
Vicia ervilia	-	-	-	-	49.5	48	31.5	24	48	54
Cicer	-	-	-	-	1	4	-	-	-	-
Lathyrus cicera/sativus	-	-	-	-	-	-	1	5	1	4
Pulse grain fragments	×	6	-	-	-	-	-	-	-	-
Linum cf. *bienne*	-	-	-	-	-	-	7	10	90	4
Vitis	-	-	3	6	3	4	-	-	1	4
Ficus	-	-	-	-	-	-	-	-	1	4
Celtis	-	-	-	-	-	-	1	5	-	-
Pistacia nutshell fragments	xx	94	xx	100	xx	96	xx	100	xx	92
Amygdalus nutshell fragments	x	6	x	24	x	35	x	10	x	25
Quercus acorn fragments	-	-	-	-	-	-	-	-	-	-
Lathyrus aphaca type	2	6	2	6	-	-	1	5	-	-
Vicia sp.	13.5	32	16.5	41	182.5	74	67.5	72	45.5	58
Vicia seed fragments	x	19	xx	41	x	17	x	10	xx	13
Astragalus	-	-	-	-	-	-	-	-	1	4
Medicago radiata	-	-	-	-	-	-	-	-	1	4
Trigonella astroites type	-	-	-	-	-	-	-	-	-	-
Unident. Leguminosae	-	-	-	-	2	9	-	-	3	4
Lolium rigidum/perenne	1	6	1.5	12	13.5	22	5	19	-	-
Bromus	-	-	-	-	1	4	-	-	-	-
Phalaris	-	-	-	-	-	-	-	-	-	-
Cynodon	-	-	-	-	-	-	-	-	-	-
Echinaria	-	-	-	-	1	4	2	10	6	4
Stipa grains	-	-	-	-	5	4	-	-	-	-
Stipa awn fragments	x	19	x	18	x	4	-	-	-	-
Agrostis type	-	-	-	-	-	-	-	-	-	-
Unident. Gramineae	2.5	12	2	6	4.5	13	0.5	5	5	13
Anagallis arvensis	-	-	-	-	-	-	-	-	-	-
Anchusa	-	-	-	-	-	-	-	-	-	-
Chrozophora	-	-	-	-	2.5	9	2	5	-	-
Rumex	-	-	-	-	-	-	-	-	-	-
Polygonum corrigioloides type	-	-	-	-	-	-	-	-	-	-
Scirpus maritimus	1	6	2	12	20	41	7	24	3	8
Vaccaria	-	-	-	-	-	-	0.5	5	-	-
Thymelaea	-	-	-	-	-	-	-	-	-	-
Verbena type	-	-	-	-	-	-	-	-	-	-
Ziziphora	-	-	-	-	-	-	-	-	1	4
Galium	-	-	-	-	3	9	-	-	-	-
Unident. Umbelliferae	-	-	-	-	-	-	-	-	-	-
Plantago lagopus type	-	-	-	-	-	-	-	-	-	-
'Lycium type'	1	6	-	-	22	4	-	-	-	-

Table 3 (Continued).

c 20		c/lr 8		lr 5		Pott.Neol. 9		Sub-phase Number of samples
Σ	%fr	Σ	%fr	Σ	%fr	Σ	%fr	
1.3	10	-	-	-	-	3	11	*Triticum boeoticum*
-	-	-	-	-	-	-	-	*Triticum boeoticum/monococcum*
2.6	25	1	13	-	-	1	11	*Triticum monococcum*
-	-	-	-	-	-	-	-	*Triticum dicoccoides*
-	-	-	-	-	-	-	-	*Triticum dicoccoides/dicoccum*
-	-	-	-	-	-	-	-	*Triticum monococcum/dicoccum*
35	65	2	25	-	-	16	67	*Triticum dicoccum*
-	-	-	-	-	-	3	11	*Triticum aestivum/durum*
-	-	-	-	-	-	5	22	*Triticum* sp.
79	35	12	63	-	-	1679	89	*Triticum* spikelet forks
-	-	-	-	-	-	1	11	*Triticum* rachis internodes
1	5	-	-	-	-	1	11	*Hordeum* (cf.) *spontaneum*
1	5	-	-	-	-	3	22	*Hordeum* rachis internodes
-	-	-	-	-	-	5	22	Unident. rachis internodes
x	10	x	13	-	-	xx	89	Cereal grain fragments
27.5	15	3	13	37	80	-	-	*Pisum*
59	35	1	13	93	80	149.5	67	*Lens*
92.5	75	1	13	3.5	40	45	78	*Vicia ervilia*
3	15	-	-	-	-	6	33	*Cicer*
-	-	-	-	-	-	-	-	*Lathyrus cicera/sativus*
-	-	-	-	-	-	-	-	Pulse grain fragments
-	-	-	-	-	-	-	-	*Linum* cf. *bienne*
-	-	-	-	-	-	-	-	*Vitis*
-	-	-	-	-	-	7	22	*Ficus*
1	5	-	-	-	-	-	-	*Celtis*
xxx	80	xx	75	x	20	x	89	*Pistacia* nutshell fragments
xx	50	x	13	-	-	x	11	*Amygdalus* nutshell fragments
x	15	-	-	-	-	-	-	*Quercus* acorn fragments
-	-	-	-	-	-	-	-	*Lathyrus aphaca* type
121	55	3	13	-	-	25.5	67	*Vicia* sp.
xx	10	xx	13	x	20	xx	89	*Vicia* seed fragments
-	-	-	-	-	-	-	-	*Astragalus*
-	-	-	-	-	-	-	-	*Medicago radiata*
-	-	-	-	-	-	1	11	*Trigonella astroites* type
1.5	15	0.5	13	-	-	-	-	Unident. Leguminosae
4	10	-	-	-	-	3.5	33	*Lolium rigidum/perenne*
2	5	-	-	-	-	0.5	11	*Bromus*
-	-	-	-	-	-	2	22	*Phalaris*
-	-	-	-	-	-	3	11	*Cynodon*
-	-	-	-	-	-	1	11	*Echinaria*
-	-	-	-	-	-	-	-	*Stipa* grains
-	-	-	-	-	-	x	22	*Stipa* awn fragments
1	5	-	-	-	-	-	-	*Agrostis* type
1.5	10	1	13	-	-	6.5	33	Unident. Gramineae
-	-	-	-	-	-	2	11	*Anagallis arvensis*
1	5	-	-	-	-	-	-	*Anchusa*
4	5	-	-	-	-	-	-	*Chrozophora*
3	10	-	-	-	-	-	-	*Rumex*
-	-	1	13	-	-	-	-	*Polygonum corrigioloides* type
5.5	20	-	-	-	-	6	56	*Scirpus maritimus*
-	-	-	-	-	-	-	-	*Vaccaria*
-	-	-	-	-	-	4	44	*Thymelaea*
-	-	-	-	-	-	1	11	*Verbena* type
2	5	-	-	-	-	-	-	*Ziziphora*
6	20	1	13	-	-	1.5	22	*Galium*
1	5	-	-	-	-	-	-	Unident. Umbelliferae
1	5	-	-	-	-	-	-	*Plantago lagopus* type
-	-	-	-	-	-	-	-	'Lycium type'

Table 4. Western excavation area. Total numbers of seeds, etc. (Σ) and sample frequencies expressed as percentages (% fr) per sub-phase. x Present; xx Frequent; xxx. Very frequent.

Sub-phase Number of samples	g 15		ch 20		cp 14		c 47	
	Σ	%fr	Σ	%fr	Σ	%fr	Σ	%fr
Triticum boeoticum	0.6	13	7	10	1	7	-	-
Triticum boeoticum/monococcum	0.3	7	-	-	1	7	1	2
Triticum monococcum	2.3	20	3	10	1.5	14	3.8	12
Triticum dicoccoides	2.5	13	5	10	-	-	-	-
Triticum dicoccoides/dicoccum	1.6	13	-	-	-	-	-	-
Triticum monococcum/dicoccum	-	-	-	-	2	7	-	-
Triticum dicoccum	-	-	7	35	2	7	1	4
Triticum sp.	-	-	-	-	-	-	1	2
Triticum spikelet forks	131	73	797	95	243	93	181	74
Hordeum (cf.) *spontaneum*	-	-	0.3	5	0.3	7	1.3	4
Hordeum rachis internodes	-	-	6	10	2	14	-	-
Unident. rachis internodes	-	-	3	5	-	-	-	-
Cereal grain fragments	xx	13	xx	80	xx	57	xx	47
Pisum	14.5	33	26	30	2	14	15.5	23
Lens	4	20	6.5	15	19.5	43	15.2	27
Vicia ervilia	13.5	40	42	70	55.5	43	3731	40
Cicer	-	-	2.5	10	-	-	4.5	10
Lathyrus cicera/sativus	-	-	1	5	-	-	-	-
Pulse grain fragments	x	7	x	25	x	7	x	4
Linum cf. *bienne*	-	-	-	-	-	-	0.3	2
Vitis	-	-	1	5	2	7	-	-
Celtis	-	-	-	-	-	-	1	2
Pistacia nutshell fragments	xx	87	xx	95	xx	93	xx	83
Amygdalus nutshell fragments	x	13	x	15	x	14	x	19
Lathyrus aphaca type	-	-	8	10	14	14	-	-
Lathyrus hirsutus	-	-	1	5	-	-	-	-
Vicia sp.	156	80	105	85	72.5	43	21	13
Vicia seed fragments	xx	13	x	20	xx	36	xx	28
Medicago radiata	-	-	-	-	1	7	-	-
Medicago	-	-	2	5	-	-	-	-
Melilotus	-	-	4	5	-	-	-	-
Unident. Leguminosae	2.8	20	1	5	1	7	1	2
Lolium rigidum/perenne	4.5	20	9.8	40	5	21	3.9	15
Bromus	-	-	2.5	10	-	-	1	2
Stipa awn fragments	x	13	x	10	x	7	-	-
Unident. Gramineae	8	13	7	20	2	7	2.5	9
Ranunculus arvensis type	-	-	-	-	-	-	1	2
Adonis	-	-	1	5	-	-	-	-
Unident. Compositae	-	-	1	5	1	7	-	-
Lithospermum tenuiflorum	-	-	2	10	-	-	-	-
Unident. Boraginaceae	-	-	-	-	1	7	-	-
Chrozophora	2	7	5	5	-	-	-	-
Unident. Cruciferae	-	-	1	5	-	-	1	2
Scirpus maritimus	3	13	2	10	3	21	-	-
Silene	-	-	1	5	-	-	-	-
Thymelaea	-	-	-	-	-	-	1	2
Unident. Labiatae	1	7	-	-	-	-	-	-
Ziziphora	2	13	58	10	-	-	-	-
Galium	2	13	1.5	10	2	14	0.5	2
Malva	-	-	1	5	-	-	-	-
Unident. Umbelliferae	-	-	1	5	-	-	-	-
Helianthemum	-	-	-	-	-	-	1	2
Fumaria	-	-	1	5	-	-	-	-
Verbascum	1	7	-	-	-	-	-	-
'*Lycium* type'	-	-	1494	90	23	36	1.5	4

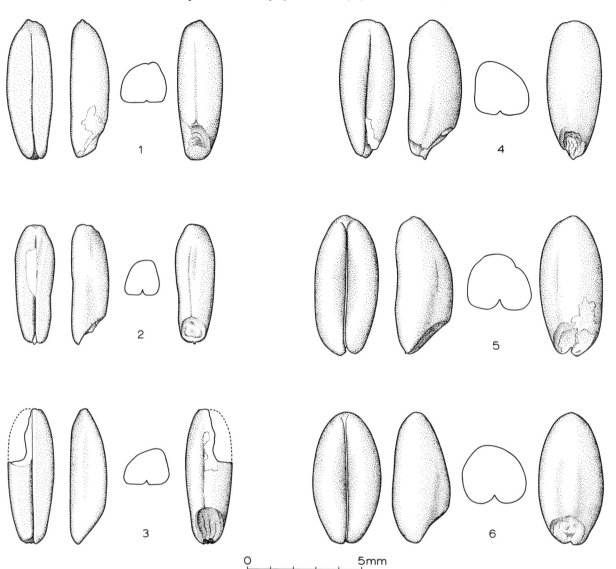

Fig. 7. 1. *Triticum dicoccoides* (70/R 14-0); 2,3. *T. dicoccoides* (70/R 15-2); 4. *T. dicoccum* (70/R 4-1); 5,6. *T. dicoccum* (70/R 5-8). After van Zeist (1972: fig. 4).

Table 5. Dimensions (in mm) and index values of *Triticum dicoccoides*(-type) grains.

Sample no.	Sub-phase	Square	L	B	T	100 L/B	100 T/B
86/90	?	27L	6.4	1.6	1.5	400	94
86/130	ch	20M	5.3	2.0	1.3	265	65
86/130	ch	20M	5.9	2.4	2.1	246	88
70/R-15	ch	27L	c.6.2	2.2	c.1.6	282	73
70/R-15	ch	27L	6.0	2.4	1.9	250	79
70/R-15	ch	27L	6.0	2.0	1.6	300	80
70/R-15	ch	27L	5.2	1.8	1.6	289	89
70/R-14	ch	27L	6.2	2.0	1.8	310	90
87/36	bp	27L	6.0	2.3	1.9	261	83
70/R-19	bp	27L	6.6	2.2	c.1.6	300	73
70/R-18	bp	27L	c.5.8	1.7	1.5	341	88

Table 6. Dimensions (in mm) and index values of *Triticum dicoccum* grains from aceramic and pottery levels.

Sample no.	Sub-phase	Square	L	B	T	100 L/B	100 T/B
Aceramic:							
70/R-5	c	27L	6.0	3.0	2.7	200	90
70/R-5	c	27L	5.0	2.9	2.2	172	76
70/R-5	c	27L	6.2	2.8	2.6	221	93
70/R-4	c	27L	6.2	2.7	2.4	230	89
70/R-4	c	27L	5.3	2.4	c.2.2	221	92
70/R-4	c	27L	6.0	2.3	1.8	261	78
86/85	ch	20L	5.7	2.2	1.8	259	82
86/79	ch	20L	4.6	2.3	2.1	200	91
Pottery section:							
87/23		27K	5.9	3.0	2.6	197	87
87/23		27K	6.0	2.6	2.1	231	81
87/23		27K	5.6	2.8	2.3	200	82
87/27		27K	5.4	2.7	1.9	200	70
87/21		27K	5.8	2.8	2.3	207	82
87/21		27K	5.2	2.7	2.3	193	85

of both types of wheat grains are illustrated in figure 7. The greater slenderness of the wild emmer-type grains compared to those of the domestic type finds expression in the L/B index values, ranging from 246 to 400 (mean 295) in *T. dicoccoides* (table 5) and from 172 to 261 (mean 221) in *T. dicoccum* (table 6).

On second thought, doubt has arisen on the attribution of the slender emmer wheat grains to *T. dicoccoides*, because no spikelet forks with an intact articulation scar have been found. The Çayönü emmer wheat-type spikelet forks show the features of the semi-tough central rachis of *T. dicoccum*. There are no signs of natural disarticulation of the ear as one would expect in wild emmer wheat.

It is striking that in the lowermost layers which yielded emmer wheat kernels, in the Grill-house sub-phase, no typical *T. dicoccum*-type grains were found, whereas in the upper half of the aceramic occupation, in the Cobble-paved house and younger sub-phases, *T. dicoccoides*-type grains are absent. Thus, in the course of the occupation one type of emmer wheat was replaced by another. Should this be explained in terms of the gradual replacement of the wild type by the domestic type? In the 1972 publication it was suggested that the Çayönü farmers had started to grow wild emmer wheat which in the course of time developed into the domestic type. However, the spikelet remains suggest rather that from the beginning on, domestic, that is to say semi-tough rachised emmer wheat was cultivated. One could speculate that changes in the shape of the grain lagged behind the transition from the brittle to the semi-tough type of rachis. It was continued cultivation which eventually resulted in the development of kernels typical of domestic emmer wheat.

For comparison with the aceramic grains, the dimensions of emmer wheat grains from pottery

Neolithic levels are available (table 6). There are no significant differences in the dimensions and index values between the aceramic and pottery Neolithic specimens. Unfortunately, only a few grains were suitable for measurement.

3.1.2. *Einkorn wheat*

Wild as well as domestic-type einkorn wheat remains have been established for Çayönü. The domestic einkorn (*Triticum monococcum*) grains are of the one-seeded type (one grain develops in a spikelet); the kernels are laterally compressed, with longitudinally curved ventral and dorsal sides (fig. 8:1,2). Of wild einkorn wheat (*Triticum boeoticum*) both the one-seeded and the two-seeded type are represented. The grains of one-seeded wild einkorn (*T. boeoticum* ssp. *aegilopoides*) show, as in those of the domestic form, a longitudinally curved ventral side, but the dorsal side is longitudinally straight or only slightly curved. One-seeded wild einkorn grains differ further from those of domestic einkorn by a yet greater lateral compression (fig. 8:3). The breadth does not exceed 1.2 mm. Because of damage and deformations it was not always possible to distinguish between wild-type and domestic-type one-seeded einkorn grains.

A few grains and grain fragments of two-seeded wild einkorn (*T. boeoticum* ssp. *thaoudar*) were recovered. In two-seeded wild einkorn the ventral side is flat, while the dorsal side is longitudinally straight or only slightly curved (fig. 8:4). The grains are not laterally compressed. In tables 3 and 4 no distinction is made between the two wild einkorn types. The dimensions of einkorn grains are shown in table 8.

No spikelet remains of the shattering (wild) type could be established. A rachis internode fragment which originally had been attributed to einkorn wheat (van Zeist, 1972: fig. 5:7) turned out to be of wild barley (see below). The distinction between spikelet forks of emmer wheat and einkorn wheat is to some extent arbitrary.

Table 7. Width of spikelet forks measured across the articulation scar (in mm). N = number of measurements.

Sample no.	Sub-phase	Square	N	Mean dimension
Triticum dicoccum type:				
85/89	c	27L	5	1.92
86/39	cp	29-30N	4	1.75
87/13	ch	20L	4	1.69
86/30	ch	20M	5	1.92
87/4	bp	27L	7	1.78
87/47	bp?	27M	6	1.86
Total (N = 31): min. 1.4 mm, mean 1.82 mm, max. 2.2 mm.				

Triticum monococcum type:

Total (N = 14): min. 1.0 mm, mean 1.23 mm, max. 1.4 mm.

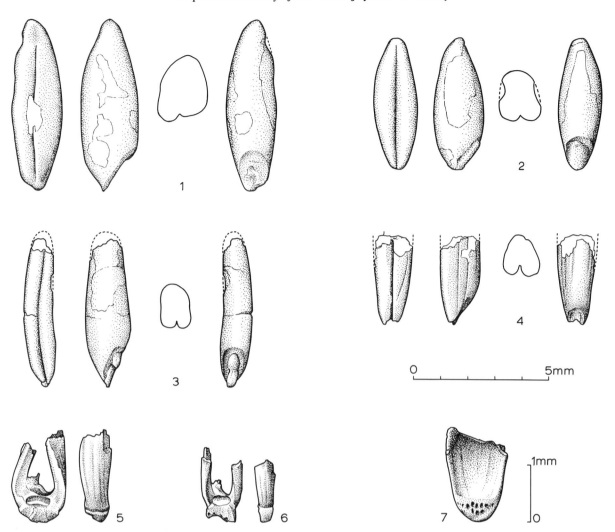

Fig. 8. 1. *Triticum monococcum* (70/R 19-3); 2. *T. monococcum* (70/R19-8); 3. *T. boeoticum* ssp. *aegilopoides* (70/R 18-0); 4. *T.* cf. *boeoticum* ssp. *thaoudar* (70/R 15-0); 5. Emmer wheat spikelet fork (70/R 15-0); 6. Einkorn wheat spikelet fork (70/R 15-0); 7. *Hordeum spontaneum* rachis internode fragment (70/R 15-0). After van Zeist (1972: fig. 5).

Table 8. Dimensions (in mm) and index values of wild and domestic einkorn grains.

Sample no.	Sub-phase	Square	L	B	T	100 L/B	100 T/B
Triticum monoccocum:							
70/X-5	c/lr	29-30L	6.2	2.0	c.2.4	310	120
86/77	cp?	19N	4.5	1.8	1.9	250	106
70/R-15	ch	27L	c.5.6	1.6	1.9	350	119
86/79	ch	20L	3.8	1.8	1.9	211	106
70/R-19	bp	27L	6.2	1.8	2.2	344	122
70/R-19	bp	27L	4.8	1.6	1.8	300	113
70/G-8	g	19N-O	c.5.6	1.8	2.1	311	117
Triticum boeoticum ssp. *aegilopoides*:							
70/R-5	c	27L	5.0	1.2	1.9	417	158
70/R-18	bp	27L	c.5.6	1.1	1.6	509	145
Triticum boeoticum ssp. *thaoudar*:							
87/45	cp	29-30N	4.1	1.3	1.4	315	108

Fairly narrow spikelet forks with the glumes at right angles on the base of the spikelet have been assigned to *T. monococcum* (figs 8:6 and 9:3). For characteristic einkorn and emmer wheat spikelet forks the width, measured across the articulation scar, has been determined (table 7).

In contrast to wild emmer wheat, wild einkorn wheat has weedy characters and consequently the species could have maintained itself as a weed in or near the Çayönü arable fields. Moreover, two-grained wild einkorn is found in massive stands on basaltic soils to the south of Çayönü. On the other hand, the absence of wild-type (shattering) spikelet remains and the large proportion of wild-type einkorn kernels with respect to that of the domestic type make one wonder whether, as is suggested for emmer wheat, einkorn wheat cultivation had also not yet resulted in the full replacement of the morphologically wild by the morphologically domestic grain type. Admittedly, this hypothesis is, to some extent, invalidated by the fact that wild einkorn-type grains do not disappear in the upper aceramic levels for

which a satisfactory floral record has been obtained (and have been attested also for pottery Neolithic levels). Thus the relatively large proportion of wild einkorn wheat remains puzzling.

3.1.3. *Barley*

The paper on the 1970 season (van Zeist, 1972) mentions only two fragments of barley (*Hordeum*) grains. Samples from the 1985-1987 seasons yielded some more barley remains, grains as well as rachis internode fragments. One fairly well-preserved kernel was recovered from a 1970 sample that had been collected after the first author had departed from the site. However, compared to the number of *Triticum* remains, those of *Hordeum* are rather insignificant.

In the same 1972 paper it was assumed that the barley remains are of the wild type, *Hordeum spontaneum*, which occurs in massive stands in the Çayönü region (Harlan & Zohary, 1966). Moreover, wild barley has weedy characters and grows in segetal habitats, e.g.

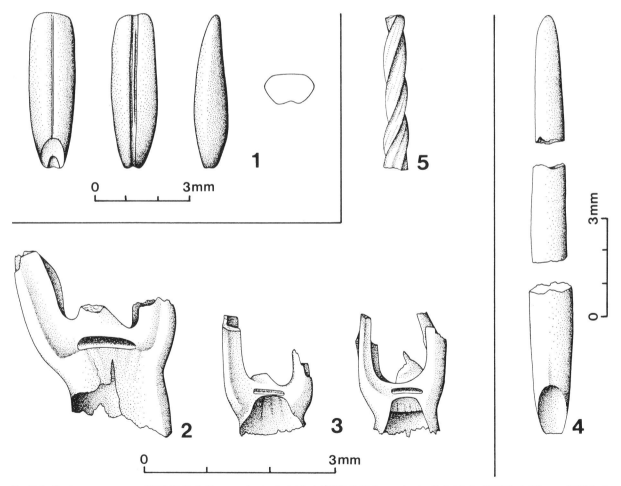

Fig. 9. 1. *Hordeum spontaneum* (70/R 19-8); 2. Emmer wheat spikelet fork (87/4); 3. Einkorn wheat spikelet forks (86/27); 4. *Stipa* sp. (87/36); 5. *Stipa* awn fragment (87/36).

Table 9. Dimensions (in mm) and index values of two *Hordeum spontaneum* grains.

Sample no.	Sub-phase	Square	L	B	T	100 L/B	100 T/B
85/89	c	27L	5.9	1.5	0.9	393	60
70/R-19	bp	27L	4.8	1.4	1.0	343	71

alongside fields. The assumption of wild barley was mainly based upon the minimal representation. If the species had been cultivated, one would have expected more remains than two grain fragments. As the continuation of the botanical analyses resulted, among other things, in a slightly better representation of barley, the question arises whether the species could have been cultivated, if only as a minor crop. In principle the kernels as well as the rachis internode remains should allow a differentiation between *Hordeum spontaneum* and the domesticated two-rowed form *H. distichum*. Charred rachis internodes of the wild form should show an intact articulation scar. Not only the articulation scar but also the base of the internodes of brittle-rachised barley may be expected to be undamaged. In tough-rachised *Hordeum distichum* the ears stay intact until threshed, and in threshing, the rachis usually does not break up at the joints between the internodes. In charred archaeological material, two or more internodes may occur still adhering together or, what is more usual, internodes are found with the basal part of the next internode attached.

The kernels of both barley species are rather similar, but those of *Hordeum spontaneum* are characterized by their flatness; they are markedly thinner than those of *H. distichum*. Although modern grains of the two barley species can satisfactorily be separated, in charred archaeological specimens the distinction is often less clear, due to deformations through carbonization. Reasonably well-preserved barley grains from Çayönü remind one of those of *H. spontaneum* (fig. 9:1). The rachis internodes are strongly fragmented, but in various specimens an intact base can be observed (fig. 8:7). Thus, the grains as well as the rachis internode remains point to *H. spontaneum* and there is no convincing evidence of *H. distichum*. It goes without saying that various fragments did not allow a species determination. Two grains were suitable for measuring (table 9), but admittedly there is not much to compare with. The only other wild barley dimensions published so far are those for late-Palaeolithic Mureybit (van Zeist & Bakker-Heeres, 1984(1986)b: table 7).

One could hypothesize that morphologically defined wild barley had been cultivated. However, in that case one would have expected the domestic form in the upper aceramic occupation deposits, as it is assumed that the latter developed from the wild form as a result of cultivation. It is very unlikely that barley was cultivated at aceramic Çayönü. It is also doubtful whether it was gathered purposely, as in that case the remains should have been more numerous.

3.2. Pulses

3.2.1. *Field pea*

As usual, also the Çayönü peas show a fairly large variation in shape: more or less spherical seeds as well as those with one or two flat or indented sides occur. The seed coat has nearly always disappeared, but in a few specimens from Basal-pit samples (70/R-19 and R-20) a rough surface could be observed. A hand-picked mineralised pea from a Cell-house sub-phase context (70/U-9) has a smooth surface and it is the only specimen in which the oval hilum has been preserved. As a rough seed coat is characteristic of wild pea, one wonders whether the peas that were initially cultivated were still of the wild type (*Pisum humile*), while peas with a smooth surface, characteristic of the domestic form (*Pisum sativum*), evolved only later. A similar development is suggested for the Çayönü emmer wheat (3.1.1). It cannot be ascertained whether the pods with the rough seed-walled (wild-type) peas were already of the non-dehiscent type (seed-retaining when mature).

The greatest dimension of peas from various levels is shown in table 10. As appears from this table, there are no significant differences in size between sub-phases. The mineralized specimen (not included in table 10), which measures 5.4 mm, is larger than any other measured Çayönü pea (maximum dimension 5.0 mm), suggesting that carbonization resulted in a decrease in size.

3.2.2. *Lentil*

As was to be expected, the Çayönü lentils are of the small-seeded type; only occasional seeds measure more than 3.0 mm in diameter. No increase in size is observed for the upper aceramic levels, and also the pottery Neolithic lentil seeds are very small on average (table 11). It cannot be ascertained whether lentil from the lower levels still had wild-type characters such as those recognized in emmer wheat and field pea.

3.2.3. *Bitter vetch*

The seeds of bitter vetch (*Vicia ervilia*) are obliquely pyramidal, with a triangular base on which the radicle

Table 10. Largest diameter of charred *Pisum* seeds. N is number of seeds measured.

Sample no.	Sub-phase	Square	N	Min.	Mean	Max.
70/U-5	lr	28-29/K-L	8	3.8	4.14	4.7
70/U-9	c	28-29/K-L	6	3.9	4.24	5.0
70/R-10	cp	27L	10	3.1	3.94	5.0
Various samples 70/R-19	ch	Various squares	11	2.8	4.04	4.6
R-20	bp	27L	9	3.8	4.16	4.6

Table 11. Diameter of *Lens (culinaris)* seeds. N is number of seeds measured.

Sample no.	Sub-phase	Square	N	Min.	Mean	Max.
Various samples	Pottery	27K	74	2.1	2.71	3.3
70/U-5	lr	28-29/K-L	22	2.1	2.59	3.0
70/S room	c	27-28/K-L	20	2.1	2.42	2.8
70/U-9	c	28-29/K-L	15	2.2	2.60	3.0
Various samples	cp	Various squares	8	2.4	2.90	3.4

Table 12. *Vicia ervilia* samples. Cell-plan sub-phase.

Square, feature			17N, house DF				17M
Sample number	85/40	85/24	85/46	85/51	85/56	85/57	85/58
Vicia ervilia	550	375	625	618	110	595	800
Lens	-	-	1	1	0.5	-	1
Pisum	4	1.5	1	-	0.5	1	1
Triticum boeoticum/monococcum	-	-	1	-	-	-	-
Triticum spikelet forks	-	3	2	4	5	8	40
Cereal grain fragments	-	-	x	x	-	-	xx
Pistacia fragments	-	x	x	x	x	x	x
Amygdalus fragments	-	-	-	x	-	-	-
Lolium	-	-	-	-	0.3	1	-

Table 13. Dimensions of *Vicia ervilia*. Cell-plan sub-phase. N = number of measurements.

		L	B	T
85/40	Min.	1.6	1.7	1.5
N = 35	Mean	2.06	1.95	1.93
	Max.	2.4	2.2	2.2
85/46	Min.	1.9	1.8	1.6
N = 35	Mean	2.19	2.07	2.00
	Max.	2.6	2.5	2.2
85/51	Min.	1.7	1.5	1.5
N = 40	Mean	2.19	2.09	2.01
	Max.	2.8	2.5	2.4
85/57	Min.	1.8	1.6	1.8
N = 25	Mean	2.14	2.01	1.99
	Max.	2.6	2.4	2.2
85/58	Min.	1.7	1.7	1.6
N = 40	Mean	2.21	2.07	2.04
	Max.	2.6	2.5	2.5

is located, and a round apex. As for the position of the measurements (table 13), length and breadth have been determined in the plane of the triangular base, while the thickness is the distance between the base and the apex. Bitter vetch seeds have been retrieved from a great number of samples.

Although in southeastern Turkey bitter vetch is found in the wild (Zohary & Hopf, 1973) and moreover weedy forms of this species occur, one may safely assume that bitter vetch formed part of the crop-plant assortment of the Çayönü farmers and that it was grown for human consumption. Convincing evidence of the intentional cultivation of bitter vetch and of its use as a human food plant is provided by the deposit of these seeds in the fill of house DF (square 17N) of the Cell-house sub-phase. The analyses of the samples from this deposit are presented in table 12. The contextual aspects of this find will be discussed in 5.3. In this section the following may be remarked.

The bitter vetch supply must have been conspicuously

pure. The cereal remains and the nutshell fragments are probably of secondary origin, that is to say, that they had been mixed in during or after carbonization. The bitter vetch crop had been thoroughly cleaned of field-weed seeds and other contaminants. It is evident that this only makes sense if the bitter vetch was intended for human consumption. Before food preparation, the poisonous substance in the seeds had to be removed, which could simply be done by soaking in water. At present bitter vetch is grown only for stock feed.

The Çayönü bitter vetch seeds (table 13) are, on average, about 20% smaller than those from pottery Neolithic Erbaba, in southwestern Turkey (van Zeist & Buitenhuis, 1983: table 13).

3.2.4. *Other pulses*

Field pea, lentil and bitter vetch were commonly cultivated at aceramic Çayönü. Chickpea (*Cicer* sp.) and grass pea (*Lathyrus cicera/sativus*), on the other hand, must have played a much more modest role in the diet of the inhabitants of the site. The few wedge-shaped grass-pea seeds (fig. 10:6) are no evidence of intentional cultivation. The seeds may have been collected in the wild or the species may have occurred as a field weed. Dimensions of these seeds are: 3.6×4.5×3.4, 4.4×4.8×3.8 and 3.8×4.4×3.8 mm.

Chickpea (fig. 10:5) is slightly better represented. As with grass pea, it cannot be determined whether the seeds are of the domestic form, *Cicer arietinum*, or of a wild species, e.g. *Cicer reticulatum*, which is assumed to be the wild ancestor of cultivated chickpea and which is native to SE Anatolia (Ladizinsky & Adler, 1976). Chickpea is only scarcely recorded for the (early-) Neolithic of the Near East. Four fairly well-preserved seeds measure: 3.8×3.6×3.2, 5.4×4.1×3.2, 4.4×4.2×3.5 and 3.5×3.4×2.9 mm.

3.3. Linseed (flax)

Virtually all linseeds are from the 1970 campaign. Subsequent sampling for floral remains yielded only one linseed fragment. Moreover, by far the majority of the seeds (90 specimens) were retrieved from one sample attributed to the Cobble-paved house sub-phase (square 27L). The Çayönü linseeds are characterized by the beak (fig. 11:1). The dimensions of the seeds that were suitable for measurement (most seeds are damaged) are shown in table 14. As for the original size, it has been determined experimentally (Helbæk, 1959) and empirically (van Zeist & Boekschoten-van Helsdingen, 1991) that through carbonization linseeds shrink in length and breadth. Mean dimensions of 2.45×1.30 mm for the charred seeds point to an original size of at least 2.8×1.6 mm. The concentration in one of the samples suggests that the seeds in question had been collected purposely or perhaps had been cultivated. This brings us to the species identity of the linseeds. The earliest firm evidence of linseed cultivation is attested for Ramad in western Syria, dated to 7190-6700 BC (calibrated ^{14}C dates). On the basis of the size of the seeds it was concluded that domestic *Linum*

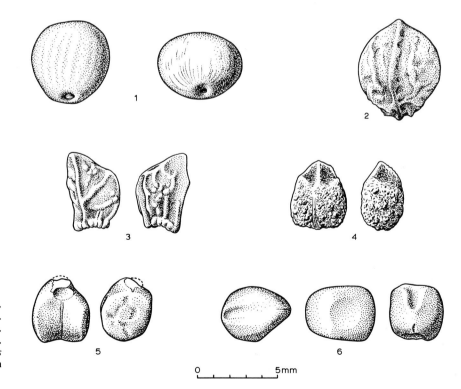

Fig. 10. 1. *Pistacia* sp. (70/R 4-1); 2. *Celtis* cf. *tournefortii* (70/S room); 3. *Anchusa* sp. (70/R 5-8); 4. *Chrozophora tinctoria* (70/R 14-2); 5. *Cicer* sp. (70/U 9-5); 6. *Lathyrus cicera/sativus* (70/R 10-3). After van Zeist (1972: fig. 7).

0   5mm

Table 14. Dimensions of linseeds, *Linum* cf. *bienne*, in mm (N = 12). Sample 70/R-10.

	L	B	100 L/B
Min.	2.2	1.1	172
Mean	2.45	1.30	189
Max.	2.6	1.5	207

Table 15. Dimensions of *Pistacia* nutshells in mm (N = 17). Sample 70/R-4.

	L	B	T	100 L/B
Min.	3.8	4.1	3.1	76
Mean	4.47	5.03	3.50	89
Max.	5.3	5.6	3.8	97

usitatissimum is concerned there. The length of the Ramad linseeds ranges from 2.8 to 3.6 mm, which after correction for shrinkage gives a measurement of 3.2 to 4.1 mm, with a mean value of 3.61 mm (van Zeist & Bakker-Heeres, 1975a, 1982(1985)). The size of the Ramad linseeds is typical of rain-fed agriculture. Under conditions of irrigation larger-sized seeds are obtained (Helbæk, 1959).

It is evident that the Çayönü linseeds cannot possibly be of *Linum usitatissimum*; they are much too small. An original mean size of 2.8×1.6 mm (see above) conforms to that reported for *Linum bienne* Mill. (syn. *L. angustifolium* Huds.): length 2-3 mm (Zohary, 1972), 2.4-2.7 mm (Helbæk, 1959), 2.6-2.7 mm (our own measurements). *L. bienne*, which is identified as the wild progenitor of *L. usitatissimum* (cf. Zohary & Hopf, 1988: p. 115), is widely distributed over the Mediterranean basin and the Near East. The species identification of the Çayönü linseed (*Linum* cf. *bienne*) remains uncertain as various other Near Eastern wild flax species have seeds which do not exceed 3 mm (cf. van Zeist & Bakker-Heeres, 1975). Flax remains (fragments of seeds and capsules) in samples from the earliest horizon of Ali Kosh, the Bush Mordeh phase, in southwestern Iran, are attributed by Helbæk (1969) to *Linum bienne*, on the basis of the dimensions.

The question of whether flax was gathered in the wild or cultivated must remain unanswered. It is tempting to hypothesize that the Çayönü farmers had pioneered with flax cultivation, but this is sheer speculation. We don't even know whether linseeds played a more than marginal role in the Çayönü plant husbandry. The find of 90 seeds in one of the samples may have been of an accidental nature.

3.4. Nuts and fruits

3.4.1. *Pistacia*

Pistacia nutshell fragments show the highest sample frequencies of all plant remains (tables 3 and 4). The 1970 campaign yielded intact nutshells (fig. 10:1); in some of them remains of the fruit flesh and fruit wall are still present. Intact nutshells were found only in Cell-house sub-phase samples from square 27L. Here large quantities of, mainly broken, pistachio nutshells (and almond fruit-stone remains, see 3.4.2.) must have been

dumped in a large ash pit (see discussion in 5.3).

The nutshells are more or less laterally flattened and broader than long, which finds expression in L/B index values of less than 100 (table 15). In the 1972 publication it is stated that the remains are of *Pistacia atlantica* and not of *P. khinjuk*. However, it may be better not to specifically attribute the nutshells to one species or the other. Both *Pistacia* species are constituents of the present natural forest cover of southeastern Anatolia (see 1.1.2).

Appreciable quantities of *Pistacia* nutshell remains are reported for late-Palaeolithic (Abu Hureyra, Mureybit) and early-Neolithic sites in the Near East. The highly nutritive fruits (the fruit flesh is rich in fats) must have contributed essentially to man's diet. Fruits of wild *Pistacia* species are at present still sold in the markets (Zohary, 1972: p. 298).

3.4.2. *Amygdalus*

Only fragmented almond fruit-stones have been recovered. This is in itself no surprise, as the stones have to be broken to obtain the kernels which are rich in fats. Fresh wild almonds have a bitter taste and, moreover, are poisonous because in crushing and chewing the seed, prussic acid (hydrogen cyanide) is formed (cf. Zohary & Hopf, 1988: p. 161). The thick-walled fragments have a pitted and grooved surface and in some of them the lateral keel can be observed (fig. 12). The shape of some large fragments indicates that the stones must have had a length of about 2 to 2.5 cm.

The majority of the almond fruit-stone fragments derive from Cell-house sub-phase levels in square 27L, the same area which yielded the greatest quantities of *Pistacia* nutshell remains (see 3.4.1). Post-1970 field-work campaigns produced only modest numbers of almond remains. Was almond for one reason or another not much appreciated or did it not commonly occur in the area? Various *Amygdalus* species are reported for SE Anatolia (Davis, 1972: vol. 4).

3.4.3. *Vitis vinifera*

From various levels one or a few grape pips were recovered, some of them in a rather battered state. The shape of the better preserved pips conforms to that of the wild type, viz. fairly squat with a short stalk. Wild grape

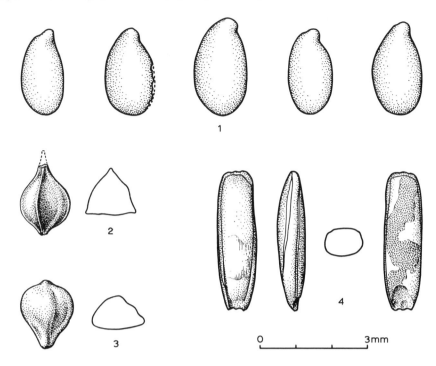

Fig. 11. 1. *Linum* cf. *bienne* (70/R 10-3); 2. *Rumex* sp. (70/S room); 3. *Scirpus maritimus* (70/R 20-9); 4. *Lolium perenne/rigidum* (70/R 15-0). After van Zeist (1972: fig. 6).

vines may have been found in thickets along streams. The berries are small and have an acid taste. Wild-type grape pips are reported for Chalcolithic Korucutepe, c. 50 km NNW of Çayönü (van Zeist & Bakker-Heeres, 1975b: fig. 7).

3.4.4. *Other fruits and nuts*

Only one *Ficus* (fig) pip was recovered from aceramic Çayönü; pottery Neolithic levels yielded a few more seeds (for illustration, see Ramad, fig. 30:1). It is likely that wild fig was hardly collected by the aceramic inhabitants, as otherwise more seeds might have been expected, seeing that each fruit contains a considerable number of pips. The distribution area of wild fig, *Ficus carica* s.l., includes SE Anatolia (Davis, 1982: vol. 7).

Two hackberry (*Celtis*) fruit stones were suitable for measurement: 6.3×5.3×5.0 and 6.2×5.6×5.4 mm. The stones are characterized by four longitudinal ridges and a rugose (coarsely wrinkled) surface structure (fig. 10:2). Most likely the Çayönü stones are of *Celtis tournefortii* Lam. (Davis, 1982: vol. 7). One wonders whether the scarce representation of this edible fruit could indicate that hackberry was not common in the area.

A few acorn (*Quercus* sp.) fragments suggest that these fruits were gathered by the inhabitants of the site, probably for human consumption. There is convincing evidence that in prehistoric Europe acorns played a part in man's diet. The tannin which causes the bitter taste can be removed by roasting the peeled acorns.

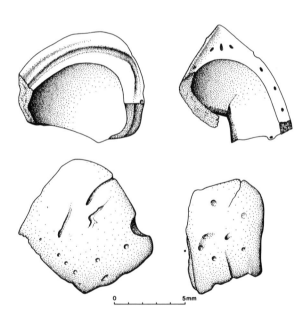

Fig. 12. *Amygdalus* nutshell fragments (70/R 4-1).

3.5. Gramineae

3.5.1. *Lolium*

The *Lolium* grains are dorso-ventrally compressed, and oblong in outline (fig. 11:4). In some *Lolium* grains parts of the finely papillose enveloping bracts are still preserved. Most of the caryopses are broken; a few almost intact grains measure 4.0x1.4, 3.6x1.3 and 3.4x1.2

mm. *Lolium perenne* L. as well as *L. rigidum* Gaudin come into consideration. The latter is at present a common arable field weed (e.g. Jansen, 1986).

3.5.2. *Stipa*

One sample (87/36) yielded broken caryopses of *Stipa*, amounting to about five complete specimens. Figure 9:4 shows a reconstructed linear-cylindrical fruit. In addition, twisted *Stipa* awn fragments (fig. 9:5) were retrieved from a few samples. *Stipa* caryopses are reported for late-Palaeolithic Abu Hureyra (Hillman et al., 1989) and early-Neolithic Ganj Dareh Tepe (van Zeist et al., 1984(1986)). Körber-Grohne (1987) describes and illustrates *Stipa* awn remains from a Middle-Neolithic site in Germany.

Feathergrass species are typical of steppe vegetation.

3.5.3. *Echinaria*

Small, squat fruits, tapering at the embryo end and rounded-truncated at the apical end (fig. 13:1). *Echinaria capitata* (L.)Desf. is reported for SE Anatolia (Davis, 1985; vol. 9). This grass is only scarcely represented at Çayönü.

3.5.4. *Other Gramineae*

Agrostis-type: small (1.2 mm long), slender caryopsis. *Bromus*: only broken grains.

3.6. Leguminosae

3.6.1. *Vicia sp. (vetch)*

Vetch seeds are among the most abundant plant remains at Çayönü (tables 3 and 4). The shape of the seeds is variable. Where (part of) the seed coat is preserved it shows a finely granular surface pattern. The dimensions of seeds in various samples are shown in table 16. The majority of the vetch seeds measure between 2.0 and 2.5 mm. Only occasional seeds are bigger than 3 mm. More

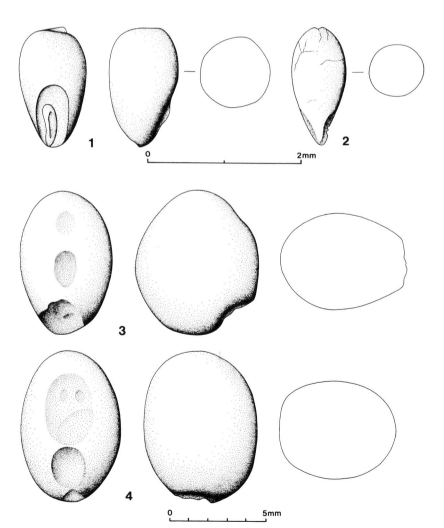

Fig. 13. 1. *Echinaria capitata* (87/45); 2. *Ziziphora* sp. (87/14); 3,4. *Lathyrus aphaca* type (86/6).

Table 16. Largest dimension (in mm) of *Vicia* sp. seeds. N is number of seeds measured.

Sample no.	Sub-phase	Square	N	Min.	Mean	Max.
86/6	cp	19L	14	1.7	2.27	3.3
70/R-15	ch	27L	13	1.8	2.25	2.5
70/R-19	bp	27L	15	1.8	2.10	2.5
70/R-20	bp	27L	25	1.4	1.96	2.4
87/36	bp	27L	11	1.3	1.94	2.3
70/G-8	g	27L	40	1.5	2.05	2.4

Table 17. Dimensions in mm and index values of '*Lycium* type's seeds.

		L	B	T	100 L/B	100 T/B
87/14	Min.	2.4	1.9	0.8	68	23
N=50	Mean	3.35	2.79	1.16	122	43
	Max.	4.2	3.8	1.7	154	72
87/13	Min.	2.3	1.8	0.6	94	21
N=20	Mean	3.06	2.55	1.04	121	43

than one *Vicia* species may be represented. Vetch species may have expanded as weeds of cultivation. One wonders whether the great numbers of vetch seeds could be interpreted as evidence of the intentional collecting of these seeds for human consumption.

3.6.2. *Lathyrus aphaca type*

The seeds are elliptic in outline (fig. 13:3,4). Hilum and radicle are not preserved. Five seeds from sample 86/6 (Cobble-paved house sub-phase) measure 3.65(3.4-4.0)x3.09(3.0-3.2)x2.64(2.4-3.0) mm. At first we wondered whether a small, somewhat aberrant form of *Pisum* or perhaps a large-seeded *Vicia* was concerned. *Lathyrus aphaca* is a common weed of arable fields.

3.6.3. *Lathyrus hirsutus type*

One *Lathyrus hirsutus*-type seed, largest diameter 3.0 mm, was recovered. Characteristic of this type is the rugulose (wrinkled) seed coat.

3.6.4. *Other Leguminosae*

Astragalus: Ramad, fig. 28:7,8.
Medicago sp.: Ramad, fig. 28:4.
Medicago radiata: Ramad, fig. 28:1.
Melilotus: Ramad, fig. 28:9,10.

3.7. Other wild plant taxa

3.7.1. *Chrozophora*

This type is listed as cf. Boraginaceae in the 1972 publication. The squat, angular seeds are pointed at the upper end and have a warty surface (fig. 10:4). The dimensions of 6 specimens are 4.2(3.9-4.4)x3.5(3.4-3.6)x2.8(2.7-3.0) mm. *Chrozophora tinctoria* (L.)A.Juss (Euphorbiaceae family) occurs as a field weed.

3.7.2. *Galium*

Hemispherical fruits, with a round concavity on the ventral side. The size of 12 specimens varies from 1.1 to 2.9 mm. Probably more than one species is represented.

3.7.3. '*Lycium type*'

Well-preserved specimens of this enigmatic seed type are flat, broadly obovate in outline, and obliquely pointed at the base (fig. 14). Due to carbonization the seeds are often misshapen. In a few specimens parts of a pitted seed wall have been preserved (fig. 14:4). The dimensions of this seed type are presented in table 17. The wide range in the L/B and T/B index values illustrates the variation in the shape of the seeds.

In the course of the investigation various possible identifications have been considered, such as *Sorbus*, *Pyrus*, a member of the Solanaceae family, *Ranunculus* and *Lonicera*. However, none of them fits. This type was found in a fair number of samples, particularly in those from the Channelled-house sub-phase (table 4). In one sample (87/14, table 23) more than a thousand specimens were counted, which makes one suspect that the seeds, or the fruits which may have contained the seeds, were collected purposely.

3.7.4. *Scirpus maritimus* L.

Nutlets obovate in outline, tapering towards the base. The ventral side is more or less flat, while the dorsal side is roof-shaped (fig. 11:3). Surface is smooth. Dimensions of 8 fruits are 1.8(1.6-2.0)x1.4(1.3-1.6) mm. Although the scientific and colloquial names of this species suggest that it is confined to more or less saline habitats, Near Eastern *Scirpus maritimus* (sea club-rush) occurs also in fresh-water swamps.

Due to an incorrect identification of modern cyperaceous fruits from a Turkish provenience, in the 1972 publication this fruit is indicated as *Cyperus* sp. Near Eastern *Scirpus maritimus* nutlets are markedly smaller than those from western Europe.

3.7.5. *Ziziphora*

Fruits obovate in outline, apical end rounded, pointed at the base, with conspicuous basal depressions (fig. 13:2). One sample (87/14, table 23) yielded a comparatively great number of *Ziziphora* fruits. Dimensions of 8 specimens: 1.4(1.3-1.5)x0.6(0.5-0.7) mm.

A few *Ziziphora* species are reported for Turkey (Davis, 1982: vol. 7).

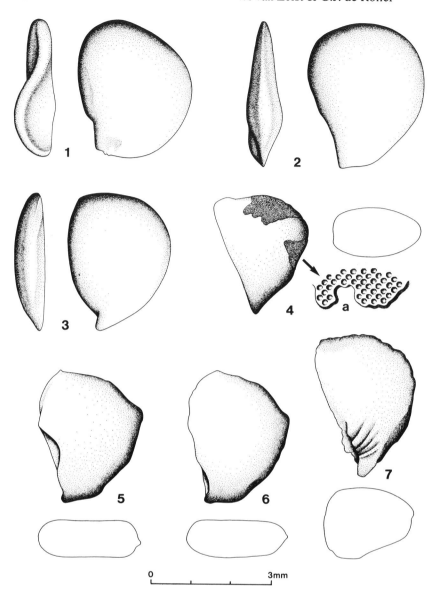

Fig. 14. '*Lycium* type'. 1-4,7. (87/14); 5,6. (86/71). Detail of wall surface structure is shown in 4a.

3.7.6. *Rarely occurring types*

Adonis: Ramad, fig. 30:9,10.
Anchusa: this publication, fig. 10:3, 5.0×3.3×2.6 mm.
Fumaria: Korucutepe, fig. 8:6; Ramad, fig. 30:2.
Helianthemum: cf. Ramad, fig. 23:6.
Lithospermum tenuiflorum: Ramad, fig. 22:6,7.
Malva: Ramad, fig. 24:11.
Plantago lagopus type: Ramad, fig. 30:5.
Polygonum corrigioloides type: Ramad, fig. 30:7.
Ranunculus arvensis type: cf. Korucutepe, fig. 8:8.
Rumex: this publication, fig. 11:2, c. 2.0×1.4×1.4 mm.
Silene: Ramad, fig. 23:4.
Thymelaea: Ras Shamra, fig. 7:12.
Vaccaria: half a seed; Ramad, fig. 24:3.
Verbascum: seed, 0.8×0.6 mm; van Zeist & Waterbolk-van Rooijen, 1985: fig. 4:4.

4. CHARCOAL ANALYSIS

No systematic charcoal identifications have been carried out. The majority of the identifications of material of the 1970 campaign are for hand-picked samples, partly destined for radiocarbon dating. In most of the flotation samples the wood charcoal pieces are too small or of too poor quality to allow an identification. The structure of the wood had often seriously been affected by carbonization. The sample frequencies of the wood types demonstrated are shown in table 18; unidentified charcoal is not included. Almost all samples which yielded identifiable charcoal are from the eastern excavation area.

The *Tamarix* identifications reported in the 1972 publication turned out to be incorrect. In preparing the present publication some doubt arose whether the

Table 18. Charcoal analysis. Numbers of samples in which the wood types identified were found.

Sub-phase	*Quercus*	*Pistacia*	Rosaceae	*Fraxinus*	*Lycium*
Cell house	1	1	-	2	-
Cobble-paved house	2	-	1	6	-
Channelled house	8	4	-	2	1
Basal pits	2	2	4	-	-
Grill house	1	1	1	-	-

Fraxinus identifications made in the field in 1970 are all correct, but this could not be checked any more. On the other hand, in one of the samples examined later, within the framework of the present study, *Fraxinus* could be established. The Rosaceous wood could be of *Amygdalus*, but other taxa, e.g. *Crataegus*, come also into consideration.

From table 18 it is clear that the data are altogether insufficient for a quantitative evaluation. They show us only which kinds of timber were used by the occupants of the site, either for building purposes or for fuel. *Pistacia atlantica* as well as *P. khinjuk* are to be expected in the surroundings of aceramic (early-Holocene) Çayönü. Apparently the fact that *Pistacia* yielded much appreciated fruits did not prevent people from cutting down trees. May we assume that there were enough? *Quercus brantii* is a deciduous oak species which very likely was found in the Çayönü area, but more oak species come into consideration, e.g. *Quercus boissieri* and *Q. infectoria*. *Fraxinus rotundifolia* is a constituent of the mixed-oak woodland which under the present climatic conditions would be the natural vegetation of the Çayönü area (1.1.2). It may be too far-fetched to see in the *Lycium* wood charcoal identification an indication that the enigmatic '*Lycium*-type' seed (see 3.7.3) is indeed of *Lycium*.

No coniferous wood has been demonstrated for Çayönü; *Juniperus oxycedrus* forms part of the SE Anatolian mixed-oak woodland.

5. THE INTERPRETATION OF THE DATA

5.1. Introduction

From the discussion of the plant remains in section 3 it will be clear that emmer wheat and einkorn were the predominant if not the only cereals of aceramic Çayönü, while lentil, field pea and bitter vetch were the main pulse crops. Wild vetch seeds (*Vicia* sp.) could have been an additional source of vegetable proteins. Wild pistachio (*Pistacia atlantica/khinjuk*) must have been a major source of vegetable fat, supplemented by wild almond and perhaps by linseed (*Linum* cf. *bienne*). Shifts in the frequencies of food-plant remains could have

been the result of dietary changes, e.g. the (partial) replacement of one crop plant by another or changes in meat consumption, implying an increased or decreased proportion of animal protein and fat in the diet. Changes in plant husbandry should find expression in changes in the archaeological plant record through time (the chronological distribution). In addition, the spatial (horizontal) distribution of plant remains could in principle inform us about the occurrence and location of crop-processing/food-preparation areas (so-called activity areas), whether or not associated with archaeological features.

It is evident that for determining possible differences in the spatial and chronological distribution of plant remains, a fairly equal representation of the various occupation phases and areas excavated in the archaeological plant record is desirable. Unfortunately, as appears from figure 6, in which the numbers of samples which yielded plant remains other than wood charcoal are shown per square, this condition cannot in any sense be met. Only for a few squares is an appreciably great number of samples available. Moreover, as has already been mentioned, many of the samples yielded only insignificant numbers of plant remains. The above implies a considerable limitation to the degree of detail to be obtained from the Çayönü plant record. In general only some broad generalizations can be made.

5.2. The chronological and spatial distribution of plant remains

For a discussion of the chronological and spatial distribution of plant remains only species and groups of species which are more than occasionally represented in the archaeological record are of importance. Seed and fruit types which occur in one or a few samples with one or two specimens are not of relevance for speculations of this kind. This means that many of the taxa demonstrated for Çayönü can be left out of consideration here. As a matter of fact, food plants are by far the best represented taxa.

The chronological distribution of plant remains will be discussed at the level of the occupation sub-phase. Thus, the samples attributed to a particular sub-phase are taken together. This procedure generally results in satisfactorily greater numbers of samples and quantities of plant remains, although the upper aceramic occupation sub-phases remain poorly represented. To facilitate the comparison between sub-phases, in table 19 the mean numbers of seeds and fruits per 10 samples are shown. Moreover, morphologically wild and domestic forms of both einkorn wheat and emmer wheat have been lumped (tables 3 and 4 provide the more detailed information). As for the spatial distribution, the western and eastern areas will be compared with each other, again at the level of the occupation phase (table 20). The highly uneven distribution of the samples examined prevents the establishment of possible plant distribution

Table 19. Densities of seeds etc. in occupation sub-phases, expressed as mean numbers per 10 samples.

Sub-phase	r	g	b	ch	cp	c	c/lr	lr
Excavation area	E	E+W	E	E+W	E+W	E+W	E	E
Number of samples	16	32	23	41	38	67	8	5[6]
Triticum boeoticum + monococcum	-	1.3	4.0	3.1	1.8	1.3	1.3	-
Triticum dicoccoides + dicoccum	-	1.6	5.9	5.4	0.8	5.4	2.5	-
Triticum spikelet forks	5.6	77.2	108.3	225.6	130.8	38.8	15.0	-
Hordeum grains	-	0.3	0.8	1.0	0.1	0.3	-	-
Hordeum rachis internodes	-	0.9	0.9	2.0	0.8	0.1	-	-
Σ Cereal grains	x	3.5	11.1	9.6	3.2	7.2	3.8	-
Pisum	-	4.8	13.5	11.3	9.1	6.4	3.8	74.0
Lens	0.6	1.3	0.7	2.0	6.7	11.1	1.3	186.0
Vicia ervilia	-	4.2	21.5	17.9	27.2	25.0[4]	1.3	7.0
Σ Pulses	0.6	10.3	36.1	32.3	43.3	43.7[5]	6.3	267.0
Pistacia[1]	xx	xx	xx	xx	xx	xxx	xx	x
Amygdalus[1]	x	x	x	x	x	xx	x	-
Vicia sp.	8.4	53.9	78.5	42.1	31.1	21.2	3.8	x
Other Leguminosae	1.3	1.5	0.9	4.1	5.5	0.4	0.6	-
Lolium	0.6	1.9	5.9	3.6	1.3	1.2	-	-
Other Gramineae	1.6	3.1	5.0	2.9	3.4	1.2	1.3	-
Scirpus maritimus	0.6	1.6	8.7	2.2	1.6	0.8	-	-
Chrozophora	-	0.6	1.1	1.7	-	0.6	-	-
Galium	-	0.6	1.3	0.4	0.5	1.0	1.3	-
Ziziphora	-	0.6	-	14.1[2]	0.3	0.3	-	-
'Lycium type'	0.6	-	9.6	81.0[3]	6.1	0.2	-	-

[1]Estimates based upon quantities of nutshell remains and sample frequencies: x present; xx fairly common; xxx common
[2]Almost all seeds in one sample (see table 23)
[3]Sample with 1170 seeds (see table 23) not included (N = 40)
[4]*Vicia ervilia* deposit samples (see table 12) not included (N = 60)
[5]Corrected for *Vicia ervilia* deposit samples
[6]Four samples from inside same house (see table 24)

patterns, apart from a few concentrations of plant remains to be discussed below (5.3).

The volumes of soil floated have not always been recorded, but an overall average of 25 litres of soil per sample should be a fair estimate. This would imply that the numbers shown in tables 19 and 20 are those per 250 litres of soil, on average.

Our initial intention was to discuss the chronological and spatial distribution of the floral remains separately, but this turned out to be unpractical.

From the lowermost occupation levels, the Rounded-hut sub-phase, conspicuously few cereal- and pulse-crop remains were recovered: some chaff and cereal grain fragments and one lentil. These numbers are so low that one could question whether wheat and lentil had been food plants of the earliest occupants of the site. Could the few remains of these plants have originated from the overlying Grill-house sub-phase levels, in which, as is shown in table 19, cereals and pulses are, on average, better represented? This suggestion is invalidated by the fact that Grill-house sub-phase samples from the eastern area, to which the rounded huts are confined, yielded relatively few cereal remains and only one pulse-crop seed (table 20). Thus, these levels were not a particularly rich source of charred plant

remains to be carried downwards (through the activity of man and burrowing animals). By far the most numerous plant remains in the rounded hut samples are *Pistacia* nutshell fragments, suggesting that pistachio fruits were gathered intensively. Admittedly, the tough nutshells may have had better chances of being preserved than cereal grains and pulse seeds, but if the latter had played a substantial role in the food economy, a better representation in the archaeological plant record might be expected. Thus, it is tempting to speculate that cereals and pulses, wild or cultivated, were at most of minor importance in the earliest stages of occupation. Be this as it may, the floral record of the Rounded-huts sub-phase is clearly distinct from that of the succeeding sub-phases.

The Grill-house sub-phase plant record, taken as a whole, provides a picture comparable to that of most other occupation sub-phases in that pulse-seed frequencies are higher than those of cereal grains. There are two differences: distinct *Triticum dicoccum*-type grains are lacking and the seed densities (numbers of plant remains per volume of soil) are still comparatively low. The low mean seed density is mainly accounted for by the eastern area (table 20). Pulse-crop and wild vetch (*Vicia* sp.) frequencies show the most conspicuous

Table 20. Densities of seeds etc. in occupation sub-phases itemized per excavation area, expressed as mean numbers per 10 samples.

Sub-phase	g		ch		cp		c	
Excavation area	E	W	E	W	E	W	E	W
Number of samples	17	15	21	20	24	14	20	47
Triticum boeoticum + monococcum	0.6	2.1	1.2	5.0	1.5	2.5	2.0	1.0
Triticum dicoccoides + dicoccum	0.5	2.7	4.9	6.0	0.4	1.4	17.5	0.2
Triticum spikelet forks	68.2	87.3	61.0	398.5	105.8	173.6	39.5	38.5
Hordeum grains	0.5	-	1.9	0.2	-	0.2	0.5	0.3
Hordeum rachis internodes	1.8	-	1.0	3.0	0.4	1.4	0.5	-
Σ Cereal grains	2.2	4.9	8.0	11.2	1.9	5.6	20.0	1.7
Pisum	0.6	9.7	9.8	13.0	13.5	1.4	13.8	3.3
Lens	-	2.7	0.7	3.3	2.5	13.9	29.5	3.2
Vicia ervilia	-	9.0	15.0	21.0	20.0	39.6	46.3	14.5[4]
Σ Pulses	0.6	21.3	26.0	39.0	36.5	55.0	91.0	22.0[5]
Pistacia[1]	xx	xx	xx	xx	xx	xx	xxx	xx
Amygdalus[1]	x	x	x	x	x	x	xx	x
Vicia sp.	9.7	104.0	32.1	52.5	19.0	51.8	60.5	4.5
Other Leguminosae	1.2	1.9	0.5	8.0	2.1	11.4	0.8	0.2
Lolium	0.9	3.0	2.4	4.9	-	3.6	2.0	0.8
Other Gramineae	1.2	5.3	1.2	4.8	4.6	1.4	2.3	0.7
Scirpus maritimus	1.2	2.0	3.3	1.0	1.3	2.1	2.8	-
Chrozophora	-	1.3	1.0	2.5	-	-	2.0	-
Galium	-	1.3	-	0.8	-	1.4	3.0	0.1
Ziziphora	-	1.3	-	29.0[2]	0.4	-	1.0	-
'Lycium type'	-	-	-	170.5[3]	-	16.4	-	0.3

[1]Estimates based upon quantities of nutshell remains and sample frequencies: x present; xx fairly common; xxx common
[2]Almost all seeds in one sample (see table 23)
[3]Sample with 1170 seeds (see table 23) not included (N = 19)
[4]*Vicia ervilia* deposit samples (see table 12) not included (N = 40)
[5]Corrected for *Vicia ervilia* deposit samples

differences between the eastern and western areas. Low seed densities are sometimes explained as the result of open and/or short-lasting occupation. There is a relation between the density of the botanical record and the number of people who lived in a certain area during a certain period of time: the more people there are, the more refuse is produced. Such an explanation does not apply to the Çayönü Grill-house sub-phase. Judging from the numbers of houses uncovered so far, the eastern area must have been inhabited at least as densely as the western one (see Schirmer, 1990: fig. 2). As a matter of fact, the question of what may have caused marked differences in quantity and quality of floral remains between sites and within a site has not yet really been tackled. We are still in the dark about the taphonomic processes which ultimately resulted in the incorporation and preservation of seeds, fruits, charcoal and other plant particles in occupation deposits.

Most curious is the near absence of pulses in the eastern area. Should one believe that during the Grill-house sub-phase there were marked differences in consumption pattern between the two areas? The comparatively high vetch seed densities from the Grill-house sub-phase onwards make one wonder whether these wild leguminous seeds were gathered for human consumption.

Botanical information on the Basal-pits sub-phase is available only for the eastern area. The densities of almost all categories of plant remains are significantly higher than those of the Grill-house sub-phase, which is even more pronounced if the comparison is confined to the eastern area. Morphologically defined *Triticum dicoccum* could be established with certainty for this sub-phase. Should there be a final attribution of the basal pits to the Channelled-house sub-phase, this would be supported by the botanical evidence: there is a fair resemblance between the floral records of the Basal-pit and Channelled-house samples. On the other hand, from a botanical point of view there is no serious objection to the idea of a separate Basal-pits sub-phase, with the understanding that a return to housing in subterranean huts did not result in a more primitive type of subsistence economy, i.e. with a reduced reliance on plant cultivation.

A characteristic feature of the Çayönü plant husbandry is the predominant role of pulses. In almost all occupation sub-phases, pulse-seed frequencies are much higher than those of cereal grains. It is true that wheat chaff remains are very numerous, but these represent crop-processing waste not destined for human consumption. In prehistoric and early-historical sites cereal grains are usually much more numerous than pulse seeds. This has raised the question whether cereals have better chances of being preserved in a carbonized condition than pulses. Be this as it may, one may safely assume that leguminous plants played a prominent part in the diet of the aceramic inhabitants of the site. It is striking that this tradition persisted for a very long time. Also in the pottery Neolithic samples, pulse-crop seeds outnumber cereal grains by far.

As has already been mentioned, the archaeobotanical record of the Channelled-house sub-phase shows considerable resemblance to that obtained for the Basal-pits levels. The greater density of chaff remains is wholly accounted for by the western area, with a five times higher density than in the eastern area. Could this indicate that during the Channelled-house occupation sub-phase the western area witnessed more crop-processing (dehusking) activities than the eastern one, for instance, because the former area had more the function of a living quarter than the latter one with the skull building? 'Lycium-type' shows a concentration in house DI, square 20L (see table 23 and discussion in 5.3), which accounts for the high mean density, also after correction for the sample in which more than 1000 specimens were counted. The distribution of 'Lycium-type' seeds is almost as enigmatic as the species identity of this type itself. Large numbers of seeds and a high sample frequency are characteristic of the Channelled-house sub-phase levels in the western area (table 20) and not a single specimen in contemporary levels in the eastern area.

The floral record of the Cobble-paved house sub-phase suggests a shift in the cereal-grain and pulse-seed proportions. In the previous occupation sub-phases the ratio between cereals and pulses is about 1 to 3, but in this sub-phase the density of pulse seeds is more than ten times higher than that of cereal grains, although this ratio may be somewhat exaggerated due to the abnormally low emmer-wheat density. This shift in favour of pulses is manifest in both the eastern and western areas, so that one may assume that there was a real reduction in the contribution of cereals to the diet of the Çayönü inhabitants. The reason for this change is puzzling, particularly because it is not clear which other food sources could have substituted for the carbohydrates of the cereals. In the Cobble-paved house sub-phase wild emmer-type grains have disappeared, but admittedly it is not clear how this should be interpreted (see 3.1.1). The increased lentil density is largely accounted for by the lentil seeds in the western area. The western area has a significantly higher chaff density than the eastern one, but the difference is not so large as in the preceding sub-phase.

In the Cell-house sub-phase the ratio between total cereal-grain and pulse-seed densities (about 1 to 6) is again distinctly higher than in the early stages of occupation, suggesting that the shift to a greater reliance on pulse crops established for the Cobble-paved house sub-phase was not of a temporary nature, but that it

persisted. The determination of the Vicia ervilia densities is somewhat complicated by the bitter vetch deposit in and near house DF, in the western area (see discussion in 5.3). The almost pure bitter vetch samples from this deposit are not included in the calculation of the densities of this seed type. In determining the total pulse-seed densities a correction is made for these bitter vetch samples. The mean density for emmer-wheat grains is much higher than in the previous sub-phase and again at the level of the Channelled-house and Basal-pit samples. One should perhaps be wary of concluding that there was a temporary decline in the economic importance of emmer wheat in the Cobble-paved house sub-phase. It cannot be excluded that the scarce representation of emmer-wheat grains in the latter sub-phase is due to accidental factors. Moreover, in this sub-phase wheat-chaff frequencies are certainly not low. Curiously, the eastern and western areas of the Cell-house sub-phase show greater difference in mean emmer-wheat densities (table 20) than there is between the Cobble-paved and Cell-house sub-phases (table 19). One wonders what explanation there could be for the striking differences in seed densities between the eastern and western areas of the Cell-house sub-phase. The usually low numbers of plant remains coupled with an unequal sample distribution may, in some instances, have resulted in inaccurate quantitative information. Only changes in food-plant proportions which embrace more than one occupation sub-phase should be considered as indicative of changes in plant husbandry.

It is evident that the botanical information available for the upper two occupation sub-phases is altogether insufficient for drawing conclusions. From deposits attributed to the Cell/Large-room transition sub-phase only 3 cereal grains and 5 pulse seeds were recovered in addition to a few seeds of wild plants. It is unlikely that during the final aceramic occupation sub-phase only pulse crops and no cereals would have been cultivated. Four of the five samples are from the fill of the same house (see 5.3) and the fifth sample is almost devoid of plant remains.

The scarcity of botanical information for the youngest aceramic sub-phases must, to a large extent, be due to the fact that in the upper levels of the site usually few plant remains are preserved. As a result of alternating wetting and drying of the upper layers during hundreds and thousands of years the charred plant remains have disintegrated.

In conclusion it can be remarked that the Çayönü archaeobotanical record provides evidence of some changes in the food-plant exploitation in the course of the aceramic occupation. Differences in floral remains between the eastern and western excavation areas may reflect intra-site differences in plant use and/or food-processing activities, but speculations of this kind should be considered with due reservation.

The plant husbandry of aceramic Çayönü, SE Turkey   91

5.3. The plant record in relation to context

One of the questions that are put to the archaeobotanist is whether a relation between archaeological features and the composition of plant remains recovered from these features can be established and if so, what such a relation may mean. In other words, do the plant remains provide further information on the function of the feature or, alternatively, do they correspond to what was expected by the archaeologist (and botanist) on the basis of the kind of feature? Obvious archaeological features on the function of which the archaeobotanist is expected to provide information include jars, storage pits and ovens (what has been stored or prepared?). Unfortunately, the interpretation of possible plant remains obtained from such features in terms of the function of the latter is less clear-cut than one would wish. Moreover, features of this kind are either not to be expected in aceramic Çayönü (jars) or have not been unearthed or sampled for botanical examination (only two samples are reported to derive from a bin, see below). As a matter of fact, the roasting pits from the lower occupation levels are distinct features, which, as it were, call for a botanical evaluation. This topic will be discussed in the next section (5.4).

The most obvious features at Çayönü are the house plans. In a first attempt to get some hold on possible correlations between the composition of plant remains and archaeological context, it was considered whether a comparison between houses makes sense. Are there archaeobotanical differences between houses which could be explained in terms of socio-economic differentiation within the site or, for instance, between the 'public' ritual skull building and the other, domestic buildings? This approach turned out to be unpromising. One of the reasons why this approach failed may have been that in general the samples from inside house plans have no direct bearing on the occupation of the house concerned. Inspection of the field data showed that many samples are from contexts which very likely date from before or after the occupation. The remains of a bitter vetch supply in house DF is a notable exception. It will be discussed below.

In the majority of the samples that yielded more than a few plant remains, typical refuse, such as chaff (glume bases, spikelet forks, rachis internodes) and *Pistacia* nutshell fragments, is most common, with occasionally a fair number of seeds of cultivated species or wild plants. The botanical contents are usually of mixed origin, that is to say, the plant remains derive from various household activities. Specific activity areas with respect to food plants, such as areas for storage and processing, are only seldom indicated by the archaeobotanical record. As a case in point, table 21 shows the contents of an 'ash deposit' in square 20M, attributed to the Channelled-house sub-phase (ch3). This example is illustrative of many of the Çayönü samples, with the understanding that the actual botanical composition varies considerably.

The botanical contents of an ash pit from square 27L, attributed to the Cell-house sub-phase (c3), differs from other refuse deposits in that chaff is almost absent (table 22). Apparently the dehusking of einkorn and emmer wheat was not carried out in the direct vicinity of this ash pit. *Pistacia* and *Amygdalus* nutshell remains are predominant by far. Even complete pistachio nutshells were recovered from the fill of this pit (fig. 10:1). Estimates of the numbers of whole pistachio and almond nutshells are presented in table 22. What is the explanation for the extraordinarily large quantities of nutshell remains? As for almond, one could imagine that the stones were cracked to retrieve the edible contents, after which the shells were thrown in the refuse pit which was periodically set afire. In wild pistachio, it is the fruit flesh that was consumed or from which the oil was extracted. Could it be that the fruits,

Table 21. Samples from ash deposit in square 20M, sub-phase ch3.

Sample number	1986/	30	38	29	23
Triticum monococcum		-	2	-	-
Triticum dicoccum		-	2	0.5	-
Triticum spikelet forks		87	89	66	4
Hordeum rachis internodes		-	4	-	-
Cereal grain fragments		xx	xx	-	x
Lens		-	1.5	-	-
Vicia ervilia		1	17	2	1
Pulse grain fragments		-	-	-	x
Vitis		-	-	1	-
Pistacia		x	xx	x	x
Vicia sp.		-	5	3	1
Vicia fragments		-	xx	-	-
Lolium		-	-	1	-
Unident. Gramineae		1	-	-	-
Unident. Compositae		-	1	-	-
Unident. Umbelliferae		-	1	-	-
'*Lycium* type'		1	9	27	1

Table 22. Numbers of seeds and fruits retrieved from five samples from 'ash pit' 70/R 5-8 (square 27L, sub-phase c3). In brackets numbers of samples in which taxon is represented. Of *Pistacia* and *Amygdalus*, numbers of whole nutshells are estimated (on the basis of weight).

Triticum monococcum	1	(2)
Triticum dicoccum	c. 18	(4)
Triticum spikelet forks	1	(1)
Pisum	2.5	(2)
Lens	2	(1)
Vicia ervilia	c. 27	(4)
Vicia sp.	6	(2)
Unident. Leguminosae	2	(1)
Pistacia	c. 256	(5)
Amygdalus	c. 16	(5)
Quercus	c. 0.5	(2)
Lolium	1	(1)
Galium	3	(2)
Anchusa	1	(1)
Unidentified	1	(1)

Table 23. Samples from ash deposits in Channelled-house (ch3) DI, square 20L.

Sample number	1987/ 6	13	14	29	1986 / 71	79	86
Triticum boeoticum	-	-	-	6	-	1	-
Triticum monococcum	-	-	-	-	-	1	-
Triticum dicoccum	-	-	-	-	-	1	-
Triticum spikelet forks	49	88	111	127	10	85	6
Unident. rachis internodes	-	-	-	-	-	3	-
Cereal grain fragments	xx	xx	xx	xx	-	xx	x
Pisum	-	-	16	-	-	2	3
Vicia ervilia	2	-	2.5	4	0.5	-	-
Lathyrus cicera/sativus	-	1	-	-	-	-	-
Pulse grain fragments	x	x	-	x	-	-	-
Pistacia fragments	xx	xx	-	xx	x	xx	xx
Amygdalus fragments	x	x	x	-	-	-	-
Vicia sp.	2	5	24	21	2	-	0.5
Vicia fragments	-	xx	xxx	-	-	xx	-
Lolium	-	0.5	1.5	1.3	-	4	-
Bromus	0.5	-	-	2	-	-	-
Stipa awn fragments	-	x	x	-	-	-	-
Unident. Gramineae	-	x	-	x	-	-	-
Adonis	-	-	1	-	-	-	-
Lithospermum tenuiflorum	1	-	1	-	-	-	-
Chrozophora tinctoria	-	5	-	-	-	-	-
Scirpus maritimus	-	-	-	1	-	-	1
Ziziphora	-	5	53	-	-	-	-
'Lycium type'	10	184	1170	6	43	-	2

after they had been made more tasty by roasting, were eaten as such and that the intact or cracked nuts were spat out, after which they ended up in the refuse pit?

Mention is made here of the botanical record of the seven samples from fill deposits inside Channelled-house DI (table 23). In general, the floral composition of these samples conforms to that of other samples of occupational fill. What is striking, however, is the extraordinarily large number of 'Lycium-type' seeds. In one of the samples 1170 specimens were counted. As has already been discussed (3.7.3), the species identity of this type is still puzzling. The concentration of 'Lycium-type' seeds suggests that somewhere in this particular area a supply of these seeds had been stored. Whether or not the seeds were stored in house DI remains uncertain. On the other hand, the circumstantial evidence is indicative of the intentional gathering of this seed type or alternatively of fruits which contained the seeds. From the sample with the many 'Lycium-type' seeds, a surprisingly great number of *Ziziphora* diaspores (53 specimens) were recovered. Was *Ziziphora* collected purposely?

It has been emphasized that the botanical contents of the Çayönü samples are usually of mixed origin. In only one case is there convincing evidence that we are dealing with the almost pure remains of a crop-plant supply. This concerns the bitter vetch (*Vicia ervilia*) deposit which has already been mentioned in 3.2.3. A concentration of bitter vetch seeds was observed in the fill of a room in house DF, square 17N, attributed to the Cell-plan sub-phase. The six samples from this deposit

are shown in table 12. A seventh bitter vetch sample from the adjacent square 17M is probably from the same supply. This sample from outside house DF has a greater admixture of cereal remains. The latter sample is to some extent a complicating factor. As to the bitter vetch deposit inside the house, one could imagine that in the fire catastrophe the charred remains of the food supply stored on the ground floor or on the upper storey had naturally come to rest there and had subsequently been covered by other debris. If the sample from outside the house derives from the same bitter vetch store, man must have had a hand in the spreading of the charred remains; for instance, in the process of levelling the house site after the fire.

From only one container a few soil samples have been secured for botanical examination. This concerns a bin in Large-room house BF (table 24). Modest numbers of peas and lentils recovered from these samples could be interpreted as evidence of the storage of these pulses in the bin. However, two samples from the fill in

Table 24. Samples from Large-room house BF, square 28-29/K-L.

| Sample number 1970/ | bin | | | |
	U 3(16)	U 4(16)	U 4(30)	U 5(49)
Pisum	c. 4	c. 7	1	c. 25
Lens	c. 18	c. 7	c. 8	c. 60
Vicia ervilia	1	-	-	2.5
Pistacia fragments	x	-	-	-

other parts of house BF give a similar picture. This makes one wonder whether the fill and its botanical contents in the whole of the house are of the same (secondary) origin, in which case the charred seeds in the bin would have no relation to what originally had been stored in it. On the other hand, the almost exclusive occurrence of pulses in the fill indicates that these seeds represent the charred remains of pulse-crop supplies which had been stored in house BF, perhaps on the upper floor.

5.4. 'Roasting pits'

One of the non-architectural features at Çayönü is the so-called roasting pits: circular depressions packed with cobbles interspersed with black earth and charcoal. From the features themselves it cannot be determined whether they served for roasting meat, parching glume wheat or for some other household activity. As for the way of functioning, Michael K. Davis (pers. comm.) advocates the following experimentally tested hypothesis. In a circular pit a wood fire was lit. When the fire was burning well, cobbles were gradually added along with additional fuel. When filled with cobbles, after a high temperature had been reached and before the fire was spent the pit's contents were covered over by a layer of clay. On this clay sealant, which soon became hot from the heat-battery of stone below, the foodstuff to be roasted was laid. Long use of the sealant's safe heat was possible as it remained relatively constant for more than two days. While the pits could have been reused after cooling, by removing the mass of cobbles and repeating the firing process, the pits excavated at Çayönü appeared to have been left with their contents undisturbed.

Irrespective of the question whether Davis' reconstruction of the operating of the Çayönü roasting pits is correct, one may wonder to what extent plant remains recovered from the fill of the pits could inform us on what has been roasted or otherwise prepared in these structures. The same applies to other features which presumably served for food preparation, such as ovens and fire places. The samples taken for botanical examination are usually from soil that had been deposited after the feature had fallen into disuse, implying that the plant remains that were found may have no relation with the household activity performed there. Be this as it may, the Çayönü roasting pit samples are not particularly informative in this respect.

As appears from table 25, in addition to wood charcoal, only one sample (No. 87/11) yielded more than an occasional identifiable particle. From various samples only wood charcoal was recovered. The few non-wood remains in samples 87/7, 87/16 and 86/146 may be considered as settlement noise: scattered waste in low densities. The composition of sample 87/11 is not indicative of any particular food-processing activity, but rather points to a refuse deposit. The scarcity of plant remains other than wood charcoal could indicate that meat and not vegetable food was prepared in these pits. If vegetable food was processed, it should not have been in direct contact with the fire or hot ash, which would support Davis' view of a clay cover over the fire. If parching of glume wheat spikelets had been carried out on these structures (prior to dehusking), one would have expected more chaff remains in the samples examined. This activity would have resulted in much chaff around the roasting pits.

5.5. Field weeds

The archaeobotanical evidence informs us fairly well on the crop plants of the Çayönü farmers. The abundant chaff remains indicate that the dehusking of the glume wheats was carried out on the site, as was to be expected. What can be said about other aspects of arable farming, such as the condition of the fields and harvesting methods?

In view of the present soil and climatic conditions (see 1.1.1) one may assume that the physical environment put no limitations on plant cultivation at early-Neolithic Çayönü. In principle, inferences about

Table 25. Roasting-pit samples.

Sample number	70/R18	87/7	87/16	86/147	87/11	86/146	70/R17	70/R14	70/R12
Square	27L	28L	28L	27L	27L	28L	27L	27L	27L
Sub-phase	bp	bp?	bp-ch	bp-ch	ch	ch	ch	ch	cp
Triticum spikelet forks	-	-	2	-	8	1	-	-	-
Cereal grain fragments	-	-	-	-	x	-	-	-	-
Cicer	-	-	-	-	0.3	-	-	-	-
Lathyrus cicera/sativus	-	-	-	-	1	-	-	-	-
Pistacia fragments	-	x	x	-	x	x	-	-	-
Celtis	-	-	-	-	1	-	-	-	-
Vicia	-	-	-	-	1	1	-	-	-
Vicia fragments	-	-	-	-	x	-	-	-	-
Lolium	-	-	-	-	2.5	-	-	-	-
Wood charcoal	+	+	+	+	+	+	+	+	+

Table 26. Potential segetal taxa attested for aceramic Çayönü. Total number of samples included is 230.

	Total number of seeds	Sample frequency
Vicia spp.	800	115
Astragalus	1	1
Medicago	2	1
Medicago radiata	2	2
Melilotus	4	1
Lolium cf. *rigidum*	48	35
Aegilops umbellulata[1]		
Bromus	6.5	5
Echinaria	9	4
Anchusa	1	1
Lithospermum arvense[1]		
Silene	1	1
Vaccaria pyramidata	0.5	1
Chrozophora tinctoria	15.5	6
Ziziphora	63	6
Malva	1	1
Fumaria	1	1
Rumex	3	2
Adonis	1	1
Ranunculus arvensis type	1	1
Galium	16	13
Verbascum	1	1

[1]reported by Stewart (1976)

agricultural practices of the past can be drawn from arable field weeds represented in the archaeological plant record. This approach, which starts from the flora of traditionally cultivated fields of the recent past, is commonly applied in archaeobotanical research, although the dangerous sides of its actualistic principle are well recognized. Table 26 lists the (potential) field-weed taxa established for the aceramic levels with the total numbers of occurrences (and numbers of samples in which the type is present).

At first sight it would seem that the arable weed flora is fairly well represented. However, it should be emphasized that the seed types concerned only possibly derive from weeds in the fields of the Çayönü farmers. Most of the taxa have been identified to the genus level only, implying that they could represent arable weeds but also species from other habitats. Moreover, many present-day segetal plants are believed to originate from steppe vegetations. Steppe (and forest-steppe) species would have invaded the fields of the ancient farmers to become arable weeds by adapting themselves to the particular conditions in cultivated fields. Thus, for many of the taxa listed in table 26 it is not certain whether they actually occurred as arable weeds at aceramic Çayönü. Yet irrespective of this question, most of the potential field weeds are scarcely represented in the seed record. Eleven taxa were found in one sample only, 9 of which with only one seed. Three types, viz. *Vicia*, *Lolium* cf. *rigidum* and *Galium*, occur in more than 10 of the 230 samples, wild vetch being by far the most numerous. Wild vetch seeds may have been

gathered intentionally (3.6.1), which could explain their great numbers.

The scarce representation of (potential) arable weeds in the archaeobotanical record does not necessarily imply that the fields of the Çayönü farmers were almost free of weeds, but it may rather be ascribed to the harvesting methods employed. If the cereals were reaped by cutting or plucking the individual ears or by uprooting the plants, only few field-weed diaspores are to be expected in the unprocessed corn crop. Pulses are usually harvested by uprooting. In this way only small numbers of arable weed seeds would unintentionally have been carried to the site. The burning of dung fuel is thought to have contributed substantially to the charred seed contents of archaeological deposits (Bottema, 1984; Miller & Smart, 1984). At aceramic Çayönü, sufficient firewood must still have been available, and moreover, domestic animals, mainly sheep, appeared only in the uppermost aceramic levels, in the Cell/Large room and Large room sub-phases (Lawrence, 1982).

From the above speculations it appears that the Çayönü floral record is not very informative on plant husbandry questions other than food-plant species and food-plant proportions.

5.6. The vegetation in the surroundings of aceramic Çayönü

The TAVO (*Tübinger Atlas des Vorderen Orients*) map of early-Holocene (c. 8000 BP) vegetation in the Near East (Bottema & van Zeist, 1990) indicates 'cold deciduous montane woodland with evergreens' for the Çayönü area. This map of the inferred early-Holocene vegetation, which is based mainly upon palynological evidence, had to be confined to presenting major vegetation zones, within which the plant cover may have varied locally and regionally. Moreover, as the early Holocene, comprising several thousands of years, witnessed considerable changes in the vegetation, the date of c. 8000 BP (cal. c. 7000 BC) is only a rough approximation. From the above it will be clear that for any given early-Holocene site, the vegetation may have deviated somewhat from that as defined for the whole of the vegetation zone. However, in broad outline the TAVO map mentioned above should provide a fair picture of the early-Holocene vegetation in the Near East.

The charred wood and seed evidence indicates that deciduous oak (*Quercus brantii* and perhaps other oak species), pistachio (*Pistacia atlantica*, *P. khinjuk*), and probably ash (*Fraxinus rotundifolia*) were major constituents of the arboreal vegetation in the Çayönü area. On the other hand, the absence of coniferous wood remains suggests that juniper (*Juniperus*), listed for the early-Holocene woodland zone, was not or only scarcely found near Çayönü. The density of the arboreal vegetation is uncertain. Was it woodland, that is to say, forest of well spaced trees (crown cover not more than

50%), or was it more open, a kind of transitional stage between steppe and woodland? The pollen record obtained for Lake Zeribar in western Iran (see fig. 1) suggests that in such a case grasses may have played a predominant role in the steppe-like ground cover (van Zeist & Bottema, 1977). Grasses are not abundantly represented at Çayönü, and the type which is most frequent, viz. *Lolium* cf. *rigidum*, is typical of ruderal and segetal vegetations. Other taxa which include characteristic steppe plants (Chenopodiaceae, Umbelliferae, Leguminosae, Compositae) are scarcely or not at all recorded. Because of its small, fragile seeds, wormwood (*Artemisia* spp.) is not to be expected in the charred seed record. The above does not necessarily imply that steppe plants were not common in the Çayönü area, but the chances for the seeds of these taxa of becoming incorporated in the settlement deposits in a charred condition may have been minor. One such negative factor may have been that no dung fuel was used. One typical steppe plant is recorded, viz. *Stipa* (feathergrass).

In conclusion, we are left with the uncertainty whether the Çayönü upland vegetation was of the woodland type or forest-steppe. For the farmers this difference in vegetation may have been crucial. In the case of woodland they would have been forced to cut down at least some of the trees in laying out their fields. In the case of forest-steppe it would have sufficed to clear the ground vegetation, which could easily have been done by burning.

Trees and shrubs which are characteristic of river-valley forest, such as *Populus (euphratica)*, *Platanus orientalis* and *Tamarix* sp., have not been demonstrated for Çayönü. Apparently the stream at the foot of the site was too small for the development of riverine forest. Among the taxa attested for Çayönü, *Scirpus maritimus* (sea club-rush) is characteristic of marshy habitats. Obvious habitats of sea club-rush in the Çayönü area must have been the stream valley and pools which carried water during part of the year. Also wild grape vine (*Vitis vinifera*) could have grown along the stream at the foot of the mound (if the berries had not been brought in from further away).

Speculations on the synanthropic field-weed flora are presented in the previous section (5.5).

6. CONCLUSIONS

The following conclusions with respect to the plant husbandry of aceramic Çayönü are drawn here:

– There is no firm botanical evidence of plant cultivation in the earliest occupation levels (the Rounded-huts sub-phase);

– Both cereals and pulses were grown by the Çayönü farmers. A characteristic feature of Çayönü is the predominance of pulses over cereals, which in the course of the aceramic occupation became even more pronounced;

– Cereal cultivation was confined to einkorn and emmer wheat. Morphologically defined wild emmer wheat in the early stages of the occupation was most probably cultivated. The absence of chaff remains of shattering wheats suggests that the wild-type emmer grains are of plants which already had a semi-tough rachis and which consequently may be considered as domestic. The same may have been the case with the wild-type einkorn wheat;

– The scarce barley remains (grains, rachis internodes) point to wild *Hordeum spontaneum* that was neither cultivated nor gathered purposely;

– Among the pulses, bitter vetch played a predominant role in addition to lentil and field pea;

– The great quantities of wild vetch (*Vicia* sp.) seeds suggest that these were gathered intentionally, probably for human consumption;

– *Pistacia* fruits must have been collected intensively throughout the entire aceramic occupation;

– The archaeological plant record provides only little information on agricultural practices.

7. ACKNOWLEDGEMENTS

The authors wish to express their sincere thanks to all who co-operated in the present study. Professors Halet Çambel (Istanbul) and Robert J. Braidwood (Chicago) offered ample facilities to both of us for botanical field work at Çayönü. Additional information on the site was provided by Dr. Mehmet Özdoğan, Asli Özdoğan, Erhan Biçakçi and Michael K. Davis. The sub-phase attributions used here (worked out by Biçakçi, Asli Özdoğan and Davis) were supplied by Davis. Mehmet Özdoğan placed the photos of figures 4 and 5 at our disposal. Robert and Linda Braidwood commented upon the first draft of this report, and actually rewrote the section on the stratigraphy and architecture. Additional comments on the manuscript were received from Michael K. Davis, who also provided the description of the 'roasting pits'. Dr. Gordon C. Hillman (London) advised on the identification of some seed types.

The drawings were executed by H.R. Roelink, Miriam A. Weijns and J.H. Zwier. Rita M. Palfenier-Vegter assisted in the laboratory work. Secretarial assistance in the preparation of the manuscript was rendered by Gertie Entjes-Nieborg. Sheila M. van Gelder-Ottway improved the English text.

The Prehistoric Project's field grants for excavation at the site of Çayönü, have come from the National Science Foundation, the Wenner Gren Foundation for Anthropological Research, and the National Geographic Society, as well as from the Oriental Institute of the University of Chicago and its friends. In particular, for the laboratory analysis of the botanical samples, financial support was received from the late Solomon A. Smith, II of Chicago.

8. NOTES

1. The term 'Basal pits'has been used in several recent Çayönü papers, so it is probably wise to retain the name. However, these pits are not truly basal, in the sense that they do not proceed downwards from the real base of the occupation. Until further excavation in achieved, it now seems most likely that they proceed downwards, stratigraphically, only from between the end of the uppermost grills and the beginning of the channel houses. Although they have been found to have been cut downwards into the sterile soil, it is the Round-house sub-phase that is the true basal level This comment of Robert J. Braidwood describes the situation in 1991.
2. The full record is on file at the Biologisch-Archaeologisch Instituut of the State University of Groningen, the Oriental Institute of the University of Chicago, and the Prehistory Section of Istanbul University.

9. REFERENCES

ALEX, M., 1985. *Klimadaten ausgewählter Stationen des Vorderen Orients* (= Beihefte zum Tübinger Atlas des Vorderen Orients, Reihe A (Naturwissenschaften) Nr. 14). Wiesbaden.

BOTTEMA, S., 1984. The composition of modern charred seed assemblages. In: W. van Zeist & W.A. Casparie (eds), *Plants and ancient man*. Rotterdam etc., pp. 207-212.

BOTTEMA, S. & W. VAN ZEIST, 1989. *Vorderer Orient. Vegetation im Frühholozän (ca. 8000 B.P.) 1:4 000 000* (= Karte A VI 2 Tübinger Atlas des Vorderen Orients). Wiesbaden.

BRAIDWOOD, R.J., H. ÇAMBEL, W. SCHIRMER et al., 1981. Beginnings of village-farming communities in southeastern Turkey: Çayönü Tepesi, 1978 and 1979. *Journal of Field Archaeology* 8, pp. 249-258.

ÇAMBEL, H. & R.J. BRAIDWOOD, 1983. Çayönü Tepesi: Schritte zu neuen Lebensweisen. In: R.M. Boehmer & H. Hauptmann (Hrsg.), *Beiträge zur Altertumskunde Kleinasiens*. Mainz am Rhein, pp. 155-166.

DAVIS, P.H. (ed.), 1965-1985. *Flora of Turkey and the East Aegean Islands*. Volume 4, 1972; volume 7, 1982; volume 9, 1985. Edinburgh.

HARLAN, J.R. & D. ZOHARY, 1966. Distribution of wild wheats and barley. *Science* 153, pp. 1074-1080.

HELBÆK, H., 1959. Notes on the evolution and history of *Linum. Kuml* 1959, pp. 103-129.

HELBÆK, H., 1969. Plant collecting, dry-farming, and irrigation agriculture in prehistoric Deh Luran. In: F. Hole, K.V. Flannery & J.A. Neely (eds), *Prehistory and human ecology of the Deh Luran Plain* (= Memoirs of the Museum of Anthropology, University of Michigan, No. 1). Ann Arbor, pp. 383-426.

HILLMAN, G.C., S.M. COLLEDGE & D.R. HARRIS, 1989. Plant-food economy during the Epipalaeolithic period at Tell Abu Hureyra, Syria: dietary diversity, seasonality, and modes of exploitation. In: D.R. Harris & G.C. Hillman (eds), *Foraging and farming. The evolution of plant exploitation*. London, pp. 240-268.

JANSEN, A.-E., 1986. Art und Bedeutung der Segetalflora im Weizenanbau Nordost-Syriens unter besonderer Berücksichtigung der Veränderung des traditionellen Anbausystems. *Plits* 4(6), Aichtal.

KÖRBER-GROHNE, U., 1987. Federgras-Grannen (*Stipa pennata* L. s.str.) als Vorrat in einer mittelneolithischen Grube in Schönungen, Landkreis Helmstedt. *Archäologisches Korrespondenzblatt* 17, pp. 463-466.

LAWRENCE, B., 1982. Principal food animals at Çayönü. In: L.S. Braidwood & R.J. Braidwood (eds), *Prehistoric village archaeology in south-eastern Turkey* (= BAR International Series 138). Oxford, pp. 175-199.

MILLER, N.F. & T.L. SMART, 1984. Intentional burning of dung as fuel: a mechanism for the incorporation of charred seeds into the archaeological record. *Journal of Ethnobiology* 4, pp. 15-28.

LADIZINSKY, G. & A. ADLER, 1976. The origin of chickpea *Cicer arietinum* L. *Euphytica* 25, pp. 211-217.

ÖZDOĞAN, M. & A. ÖZDOĞAN, 1989. Çayönü. A conspectus of recent work. In: *Colloque Préhistoire Levant II, Maison de l' Orient-Lyon, 20 mai-4 juin 1988*. Paris, pp. 65-74.

SCHIRMER, W., 1983. Drei Bauten des Çayönü Tepesi. In: R.M. Boehmer & H. Hauptmann (Hrsg.), *Beiträge zur Altertumskunde Kleinasiens*. Mainz am Rhein, pp. 463-476.

SCHIRMER, W., 1990. Some aspects of building at the 'aceramic-neolithic'settlement of Çayönü Tepesi. *World Archaeology* 21, pp. 363-387.

STEWART, R.B., 1976. Paleoethnobotanical Report – Çayönü 1972. *Economic Botany* 30, pp. 219-225.

WALTER, H., 1956. Vegetationsgliederung Anatoliens. *Flora* 143, pp. 295-326.

ZEIST, W. VAN, 1972. Palaeobotanical results of the 1970 season at Çayönü, Turkey. *Helinium* 12, pp. 3-19.

ZEIST, W. VAN & J.A.H. BAKKER-HEERES, 1975a. Evidence for linseed cultivation before 6000 bc. *Journal of Archaeological Science* 2, pp. 215-219.

ZEIST, W. VAN & J.A.H. BAKKER-HEERES, 1975b. Prehistoric and early historic plant husbandry in the Altinova Plain, southeastern Turkey. In: M.N. van Loon (ed.), *Korucutepe. Final report on the Excavations of the Universities of Chicago, California (Los Angeles) and Amsterdam in the Keban Reservoir, Eastern Anatolia, 1968-1970*, vol. 1. Amsterdam etc., pp. 223-257.

ZEIST, W. VAN & J.A.H. BAKKER-HEERES, 1982(1985). Archaeobotanical studies in the Levant. 1. Neolithic sites in the Damascus Basin: Aswad, Ghoraifé, Ramad. *Palaeohistoria* 24, pp. 165-256.

ZEIST, W. VAN & J.A.H. BAKKER-HEERES, 1984(1986)a. Archaeobotanical studies in the Levant. 2. Neolithic and Halaf levels at Ras Shamra. *Palaeohistoria* 26, pp. 151-170.

ZEIST, W. VAN & J.A.H. BAKKER-HEERES, 1984(1986)b. Archaeobotanical studies in the Levant. 3. Late-Palaeolithic Mureybit. *Palaeohistoria* 26, pp. 171-199.

ZEIST, W. VAN & A.M. BOEKSCHOTEN-VAN HELSDINGEN, 1991. Samen und Früchte aus Niederwil. In: H.T. Waterbolk & W. van Zeist (Hrsg.), *Niederwil, eine Siedlung der Pfyner Kultur. III, Naturwissenschaftliche Untersuchungen*. Bern etc., pp. 49-113.

ZEIST, W. VAN & S. BOTTEMA, 1977. Palynological investigations in western Iran. *Palaeohistoria* 19, pp. 19-85.

ZEIST, W. VAN & S. BOTTEMA, 1991. *Late Quaternary Vegetation of the Near East* (= Beihefte zum Tübinger Atlas des Vorderen Orients, Reihe A (Naturwissenschaften) Nr. 18). Wiesbaden.

ZEIST, W. VAN & H. BUITENHUIS, 1983. A palaeobotanical study of Neolithic Erbaba, Turkey. *Anatolica* 10, pp. 47-89.

ZEIST, W. VAN, P.E.L. SMITH, R.M. PALFENIER-VEGTER, M. SUWIJN & W.A. CASPARIE, 1984(1986). An archaeobotanical study of Ganj Dareh Tepe, Iran. *Palaeohistoria* 26, pp. 201-224.

ZEIST, W. VAN & W. WATERBOLK-VAN ROOIJEN, 1985. The palaeobotany of Tell Bouqras, eastern Syria. *Paléorient* 11/2, pp. 131-147.

ZOHARY, M., 1972. *Flora Palaestina*. Vol. 2. Jerusalem.

ZOHARY, M., 1973. *Geobotanical foundations of the Middle East*. 2 vols. Stuttgart etc.

ZOHARY, D. & M. HOPF, 1973. Domestication of pulses in the Old World. *Science* 182, p. 4.

ZOHARY, D. & M. HOPF, 1988. *Domestication of plants in the Old World*. Oxford.

A RE-ASSESSMENT OF THE *HUNEBEDDEN* O1, D30 AND D40: STRUCTURES AND FINDS

A.L. BRINDLEY & J.N. LANTING

Biologisch-Archaeologisch Instituut, Groningen, Netherlands

ABSTRACT: Re-investigations at three *hunebedden* excavated by A.E. van Giffen in 1918 are described and changes to the original conclusions given. O1 is shown to have had a minimum of seven pairs of side stones. The mounds of D40 and D30 are shown to have been constructed in several phases; the primary mounds did not completely cover the chambers. The finds have been re-analysed.

KEYWORDS: Netherlands, Neolithic, TRB culture, *hunebedden*, pottery, typochronology.

1. INTRODUCTION

In the summer and autumn of 1918, A.E. van Giffen investigated five *hunebedden* in Drenthe, namely D21 and D22 at Bronneger, D30 at Exlo, D40 at Emmen, and D53 at Havelte, and the remains of a destroyed *hunebed* in Overijssel, O1 on the estate De Eese, 6.5 km north of Steenwijk.

The excavations were made possible thanks to an 'unusually generous gift' from M. Onnes van Nijenrode, the owner of the De Eese estate (van Giffen, 1919: p. 110, *Verslag PMD Assen over 1918*, p. 7). As labourers, van Giffen used a number of Belgian internees, of whom A. van Dinter and J. Verdonckt acted as supervisors and draughtsmen. Neither van Giffen nor his fieldteam had any experience of monuments on sand. It is therefore not surprising that the weakly developed soils in and beneath the mounds of the excavated *hunebedden* were either not recognized or only recognized in a few places. The fill of the burial chambers and the contents of the large extraction pit where the chamber of O1 had stood formerly were only dug over for finds with a spade. Sieving of chamber fills was not introduced until the 1960s, during the excavations by J.A. Bakker of D26 and by J.N. Lanting of G2. There is no doubt that the limited number of small artefacts such as transverse arrowheads and amber and jet beads amongst the finds of the 1918 excavations was due to the method of excavation employed by van Giffen. Most of the very small finds must have been missed during the digging of the stony and gritty chamber fills. Many of the smaller sherds and pieces of flint must have escaped discovery as well.

This does not mean that the results of the 1918 excavations have no value; on the contrary. In a number of cases, van Giffen left parts of the mounds intact in 1918 so that further research and re-interpretation of the original data is still possible. He did this at D30 and D40

amongst others, and at O1 where in September 1985, October 1987 and September 1985 respectively several of his cuttings were re-opened and the profiles inspected again. Recently, finds from these three sites have been re-examined and new drawings prepared.

The details of the excavations and descriptions and drawings of the finds are presented together in this article. It is assumed that van Giffen's publications of D30 and D40 (van Giffen, 1925/27, II: pp. 207-230, pp. 165-207) and of O1 (van Giffen, 1924, resp. 1925/27, II: pp. 311-322, the latter however, without finds catalogue) are known to the reader. The original documentation of the 1918 excavations is housed in the B.A.I., and consists of field drawings, photographs and finds lists. Excavation notes were not made at the time. The finds from O1 and D40 are stored in the B.A.I., those from D30 in the Provinciaal Drents Museum in Assen.

2. THE DESTROYED *HUNEBED* O1

2.1. The site (fig. 1)

In July and August 1918 van Giffen excavated the remains of the destroyed *hunebed* O1 on the De Eese estate in the *gemeente* Steenwijkerwold (now *gemeente* Steenwijk). This was without doubt the *hunebed* drawn by Petrus Camper in 1781 (fig. 2) when it was still in a reasonably good state of preservation although the capstones had already been displaced (Camper, n.d.). At some stage during the first half of the nineteenth century, possibly during the 1840s, the burial chamber was demolished. Van Giffen's publication is not very satisfactory (van Giffen, 1924b; 1925/27, II: pp. 311-322) which is largely due to the fact that he had no experience whatsoever with *hunebedden* at that time. It would have been much better if

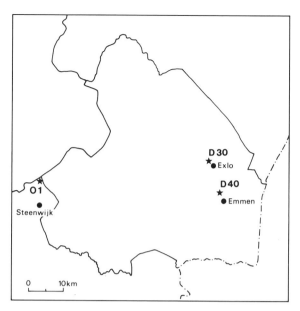

Fig. 1. Location of O1 near Steenwijk, D30 near Exlo and D40 near Emmen.

he had started with an undisturbed *hunebed* and then moved to O1, instead of the other way around as he might have developed a better understanding of the 'foundation pits', 'extraction pits' and the dimensions of chambers etc. He considered one part of the O1 mound to be an independant burial mound (No. II), apparently as a result of a mistaken interpretation of Camper's 1781 drawing. Camper's burial mound C is van Giffen's burial mound No. III, which lies 85 m to the north of the *hunebed* O1. His reconstructed groundplan is too short for a *hunebed* which according to Camper was 'fairly large', 'made of very large stones', and which according to his drawing, had at least 5 pairs of side stones.

During the excavation, van Giffen carefully mapped the remains of O1 and burial mound No. III in relation to the provincial border between Overijssel and Drenthe and actually noted down the distances to the border posts 5 and 6. In spite of this, in 1985 there was some uncertainty about the precise location of the monument. A *hunebed* and two burial mounds are shown immediately west of the border in the publication of the

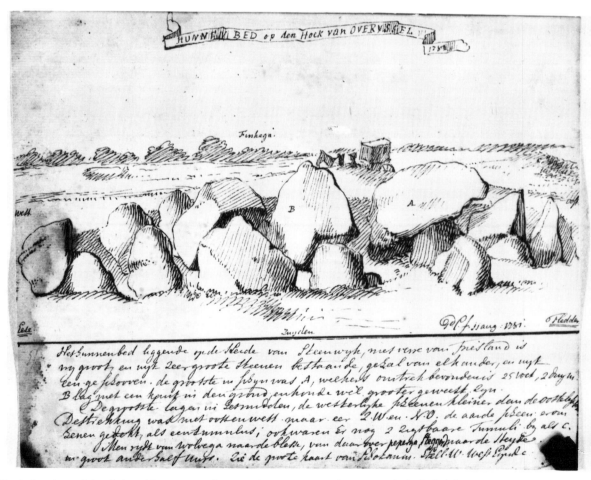

Fig. 2. Drawing and description of O1 by Petrus Camper. After a photocopy of the manuscript in the B.A.I.

excavation of a burial mound on the Drents part of the De Eese estate (Waterbolk, 1964: fig. 11). According to the list of protected monuments in the *gemeente* Steenwijk, the two burial mounds were protected, but the *hunebed* was not mentioned. There was also no visible trace on the ground of a prehistoric burial monument on the spot indicated although both the burial mounds were prominently visible. Further inspection made it clear that in fact the southern mound shown on Waterbolk's map was the mound of the *hunebed* O1 (i.e. van Giffen's mounds I and II), and the northern mound, van Giffen's mound III. The oval shape of the southern mound with the rectangular hollow in the centre already indicated this and it was confirmed by measurements in relation to the provincial border.

2.2. The 1918 excavation (fig. 3)

In advance of the excavation in 1918 a contour plan was made. The heights were taken relative to the base of the boundary post 6, and not corrected to heights NAP. It appears from this contour plan that two oblong mounds lay on both sides of the southwest-northeast oriented hollow in which the remains of the burial chamber had previously stood. Van Giffen began his excavation by putting a narrow northwest-southeast trial trench across both remaining higher areas and the hollow. This trench is shown on the published ground plan only by a dotted line marking its position on the west profile face (section face A-B). The outline of the cutting and the word *proefgrep* (trial trench) are still visible on the original field drawing of the contour plan.

After this, van Giffen excavated an extended cutting between the two hillocks. This cutting was apparently cleared in stages down to what he took to be undisturbed subsoil. In addition, he made two smaller cuttings in the northern rise which he named mound II. Numerous pits with recent filling were located in the large cutting, some of which did not reach undisturbed ground. A few of these pits were identified as the extraction pits of side or end stones, apparently on the basis of their stony

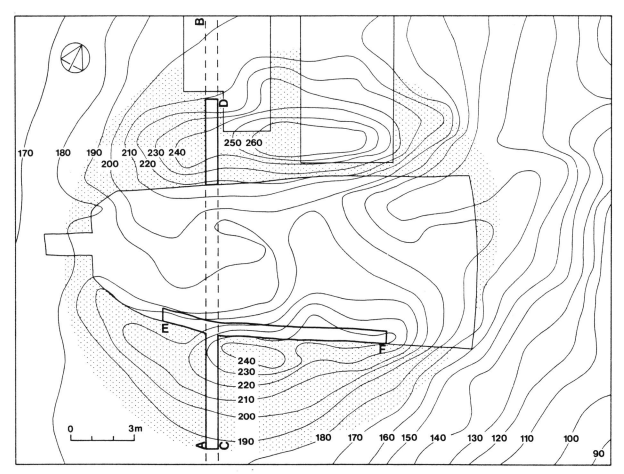

Fig. 3. Contour plan of the destroyed *hunebed* O1 in 1918. Elevations are relative to the base of borderpost 6 on the provincial border. The trial trench (interrupted line) and the three extended cuttings (thin solid line) of 1918 and the trenches of 1985 (thick line) are shown. The mound is shaded. The fig. 4 profiles are indicated by letters.

A.L. Brindley & J.N. Lanting

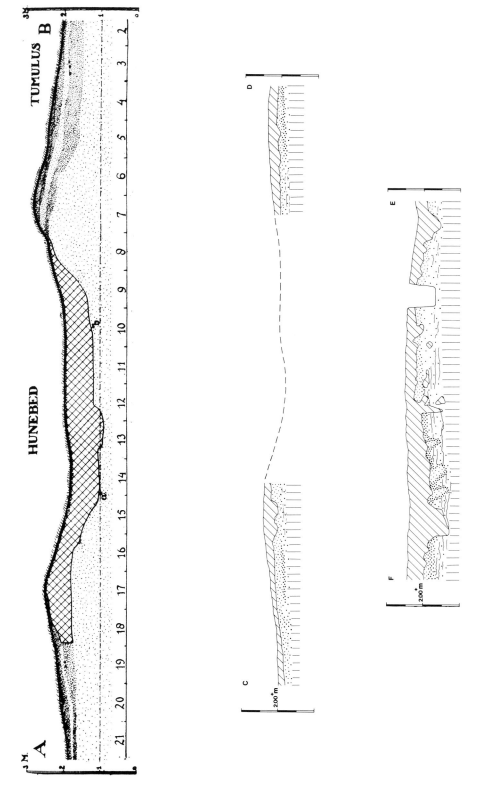

Fig. 4. Profiles through the mound, drawn in 1918 (A-B) and 1985 (C-D and E-F). The locations of these profiles are indicated in fig. 3.

filling. Other pits were marked as 'stone free' on the field drawings.

Other than this, the plans and section drawings of 1918 are not very informative.

2.3. The re-excavation of 1985 (fig. 3)

During the re-excavation in 1985, a trench about 10 metres long was excavated along the south side of van Giffen's large cutting. The profile was cleaned and drawn. The northwest-southeast oriented trial trench was also re-opened, for a length of 6 m in the southern rise, and for about 4 m in the northern one. This trench appeared to be about 0.5 m wide. The east profiles of these trenches were drawn. The levels were taken relative to the base of the border post 6, as in 1918.

Unfortunately the three profiles turned out be particularly uninformative because of the very strong secondary staining resulting from the development of the soil profile in the top of the mound. A strong humus infiltration had taken place under a thick leached horizon. A similar well-developed soil profile had been en-

countered by Waterbolk in 1956 in the burial mound in the Drenthe part of the estate, about 700 m north of O1 (Waterbolk, 1964). The original ground level and the structure of the O1 mound were no longer recognizable due to the strong discolouration which had occurred.

During the cleaning of the profile sides, several big pieces of a large bowl (No. 1) were found in undisturbed ground 2.0 m above the level of the base of the boundary post 6, at the point where the eastern profile of van Giffen's trial trench through the southern rise had been cut by the long south side of the large cutting. A small undiagnostic sherd (No. 50a), was found at a level of 1.70+ m, in the profile along the long south side of the cutting.

2.4. The reconstruction of the mound and the burial chamber

From the contour plan which van Giffen had made in 1918, and taking into consideration the location of O1 on a low elevation, the shape and height of the mound can be reconstructed without difficulty, that is, at least

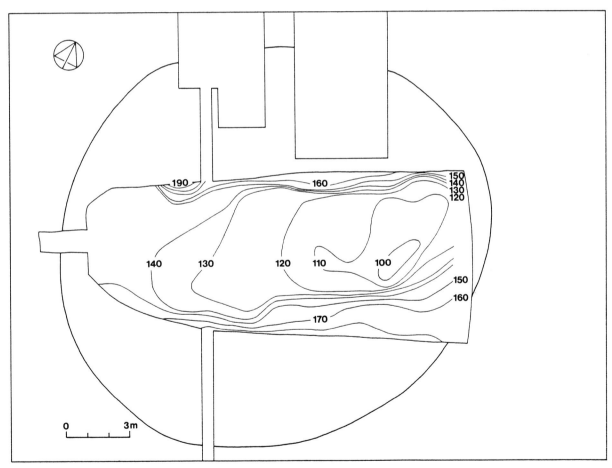

Fig. 5. Contour plan of the subsoil of van Giffen's main cutting, based on heights noted on the 1918 field drawing. The edge of the foundation pit of the megalithic structure is clearly visible. The outline of the mound is also indicated.

the mound which existed after the demolition of the hunebed (fig. 3). It can be assumed that the original ground level occurred at a depth of about 1.80 m on the south side and about 2.0 m on the north side, based on the pattern of the contours outside the mound. The mound itself was oval, oriented southwest-northeast, with dimensions of c. 21×16 m. The 1918 and 1985 profiles of the mound show that it had not been heightened by soil thrown out from the centre during the destruction of the burial chamber (fig. 4). The higher parts – 2.6 m on the north side and 2.4 m on the south side – indicate that it was 0.6 m high in these places. This means that the sherds of the previously mentioned bowl No. 1 lay in the body of the mound and that the small weathered sherd (No. 50a) lay in the old soil under the mound.

During the excavation of the central hollow, van Giffen apparently cleared out the soil, which had been dug over during the demolition of the chamber, down to the level at which the individual pits showed up in undisturbed subsoil. As a result, the base of the cutting was very irregular. A large number of levels were taken

to show this. Only a small number of these were reproduced on the published ground plan on which van Giffen used a stippled line in an attempt to indicate how the ground rose up towards the sides of the cutting. This was not very successful. However, using the many levels recorded on the field drawing it is possible to make a contour map that shows the relief quite clearly (fig. 5). In plan, the outline of a 4.5 m wide depression is visible. This appears to have a rounded end at the northeast but, unfortunately, no levels were taken for the last 1.5 m of the cutting. The depression has no clear end at the southwest. Most of this feature must have been the foundation pit, i.e. the pit which was dug by the builders of the burial chamber and in which it was constructed. The disappearance of a clear edge at the southwest end is probably due to the radical destruction of the chamber. The length of the foundation pit appears to have been at least 14 m to judge from the preserved straight edges of the southwest half. There are however, strong indications that the foundation pit had a length of at least 17 m (see below).

The side and end stones of *hunebedden* usually stand

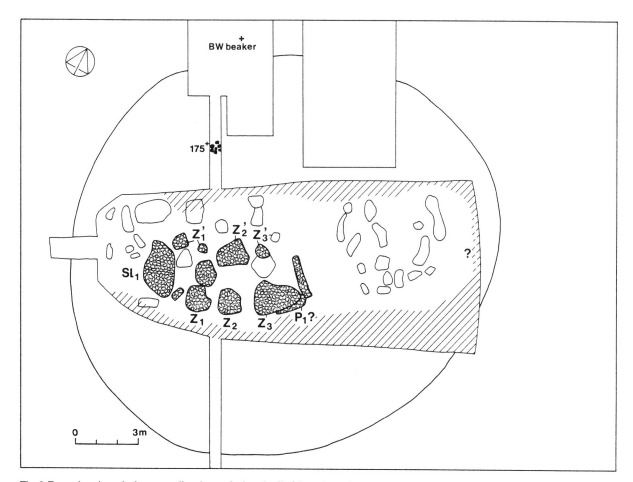

Fig. 6. Extraction pits and other recent disturbances in the subsoil of the main cutting. Redrawn after van Giffen (1925/27, Atlas: pl. 149). Pits with stony filling are shaded. Letters and numbers refer to the new identification of the extraction pits. The outline of the mound and the foundation pit are indicated.

on a base of field stones and sand in individual pits dug into the the base of the construction pit. Extraction pits, i.e. holes with a modern fill where end and side stones had formerly stood, are often still found during the excavation of destroyed *hunebedden*, if the subsoil has not been dug away to any great depth. A large number of pits in van Giffen's cutting were drawn of which several were identified as extraction pits apparently on the basis of their stony filling (fig. 6). According to van Giffen, these extraction pits belonged to a small *hunebed* of five pairs of side stones with external dimensions of c. 8x3.5 m.

The contour plan of the excavated cutting shows clearly that the subsoil in the eastern half of the foundation pit was dug out to a greater depth than in the western half. The extraction pits of the side and end stones must have been dug away as well. The features which van Giffen described as extraction pits only indicate the western half of the burial chamber, or at least in part because several pits do not fit into the picture. It is suggested here that Z1 and Z2 of van Giffen's ground plan together represent the remains of Z1, Z3 and Z4 should be adjusted to the numbers Z2 and Z3. On the south side, it is suggested that Z1' be disregarded and Z2', Z3' and, Z4'/Z5' be renumbered Z1', Z2' and, Z3'. The socket of a western portal stone is perhaps also included in van Giffen's Z4'/Z5' extraction pit. Van Giffen's extraction pit S12 may have been caused by the digging out of one of the capstones, just as pit D2. To judge by the position of S11, the length of the foundation pit must have been c. 17 m. Apparently, less than half the ground plan survived, represented by the extraction pits of one end stone, 3 pairs of side stones and possibly one portal stone. Neither the fieldstone floor of the burial chamber nor the stone packing around it had survived.

It is clear that van Giffen's reconstruction of O1 is not supported by the size of the mound, the length of the foundation pit and drawings and descriptions made by Camper. Camper's drawing shows a *hunebed* with certainly 5 but possibly 6 or 7 pairs of side stones, which according to him were *zeer groot* (very large). From this it may be taken that O1 was comparable to those Drenthe *hunebedden* which were built of very large field boulders. *Hunebedden* with 5 pairs of large side stones can be expected to have an external length of 8.5-10 m, 6 pairs as having one of 10-12 m and 7 pairs, 12-15 m. The length of the associated foundation pits can be estimated at 10-12 m, 12-14 m, and 14-17 m respectively. With a length of c. 17 m for the foundation pit, O1 could have had 7 pairs of side stones according to these calculations. Camper's drawing does not exclude this possibility. The associated length of the burial chamber proper must have been as much as 14-15 m, clearly longer therefore than the 8 m which van Giffen allowed for at O1.

Hunebedden with 7 pairs of side stones may have a stone kerb around the base of the mound. However, there were no indications of a kerb found in the excavation cutting of 1918. A small pile of fieldstones were discovered in situ during the re-opening of the trial cutting suggesting that a kerb may have been present (fig. 6).

2.5. The finds (fig. 7)

2.5.1. *The distribution of the finds*

During the 1918 excavation a surprisingly small number of finds were recovered. This is partly the natural result of the very complete destruction of the burial chamber. On the other hand, one must also consider that during the excavation insufficient attention was paid to the recovery of sherds etc. This is suggested by the fact that during the backfilling of the excavation trenches a small stone axe and various large sherds were found in the spoil heaps. As may be anticipated, the majority of the finds with known findspots came from the western half of the burial chamber. Only a few sherds came from the deeply dug away east part. Likewise, a few sherds came from the two excavated cuttings in the northern half of the mound, in part from the undisturbed subsoil according to the finds book (find number 14). During the 1985 re-excavation, several large sherds of a bowl (No. 1) were discovered in the undisturbed mound. Unfortunately it was no longer possible to discover if these sherds were deposited during the raising of the mound or belonged to an offering buried in the mound later.

Outside the edge of the mound sherds of a Barbed Wire pot of the Early Bronze Age were discovered. The sherds are apparently recorded twice in the finds book, under the numbers 2 and 29. Find number 29 does not appear on the field drawings. In the publication the pot has been given the find number 28 by mistake. The sherds apparently lay together at a depth of 0.6 m beneath the surface. No pit was visible, probably because of the strongly developed soil profile. The vessel is a beaker, not a large domestic/storage vessel. Because of its broken state, it is not clear whether it represents a burial gift, or sherds in a domestic pit.

The finds from O1 consist of a relatively small group of sherds (about 120), two stone axes, a flake from a polished flint axe, two flint flakes and an amber bead. With the exception of the sherds of bowl No. 1, two flint flakes (Nos 58 and 60), and a single undiagnostic sherd (No. 50a) which were found in 1985, all the finds stem from the 1918 excavation.

The pottery from the 1918 excavation was drawn on three occasions; a selection of sherds was drawn for the *Atlas der hunebedden* published in 1925/27; a smaller selection was drawn, possibly for a revised edition, but never published (archives of the B.A.I.); and in the 1940's, the draughtswoman J.C. Kat-van Hulten drew almost all the pottery for the B.A.I. finds register. The first two sets of drawings included two joining rim sherds of a funnel beaker with two zigzag lines below

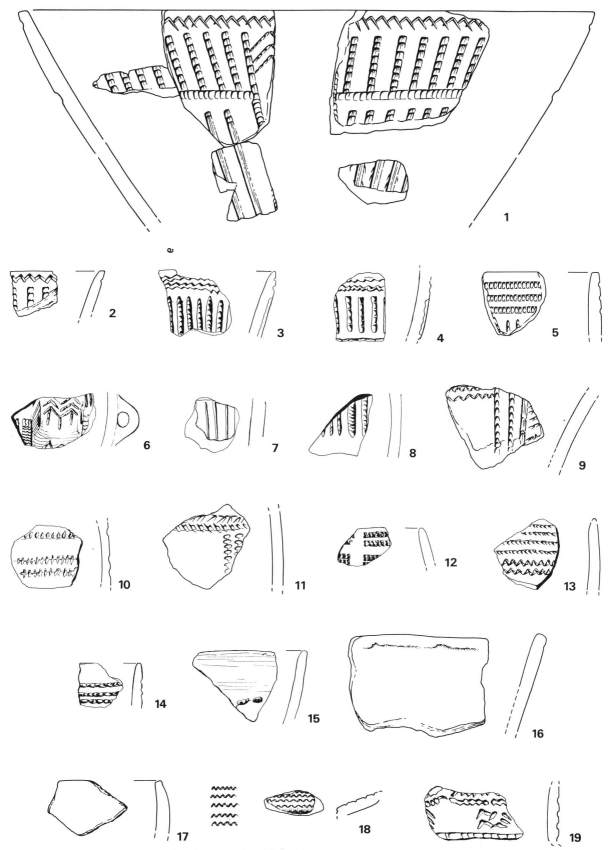

Fig. 7a-c. The finds from the destroyed *hunebed* O1. Scale: all finds 1:2.

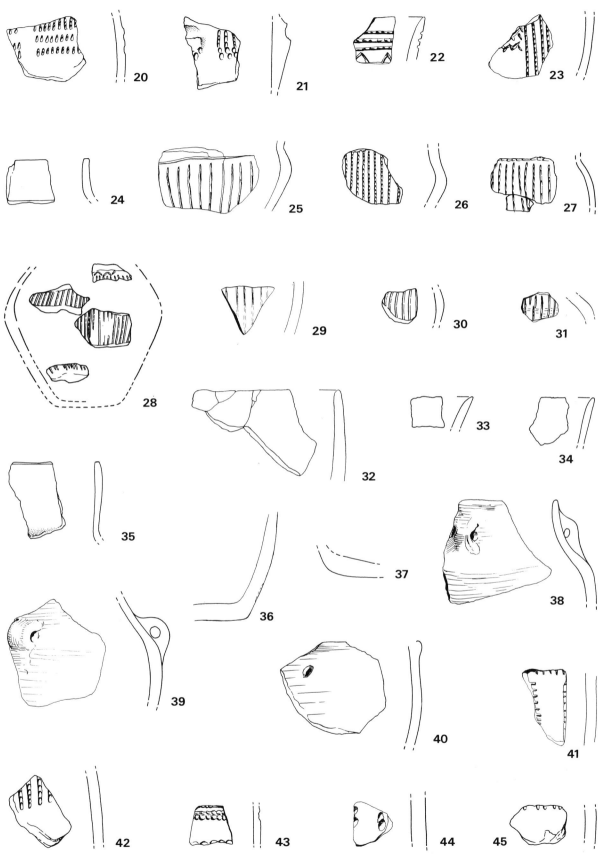

Fig. 7b.

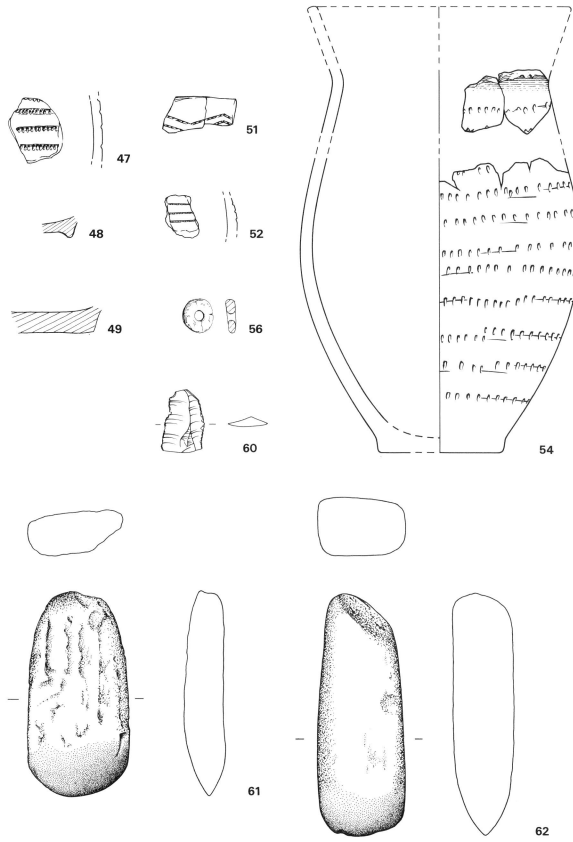

Fig. 7c.

the rim but these had apparently disappeared by the 1940's. The Kat-van Hulten drawings with some corrections and additional drawings are used in this publication.

Probably as a result of the very thorough demolition of the burial chamber which apparently involved the use of explosives, the majority of the sherds are small and in most cases each vessel is represented by a single sherd. The two notable exceptions are the bowl No. 1 and the Barbed Wire beaker (No. 54), both of which were found in undisturbed ground.

2.5.2. *Dating the construction and use of the chamber*

The majority of the finds belongs to the TRB use of the monument. Later material is represented by sherds of two or three beakers (Nos 52-54) and a post-medieval vessel (No. 55). It is unclear to what extent the surviving assemblage represents the original contents of the burial chamber. A *hunebed* with 7 pairs of side stones could have contained a relatively large assemblage of 300-400 vesssels (or more), considerably in excess of the number of sherds actually found at O1. The surviving identifiable vessels must form a relatively small proportion of the original contents, probably from a relatively restricted part of the chamber (the western part). Excluding the typologically insensitive material (funnel beakers, undecorated bowls and non-specific sherds), the following horizons (Brindley, 1986b) are represented:

Horizon 3: 7 pails and lugged bowls, 1 tureen (Nos 1-7 (possibly 8), 18);

Horizon 4: 4 bowls, 1 tureen (Nos 9-12, 19);

(Horizons 3 and 4: funnel beakers (Nos 25-36);

Horizon 5: 1 bowl, 4 tureen-amphorae, and one other vessel (Nos 13, 20-23, 14);

Horizon 6: 1 possible bowl (No. 15);

Horizon 7: 1 possible shouldered bowl (No. 24).

This suggests that the chamber was used chiefly during Horizon 3 and early Horizon 4, and during Horizon 5 but possibly not on a continuous basis. Discontinuous activity at *hunebedden* is not uncommon (e.g. G2 (Brindley, 1986a) appears to have been avoided during H6 and G1 (Bakker, 1982-83) during the latter part of H4, H5 and H6) and it may be that the surviving pottery is reflective if not wholly representative of the original pattern of use. As regards the date of construction of the tomb, the position of the large bowl No. 1 in the makeup of the original mound, whatever the precise details of its deposition, indicates that the mound cannot have been raised later than Horizon 3. The style of the bowl is similar to pails and bowls Nos 2-4, which are likely to represent the earliest material in the burial chamber.

Single vessels or very small numbers of sherds of beaker pottery are found quite frequently in *hunebedden* (i.e. Bell beaker at D9, D30, D40, G2, D54b/c; Single Grave pottery has been found at D9, D30, D32a, D54b/c, G5 etc.).

Conclusion: O1 was built no later than Horizon 3.

2.5.3. *Catalogue of finds*

In the catalogues the following terminolgy is used:

Complete: complete section of profile.

Almost complete: indication of base or rim, but actual feature missing

Incomplete: not reconstructible

Fragmentary: identification of type of vessel is evident

Restored: the pot has been conserved and the restoration covers some of the individual pottery and makes an independant assessment of the reconstruction impossible

H+number: Horizon assignation.

TRB pottery:

1. Bowl. Incomplete. Rim and body sherds of very large bowl with horizontally perforated lugs. *Tiefstich*. Upper zone, band of verticals with stacked 'M' motif over lugs. Horizontal line divides upper and lower zone. Lower zone with defined panels of vertical lines and possibly 'M' motif under lugs. H3;

2. Pail/bowl. Fragmentary. *Tiefstich*. 1 rim sherd, zigzag under rim, verticals below. H3;

3. Pail/bowl. Fragmentary. *Tiefstich*. 1 rim sherd, two small zigzags below rim, verticals below. H3;

4. Pail/bowl. Fragmentary. *Tiefstich*. 1 rim sherd, two small zigzags below rim, verticals. Horizontal line at base of upper zone. H3;

5. Pail/bowl. Fragmentary. *Tiefstich*. 1 rim sherd, 3 horizontal lines below rim, verticals and another element (?M over lug?). H3;

6. Pail. Fragmentary. *Tiefstich*. 1 body and lug sherd. 'M' motif on lug. Lower zone of defined panels with 'M' motif under lug. H3;

7. Pail/bowl. Fragmentary. *Tiefstich*. 1 body sherd, horizontal line separating upper zone and lower zone with verticals. H3;

8. Pail/bowl. Fragmentary. *Tiefstich*. 1 lower body sherd, showing verticals in panels. H3;

9. Pail/bowl. Fragmentary. *Tiefstich*. 1 body sherd with panels of chevron/'M' motif defined by three verticals, and separated by at least two small zigzag lines at the top of open spaces. H3;

10. Bowl. Fragmentary. *Tvaerstik*. Neck sherd of probable bowl (no indication of shoulder), with horizontal lines under the rim, narrow empty zone and blocks of at least two lines below. H4;

11. Bowl. Fragmentary. *Tvaerstik*. 1 body sherd, horizontal chevron (technique, opposed obliqe, pointed impressions superimposed on parallel grooved lines, see D40, pot No. 11 for comment) with spaced groups of vertical *tvaerstik*. H4;

12. Bowl. Fragmentary. *Tvaerstik*. 1 rim sherd, narrow blocks of horizontal lines. H4;

13. Bowl. Fragmentary. *Tiefstich*. 1 upper body sherd. Min. 4 horizontal lines below rim, two lines small zigzag. H5;

14. Bowl or tureen rim. Fragmentary. 1 rim sherd. *Tiefstich*. Min. 3 lines under rim. Position of lines and use of pointed *Tiefstich* suggest H5;

15. Fragmentary. 1 rim sherd of open shallow dish with some decoration. ?H6/H7;

16. Bowl. Large fragment of largish, open bowl with fairly straight sides;

17. Bowl. Rim of undecorated, slightly globular bowl;

18. Tureen. Fragmentary. 2 sherds: shoulder sherd with min. 4 concentric lines very small zigzags on shoulder and fifth line below shoulder; second very small sherd has grooved lines. H3?;

19. Tureen. Fragmentary. *Tiefstich*. 1 neck sherd, irregular small tight zigzag lines under rim, interrupted by inverted 'V' motif, stacked 'M' motif above horizontal line at base of neck. Late H3?;

20. Tureen-amphora. Fragmentary. *Tiefstich*. 1 body sherd, groups of vertical lines terminating in double line of small stabs (tear motif). H5;

21. Tureen-amphora. Fragmentary. *Tiefstich*. 1 shoulder sherd close to horizontally perforated lug, short band of verticals terminating in bone impresssions. H5;

22. Tureen-amphora or possibly small bowl. Fragmentary. *Tiefstich*. 1 rim sherd, 3 lines below rim, line of zigzag. H5;

23. Amphora type 2. Fragmentary. *Tiefstich*. 1 shoulder sherd with alternating groups of vertical lines and small zigzag. H5;

24. Shouldered bowl? Fragmentary. 1 small rim sherd. H7;

25. Funnel beaker. Fragmentary. 1 body sherd with high, fairly angular shoulder, horizontal line at base of neck and vertical *Tiefstich* on body;

26. Funnel beaker. Fragmentary. 1 body sherd with fairly angular shoulder and *Tiefstich* on body;

27. Funnel beaker. Fragmentary. 2 body sherds, rounded body, probably line at base of neck, *Tiefstich* on body;

29. Funnel beaker. Fragmentary. 1 body sherd with *Tiefstich* on body;

31. Funnel beaker. Fragmentary. 1 body sherd with *Tiefstich* on body;

28. Funnel beaker. Fragmentary. 4 small body sherds. Zigzag at base of neck and finely grooved lines on body;

30. Funnel beaker. Fragmentary. 1 body sherd. Finely grooved lines;

32. Funnel beaker. Incomplete. Rim and neck, undecorated;

33. Funnel beaker. Small fragment;

34. Funnel beaker. Small fragment;

35. Funnel beaker? Rim and neck fragment;

36. Funnel beaker. Incomplete. Base and part of side wall, undecorated;

37. Funnel beaker. Fragmentary. Small base;

38. Lugged vase. Incomplete. 1 large piece. Undecorated small vase with applied horizontally perforated lugs on the neck and a semi-angular shoulder. H5/6;

39. Lugged vessel. Incomplete. 1 shoulder sherd with lug. Undecorated. Horizontally perforated applied lug on semi-angular shoulder. H5/6;

40. Lugged vessel. Incomplete. 1 body sherd with small lug. Undecorated bowl with small horizontal unperforated and probably pinched up lug. Post firing perforation. ?H6/H7;

Miscellaneous:

41. Fragmentary. *Tvaerstik*. 1 body sherd, horizontal line above spaced verticals. H4;

42. Fragmentary. *Tiefstich*. 1 lower body sherd, vertical lines;

43. Fragmentary. *Tvaerstik*. 1 body sherd, possibly neck of tureen, horizontal lines with an empty band;

44. Fragmentary. *Tiefstich*. 1 body sherd, vertical lines, possibly pail;

45. Fragmentary. 1 body sherd, probably base of horizontal *Tiefstich* decoration;

46. Fragmentary. 1 very small coarse sherd with *Tiefstich*;

47. Fragmentary. 1 body sherd with horizontal *tvaerstik*. Probably bowl with lugs;

48. Base. Small portion of base with undecorated foot ring;

49. Base. Portion of flat base;

50. Sixty-three featureless body sherds;

51. Sherd shown in publication but no longer present and not included in inventory of B.A.I. stores. 2 lines shallow zigzag below rim of funnel beaker.

Non-TRB pottery:

52. Beaker. One featureless body sherd from largish, apparently undecorated vessel;

53. One sherd of a Bell beaker marked with the year and month of the De Eese excavation. However, the distinctive elements on the sherds (two shallow grooved lines and a line of horizontal finger nail impressions bordering an empty zone) indicate without doubt that this sherd (and possibly also No. 52) comes in fact from Havelte D53.

Fig. 8. Fragment of granite boulder with cylindrical bore hole, found in 1918. This shows that the boulders at O1 were blasted with black powder. Photograph: C.F.D. Scale 3:5.

54. Almost complete. 1 medium sized Barbed Wire beaker.

55. 10 sherds soft reddish fabric, glazed. At least two wheel thrown pots, probably *grapes*, are present. Of red fabric, with brownish-greenish glaze on the inner surface only. Most likely of late 18th – or early 19th century date.

Other finds:

56. Disc-shaped amber bead with central perforation;

57. Flint flake, ending in hinge fracture. No traces of working;

58. Irregular-shaped flake of flint with traces of cortex;

59. Flake from polished flint axe, no traces of working;

60. Distal end of blade of grey, transparent flint. Use retouch along edges. Found with large bowl (No. 1) in 1985;

61. Axe of fine crystalline stone (unidentified). Apparently made from piece of stone that had an axe-like shape, although the sides may have been shaped artificially. Only the cutting edge shows polishing;

62. Axe of fine crystalline stone (unidentified). Regular shape, but top broken obliquely. In side view, widest near top;

63. Fragment of granite with part of cylindrical bore hole for explosive powder. This indicates that at least one stone of O1 must have been blasted to pieces using black powder (fig. 8).

3. *HUNEBED* D40

3.1. The 1918 excavation

After van Giffen had investigated the destroyed *hunebed* O1 and seven burial mounds at De Eese, and subsequently the large *hunebed* of D53 and a small group of pyre mounds near Havelte, he excavated the small *hunebed* of D40 with its mound, on the Emmerveld (now Valtherbos), *gemeente* Emmen (fig. 1).

It is possible to reconstruct the sequence of events during the excavations to a large extent, using the field drawings and photographs. It is even possible to make some corrections to the location of the profiles as published. It appears that the profiles E and G lay 1 m further east than is indicated. The excavation began with the preparation of a contour plan of the mound and

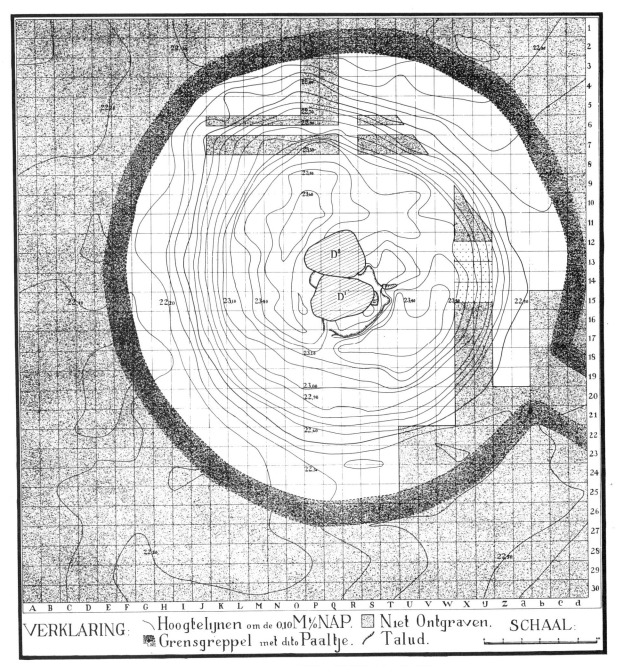

Fig. 9. Contour plan of *hunebed* D40 and surrounding area, after van Giffen (1925/27, Atlas: Pl. 127).

its immediate vicinity (fig. 9). Following this, two trenches were dug through the mound; a 1 m wide north-south trench which apparently was dug immediately down to undisturbed subsoil and which joined Sl2 in squares R2-11, and an east-west trench in front of the entrance to the chamber. This trench was initially 3 m wide but was reduced in width to 1 m about 0.5 m below the surface of the mound. At a depth of 0.75 m excavation was stopped although the subsoil had not yet been reached. The north-south trench is particularly poorly

documented; not even the profiles were drawn. It is possible that it was excavated earlier in the year to establish the potential of the site.

The mound was excavated in stages (fig. 10), beginning with the excavation by layer of the parts of squares O-T/10-17 which lay outside the chamber. Apparently van Giffen only realized after some time that he was digging away valuable data in the form of profiles which joined up with the chamber (van Giffen, 1925/27, II: p. 181). At this stage, it was still possible to

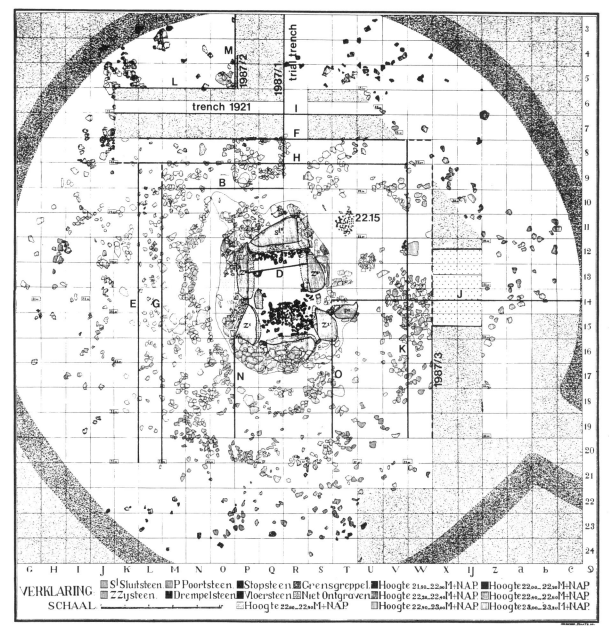

Fig. 10. Plan of the excavated area surrounding D40, after van Giffen (1925/27, Atlas: Pl. 129, with minor changes). The location of the profiles of 1918, 1921 and 1987 through the mound are indicated.

rectify this in part by drawing the profiles N, O and J. The gradual extention of the cutting made it possible to draw two parallel profiles (G and E) on the west side, three (B, H and F) on the north side, and two on the east side (K and the profile drawn only in outline between the squares W and X/15-19).

Following this, the edge of the mound was investigated, leaving on the north side a T-shaped piece of mound of which the profiles L and M were drawn, on the east side a broad baulk of which the profile on the west side was drawn only in outline, and on the southeast side several small pieces of the edge of the mound. A large number of field stones were found both in and under the mound. These were very carefully recorded in three dimensions (fig. 10). According to van Giffen these stones lay more-or-less concentric to the chamber on a slope, forming a sort of capping. In places, mainly on the northwest side of the mound, van Giffen came across fragments of stone in and on the podsolized soil of the mound. He interpreted this as an artificial surface deliberately laid as a path along which the capstones had been dragged.

Van Giffen claimed to have recognized two phases during which the mound was raised, on the first of which rested the stone capping and on the second, the 'path'. Both episodes should date therefore to the construction of the *hunebed*. The northwest side was subsequently raised again. Van Giffen did not exclude completely the possibility that this was a result of drifting sand.

There was a surprising number of finds, chiefly sherds of TRB pottery and flint but also sherds of a late Bell beaker in the body of the mound in ground which had not recently been disturbed. A large proportion of this material lay southeast of the entrance to the burial chamber, in squares V15-17, under the stone capping according to van Giffen (1925/27, II: p. 201).

The burial chamber was similarly investigated. Unexpectedly, the filling of the chamber consisted largely of recently silted in material. The floor was destroyed in places. The number of finds in the chamber was, not surprisingly, small. A large stone (1.0×0.85×0.4 m) was found in the modern filling which van Giffen identified as the missing portal stone P2. In other respects, the chamber was intact, although a substantial amount of the drystone infill between the side, end and cap stones was missing.

The chamber is oriented approximately north-south and consists of two pairs of side stones, two end stones and two capstones. Because the southern end stone is much shorter than the adjacent side stones, there is a gap between it and the capstone D1. The entrance lay in the centre of the east side and was originally flanked by two portal stones. The northern portal stone P2 was later removed and was found by van Giffen in the chamber. Apparently van Giffen left the stone there as the position of P2 is indicated now by a concrete marker.

A large portion of the stone packing was still intact around the chamber, although in places stones had been removed. The chamber and stone packing stood in a more-or-less oval foundation pit of 6.5×4.5 m. Undisturbed subsoil under the chamber lay about 0.30 m below the level of the old ground surface. According to van Giffen, the side and end stones stood in a trench which was 0.7 m deep.

3.2. Supplementary excavation in 1921

Apparently during the writing up of the excavation van Giffen ran into a problem relating to the construction of the mound. Therefore in 1921 it was decided to carry out a supplementary excavation. A trench was dug in the cross baulk of the T-shaped piece of the mound, of which profile I was drawn (fig. 10). It is not clear why van Giffen wanted to see a 'new' profile and did not just re-open one of the sections which he had already drawn.

3.3. The re-excavation of 1987

The discovery of what was apparently part of the original chamber contents in the mound and beneath the concentric stone capping clearly indicates that van Giffen's dating of the phases of construction of the mound cannot be correct. The chamber contents included sherds of a late Bell beaker. The clearing out of the chamber cannot therefore have taken place earlier than during the Late Neolithic. The stone capping can only have been laid down after that had taken place and subsequently the second heightening, on top of which lay the 'path' along which the capstones were supposed to have been dragged. Only van Giffen's first phase, the mound under the stone capping, can be the *hunebed*'s original mound. The other heightenings and, therefore, also the 'path' are clearly later.

In order to test this, it was decided to carry out a small re-excavation. This took place in October 1987. Three profiles were re-opened, namely a 8.5 m long piece of the western profile of van Giffen's north-south oriented trial cutting (profile 1987/1), a c. 1 m long piece of profile M (1987/2) and a 3.5 m long part of the profile which had only been drawn in outline, on the edge of squares W and X (1987/3). These profiles were drawn and photographed and soil samples for pollen analysis collected. Samples for dating purposes were not found.

The re-excavation showed that the mound had been constructed in three clearly recognizable phases, separated by well-developed soil horizons. A soil horizon is also clearly visible under the mound. Furthermore, it appears that the stones in the mound did not lie on the slope of the primary mound, but on the slopes of periods 1 and 2. An analysis of the orginal three dimensional records had already shown that the stones lay at different levels.

The height NAP was not remeasured in 1987. The levels were taken relative to the top of the portal stone P1 whose height in metres NAP had already been established in 1918. There are strong indications, however, that the levels of 1918 are about 1 metre too high.

3.4. The construction of the mound on the basis of the evidence from 1918, 1921 and 1987 (fig. 11)

It appeared in 1987 that the old ground surface below the primary mound was clearly visible in profiles 1987/1 and 1987/3 as a 0.10 m thick light grey coloured layer, with some local secondary infiltration veins. Van Giffen's observation that the old ground surface had been desodded because no humic layer was visible is therefore incorrect (van Giffen, 1925/27, II: p. 179). The old ground surface was not recognizable in profile 1987/2 because it lay too near the edge of the mound. This old ground surface was observed and drawn in 1918 in profile B, just north of the chamber, as a thin grey layer. In profiles H, squares O-V, and K, squares 15-19, it was drawn as the lowest, thick band of

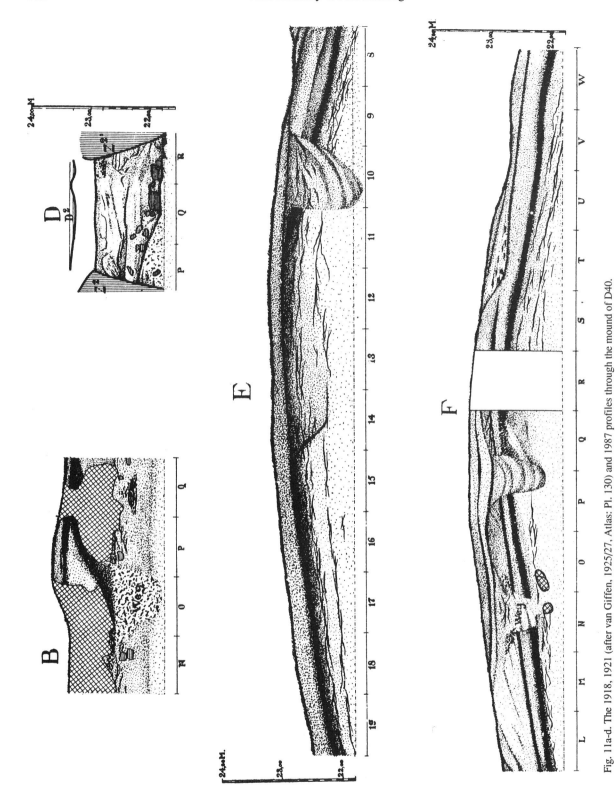

Fig. 11a-d. The 1918, 1921 (after van Giffen, 1925/27, Atlas: Pl. 130) and 1987 profiles through the mound of D40.

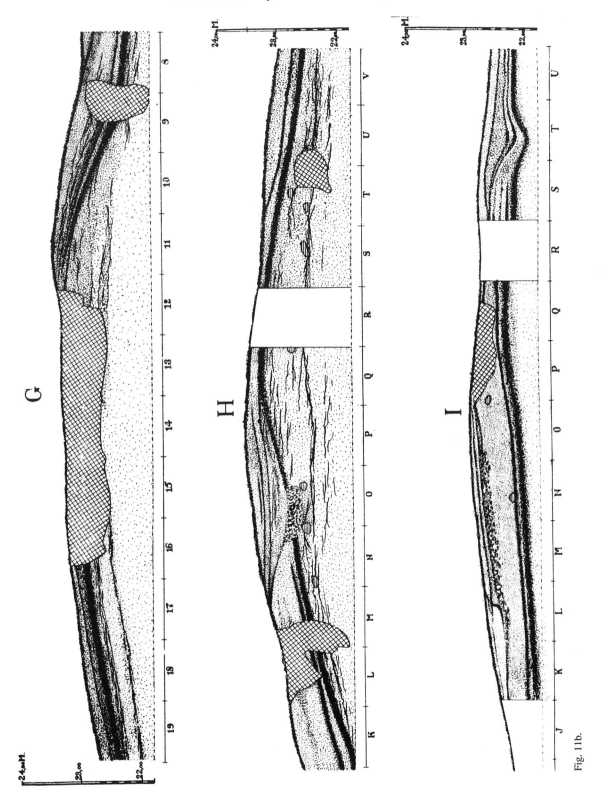

Fig. 11b.

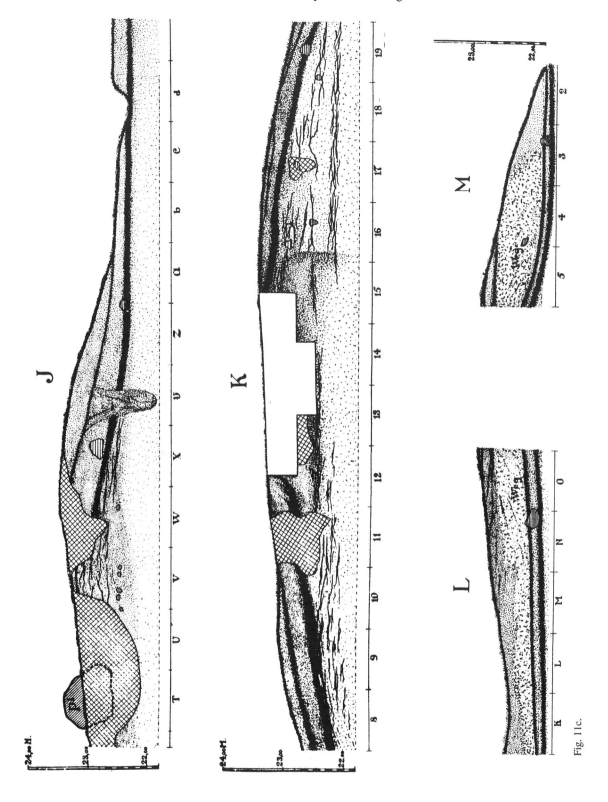

Fig. 11c.

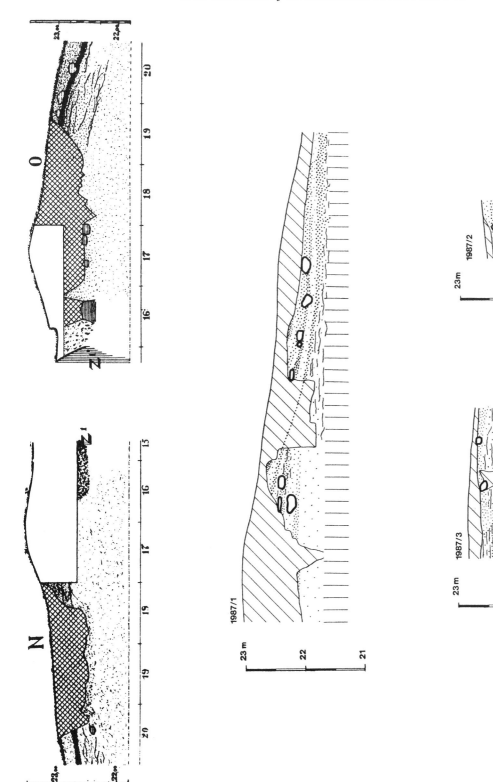

Fig. 11d.

infiltration veins. The old ground surface under the mound occurred at c. 22.20 m +NAP.

The primary mound was constructed of yellow sand. In its upper part a clear soil development was present in the form of a dark grey humus layer, c. 0.15 m thick and with locally strongly developed veins. This soil is also visible in places in the 1918 profiles, for example in the profiles E, G, and F as the lowest, curved band of infiltration veins, in the profiles H and K as the lowest but one of the bands of infiltration veins and in profile B as the thin band of infiltation veins halfway up. Large and small stones occur scattered in and over this soil horizon.

Although the primary mound was only well documented in a couple of profiles, the dimensions and height can still be extrapolated. The mound extended to about 6 m outside the north and south sides of the chamber and therefore had a diameter of about 15 m. The mound may have been slightly oval (oriented north-south) and was 0.6 m high at the time that topsoil formed on its surface. It therefore did not reach the tops of the side and end stones, which stood about 1 m above the old ground surface (with the exception of S11). The fact that the top of the mound reached only c. 22.80 m+ appears to contradict the maximum height of 22.95 m+ given by van Giffen (1925/27, II: p. 178) for the stone packing around the chamber. After some checking, it was established that this level related to loose stones above the stone packing. The in situ stone packing did not extend above c. 22.75 m+.

As a corollory one must ask whether the small primary mound consisted of more than the sand which originated from the digging of the foundation pit.

The volume of the mound can be established without much difficulty as 30 to 35 m³. The contents of the foundation pit must have had a maximum volume of 15 m³ based on the dimensions given by van Giffen (see above). A large part of the mound's soil must therefore have been obtained from elsewhere.

The period 2 mound was made up of brownish sand, and was strongly veined. It had a well developed soil on its upper surface with a thin black humus layer, thick grey leached horizon and a thin brown horizon. It is easily recognizable in the 1918 and 1921 profiles. The edge of the mound is only visible in these profiles on the north side. On the south side, where the mound had not been raised further, the contour map can be used to establish the position of the edge of the period 2 mound. The period 2 mound appears to have had a diameter of about 19 m at the time of consolidation. By extrapolation, the mound at that time had a height of c. 1.5 m, through which only the tops of the capstones may have protruded.

The period 3 mound is eccentric to the periods 1 and 2 mound, and is limited more-or-less to the northwest side of the mound. This restricted heightening is visible on the contour plan (fig. 9). The profile 1987/2 shows that the mound consisted of yellow-grey sand with flecks of humus and above, yellow sand. On its upper surface a narrow grey leached horizon and a brown infiltration layer had developed. Van Giffen had seen the same soil horizon in 1921 in profile 1. Large field stones were found in various places in the period 3 mound. This eliminates the suggestion that this was a natural raising formed of blown sand as van Giffen (1925/27, II: p. 189) thought possible. It must have been added artificially. In the upper surface of the period 3 mound are the stones and stone fragments which according to van Giffen belonged to the cobbled track along which the capstones had been dragged.

3.5. The finds (fig. 12)

3.5.1. *The distribution of the finds and the dating of the phases*

Van Giffen established that the chamber was for the most part filled with recently deposited/silted sand, and that the floor had been partially destroyed. Only where the floor was still intact were some of the original contents of the chamber present. The remainder apparently lay in and under the mound, where, above all in squares V16/17, artefacts were recovered. These occurred quite deep down, just above the old ground surface. Together with the finds in these squares were also the sherds of a late Bell beaker. It appears that a large part of the chamber contents were deposited in a pit at the edge of the primary mound, in squares V16/17. The depth at which other sherds were found suggests very strongly that they lay on the slope of this mound. The good condition of all the sherds also suggests that they were covered over immediately by the period 2 mound. It is likely that a clearing of the chamber took place before its re-use and the associated raising of the mound. Because of the presence of the Bell beaker sherds, this must have taken place at the earliest at the end of the Neolithic, or more probably in the Early Bronze Age or the beginning of the Middle Bronze Age. Much later is not likely because of the absence of a podsol profile under the period 2 mound.

In addition to the chamber contents there appear to have been several separate deposits in the primary mound. Outside the chamber, behind end stone S11 and in the stony backfill, portion of a funnel beaker (No. 25) was found. It seems likely that this became incorporated in the backfill during the construction of the chamber rather than being deposited as a formal offering. About two metres north of the same end stone, and within the primary mound, the complete lower half of a funnel beaker (No. 30) was found. It is possible that the neck was inadvertently dug away during the 1918 excavations. No trace of a pit was recorded but stones are noted in the finds book. About two metres northwest of the same end stone, a shouldered bowl (No. 38) was found within the primary mound. Again, no pit was recognized.

The date of the period 3 mound is very unclear, it may belong to the Iron Age.

3.5.2. *Dating the construction and use of the chamber*

The finds consist of a small collection of pottery and three pieces of flint. 680 sherds of pottery were found during van Giffen's excavations but only a small number of these was discovered in the chamber. The majority were found in two locations, in a pit at the foot of the first period mound (find numbers 25-28) and in a patch of recently disturbed ground just outside the entrance (find number 21). In some cases, large portions of individual pots were recovered intact or more-or-less so. Some of the sherds are also large but the majority are medium to small in size. The vast majority of the pottery, representing about 60 vessels, belongs to the TRB family. Sherds of a *Wellenband* pot, a potbeaker, and an almost complete Bell beaker are also present. A complete *Kümmerkeramik* pot clearly marked 'Emmen' and with the year and month of the excavation has also been discovered in the collections of the B.A.I. Its relationship to the excavation is, however, unclear. Sub-recent material is represented by one sherd of glazed 16th/17th-century fabric and a spindle whorl of the same date. The apparent absence of small items such as arrowheads and amber beads may be due to the fact that sieves were not employed during the excavation.

TRB pottery
The TRB assemblage consists of an estimated original total of 60-80 pots. This figure is based on the number of reconstructible pots (thirty-eight) and the number of decorated sherds (fifteen) plus a notional twenty for undecorated and featureless vessels. With a few exceptions, the pots are represented by small numbers of sherds. Complete profiles of only two pots (Nos 2 and 25) are preserved although others are reconstructible. However, because of the regularity of the ornament, the reconstruction of the full decorative scheme is possible in many cases. All the TRB pottery can be identified to a single horizon, Horizon 3. The pottery without diagnostic features (the funnel beakers) can also be accomodated within this horizon. The assemblage consists of six pails (Nos 1-6), seven bowls (Nos 7-12, 42) plus lugs of two others (Nos 41B and C), six tureens (Nos 13-18) plus the lug of another arguably shouldered pot (No. 41A), and nineteen funnel beakers (Nos 19-37). There are also sherds of a weakly shouldered bowl (No. 38). The fifteen small, decorated but otherwise featureless sherds (No. 39) represent other vessels, probably funnel beakers but possibly other types as well. The assemblage does not appear to have included biberons, collared flasks, baking plates, type 1 amphorae or, surprisingly, undecorated bowls, all of which can occur in Horizon 3 assemblages.

The finds indicate that the tomb was built and used exclusively while pottery of Horizon 3 was in use. Aspects of the vessels themselves indicate that the tomb was constructed at the beginning of this horizon and continued in use throughout its duration. Most of the

parallels suggested for the small tureen No. 13 appear to belong to late Horizon 2 or early Horizon 3 contexts. The bowls Nos 9 and 12 and tureen No. 15 have elements which suggest the end of Horizon 3.

Conclusion: D40 was constructed at the beginning of Horizon 3 and remained in use for a period estimated as lasting about 100 years between 3300 and 3200 cal BC. It was not subsequently re-used during TRB times.

Non-TRB pottery
Finds of Bell beakers and other Late Neolithic/Early Bronze Age pottery, either complete or represented by small numbers of sherds are relatively frequent at *hunebedden*. The Bell beaker appears to have been the last deposit in the burial chamber as its condition (practically complete) and its association with the larger part of the redeposited chamber contents indicates. Sherds of *Wellenband* pottery have been found at a number of *hunebedden*, e.g. Ostenwalde 1 (Fansa, 1978) and Havelte D53 (van Giffen, 1925/27, Atlas: Pl. 154:72 and 80) as have potbeaker sherds e.g. Bronneger D21/22 (van Giffen, 1925/27, Atlas: Pl. 154:87 and 89) and Annen D9 (de Groot, 1988).

3.5.3. *The catalogue*

TRB pottery:
1. Pail. Incomplete. *Tiefstich* of three different types. 4 lugs in spaced pairs. Very regular upper and lower zone with defined panels including zippers. H3;
2. Pail. Complete profile. 4 lugs in spaced pairs. Pointed *Tiefstich*. Upper and lower zone and lower zone panels defined by grooved line. 2 horizontal lines immediately below horizontal groove forming horizontal element in lower panels. This last combination is not common but three examples occur at Gross Berssen (Nos. 15, 20 and 144, Schlicht, 1972) and at least one other example (although of not fully continuous lines) occurs at Drouwen D19 (Bakker & Luijten, 1990: pl. 2 d). H3;
3. Pail. Incomplete. Lugged (one present). *Pseudo-tvaerstik*; although *Tiefstich* lines are deeply indented, well-marked guideline is clearly visible. Verticals in upper zone, well-defined panels in lower zone. H3;
4. Pail. Fragmentary. *Tiefstich* and *tvaerstik*. Horizontal lines in lower panels. H3;
5. Pail/bowl. Fragmentary. Upper zone consists of band of verticals bordered by double line of small zigzag. H3;
6. Pail. Incomplete. Lower body consists of panels with horizontal lines, vertical lines and chevron/'M' motif. H3;
7. Lugged bowl. Fragmentary. Lower zone only, apparently consisting of defined panels with vertical lines, hatched strips and chevrons/'M' motif, layout similar to Pail No. 1. H3;
8. Lugged bowl. Fragmentary. Regular, neatly executed ornament. *Tiefstich* and *tvaerstik*. H3;
9. Lugged bowl. Fragmentary. *Tvaerstik*. 3 discontinuous horizontal lines below rim, band of verticals. H3;
10. Lugged bowl. Fragmentary. *Tvaerstik*. Two horizontal lines below rim, verticals below. H3;
11. Lugged bowl. Incomplete. Low, unperforated lugs and decorated footring. Variant of *tvaerstik*, consisting of regularly grooved line with dots superimposed on it. 3 horizontal lines below rim, band of verticals. Horizontal zipper. Lower zone, groups of verticals, 'M' motif below lugs. Bakker & Luijten (1990) have recently drawn attention to this technique and its apparently limited distribution. Other versions of the technique include the use of an obliquely impressed pointed implement (O1, No. 11, above) and a single

118   A.L. Brindley & J.N. Lanting

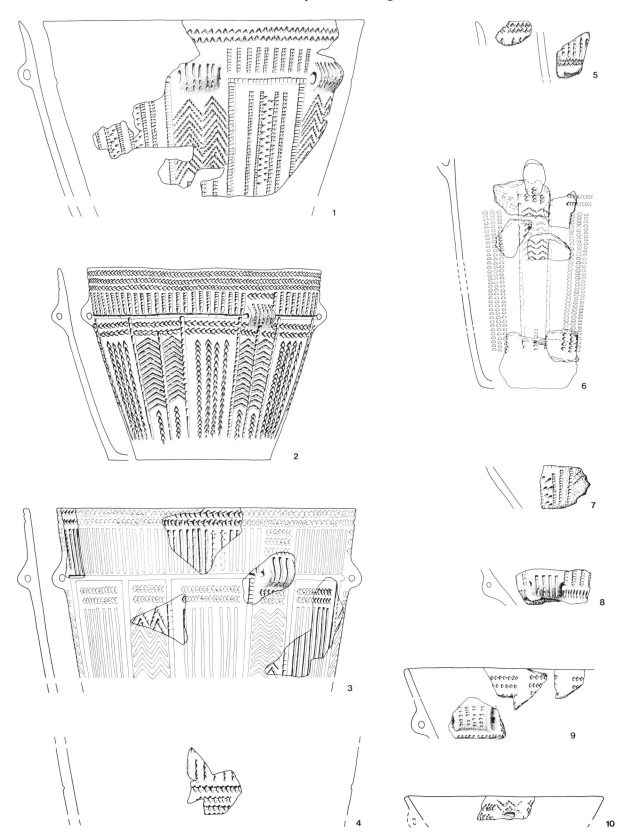

Fig. 12a-e. The finds from D40. Scales: pottery 1:3, other finds 1:2.

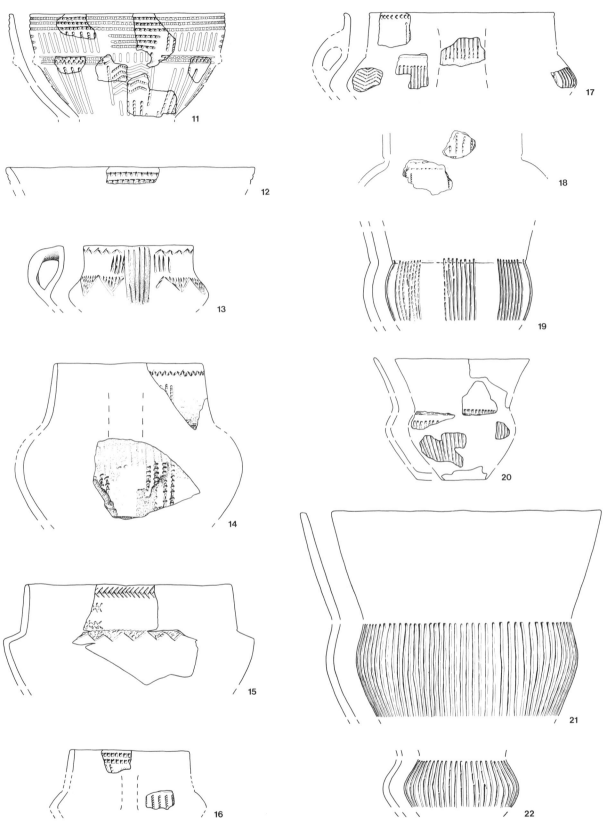

Fig. 12b.

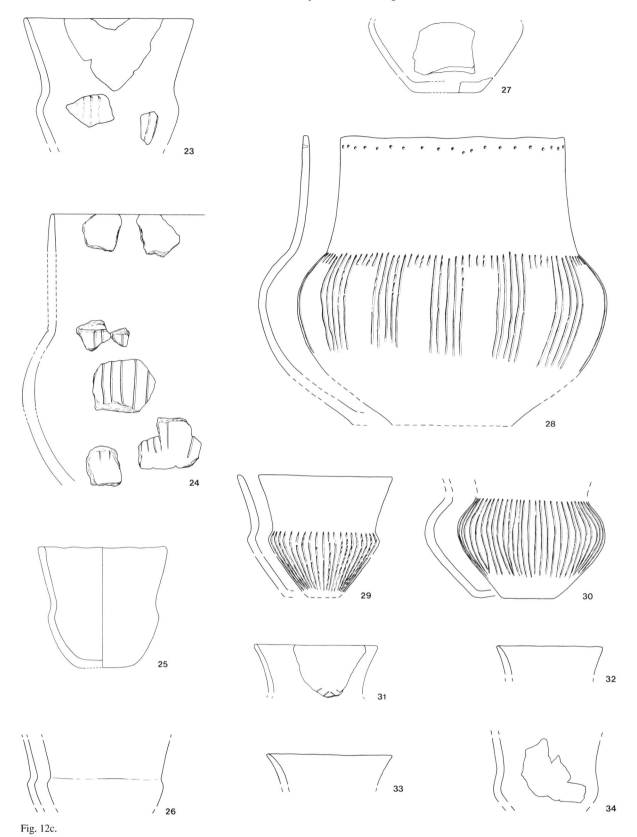

Fig. 12c.

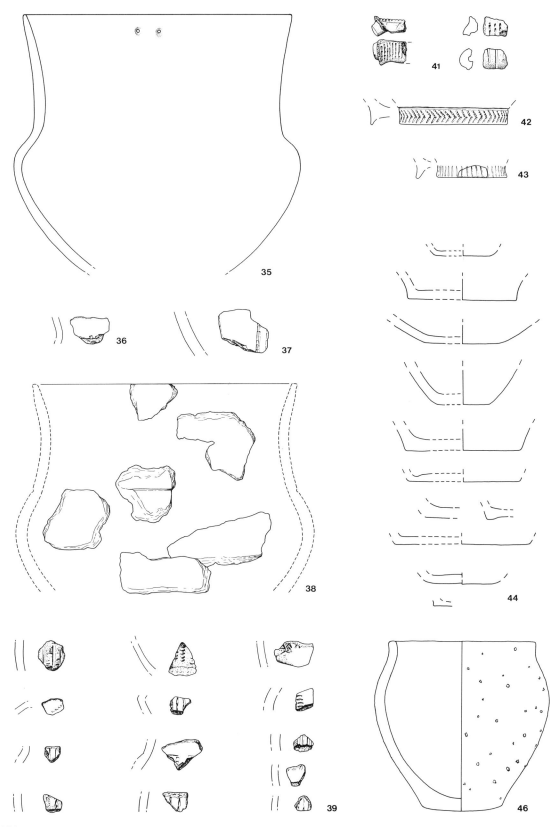

Fig. 12d.

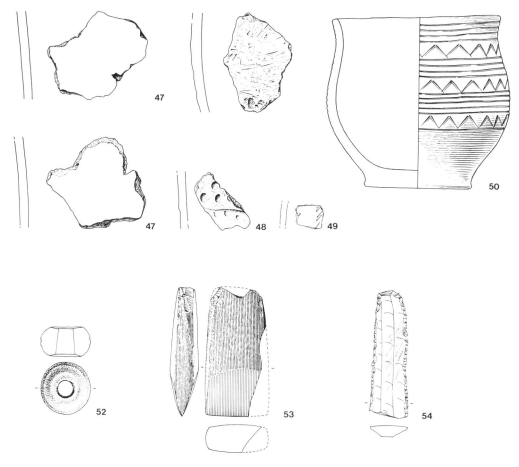

Fig. 12e.

grooved line with the rounded impressions on either side (e.g. Emmeln Nos 166 and 167, Schlicht, 1968; Bronneger, Knöll, 1959: Tafel 15, 14). H3;

12. Bowl. Fragmentary. Horizontal *tvaerstik*. H3.

In addition to the above, the decorated footring, No. 42, probably represents a finely decorated bowl, probably of Horizon 3.

13. Tureen. Incomplete. Zigzag below rim, filled triangles on shoulder and vertical incised lines on high, thick handle. True hatched triangles are uncommon on tureens and jugs in this area although they do occur on a small tureen from Tinaarlo which also has a high handle. Parallels for the combination of hatched triangles, high placed handle and small and slightly crude shape become more frequent as one moves eastwards (e.g. Kleinenkneten 1, Knöll, 1959: Tafel 2, 15; Kleinenkneten 2 (filled triangles), Knöll, 1959: Tafel 3, 3; Kleinenkneten, Fansa, 1982: Tafel 9, 7; Sögel, Kr. Aschendorf-Hümmling (reversed triangles), Knöll, 1959: Tafel 4, 8; Himmelpforten, Kr. Stade (Grave 5, hatched triangles), Knöll, 1959: Tafel 8, 15; Dötlingen, Kr. Oldenburg, Fansa, 1982: Tafel 25, 1962) and amongst the pottery of the Altmärk (e.g. Düsedau, Kr. Osterburg, Preuss, 1980: Tafel 8, 1 and 2 and Eichholz, Kr. Zerbst, Preuss, 1980: Tafel 48, 1). H3;

14. Tureen. Fragmentary. Rim and neck ornament suggests Horizon 3 tureen; lower sherd is unusually rounded for a tureen but has indications of a wide strap handle. H3;

15. Tureen. Fragmentary. Slightly cylindrical neck and short sharp shoulder. Single line of skating technique below rim, blocks of two lines of incised lozenge, possibly only in vicinity of handle. Filled

triangles on shoulder. Shape and general style of decoration is similar to tureen from D26 (Bloemers et al., 1981: p. 46). H3;

16. Tureen. Fragmentary. H3;

17. Tureen. Incomplete. Groups of verticals on neck, panels of verticals and zigzag on shoulder. Wide strap handle. H3;

18. Tureen. Fragmentary. Very friable fabric. H3;

19. Funnel beaker. Incomplete. Undecorated neck, horizontal line at base of neck, pendant groups of verticals on body;

20. Funnel beaker. Incomplete. Undecorated neck, line of stab marks at base of neck, fine grooves running over shoulder onto body;

21. Funnel beaker. Incomplete. Undecorated neck. Vertical lines on body;

22. Funnel beaker. Incomplete. Body, vertical lines;

23. Incomplete. 12 sherds. Undecorated, slightly flaring neck with faintly scored lines on body;

24. Funnel beaker. Incomplete. Undecorated neck, groups of grooved lines on body;

25. Funnel beaker. Complete profile, undecorated;

26. Funnel beaker. Incomplete. Faint line at base of vertical neck;

27. Funnel beaker. Incomplete. Undecorated;

28. Funnel beaker. Incomplete. Line of dots below rim, alternating groups of long and short lines on body;

29. Funnel beaker. Almost complete. Undecorated neck, vertical lines on body;

30. Funnel beaker. Incomplete. Almost complete body with vertical lines;

31. Funnel beaker. Incomplete. Very faintly scored double zigzag at base of neck;

32. Funnel beaker. Fragmentary;

33. Funnel beaker. Fragmentary;
34. Funnel beaker. Incomplete;
35. Funnel beaker. Incomplete. Large, thick-walled, with repair hole below rim;
36. Funnel beaker. Fragmentary;
37. Funnel beaker. Fragmentary.

Miscellaneous TRB pottery:
38. Shouldered bowl. Incomplete. Coarse friable fabric. Single line at base of neck. A fairly similar vessel in both shape and fabric is present at Exlo (No. 67, this paper) and another is known from Eext (1923/1 2k, van Giffen, 1944b: fig. 7);
39. Fifteen small decorated but otherwise featureless body sherds, probably representing fifteen different vessels, chiefly funnel beakers. H3;
40A-E. 5 thumb-sized rim sherds (not illustrated);
41A-D. 4 lugs. A, portion of horizontal strap handle with vertical *Tiefstich* lines. The slight but distinctive curve at the end of the lug suggests either it is half of a double vertical handle (cf. Emmeln, Schlicht, 1968: 21/Tafel7:7) or that it is a horizontal lug (cf. Emmeln, Schlicht, 1968: 3/Tafel4:4). Double handles tend to have more complex ornament than horizontal handles and to occur on tureens while horizontal handles occur on amphorae; B, portion of horizontally pierced lug with vertical *Tiefstich* lines; C, horizontally pierced lug with vertical grooves (not illustrated); D, fragment of horizontally pierced lug, very worn (not illustrated);
42. Base with footring. Decorated, probably H3;
43. Base with footring. Decorated;
44. Bases. Twenty-one sherds of twelve bases, eleven illustrated;
45. Body sherds. Small (thumb-sized and under), undecorated and featureless.

Non-TRB pottery:
46. *Kümmerkeramik*. Complete. There is some doubt as to the origin of this pot. It is not referred to in the report or shown in the photographs which, as it is more-or-less complete although undecorated, is surprising. However, it is clearly marked 'Emmen' with the year and month of the Emmen excavation;
47. *Wellenband* pot. Fragmentary. 5 undecorated featureless sherds with little curvature;
48. Pot beaker. Fragmentary. 1 sherd with finger tip rustication;
49. Fragmentary. 2 small sherds with finger nail impressions in vertical lines. The sherds are from a coil built vessel. Originally thought to be part of an EGK beaker, both the break pattern and the vertical layout of the impressions suggest that this identification is incorrect;
50. Bell beaker;
51. Sherd of wheel-thrown pottery, with dark brown glaze with abundant pale coloured small pits. Probably 16th/17th century (not illustrated);
52. Spindlewhorl with dark brown glaze. Probably 16th/17th century.

Flint:
53. Heavily damaged small axe of light grey flint whose form and dimensions are reconstructible. Part of top and large fragment of one of the sides missing. Polished on all sides. Originally longer, cutting edge shows evidence of intensive resharpening;
54. Very regularly shaped flint blade, proximal end missing. Steep retouch along both sides and around distal end. Light grey flint. Found in square K4, at the edge of the mound, depth 25 cm. Late Neolithic?;
55. Large flake of light grey coloured flint without traces of use (not illustrated).

4. *HUNEBED* D30

4.1. The 1918 excavations

After finishing the excavation of D40, van Giffen and his field team moved to the *hunebed* D30, northwest of Exlo, *gemeente* Odoorn, which was excavated between 23rd September and 10 October 1918 (fig. 1).

The excavation also began here with the preparation of a contour plan of the mound and its immediate vicinity (fig. 13). After this, the mound was almost completely excavated down to the undisturbed yellow subsoil. As a result of the experience he had gained at D40, van Giffen this time laid out several profiles more-or-less at right angles to and joining the chamber (fig. 14). Unfortunately, the drawings and photographs do not show clearly how he carried this out. It is certain that a wedge-shaped piece of the mound between profiles II and III was left until near the end of the excavation, but in the end, only the baulk with profile I was left standing. The 1918 profile drawings give little information about the construction of the mound. A soil horizon under the mound was not observed and van Giffen (1925/27, II: pp. 213-214) therefore concluded that the old ground surface had been desodded. The structure of the mound was equally poorly observed, apart from the heavy podsol profile in the upper part. Large numbers of field stones were discovered in and under the mound which, especially on the south side of the chamber, formed a cobbled surface on a slope. There is less recognizable coherence on the north side. It appears that the cobbling was present there originally but later was mostly disturbed. There is a surprisingly large quantity of stones in front of the entrance. All the stones were recorded three dimensionally. In van Giffen's opinion, the stones were a sort of cobbling (*plaveisel*) on the slope of the first period mound which was enlarged and heightened shortly afterwards (van Giffen, 1925/27, II: pp. 213-214).

The chamber of D30 is oriented NNW-SSE, and consists of four pairs of side stones, two end stones and now two capstones. The other two capstones were already missing in 1818 (van Giffen, 1925/27, II: pp. 208-209). The entrance in the middle of the south side was flanked by a pair of portal stones. At present there is a third stone of capstone dimensions in the chamber, resting partly on the ground and leaning against side stones Z3' and Z3. This is not an original capstone, but a stone dumped in the chamber by foresters. At the beginning of the 1918 excavations, the capstone D2 was lying in the chamber but was replaced, however, by van Giffen. A large stone of the same height was set between Z1' and Z2' to give this capstone extra support.

In 1918, the floor of fieldstones appeared to be fairly intact. The internal measurements of the chamber at floor level are 6.2×2.5 m with a depth of c. 0.7 m; the external measurements are c. 7.5×3.5 m. The chamber stands in a more-or-less oval foundation pit of 9×4.5 m. The stone packing around the chamber appeared in places to be intact in 1918. In one or two places, a few stones had been dug out by 'stone diggers'.

The fill of the chamber was excavated in horizontal spits. The stones encountered during this operation

A.L. Brindley & J.N. Lanting

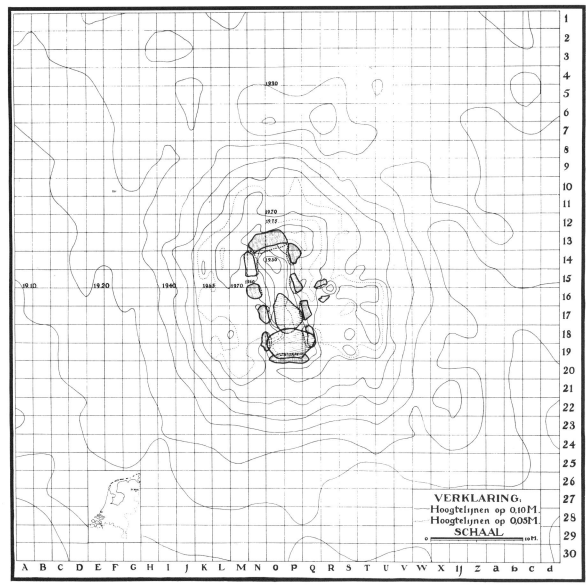

Fig. 13. Contour plan of *hunebed* D30 and surrounding area, after van Giffen (1925/27, Atlas: Pl. 134).

were, according to van Giffen, arranged in several man-made floors. These stones were also recorded three dimensionally.

4.2. The re-excavation of 1985

In 1985 profile I was re-opened, insofar as that was possible. Unfortunately, the former owner, the Province of Drenthe (D30 has since been transferred to the State; see *Jaarverslag Rijksdienst voor het Oudheidkundig Bodemonderzoek* 1990: p. 111), had erected a signpost on the only remaining intact part of the mound. As a result, part of the profile was not accessible.

The re-excavation showed that a clear soil was present under the mound, represented by c. 0.10 m

thick, light grey layer above an orange-yellow illuvial horizon (fig. 15). Fieldstones are present in places in the subsoil. The mound consists of orange-yellow sand with humic flecks. A light grey band, about 0.10 m thick, was visible in the body of the mound, rising up towards the chamber. This was apparently a soil which had formed on the slope of the primary mound. Unfortunately in 1985 it appeared that profile I was 20-25 cm lower in the vicinity of the chamber than it had been in 1918 as a result of erosion caused by the trampling of visitors. Furthermore, there was a shallow recently dug hole just above the band of humus in the body of the mound. As a result, the orange-yellow body of the mound was only just still visible above the rising light grey horizon.

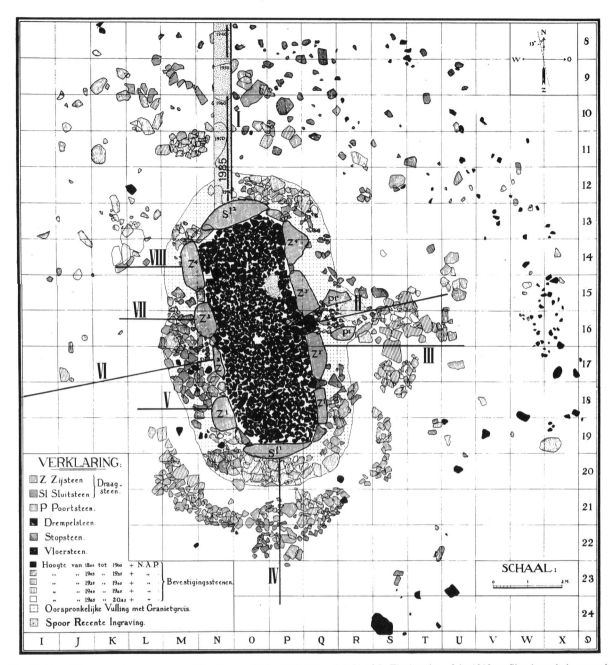

Fig. 14. Plan of the excavated area around D30, after van Giffen (1925/27, Atlas: Pl. 136). The location of the 1918 profiles through the mound and the redrawn profile of 1985 are indicated.

The height NAP was not remeasured in 1985. By comparing the levels taken in 1918, of the tops of the side, end and portal stones with those taken in 1985, the height NAP of the 1985 reference level could be established to an acceptable level of accuracy. There are strong indications that in this case the levels of 1918 are about 1 metre too low.

4.3. The reconstruction of the mound on the basis of the evidence from 1918 and 1985.

According to the contour plan of 1918, D30 was located on a slight rise (fig. 13). North of the mound the ground surface rises above 19.30; south of it, it is below 19.10 m +NAP. In 1985, the height of the old ground surface below the mound immediately north of S12 was recorded at c. 19.20 +NAP. This surface also rises towards the north.

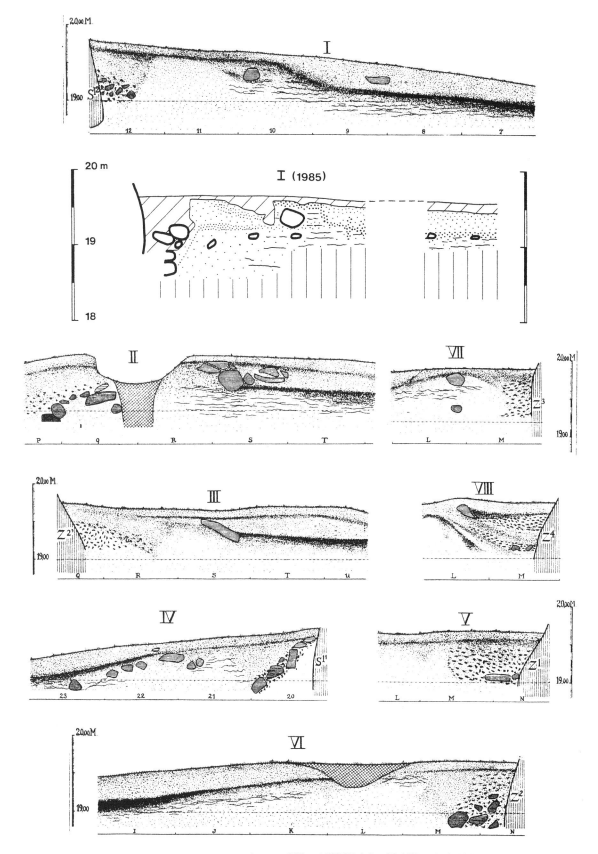

Fig. 15. The profiles through the mound of D30 of 1918, after van Giffen (1925/27, Atlas: Pl. 137) and of 1985.

As previously stated, in 1985 clear indications of a primary mound around the burial chamber were visible in profile I. In the upper part of this, a soil horizon is present which could be traced over a distance of one metre. The edge of the mound lies two metres north of Sl2. By extrapolation, it must have had a maximum height of 0.75 m by Sl2, i.e. of 19.95 +NAP. That means that a good part of Sl2 (top, 20.30 +NAP) must have protruded above the primary mound. The stone packing around the chamber which according to van Giffen reached to 19.55 +NAP must have been covered by this mound. The 1985 profile shows a large fieldstone lying on the old ground surface at the edge of the primary mound (fig. 15). This stone had already been recorded in 1918, in square O-10, just on the boundary with square O-11 (fig. 14). At that time only a small part of this stone projected from the baulk. According to the 1918 field drawings, this stone must be one of the stones which made up the cobbling on the north side of the mound and which apparently was still lying in situ. The edge of the cobbling on the south side of the chamber also lies about 2 m from the outer edge of the end stone, and according to profile IV, on a slope with a similar angle to that in profile I (fig. 15). Van Giffen's primary mound with cobbling therefore exists. According to the soil on its surface, this small primary mound must have existed for a longish period of time. The primary mound was oval in shape, c. 11x8 m, with cobbling around and on its edge.

During a later phase, the mound around the burial chamber was enlarged to an approximately circular mound with a diameter of c. 16 m. There is scarsely any question of heightening, because at the time that the podsol horizon was developing on its upper surface, the mound was still not much higher than c. 19.80 +, i.e. even lower than 19.95 + which was allowed to the period 1 mound. Probably the mound had already started to deteriorate and some of the top of the mond had silted into the chamber. Considering the type of soil profile in the upper part of the period 1 mound, the period 2 mound must have been raised at the latest in the Early Bronze Age.

The question remains to what extent soil for the construction of the period 1 mound must have been brought from elsewhere. The volume of the foundation pit is equal to c. 22 m³. The volume of sand in the period 1 mound can be calculated as c. 18 m³. Some of the sand from the foundation pit was replaced in the pit with the stones and the granite grit. It therefore appears that in this case the primary mound consists entirely of sand from the foundation pit.

4.4. The finds (fig. 16)

4.4.1. *The distribution of the finds*

The fill of the chamber of D30 appeared in 1918 to be relatively undisturbed. Most of the finds were discove-red in the chamber, in a relatively thin layer between van Giffen's fourth (i.e. lowest) layer of stones in the fill of the chamber and the actual floor. The four floors of stone in the chamber fill appear to be due to the manner by which the chamber was excavated, rather than being actual paved floors laid down the the users of the chamber. Without doubt the stones are for the most part drystone walling which fell into the chamber and which became imbedded in the sand and soil which spread in from the mound.

Both in the body of the mound and under it a few finds were discovered, which must be construction offerings. These are:

Finds number 8: sherds found in ground disturbed by the digging out of the stone packing, squares Q14/15, depth 19.20-19.30 +;

Finds number 44: sherds found in square R12, above, below and between stones, depth 19.00-19.15 +;

Finds number 45: sherds found under the stones, in square Q22, depth 19.05 +;

Finds number 46: sherds found between the stones, south of the entrance in square S17, depth 19.55 +.

In the description of some of the finds, van Giffen (1925/27, II: p. 227) gives incorrect depths for the finds numbers 44 and 45, as 18.50 and 18.65 +NAP respectively. This appears to be a simple mistake in calculation. During the excavation only relative depths were recorded. After a long list of finds from the chamber which all were found lower than the reference point, the three finds groups 44, 45 and 46 which were found higher than the reference level, occur. Van Giffen did not notice this and subtracted the recorded depths from the NAP depth of the reference point, instead of adding them.

The location of the finds numbers 44 (funnel beaker No. 52) and 45 (bowls Nos 1-3) is of particular interest. Both of these finds occurs at the edge of the primary mound. The original ground surface in square R12 occurred at a depth of c. 19.20 +. The sherds therefore apparently lay in a shallow pit, amongst some stones. The old ground surface must have been at about 19.10 + in square Q22. The finds lay more-or-less on the old ground surface.

It is unfortunate that the details of the finding of both these finds groups can not be more precisely reconstructed. The earliest pottery at the site is involved and establishing the relationship of the pottery to the burial chamber and the period 1 mound would be interesting. Considering the distribution of the finds groups 44 and 45 it does not seem unlikely that both groups were offerings on the edge of the primary mound, although it is possible that they had already been buried at the time the mound was being built.

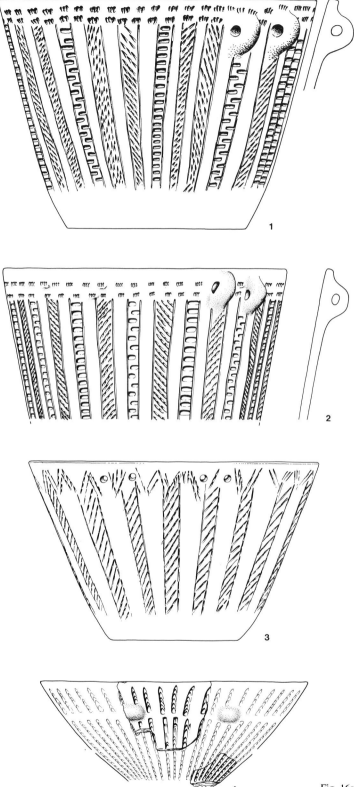

Fig. 16a-g. The finds from D30. Scales: pottery 1:3, flint 1:2.

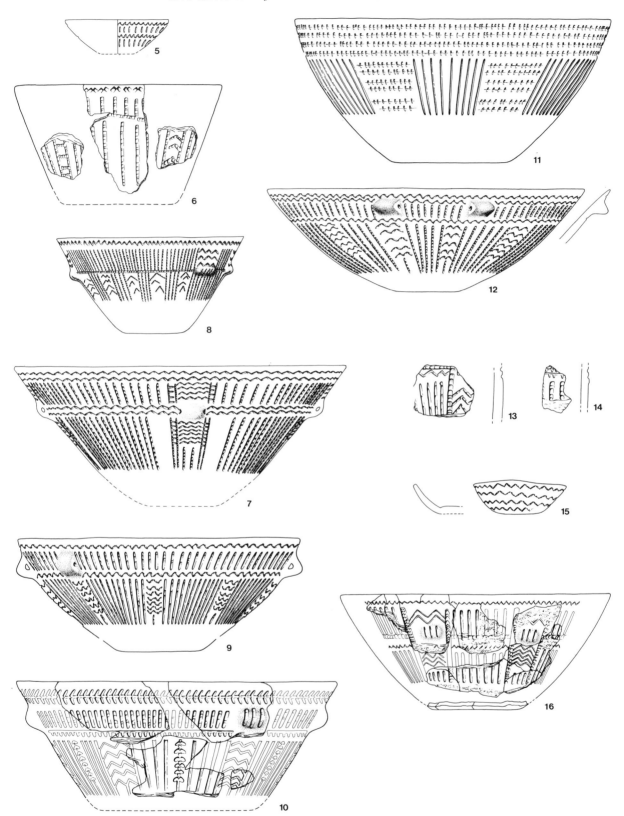

Fig. 16b.

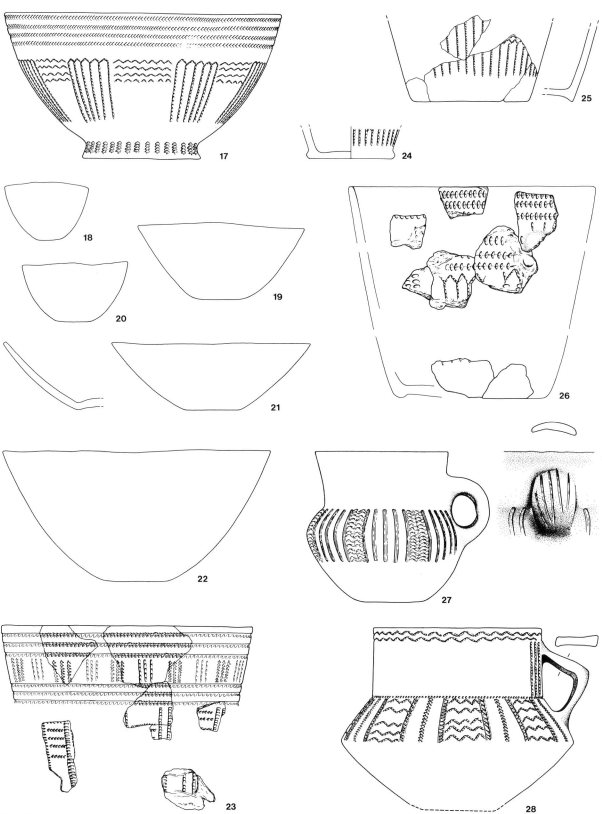

Fig. 16c.

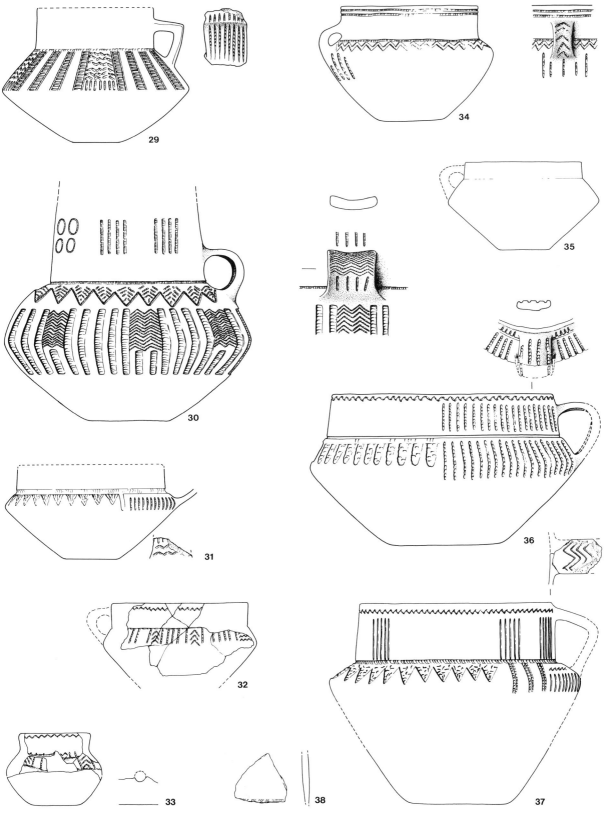

Fig. 16d.

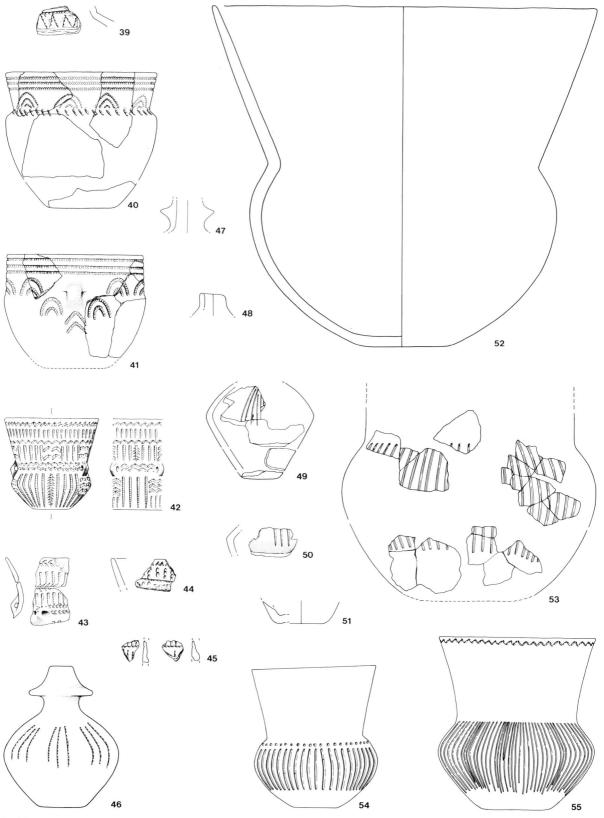

Fig. 16e.

Fig. 16f.

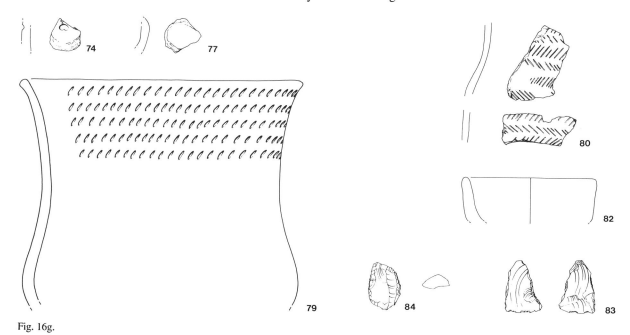

Fig. 16g.

4.4.2. *Dating the construction and use of the chamber by the finds*

The finds consist of 1500 sherds of pottery and a few pieces of flint. The vast majority of the pottery, representing about eighty vessels, belongs to the TRB. Sherds of three Single Grave beakers and what is possibly a *Kümmerkeramik* bowl are also present. Sherd sizes vary from relatively large fragments to very small. In some cases, large portions of individual vessels can be identified and joined. Although there is no evidence that the chamber was disturbed significantly at any time, it is not now possible to reconstruct the original position of individual vessels because a significant amount of the pottery is unnumbered and in some cases individual sherds bear more than one number.

The TRB Pottery:
About 80 vessels are recognizable; in addition, there are sherds which cannot be assigned to individual pots, of which some probably represent vessels not otherwise identified. Excluding the funnel beakers, collared flasks, undecorated bowls and miscellaneous vessels which may belong to several horizons, the pottery represents the following horizons (Brindley, 1986b):
  Horizon 1: three bowls (Nos 1-3), and possibly also two funnel beakers (Nos 52 and 53). Minimum total: 3-5. All this material was found outside the chamber;
  Horizon 2: one bowl (No. 4), three jugs (Nos 27-29), and possibly the funnel beaker No. 54 on the basis of its general similarity to jug No. 27. Minimum total: 4-5;
  Horizon 2/3: Jug No. 30, tureens Nos 31 and 34. Minimum total: 3;
  Horizon 3: Bowls Nos 5-10, 12-16, pails Nos 23, 26

(late), tureens Nos 32, 33, 36, 37 (39), lugged beakers Nos 42-45. Minimum total: 22;
  Horizon 4: Bowls No. 11, tureens Nos 40, 41. Minimum total: 3;
  Horizon 4/5: Bowl No. 17. Minimum total: 1.

The remainder of the pottery is consistent with this pattern of use. The open undecorated bowls Nos 19-22 probably represent Horizons 2 or 3; in form they are similar to the decorated bowls of these horizons rather than the more globular and rounded bowls of the later horizons. Undecorated bowls occur in several small Horizon 2 and 3 assemblages e.g. Zeijen flat grave E under tumulus II (Bakker, 1979: fig. B14) and Eext stone cist D13a (van Giffen, 1944b: fig. 7, pots 2g and f). The funnel beakers lack the large zigzags which although not closely datable, appear to occur chiefly with late 3 and Horizon 4 pottery. Likewise, none of the funnel beakers have the short, high and sharp shoulder which occurs commonly with large zigzag motifs. The shouldered bowl which bears a general resemblance to Horizon 7 shouldered bowls, is paralleled by a vessel from Eext stone cist D13a (van Giffen, 1944b: fig. 7, pot 2k) where the associations point to an early Horizon 3 date. The small perforated beaker No. 68 has its closest parallels at Bronneger D21/22. Although the context precludes a precise assignation, material of Horizons 1-3 is present.

In general, the Horizon 3 pottery appears to represent the earlier rather than the later aspect of this horizon.

It is unfortunately not clear whether the bowls Nos 1-3 were deposited prior to the construction of the *hunebed* or were deposited at its edge subsequently. The earliest identifiable material in the chamber belongs to

Horizon 2 and apparently marks the beginning of use of the chamber on a continuous if not necessarily regular basis as shown by the presence of pottery representing Horizons 2, 2/3 and 3. Subsequently, the frequency of deposition seems to have declined. Very little Horizon 4 material is present and the latest pottery appears to bowl No. 17. It has been suggested that Horizons 1 and 2 were of relatively short duration, each lasting approximately fifty years (Brindley, 1986b: pp. 104-105). It is possible that the *hunebed* was constructed at or around 4700 BP/3400 cal BC, on the basis of similarities between the bowls of Horizon 1 and products of the Fuchsberg Style of North Group TRB for which a number of radiocarbon dates are available. A slightly later date is suggested by the Horizon 2 pottery from the chamber. The time lapse is possibly only significant in terms of the pottery itself. In either case, D30 must be considered as one of the oldest *hunebedden* of the West Group. Pottery of Horizon 1 is also known from Bronneger D21, Emmen D43a (see Brindley, 1986b: fig. 3), Heveskesklooster G5 and the stone cist of Heveskesklooster (unpublished). *Hunebedden* which were built during Horizon 2 include Tinaarlo D6e/f (van Giffen, 1944a), Hooghalen D54b/c and Rijs F1 (van Giffen, 1924a). None of these *hunebedden* has a large quantity of early pottery. After Horizon 3, D30 was only sporadically used. The abandonment of *hunebedden* either permanently or for several hundred years is not uncommon.

The three Single Grave beakers Nos 79-81 and the crude bowl No. 82 represent later activity; the former can be dated to 4100-4000 BP/2700-2500 cal. BC and the later to somewhere between 3500-3000 BP/1900-1200 cal. BC. Finds of small numbers of Single Grave beakers occur quite regularly at *hunebedden*. In some instances where Horizon 7 pottery is also present, this material can be viewed as representing some form of continuity of practice. In this case, however, a gap of several hundred years appears to exist between the latest TRB use and the deposition of the Single Grave beakers.

4.4.3. *Catalogue*

1. Pail. Incomplete. 4 lugs in 2 pairs. 2 lines of maggot-shaped impressions below rim and continuous over lugs. Separate vertical strips (ladders, zippers, hatched diagonally and vertically) and horizontal maggot zippers. Each ladder or zipper strip alternates with two hatched strips in regular pattern. H1;

2. Pail. Incomplete. 2 pairs of lugs. 2 lines maggot impressions which do not run over lugs. Separate ladder, zipper and hatched strips in regular pattern. H1;

3. Bowl. Incomplete, restored. Very irregular zigzag ladder below rim. 2 horizontal sets of four perforations for 2 pairs of some form of lugs. Separate alternating diagonally hatched strips and possibly one ladder strip. Good parallels for the irregular ornament below the rim do not seem to be as easy to find as the line of impressions on Nos 1 and 2 but nevertheless, they appear to have been quite widespread. The motif occurs for instance on a Fuchsberg Style bowl from Flensburg illustrated by Schwabedissen (1979: Abb. 3, 1) and on a bowl from Samswegen (Preuss, 1982: Tafel 38, 10). H1;

4. Dish. Incomplete. Slightly curved profile. Low, perforated, undecorated lugs. *Tiefstich*, irregularly executed. Upper zone, undefined, broad pointed *Tiefstich* verticals. Lower zone: pointed *Tiefstich* verticals. H2;

5. Dish, miniature. Almost complete profile. *Tiefstich*. No lugs. H3;

6. Dish. Almost complete profile. Broad *Tiefstich*. Zigzag below rim. Upper zone of verticals separated from lower zone by horizontal *Tiefstich* line. Lower zone: chiefly verticals, but including at least one ladder and one vertical chevron, possibly below a lug. H3;

7. Dish. Incomplete/restored. *Tiefstich*. 4 equally spaced perforated lugs. Upper zone defined by double line of zigzag, with continuous band of verticals except above and below lugs where panel of zigzag defined by ladders is present. H3;

8. Complete profile. *Tiefstich*. 2 pairs lugs. *Tiefstich*. Zigzag below rim. Band of verticals except over lugs where replaced by four zigzags. Horizontal line separates lower zone with alternating groups of vertical lines, multiple 'M' motif and irregular pointed arches. H3

9. Dish. Almost complete profile. *Tiefstich*. 4 horizontal lugs, slightly curved walls. Upper zone: verticals defined by single zigzags. Lower zone: groups of vertical grooves and 'W' motif. H3;

10. Dish. Broad *Tiefstich* and *tvaerstik*. 4 unperforated lugs. Slightly curved wall. Coarse *tvaerstik* defining upper zone of vertical *Tiefstich*. Lower zone: alternating multiple 'M' motif and single *tvaerstik* lines repeating. The type of *tvaerstik* employed is reminiscent of the Horizon 2 type rather than the later type. H3;

11. Bowl. Almost complete profile. *Tvaerstik*. 4 continuous lines under rim, alternating groups of horizontal lines and vertical grooved lines below. H4;

12. Lugged bowl. Complete profile. *Tiefstich*. 4 horizontally perforated undecorated lugs close to rim. 2 zigzag lines below rim, upper zone of verticals separated by single zigzag line from lower zone of alternating groups of 'M' motif and verticals which do not relate to the position of the lugs. H3;

13. Dish. Fragmentary (1 sherd). Lower zone, defined panels with verticals below single zizag and panel of 'M' motif, probably below lug. H3;

14. Bowl. Fragmentary (1 sherd). 2 lines horizontal *tvaerstik* above vertical *Tiefstich*. Upper zone of bowl. H3;

15. Miniature bowl, base sherd missing. 3-4 lines of untidy *Tiefstich* zigzag continuous around body. H3;

16. Lugged bowl. *Tiefstich* and grooves. 4 unperforated lugs with incised lines. Zigzags below rim, band of vertical grooves and large 'M' motif over lugs. Lower zone: defined panels of 'M' motif and grooved lines. H3;

17. Bowl. Complete profile. Footring with vertical *tvaerstik*. 4 *tvaerstik* lines below rim. Alternating groups of 4 horizontal zigzags and 6 *Tiefstich* lines pendant from zigzag. Although the ornamental scheme belongs to the Anlo-Uddelermeer Style of Horizon 5, the use of *tvaerstik* is more characteristic of Horizon 4;

18. Small hand bowl. Complete profile. Neatly made;

19. Small bowl. Complete profile;

20. Bowl. Complete;

21. Bowl. Reconstructible;

22. Bowl. Complete profile;

23. Pail. Fragmentary. *Tiefstich* and *tvaerstik*. Three lines *Tiefstich* overlain by small half bone impressions defining upper zone of alternating 3 *Tiefstich* lines and 3 *tvaerstik* lines. Lower zone includes vertical and horizontal lines. H3;

24. Pail. Fragmentary (1 sherd). Complete base with close-set vertical *Tiefstich* lines and indications of panel of zigzag/chevron. The pinched out foot is reminiscent of the small Hooghalen pails;

25. Pail. Fragmentary, lower part only. Decoration of vertical *Tiefstich* with panels of zigzag extends close to base. H3;

26. Pail. Incomplete. *Tiefstich* and *tvaerstik*. Upper zone of *Tiefstich* verticals defined by three lines *tvaerstik*. Lower zone of wide panels defined by *tvaerstik* containing spaced groups of *Tiefstich* lines pendant from zigzag. H3;

27. Jug. Complete profile. Rounded profile, vertical neck, rounded short shoulder, thick crude handle. Shallow grooves. Undecorated neck. Body decorated to below mid-belly with vertical chevrons

(single or paired). Vertical lines on handle. There are good parallels for this type of jug at Bronneger (Knöll, 1959: Tafel 1, 12) and Glimmen G2 (Brindley, 1986a: fig. 40, 238). H2;

28. Jug. Almost complete profile. Angular profile, tall, cylindrical neck, long shoulder and wide angular strap handle from mid-neck to mid-shoulder. *Tiefstich* and *tvaerstik*. Neck: two zigzag lines below rim, 3 verticals on either side of handle, *tvaerstik* line at base. Shoulder: alternating panels of multiple zigzag defined by narrow ladders with 2 vertical *tvaerstik* lines in between. Upper portion of handle badly worn but probably decorated. There are close parallels for this type of jug at Heveskesklooster G5 (unpublished), Hooghalen D54b/c (Brindley, 1986b: fig. 9; recent examination has shown that this jug also has panels on the shoulder) and Zeijen (Bakker, 1979: fig. B13, 33 and B14, 20a). H2;

29. Jug. Body only. Angular profile. *Tiefstich*. Ladders and panels of zigzag defined by ladders. Angular strap handle with vertical lines on upper part. H2;

30. Jug. Restored. Almost complete profile (rim missing). *Tiefstich*. Tall, slighty conical neck and slightly convex shoulder. Complex ornament. Neck: two groups of 4 verticals on either side of the handle. Two groups of 4 'O' motif opposite handle. Shoulder: hatched triangles pendant from line at base of neck and on and running over shoulder to mid belly, alternating panels of multiple 'M' motif and 4 broad vertical *Tiefstich* verticals. Handle: multiple zigzag and vertical lines. Some sherds marked 38. This jug belongs to be compared to vessels which share its general proportions and size including an unpublished vessel from Bronneger D21/22, four vessels from Emmeln (Nos 2, 3, 16 and 17, Schlicht, 1968), possibly one from Gross Berssen (No. 209, Schlicht, 1972) and from further east in the Altmark, a vessel from Niedergörne, Kr. Stendal (Stolle et al., 1988: Abb. 4) and various other sites shown by Preuss (1982: e.g. Tafel 26, 1; Tafel 51, 50; Tafel 51, 1). These vessels show considerable variation in the form, number and arrangement of the lugs or handles, which range from small horizontally pierced bosses at the base of the neck, to large horizontal handles, to large angular vertical strap handles, and in number from single handles to one pair and two pairs. These vessels share angular profiles and long necks and often have long shoulders and are relatively large. Because of the number and variety of handles, they may be treated with jugs, tureens and amphorae, but are clearly outside the norms of these classes. They are more consistent when treated as a group and should perhaps be considered as a type apart. The decoration indicates that they were current during Horizons 2 and 3. H2/3;

31. Tureen. Incomplete. Angular profile (body only). *Tiefstich*. Wide and probably angular strap handle attached to junction of shoulder and body. Grooved line at base of neck. Vertical *Tiefstich* adjacent to handle, filled triangles on remainder of shoulder. Vertical lines and some zigzag on handle. Neatly finished. Slightly hollow base. There is a similar but larger tureen at Bronneger D21/22 (B.A.I. store). H3;

32. Tureen. Incomplete. Handle missing. Angular profile. *Tiefstich*. Vertical neck. Zigzag line under rim, line at base of neck. Shoulder: alternating panels of zigzag, chevron 'V', vertical lines adjacent to handle and small zigzag panels. A good parallel for this tureen and No. 33 is the pot from Emmen D43a (Knöll, 1959: Tafel 3, 11). H3;

33. Miniature tureen. Incomplete. *Tiefstich*. Line of zigzag below rim. Chevrons, panels of zigzag and *Tiefstich* lines on shoulder. H3;

34. Tureen. Restored. Complete profile. *Tiefstich*. Angular profile with small thick handle. Vertical neck: 2 horizontal lines below rim. Line at base of neck and empty triangles (or zigzag line) on shoulder. Vertical lines on either side of the handle. Inverted 'V' motif on handle. H2/3;

35. Tureen. Complete profile. Undecorated. Short, slightly conical neck, sloping shoulder and evidence for handle;

36. Tureen. Restored, in poor condition. Complete profile. Angular profile. Deeply impressed *Tiefstich*. Steeply sloping shoulder. Strap handle. 1 line zigzag below rim. Line at base of neck. Verticals on neck and shoulder on either side of handle, cover part of vessel circumference. Pseudo-triangles on remainder of shoulder. Strap handle has some vertical decoration. H3;

37. Tureen. Incomplete. *Tiefstich*. Slightly conical neck, rounded shoulder. Wide, flat handle with zigzag. Neck: zigzag line below rim, well-spaced groups of 5 verticals, wide *Tiefstich* line at base. Shoulder: rusticated triangles and double line of *Tiefstich* by handle. H3;

38. Tureen. Fragmentary. Line of *Tiefstich/tvaerstik* at base of neck. Possibly from tureen similar to No. 37. H3;

39. Tureen. Fragmentary. 1 shoulder sherd with rusticated triangles outlined with fine pointed *Tiefstich*. H3;

40. Tureen. Complete profile, except for lug. Almost vertical neck, very small shoulder. *Tvaerstik*. 3 horizontal lines below rim. Stacked 'U' motif on neck. *Tiefstich* stabs on shoulder. H4;

41. Tureen. Almost complete profile, except for lug. *Tiefstich*. Very slight, unmarked shoulder. 3 lines below rim, stacked, inverted 'U' motif. Below handle, stacked, inverted 'V' with inverted 'U' on either side. Early H4;

42. Lugged beaker. Complete profile. Slightly open straight neck, 4 small lugs in widely spaced pairs, angular shoulder. Two zones on neck, defined by zigzags and filled with *Tiefstich* vertical lines only in upper zone, vertical *Tiefstich* lines with at least one vertical *tvaerstik* and multiple 'M' motif over lugs. Decoration repeated over entire body. Simple lines on lugs. H3;

43. Lugged beaker. Incomplete. *Tiefstich*. Skating, verticals and small lug. H3;

44. Lugged beaker. Fragmentary. Zigzag motif below rim, verticals, horizontal pointed *Tiefstich* line with verticals and 'M' or chevron motif in lower zone. H3;

45. Lugged beaker. Fragmentary (3 sherds). *Tiefstich*. Neck of small lugged beaker. H3;

46. Collared flask. Restored. Short neck and round body. Neck apparently drawn up round stick, collar possibly applied. Group of 3 *Tiefstich* lines to edge of shoulder;

47. Short-necked collared flask. Collar and neck sherd;

48. Collar piece. Appears to have been initially pinched out and then enhanced by applied collar.

49. Collared flask. Incomplete. Angular body of collared flask with groups of incised lines;

50. Collared flask. Fragmentary. Angular shoulder of collared flask with groups of incised lines;

51. Base sherd, rough on inside. Probably collared flask;

52. Funnel beaker. Restored. Almost complete. Undecorated, with flaring neck and small base. Possibly H1;

53. Funnel beaker. Incomplete. Rounded body with vertical grooves. There is a general similarity between the undecorated sherds of the lowest parts of bowl No. 1 (Findspot 45) and No. 53. Both pots are also decorated with grooved lines (although, of course, No. 1 has additional techniques). Possibly H1;

54. Funnel beaker. Complete profile. Undecorated neck, line of stabs at base of neck, shallow *Tiefstich* on body. There is a general similarity between this pot and jug No. 27;

55. Funnel beaker. Partially restored. Almost complete profile. Slightly flaring neck and rather angular shoulder. Line of zigzag below rim. Fine *Tiefstich* to mid-belly. Probably H3;

56. Funnel beaker. Almost complete. *Tiefstich*. Two small zigzag lines below rim, line at base of neck, fine lines all-over-body to close to base. Probably H3;

57. Funnel beaker. Restored. Complete profile. *Tiefstich*. Undecorated neck, short body with pronounced shoulder, fine irregularly incised lines close to base;

58. Funnel beaker. Complete profile. Undecorated neck, highish rounded shoulder. Regular *Tiefstich* to mid-belly;

59. Funnel beaker. Complete profile. Assymetrical profile. Undecorated. Poorly finished;

60. Funnel beaker. Complete profile. Undecorated neck, unaccentuated shoulder. Scored lines. Not well-finished;

61. Funnel beaker. Undecorated neck. Fine fabric.;

62. Funnel beaker. Lower part of body of undecorated funnel beaker;

63. Funnel beaker. Undecorated neck, rounded body. *Tiefstich*. Two horizontal lines opposed *Tiefstich* with vertical lines below. See G2: 68 for good parallel;

64. Funnel beaker. Incomplete. *Tiefstich*. Undecorated neck, rounded body with vertical lines;
65. Funnel beaker. Reconstructed profile. *Tiefstich*. Undecorated neck. Horizontal line at base of neck, vertical lines on body;
66. Funnel beaker. Fragmentary. Fine *Tiefstich* lines on body. Coarse, densely gritted fabric.

Miscellaneous:
67. Shouldered bowl. Reconstructible profile. Flat-topped rim, wide with low neck, slight shoulder, with faintly scored line at base of neck. Not well finished. Has split along lines of manufacture. A similar vessel is known from the stone cist at Eext (van Giffen, 1944b: afb. 7, 2k) where the associations indicate a H2/3 date;
68. Beaker with rim perforations. Reconstructed. Asymmetrical profile. Slightly flaring neck and unaccentuated, slack, shoulder and body. Two pairs of post-firing perforations, one pair preserved. Several parallels for this vessel were found at Bronneger D21/22 (e.g. Knöll, 1959: Tafel 12, 11).

These vessels appear to form a distinct group characterized by the following, unaccentuated profile with short open neck usually with two to four perforations under the rim, and biconical body, sometimes with lugs at the top of the shoulder. Although the Bronneger and Exlo examples are undecorated, ornamented versions also occur, e.g. Buinen D28 (van Giffen, 1943: afb. 30, 82). As a group, they may have escaped attention as either unusual forms of funnel beakers or amphorae;
69. Neck sherd with line of *Tiefstich* at base of neck. ?Tureen;
70. Sherds of coarse, tureen-like vessel including thick handle and body sherds. Not illustrated;
71. Undecorated, straight neck with line of decoration at base of neck and fracture point. Thickening indicates lug at base of neck;
72. Large shoulder fragment of biconical vessel. Min. 3 grooved vertical lines;
73. Reconstructible profile. Short vertical neck, rounded body. 4 large perforations in two opposed pairs;
74. Fragmentary. 1 body sherd;
75-76. 2 base sherds, not illustrated;
77. 5 very small rim sherds of different pots. 4 of the sherds are too small for further identification (not illustrated);
78. 100 undecorated featureless sherds, mostly very small (not illustrated).

Other pottery:
79. Incomplete. Large Single Grave beaker. Diagonal stabs in 5 horizontal undefined bands;
80. Incomplete. Single Grave beaker. 1d type;
81. Lower undecorated portion of Single Grave beaker (not illustrated);
82. Bowl. Incomplete, base missing. Crude. Possibly *Kümmerkeramik*.

Flint:
83. Transverse arrowhead of dark grey flint;
84. Small scraper with retouch along two sides;
85. Flint flake with some retouch near distal end. Dark grey flint;
86. 27 pieces of struck flint, flakes and pieces without traces of working (not illustrated).

5. ASPECTS OF TRB POTTERY FROM *HUNE-BEDDEN*

Pottery in ceremonial contexts is either custom made for ritual purposes or selected from the range of domestic wares. Highly decorated pottery such as the TRB pottery is usually considered to have been made with ceremonial purposes in mind. There are, however, grounds for suggesting that the pottery found in *hunebedden* was selected from a domestic range of pottery which was highly decorative and included an unusually wide range of distinct forms. *Hunebed* inventories include not only fine pottery in the sense of well-finished and competently decorated vessels; they also include both very poorly made and finished specimens and relatively large funnel beakers as well as undecorated and sometimes not particularly well-finished bowls. The poorly made specimens include funnel beakers, tureens, bowls etc. which may have irregular bases and rims, or be markedly asymmetrical in profile, have uneven, unsmoothed walls, on occasion even showing horizontal lines along the coils, be badly fired and have crude decoration. These vessels are clearly not chosen for their aesthetic appeal or competent workmanship. They are the products of very poorly skilled individuals. The 'incompetent' pieces are usually limited in number but poorly finished pottery appears to be present on a regular basis in *hunebedden*, during all horizons (except Horizon 5 perhaps). It appears therefore, that not all pottery was selected for deposition on the basis of its quality. The more competently made pottery includes some very finely finished and decorated pieces, but the vast majority of the pottery is chiefly characterized by the large amount of basically simple and repetitive decoration. Once the basic pattern and the technique is understood, the decoration can be applied to a Horiozon 3 bowl in less than 15 minutes by a novice (authors' test). The most striking aspect of this pottery is the wide range of pots which bear decoration rather than the decoration itself. That this range was perpetuated over large distances is surprising. Large funnel beakers whose size suggests that they had a domestic function were found at Havelte D53, with a rim diameter of 28 cm and a height of 30 cm, at Exlo (No. 52), with rim diameter of 31 cm and height of 27 cm and fragments at Heveskesklooster G5, with a rim diameter of about 29 cm (unpublished). More frequently found are the relatively thick-walled funnel beakers with rim diameters of c. 20 cm (Emmen D40, Nos 21, 24 and 28, and 35; Glimmen G2, Nos 97 and 99 (Brindley, 1986a: fig. 31), Hooghalen D54b/c, Nos 182, 184 and 185 (Brindley manuscript, 1993), at least three further examples at Heveskesklooster G5. Neck sherds of similarly large funnel beakers have also been noted at Papeloze Kerk D49 (Brindley manuscript, in prep.).

These two aspects (the inclusion of pottery displaying a wide range in competency and skill together with undecorated bowls and very large beakers) suggest that the pottery stems from the personal property of individuals. It is questionable whether much of it is ceremonial in origin. Settlement pottery from Midlaren, Elspeet and Laren (Bakker, 1979: figs B1, B6 and B7; B9 resp. B10) does not appear to include significantly more decorated pottery of lower quality.

It is usually assumed that the quantity of pottery in some of the western *hunebedden* is the result of long and intensive use. The large inventories of Emmeln (1220 pots) and Havelte D53 (660 pots) are frequently cited as examples of this. However, a recent survey has shown that at individual *hunebedden* neither protracted use of the chamber nor necessarily intensive use can be assumed and is probably not the rule.

In the following discussion, estimates of duration are based on the dates indicated for each horizon by Brindley (1986b: pp 104-106). Based loosely on the amount of typological development within a horizon and not solely on the quantity of material known, the following time brackets have been calculated.

Horizon 1	c.	50 years	3400-3350 BC
Horizon 2		50 years	3350-3300 BC
Horizon 3		100 years	3300-3200 BC
Horizon 4		150 years	3200-3050 BC
Horizon 5		100 years	3050-2950 BC
Horizon 6		50 years	2950-2900 BC
Horizon 7	c.	50 years	2900-2850 BC

Although crudely arrived at, given the detail of the typological developments involved and the constraints of the dates for Horizons 1, 5 and 7, it is very likely that these provide a strong chronological framework. It is possible that Horizons 1 and 7 may extend slightly beyond the earlier and later limits respectively. In the following discussion, it is assumed that this chronology is accurate enough to allow for the dating of the construction and use of individual *hunebedden* based on the type of pottery found in them.

Of the three inventories catalogued above, only D30 and D40 are sufficiently well-preserved for a general reconstruction of the manner and frequency with which they may have been used. In addition, the following inventories are considered to be relatively complete: Emmeln, Gross Berssen, G1, G2, D9, D32a, D9, D43 and D53 and cited in a more general manner, D28, D43a, D32d and O2. The finds from D54b and D54c are also included although the two inventories have been mixed up since being excavated.

It is clear that *hunebedden* in continuous use over long periods of time (i.e. more than three of the seven horizons of ceramic development) are the exception rather than the rule. Since *hunebedden* ceased to be constructed during Horizon 4 (or possibly at its start), it is technically possible for each to have been in use during four horizons or a minimum of 350 years.

Of the *hunebedden* surveyed here:

Emmeln 2 is the best known exception with over 1220 identified vessels, 959 of which are illustrated in the excavation report. The majority of the illustrated pottery can be easily assigned to Horizons 3 (early) to 5 (including quite late looking bowls) with a small quantity indicating Horizons 6 and 7;

Gross Berssen 7. The catalogue indicates Horizons 3-4 and a few Horizon 5 pots, all in the Heek-Emmeln Style.

Noordlaren G1. The majority of the pottery from this *hunebed* belongs to only two, separate, horizons. The first period of use is shown by the Horizon 3 pottery, including some early looking vessels and some late Horizon 3/early Horizon 4 pots. There is one Horizon 6 pail. The second period of use occurred during Horizon 7;

Glimmen G2. This hunebed was constructed at the very end of Horizon 2 and used continuously until the quite late during Horizon 5. It was used once during Horizon 6 and for re-used during Horizon 7 for a second period;

Annen D9. This *hunebed* was possibly constructed during Horizon 3 (2 sherds which may be residual as they suggest a fairly early stage in that horizon), but was more probably constructed at the beginning of Horizon 4. After Horizon 4 it was apparently used once or twice during Horizon 5;

Odoorn D32a. This *hunebed* was in use throughout Horizons 3 and 4. It was abandonned at the beginning of Horizon 5 which is repesented by a small number of vessels (not Anlo-Uddelermeer Style);

Emmen D43. The inventory consists of the now unseparatable contents of two chambers within one kerb. Apart from the Horizon 1 sherds in a pit outside the burial chambers, the inventory includes a small amount of Horizon 2 pottery, and a very small amount of Horizon 5 (Heek-Emmeln Style) pottery. The majority of the pottery belongs to Horizons 3 and 4;

Havelte D53. The inventory includes a small quantity of mature Horizon 3 pottery and was in continuous use up to and including Horizon 7;

Buinen D28. The inventory includes some early Horizon 3 pottery. The majority of the pottery belongs to late Horizon 3 and early Horizon 4. There are also several Horizon 5 pots. There are no tureen-amphorae or bowls with block patterns which indicate the more mature Horizon 4 Style.

Emmen D43a. The inventory includes one Horizon 1 jug, possibly of a late form (no decoration below the shoulder). According to Molema (pers. comm.), 4 pots could be assigned to Horizon 2, 13 to Horizon 3 and 18 to Horizon 4, with a single pot attributable to Horizon 5;

Odoorn D32d. Includes several developed Horizon 2 pots, and a small number of Horizon 5 pots (both Heek-Emmeln and Anlo-Uddelermeer Styles). The vast majority of the pottery belongs to Horizons 3 and 4. There is one Horizon 7 bowl (B. Kamlag, pers. comm.).

Mander O2 has pottery exclusively of Horizons 3 and 4 (A. Ufkes, pers. comm.).

Hooghalen 54b and c. The pottery from the two *hunebedden* cannot now be separated; however, the combined assemblage includes a small quantity of Horizon 2 pottery, and spreads across Horizons 3 and 4. There is a limited amount of Horizon 5 pottery, a few examples of Horizon 6 pottery, sufficient only to show sporadic visits, and a second phase of activity represented by a relatively large quantity of Horizon 7 pottery. The combined assemblage also shows at least one monument was not in continuous use and the other monument was either abandonned fairly early (i.e. around Horizon 5) or also re-used after abandonment (Brindley manuscript, 1993).

Against this background, the single horizon assemblage of D40 no longer appears in any way unusual and the relatively early abandonment of both D40 and D30 can be easily paralleled at other monuments.

The sometimes dramatic figures given for the number of pots in individual *hunebedden* can be seen to be relatively consistent when viewed against their likely timescale. Excluding D43 and Hooghalen D54b/c and including only the episodes of concentrated activity, approximate figures for the duration and use of the above *hunebedden* can be found in the table below.

Taken by themselves, these figures suggest that whereas *hunebedden* may have been used on a regular basis, they were not used on a frequent basis, not even, apparently, on an annual one. The 'service sets' (Brindley, 1986a: p. 35) indicate that more than one pot might be deposited at once (i.e. assuming that the 'service sets' are not the result of pots being placed together on several occasions as the products of a single potter deposited at a favoured or 'own' place within a communal tomb), and that as many as five or six pots might sometimes have been left in a single act. Assuming that this happened, the frequency of activity is likely to be even less than suggested by the table above and variation in the number of pots used on any one occasion is likely to cancel out any differences between the apparent frequency of activity between monuments as shown by the figures in the last column of the table.

No 'service sets' were recognized at either D30 or D40 (the two Horizon 1 Exlo pails are not considered in this context). Examining the pottery from these two sites from a stylistic point of view, however, it is noticable

that groups of pots seem to share a particular stage of development even within a Horizon. It is possible to distinguish putative groups in D30 as follows (not all the pottery can be placed in groups):

Horizon 2
– Nos 27, 54
– Nos 4, 28, 29
Horizon 3
– Nos 5, 7, 9, 12, ?30, 55, 56
– Nos 8, 32, 33, ?15
– Nos 6, 34, 36
– Nos 10, 42 (43-45)
– Nos 16, 37
– Nos 11, 40, 41.

This approach is less successful when applied to the pottery from D40, partly due to the more fragmented condition of the pots and partly because the pottery stems from a shorter period and therefore displays less typlogical variation. However, suggested groups amongst the pails and bowls are
– Nos 5, 7 (?8)
– No. 1
– Nos 3, 4
– Nos 9, 11
– Nos 2, 6.

This apparent stylistic clustering may be the result of groups of pots being deposited at intervals of time.

It appears that D30 and D40 were used on a possibly infrequent basis for a limited amount of time and went out of use at a relatively early stage within the chronological framework of the TRB. In both these aspects, they are well within the behaviour indicated at other *hunebedden* by the range of pottery they contain.

The relevance of these conclusions is not limited to the contents of the individual *hunebedden*. Because of their prominence in the landscape, *hunebedden*, despite the distinct factors which limit their distribution in the Dutch landscape, are frequently discussed in relation to

Hunebed	Horizon(s)	Years	Pots	Per year
Emmeln 2	early 3, 4, 5, 7	400	1220	3
Gross Berssen 7	3, 4 and part of 5	275	325	1.2
Noordlaren G1	3, 7	300	150	0.5
Glimmen G2	(2/3), 3, 4, 5, 7	400	400	1
Annen D9	4	150	80	0.5
Odoorn D32a	3, 4	250	160	0.6
Havelte D53	3 (mid), 4, 5, 6, 7	400	660	1.7
Emmen D40	3	100	80	0.8
Exlo D30	1, 2, 3, part of 4	275	80	0.3
Buinen D28	3, 4	200	-	
Emmen 43a	1, 2, 3, 4	350	-	
Odoorn 32d	2, 3, 4, 5	400	-	
Mander O2	3, 4	250	-	

settlement distribution. Bakker has already indicated that the Hondsrug distribution is more likely to be related to roads and soil types than to the monuments functioning as territorial markers (Bakker, 1980). However, it may be possible to relate the changing fortunes of individual *hunebedden* to changes in territories. On the basis of the chronology of the pottery, it seems probable that only during a very limited time during the later part of Horizon 3 and the earlier part of Horizon 4 were all *hunebedden* in use. By plotting *hunebedden* use by horizon it may be possible to come up with a picture of changing land ownership. Even during the period of maximum activity (late in Horizon 3), some *hunebedden* were on the wane while others were only being constructed or were still in their earliest stages of use.

6. ACKNOWLEDGEMENTS

We wish to thank Mrs J.C. Kat-van Hulten (O1), Miss M.A. Weijns (O1, D30) and S.W. Jager (D40) for the illustrations of the finds, and J.H. Zwier for (re)drawing the plans and sections of these *hunebedden*. We are moste grateful to the heirs van Karnebeek, owners of the De Eese estate and their local manager A. Boschloo for permission to re-excavate O1; to the Province of Drenthe represented by Ir. B. Volbeda and S. Smid of Provinciale Waterstaat for permission to re-excavate D30 and to the State, represented by Ir. J.J. Kalb and L. Dijk of Staatsbosbeheer Drenthe for permission to re-excavate D40.

7. REFERENCES

BAKKER, J.A., 1970. Diepsteekceramiek uit Hooghalen, gem. Beilen. *Nieuwe Drentse Volksalmanak* 88, pp. 185-211.

BAKKER, J.A., 1979. *The TRB West Group. Studies in the chronology and geography of the makers of hunebeds and Tiefstich pottery.* Amsterdam,

BAKKER, J.A., 1980. Einige Bemerkungen über die niederländischen Groszsteingräber und deren Erbauer. *Nachrichten aus Niedersachsens Urgeschichte* 49, pp. 31-59.

BAKKER, J.A., 1982-83. Het hunebed G1 te Noordlaren. *Groningse Volksalmanak*, pp. 115-149.

BAKKER, J.A. & H. Luijten, 1990. 'Service sets' and other 'similarity groups' in Western TRB pottery. In: *La Bretagne et l'Europe préhistoriques. Mémoire en hommage à Pierre-Roland Giot* (=

Revue Archéologique de l'Ouest, Supplement no. 2). Rennes, pp. 173-187.

BLOEMERS, J.H.F., L.P. LOUWE KOOIJMANS & H. SARFATIJ, 1981. *Verleden Land. Archeologische opgravingen in Nederland.* Amsterdam, 1981.

BRINDLEY, A.L., 1986a. Hunebed G2: excavation and finds. *Palaeohistoria* 28, pp. 27-92.

BRINDLEY, A.L., 1986b. Typochronology of TRB West Group pottery. *Palaeohistoria* 28, pp. 93-107.

CAMPER, P., n.d. De Hunnen Bedden van Drenthe. Manuscript, University Library Amsterdam, MS II G 53.

FANSA, M., 1978. Die Keramik der Trichterbecherkultur aus dem Megalithgrab I van Ostenwalde, Kreis Aschendorf-Hümmling. *Neue Ausgrabungen und Forschungen in Niedersachsen* 12, pp. 33-77.

GIFFEN, A.E. VAN, 1919. Mededeling omtrent onderzoek en restauratie van het grote hunebed te Havelte. *Nieuwe Drentse Volksalmanak* 37, pp. 109-139.

GIFFEN, A.E. VAN, 1924a. Het hunebed te Rijs in Gaasterland. *De Vrije Fries* 27, pp. 307-325.

GIFFEN, A.E. VAN, 1924b. Het verstoorde hunebed op De Eeze bij Steenwijk. *Verslagen en Mededelingen van de Vereniging tot Beoefening van Overijsselsch Regt en Geschiedenis* 41, pp. 56-71.

GIFFEN, A.E. VAN, 1925-27. *De hunebedden in Nederland.* Utrecht, 2 volumes, with atlas.

GIFFEN, A.E. VAN, 1943. Het ndl. hunebed (DXXVIII) te Buinen, gem. Borger (een bijdrage tot de absolute chronologie der Nederlandse hunebedden). *Nieuwe Drentse Volksalmanak* 61, pp. 115-136.

GIFFEN, A.E. VAN, 1944a. De twee vernielde hunebedden DVIe en DVIf bij Tinaarlo, gem. Vries. *Nieuwe Drentse Volksalmanak* 62, pp 93-112.

GIFFEN, A.E. VAN, 1944b. Een steenkeldertje, DXIIIa, te Eext, gem. Anlo. *Nieuwe Drentse Volksalmanak* 62, pp 117-119.

GROOT, D.J. DE, 1988. Hunebed D9 at Annen (gemeente Anlo, provincie of Drenthe, the Netherlands). *Palaeohistoria* 30, pp. 73-108.

KNÖLL, H., 1959. *Die nordwestdeutsche Tiefstichkeramik und ihre Stellung im nord- und mitteleuropäischen Neolithikum.* Münster.

PREUSS, J., 1980. *Die altmärkische Gruppe der Tiefstichkeramik.* Berlin.

SCHLICHT, E., 1968. *Die Funde aus dem Megalithgrab 2 von Emmeln, Kreis Meppen.* Neumünster.

SCHLICHT, E., 1972. *Das Megalithgrab 7 von Gross Berssen, Kreis Meppen.* Neumünster.

SCHWABEDISSEN, H., 1979. Zum Alter der Grossteingräber in Norddeutschland. In: H. Schirnig (ed.), *Grosssteingräber in Niedersachsen.* Hildesheim, pp. 143-160.

STOLLE, T., N. BENECKE & J. BERAN, 1988. Zwei Siedlungsgruben der Tiefstichkeramik mit zahlreichen Tierresten von Niedergörne, Kreis Stendal. *Jahresschrift für Mitteldeutsche Vorgeschichte* 71, pp. 37-55.

TAAYKE, E., 1985. Drie vernielde hunebedden in de gemeente Odoorn. *Nieuwe Drentse Volksalmanak* 102, pp. 125-144.

WATERBOLK, H.T., 1964. Ein Grabhügel auf dem Gut "De Eese", Gem. Vledder, Prov. Drenthe. *Palaeohistoria* 10, pp. 71-86.

IMPORT VAN NOORDELIJKE VUURSTEEN
ENKELE VOORLOPIGE CONCLUSIES MET BETREKKING TOT SIKKELS IN NOORDWEST-EUROPA

J.R. BEUKER

Drents Museum, Assen, Nederland

ABSTRACT: Since 1985 a research programme concerning the importation of northern flint has been carried out at the Drents Museum in Assen. One of the results sofar is the discovery that red Heligoland-flint was used during prehistory in large parts of northwestern Europe. Study of axes, sickles, daggers and large blades of flint in the collection of the Drents Museum shows clearly that the sickles in this collection share some common characteristics, something that was not found (or only to a lesser extent) with the axes, daggers and blades. To obtain a larger sample it was decided to study a number of sickles from the provinces of Groningen, Friesland and Noord-Holland. The situation there appeared to be identical to that in Drenthe.

According to Kühn (1979: p. 64) flint sickles can be divided into A-types and B-types. Except for one, all the Dutch examples studied are of the first type. Remarkably the only B-sickle appeared to be made of flint lacking the common characteristics of the A-sickles. Therefore it was decided to study a sample of sickles from Schleswig-Holstein, a region where both types are present in rather large numbers. Of the 13 German A-sickles studied 11 had the same characteristics as the Dutch examples. However, none of the B-sickles had these characteristics. The distribution of the semi-finished B-sickles in Schleswig-Holstein justifies the conclusion that these implements were fabricated on the east coast. Flint of good quality can be collected there from Weichsel-moraines. The common characteristics of the A-sickles show that they were not made of moraine flint. The flint for these objects must have been collected at a primary outcrop. Within the distribution area of the sickles Heligoland was the only outcrop of economic importance. In addition to the red flint other flint types are also found there, for instance flat flint modules that are very suitable for making bifacially worked objects such as daggers and sickles. When we compare this flint with the A-sickles there is a striking resemblance of various characteristics.

KEYWORDS: Northern flint, flint sickles, Drenthe, Friesland, Groningen, Noord-Holland, Schleswig-Holstein, Heligoland.

1. INLEIDING

Vuursteen van een kwaliteit die geschikt is om er grote werktuigen als neolithische bijlen, dolken, klingen en sikkels van te maken ontbreekt in Noord-Nederland en grote delen van Noordwest-Duitsland. Deze voorwerpen werden in de prehistorie geïmporteerd uit streken waar de grondstof wel aanwezig was.

Sinds 1985 wordt vanuit het Drents Museum te Assen onderzoek verricht naar de herkomst van deze 'exotische' werktuigen.[1] Dit leidde o.a. tot de ontdekking dat in het Neolithicum, de Bronstijd en de IJzertijd een rode vuursteensoort werd gebruikt die afkomstig is van het eilandje Helgoland in de Duitse Bocht. Het betreft goed herkenbaar materiaal waarvan kan worden aangetoond dat het over een groot deel van West-Europa werd verspreid.

Om de kenmerken en daarmee het gebruik van verschillende noordelijke vuursteensoorten nog beter te kunnen bestuderen werd in 1989 met financiële steun van de Drents Prehistorische Vereniging een grote vergelijkingscollectie opgebouwd. De schrijver werd in de gelegenheid gesteld vuursteen te verzamelen in Denemarken en Noord-Duitsland (Lüneburg, Hemmoor en de Weichselienmorenes in Sleeswijk-Holstein). In de jaren ervoor waren van de op Helgoland voorkomende vuursteensoorten – naast de rode nog vier andere – al monsters verzameld. Gewapend met de vergelijkingscollectie werd een begin gemaakt met de bestudering van grote vuurstenen werktuigen.

Een inventarisatie van de collectie van het Drents Museum maakte duidelijk dat vuurstenen sikkels wat betreft materiaal afwijken van de bijlen, dolken en grote neolithische klingen. Dit gegeven was dermate interessant dat besloten werd, vooruitlopend op een groter onderzoek, een aantal sikkels nader te bestuderen. In totaal werden 91 sikkels en fragmenten uit Nederland en Sleeswijk-Holstein op de specifieke kenmerken van hun grondstof bekeken. Daarbij bleek dat sikkels die aan één uiteinde puntig uitlopen en aan het andere uiteinde afgerond of recht zijn[2], wat betreft grondstof een opvallende uniformiteit vertonen en grotendeels uit hetzelfde produktiecentrum afkomstig moeten zijn.

2. IMPORT OF LOKAAL?

Tot en met het Mesolithicum deed de noodzaak tot het importeren van vuursteen zich in Noord-Nederland en Noordwest-Duitsland niet voor. De werktuigen waren relatief klein en vuursteen van een daarvoor toereikende kwaliteit was lokaal te verzamelen. Zelfs grotere werktuigen als kern- en afslagbijlen konden daarvan worden gemaakt. Deze voorwerpen zijn niet zo goed afgewerkt als de latere bijlen en volgen vaak grotendeels de vorm van het oorspronkelijke stuk vuursteen. Hooguit kunnen motieven van bijvoorbeeld esthetische aard er toe hebben geleid dat hier 'exotische' vuursteensoorten terechtkwamen. Een voorbeeld zijn twee schrabbers van de Federmessertraditie, beide gevonden in de provincie Drenthe, en gemaakt van rode Helgoland-vuursteen.

Toen een agrarische bestaanswijze de plaats innam van een economie die gebaseerd was op jagen en verzamelen van voedsel ontstond de behoefte aan grote vuurstenen werktuigen zoals bijlen. Van de inheemse vuursteen konden dergelijke voorwerpen niet worden vervaardigd. Wel kon gebruik gemaakt worden van inheemse steensoorten als gabbro, dioriet en diabaas. Daarnaast werd vuursteen geïmporteerd.

De discussie wanneer we in Noord-Nederland nu eigenlijk met importen te maken hebben en welke werktuigen van inheemse makelij zijn heeft zich soms uitsluitend geconcentreerd op de grootte van de voorwerpen. Bakker (1979: p. 80) is over de bijlen iets genuanceerder en zegt:

> For the moment we must content ourselves with the unsatisfactory rule of thumb that specimens of the thick-bladed TRB axes with an original length of more than c. 15 cm will be imported pieces, and that this is probably also the case with the originally shorter specimens if they are well made.

Dit beeld behoeft enige aanvulling. Voor het maken van bijlen is vaak een relatief groot halffabrikaat nodig.[3] Probeert men van morenemateriaal uit Noord-Nederland een bijl van het noordelijke type te maken dan doet zich een aantal problemen voor. Zo zal men er nauwelijks in slagen een regelmatige rechthoekige dwarsdoorsnede te creëren. De bijl zal niet zo regelmatig worden en deels de natuurlijke vorm van het uitgangsmateriaal volgen. Ook is te verwachten dat er vorstsplijtvlakken blijven zitten en tot slot zal de bijl meestal relatief klein uitvallen. Een inventarisatie van de exemplaren in de collectie van het Drents Museum leerde dat er bijlen voorkomen die verschillende van de hierboven genoemde kenmerken in zich verenigen.[4] Overgangsvormen tussen deze vaak op mesolithische kernbijlen lijkende werktuigen en goed afgewerkte exemplaren zijn er niet. De meeste vuurstenen bijlen zijn dus geïmporteerd, zelfs de kleinere exemplaren.

Voor objecten als dolken en sikkels geldt in principe hetzelfde. De lokale vuursteen was ook voor het maken van deze voorwerpen volstrekt ongeschikt. Slecht ge-

vormde, kleine sikkels met bijvoorbeeld veel vorstsplijtvlakken zijn bij de inventarisatie niet aangetroffen, zodat we er vanuit kunnen gaan dat ze alle geïmporteerd zijn.

3. INVENTARISATIE VAN DE NEDERLANDSE SIKKELS

De in Nederland geïnventariseerde sikkels zijn afkomstig uit de provincies Friesland, Groningen, Drenthe en Noord-Holland (tabel 1).

Volgens Kühn (1979: p. 64) zijn de sikkels te verdelen in een type A en een type B. Hij beschrijft ze als volgt (zie ook fig. 1 en 2):

> Typ A. Kennzeichend für diesen Sicheltyp ist die nahezu gerade oder halbrund abschliessende Basis, die häufig einen Rest der Naturkruste trägt. Ein weiteres typen-spezifisches Merkmal ist das stets in Richtung der Längsachse gekrümmte Blatt. Zwar überwiegt die Blattkrümmung zur rechten Seite (von der Basis aus gesehen), es treten aber auch nach links gekrümmte Sichelblätter auf, die eine Handhabung der Sichel mit der linken Hand wahrscheinlich machen.
>
> Der Sichelrücken ist in der Seitenansicht stets bogenförmig ausgeformt, die Schneide ist bei nicht nachgeschärften Exemplaren gerade oder nur wenig, bei abgenutzten Sicheln dementsprechend stärker eingezogen. Im Querschnitt ist die Innenseite (konkave Breitseite) in der Regel nahezu flach, die Aussenseite (konvexe Breitseite) halbrund oder auch dachförmig gearbeitet.
>
> Typ B. Sicheln dieses Typs, die auch als halbmondförmige Sicheln bezeichnet werden, laufen an beiden Enden spitz aus, so dass Spitze und Basis in der Regel beliebig austauschbar sind. In der Seitenansicht ist der Sichelrücken flachbogig geformt, die Schneide ist konvex, gerade oder, wie die Schneiden der A-Sicheln, dem jeweiligen Abnutzungsgrad entsprechend eingezogen. B-Sicheln zeigen keine dem Typ A vergleichbare Krümmung des Blattes zu einer Seite.

Onder de bestudeerde Nederlandse sikkels bevindt zich slechts één B-sikkel. De overige zijn alle tot het type A te rekenen.[5] Zoals al ter sprake kwam vertonen de laatste een opvallende uniformiteit in de gebruikte grondstof. Dit heeft er waarschijnlijk ook toe geleid dat Van der Waals (1972/73: p. 180) het volgende vermeldt over twee bij Onstwedde gevonden sikkels:

> Sikkel van plaatselijk licht-doorschijnende, licht en donker gevlekte en gespikkelde vuursteen; mogelijk zelfde vuursteen als 1969/III 1, maar meer uit inwendige van de knol; geen resten van schors.

Sikkel 1969/III 1 wordt omschreven als:

> Sikkel van plaatselijk licht-doorschijnende, meer en minder melkachtig-gevlekte, grijze vuursteen; resten van de schors aan beide uiteinden.

De A-sikkels[6] bezitten wat betreft het materiaal waarvan ze zijn gemaakt een aantal kenmerken die bij andere grote voorwerpen (bijlen, dolken en grote klingen) niet of in mindere mate voorkomen. Bij de bestudering van enkele van die kenmerken is het gebruik van een binoculair of loep gewenst. In het onderstaande wordt

Tabel 1. De in Nederland bestudeerde sikkels en fragmenten.

Inv.nr.	Vindplaats	Collectie	Type
Drenthe			
1965/IV 2a	Angelslo	D.M.	-
1965/IV 2b	Angelslo	D.M.	-
17	Anloo	M. Wetterauw	-
1993/IV 9	Annen	D.M.	A
1920/V 1	Buinen	D.M.	A
12 E 46	De Groeve	G. Holtrop	A
12 G 41	Drouwen	G. Holtrop	A
1939/XII 32	Drouwen	D.M.	A
1962/VI 102	Elp	D.M.	-
1978/II 4	Emmen	D.M.	-
1981/III 101	Emmen	D.M.	-
LGV	Emmen	R. Trip	-
17 H 35	Emmen	G. Holtrop	A
17 H 31	Emmen	G. Holtrop	-
17 H 3	Emmen	G. Holtrop	-
1986/I 111	Exloo	D.M.	-
12 G 32	Gasteren	G. Holtrop	A
1988/VIII 13	Meppen	D.M.	-
1925/VI 20	Nieuw-Weerdinge	D.M.	A
1914/III 1	Nijlande	D.M.	A
1914/III 1a	Nijlande	D.M.	A
1914/III 1b	Nijlande	D.M.	A
1914/III 1c	Nijlande	D.M.	A
1991/X 26	Oosterhesselen	D.M.	A
1962/II 32	Oranjedorp	D.M.	A
1866/I 1	Paterswolde	D.M.	A
1920/I 3	Rhee	B.A.I.	A
1938/XI 1	Schipborg	D.M.	A
1992/II 2	Vries	D.M.	-
1957/VIII 48	Vries	D.M.	A
1959/XI 6	Weerdinge	D.M.	A
9	Weerdinge	F. Modderkolk	A
17 F 44	Weerdinge	G. Holtrop	B
12 B 4	Westlaren	G. Holtrop	-
1934/VII 1d	Zeijen	D.M.	A
1962/II 146	Onbekend	D.M.	A
1962/II 145	Onbekend	D.M.	A
Noord-Holland			
AN V3	Andijk	W.M.	A
And V 4	Andijk	W.M.	A
1962/X a	Andijk	W.M.	A
BOV/K16	Bovenkarspel	W.M.	A
1963/X a	Bovenkarspel	W.M.	A
1968/III a	Bovenkarspel	W.M.	A
Enk V2	Enkhuizen	W.M.	A
Enk V12	Enkhuizen	W.M.	A
ENK V19	Enkhuizen	W.M.	A
1961/IX a	Enkhuizen	W.M.	A
VEN/VI 7	Venhuizen	W.M.	A
1970/VII a	Venhuizen	W.M.	A
1937/VII 52	West-Friesland	B.A.I.	A
1969/XI n	Zwaagdijk Oost	W.M.	A
1969/XI o	Zwaagdijk Oost	W.M.	A
Friesland			
FM 1974-IX-5	Baard	F.M.	A
FM 1984-III-10	Boksum	F.M.	A
82a/N14	Hartwerd	F.M.	A
82d/N2	Hartwerd	F.M.	A
82d/N4	Hartwerd	F.M.	A
82D/N3	Hartwerd	F.M.	A
FM 1980-XI-4	Hartwerd	F.M.	A
82d/N13	Hartwerd	F.M.	A
FM 1977-XI-26	Hartwerd	F.M.	A
82D/N4	Hartwerd	F.M.	A

144   J.R. Beuker

Tabel 1 (vervolg).

Inv.nr.	Vindplaats	Collectie	Type
65/N60	Spannum	F.M.	A
65/N61	Spannum	F.M.	A
65/N62	Spannum	F.M.	A
81/N4	Witmarsum	F.M.	A
Groningen			
S84-1	De Haar	F.M.	A
S84-2	De Haar	F.M.	A
1970/X 14	Middelstum	B.A.I.	-
1970/IX 8	Middelstum	B.A.I.	A

B.A.I., Biologisch-Archaeologisch Instituut te Groningen
D.M., Drents Museum te Assen
F.M., Fries Museum te Leeuwarden
W.M., Westfries Museum te Hoorn

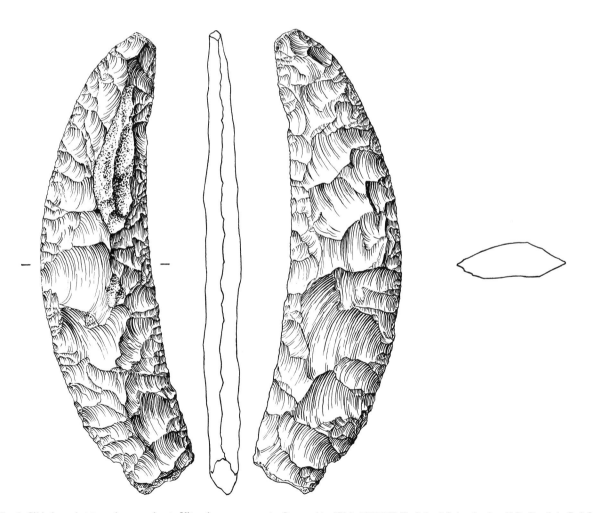

Fig. 1. Sikkel van het type A, gevonden te Uitwedsmee, gemeente Onstwedde (GM, 1959/VII.3). Schaal 3:4; tekening H.R. Roelink, B.A.I.

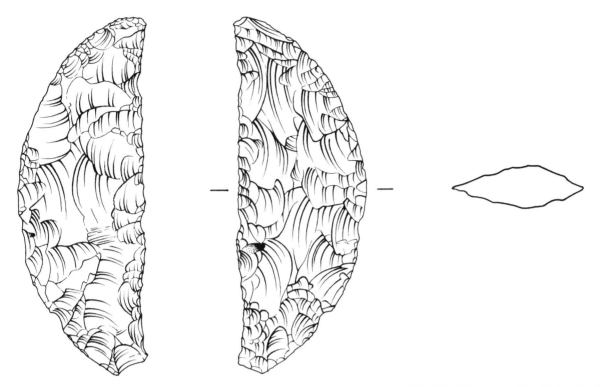

Fig. 2. Sikkel van het type B, gevonden te Weerdinge, gemeente Emmen (collectie G. Holtrop, 17 F 44). Schaal 1:1; tekening M.A. Weijns, B.A.I.

nader ingegaan op vijf van die kenmerken en op het voorkomen daarvan bij verschillende werktuigtypen.

1. Het meest opvallend is de vaak transparante matrix met daarin kleine, gemiddeld 0,2 tot 1 mm grote, vaak scherp begrensde, veelal ronde vlekjes.[7] Deze vertonen geen structuur, zijn het beste te omschrijven als melkachtig en worden vaak afgewisseld met vormloze flarden van dezelfde kleur (fig. 3 en 4). De vlekjes en flarden komen in wolken voor. Hun aantal kan de 30 per cm² overschrijden. De aanwezigheid van vlekjes is een kenmerk dat zelfs op afbeeldingen in publikaties vaak goed te zien is (zie o.a. Dorst, 1980: p. 231 en Stapert, 1988: p. 47).

Zowel bijlen, dolken, klingen als sikkels uit de collectie van het Drents Museum zijn op de aanwezigheid van deze vlekjes bekeken. Dit leverde de volgende percentages op:

Bijlen (A=189)	32%
Dolken (A=35)	49%
Klingen (A=34)	32%
A-sikkels (A=69)	100%
B-sikkel (A=1)	0%

2. Een tweede kenmerk is het voorkomen van wolken van kleine spikkels vaak met naaldvormige structuren erin. Deze spikkelwolken concentreren zich meestal rond inclusies, maar komen ook los daarvan voor. De volgende percentages zijn te geven.

Bijlen	35%
Dolken	43%
Klingen	9%
A-sikkels	93%
B-sikkel	0%

3. Afgezien van de onder 2 genoemde naaldvormige structuren, die overigens niet alleen in de spikkelwolken voorkomen, komen in de voor sikkels gebruikte vuursteen niet of nauwelijks bryozoën voor. Vertalen we de afwezigheid van bryozoën in percentages dan levert dat het volgende beeld op.

Bijlen	34%
Dolken	46%
Klingen	15%
A-sikkels	96%
B-sikkel	0%

4. Minder evident in een vergelijking van de A-sikkels met andere werktuigen is de aanwezigheid van grofkorrelige inclusies. Bij de sikkels zijn deze inclusies vaker aanwezig en zijn ze markanter dan bij andere voorwerpen. Een vergelijking levert de volgende cijfers op.

Bijlen	52%
Dolken	46%
Klingen	29%
A-sikkels	84%
B-sikkel	100%

5. Meer dan bij de andere voorwerpen bezit de vuur-

Fig. 3. Sikkel van het type A uit Weerdinge, gemeente Emmen (collectie F. Modderkolk, 9), met duidelijk zichtbare 'melkachtige'vlekjes in een transparante matrix. Ware lengte 16,4 cm.

Fig. 4. Detail van een sikkel van het type A uit Annen, gemeente Anloo (D.M. 1993/IV.9), met 'melkachtige'vlekjes en flarden.

steen waarvan de sikkels zijn gemaakt vaak een bruinige 'zweem'. We moeten daarbij direct aantekenen dat, in tegenstelling tot de voorgaande kenmerken, de kleur beïnvloed kan zijn door patinering. Hierdoor kunnen de onderstaande percentages enigszins vertekend zijn.

Bijlen	12%
Dolken	31%
Klingen	3%
A-sikkels	57%
B-sikkel	0%

Dat de sikkels (van het type A) wat betreft de vuursteen een afzonderlijke groep vormen is duidelijk op grond van het al of niet voorkomen van de bovenstaande

kenmerken. Dit blijkt overigens nog meer wanneer we berekenen bij hoeveel van de verschillende werktuigen de eerste drie bovengenoemde kenmerken – de meest markante – voorkomen.[8]

Bijlen	10%
Dolken	23%
Klingen	3%
A-sikkels	88%
B-sikkel	0%

De uniformiteit in de voor A-sikkels gebruikte vuursteen is dus groot. Dit betekent dat het materiaal niet afkomstig kan zijn uit secundaire voorkomens zoals de

Weichselienmorenes in Sleeswijk-Holstein. In dat geval zou een beeld te verwachten zijn zoals dat bijvoorbeeld bij de bijlen optreedt. Ze zijn vervaardigd van een heel scala verschillende vuursteensoorten, ieder met specifieke kenmerken.[9] De grondstof voor de A-sikkels moet dus uit een primair voorkomen afkomstig zijn.

De combinatie van kenmerken maakt de vuursteen zo typerend dat het materiaal zelfs op het oog meestal goed herkenbaar is.

Op enkele sikkels waarvan de grondstof iets afwijkt van het gemiddelde komen we in de conclusie nog terug. Markant is het feit dat de B-sikkel van een andere grondstof is vervaardigd dan de andere sikkels. Vlekjes en spikkelwolken ontbreken, de vuursteen is grijs en bevat veel bryozoën.

Een combinatie van de kenmerken, vlekjes spikkelwolken en afwezigheid van bryozoën, komt niet uitsluitend bij de sikkels voor, maar is ook te vinden bij de bijlen, dolken en in mindere mate bij de klingen. Wanneer we op het oog een inschatting maken van de vuursteen dan is 98% van de A-sikkels gemaakt van hetzelfde materiaal. Van de bijlen is 6% en van de dolken 23% gemaakt van dezelfde vuursteen. Onder de klingen bevindt zich geen enkel exemplaar dat van deze grondstof is vervaardigd.

De bovenstaande cijfers komen vrijwel overeen met de berekening van de combinatie van de kenmerken 'vlekjes, spikkelwolken en afwezigheid van bryozoën'.

4. SIKKELS IN SLEESWIJK-HOLSTEIN

Het feit dat de in Nederland gevonden B-sikkel wat betreft grondstof afwijkt van de A-sikkels is aanleiding geweest om ook in Duitsland een (beperkt en in feite te klein) sample van zowel A- als B-sikkels te onderzoeken. Omdat vooral in Sleeswijk-Holstein A- en B-sikkels deels in dezelfde gebieden voorkomen (Kühn, 1979: kaart 15 en 16; fig. 5 en 6) werden 13 exemplaren van het eerste en 8 exemplaren van het tweede type onderzocht in het Archäologisches Landesmuseum der Christian-Albrechts-Universität in Sleeswijk (tabel 2).

Van de A-sikkels bleken 11 exemplaren qua materiaal overeen te komen met de Nederlandse exemplaren. Van de B-sikkels bleek geen enkel exemplaar van de vlekjeshoudende vuursteen te zijn gemaakt.

Alhoewel niet helemaal zo evident als in Noord-Nederland – 2 van de 13 exemplaren wijken af – vormen de A-sikkels ook in Sleeswijk-Holstein grotendeels een groep (in het onderzochte sample 85%) met hetzelfde herkomstgebied. De B-sikkels zijn niet alleen van andere vuursteen gemaakt, de diversiteit van de voor deze sikkels gebruikte vuursteen is groter, hetgeen wijst op een herkomst uit een morenegebied. De verspreiding van de sikkels bevestigt dit (fig. 6). Kühn (1979: p. 64) zegt over deze verspreiding:

B-Sicheln finden sich, abgesehen vom Mittelrücken, im gesamten Schleswig-Holstein in annähernd gleichmässiger Verbrei-

Tabel 2. In Sleeswijk-Holstein bestudeerde sikkels, alle in de collectie van het Archäologisches Landesmuseum der Christian-Albrechts-Universität in Sleeswijk.

Inv.nr.	Vindplaats	
Type A		
KS 13242	Sehestedt,	Kr. Eckernförde
KS 20618	Ringsburg,	Kr. Flensburg
KS 19574	Oldersbek,	Kr. Husum
KS 10996	Weddingstedt,	Kr. Norderdithmarschen
KS 13051	Weddingstedt,	Kr. Norderdithmarschen
KS 18613a	Silberstedt,	Kr. Schleswig
KS 20902	Albersdorf,	Kr. Süderdithmarschen
KS 8976b	Nordhastedt,	Kr. Süderdithmarschen
KS 8976c	Nordhastedt,	Kr. Süderdithmarschen
KS 5500	Kampen,	Sylt
KS 10994	Weddingstedt,	Kr. Norderdithmarschen
8976a	Nordhastedt,	Kr. Süderdithmarschen
KS 15220	Schuby,	Kr. Schleswig (door Kühn als B-sikkel betiteld)
Type B		
KS 12440	Brodersby,	Kr. Eckernförde
KS 13748	Weddinghausen,	Kr. Norderdithmarschen
KS 11224	Neuratjensdorf,	Kr. Oldenburg
KS 8988	Schleswig,	Kr. Schleswig
KS 8987	Albersdorf,	Kr. Süderdithmarschen
KS 24790	Bunsoh,	Kr. Süderdithmarschen
KS 11972	Grossenrade	Kr. Süderdithmarschen
KS 4057	Rantum,	Sylt

Fig. 5. De verspreiding van sikkels van het type A naar Kühn (1979: Karte 15).

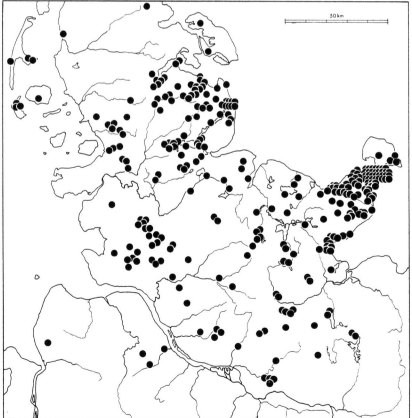

Fig. 6. De verspreiding van sikkels van het type B naar Kühn (1979: Karte 16).

tungsdichte. Eine Fundkonzentration zwischen Heiligenhafen und Grossenbrode, beide Kreis Oldenburg, deckt sich mit dem massierten Auftreten von Vorarbeiten und Halbfabrikaten desselben Sicheltyps (Karte 2). Offensichtlich wird damit eines der Produktionszentren dieses Sicheltyps innerhalb des Arbeitsgebietes angedeutet.

Ook op andere plaatsen langs de oostkust van Sleeswijk-Holstein, vooral waar steilkusten liggen en relatief veel vuursteen te vinden is, zijn halffabrikaten van (uitsluitend) B-sikkels[10] gevonden. De vuursteen die hier voorkomt is van goede kwaliteit in tegenstelling tot de vuursteen die bijvoorbeeld in Noord-Nederland in Saalienmorenes wordt aangetroffen. Deze is langer geleden afgezet en heeft een langer transport achter de rug.

De verspreiding van de A-sikkels in Sleeswijk-Holstein wijkt af van die van de B-sikkels (fig. 5). Kühn (1979: p. 64) zegt daarover:

In Ihrer Verbreitung sind die Flintsicheln des Typs A an die Altmoränengebiete des westlichen Landesteiles gebunden, wobei sich in Dithmarschen und auf der nordfriesischen Insel Sylt Verbreitungsschwerpunkte abzeichen. Vereinzelt sind Exemplare treeneaufwärts bis in das westliche Angeln gekommen. Aus dem östlichen Schleswig-Holstein ist bislang keine einzige Flintsichel des Typs A bekannt geworden. Somit spricht auch das Verbreitungsbild dieses Sicheltyps, das dem der Flintdolche des Typs I auffallend ähnelt, für die Existenz regional begrenzter Formenkreise innerhalb des Arbeitsgebietes.

Bij zijn behandeling van de door M. Malmer opgestelde dolk/sikkel-regel[11] gaat Kühn (1979: pp. 27-28) ervan uit dat produktie van A-sikkels plaatsgevonden heeft in de

Altmorenenlandschaften des westlichen Landesteiles, in denen der Flint minderwertiger und schwerer erreichbar ist als in den Jungmoränengebieten im östlichen Landesteil.

Hij komt tot de conclusie dat de Malmer-regel voor het oostelijk deel van Sleeswijk-Holstein wel en voor het westelijk deel niet geldt. Zoals reeds gezegd is de vuursteen voor de A-sikkels niet uit een morene afkomstig. Dat de Malmer-regel in westelijk Sleeswijk-Holstein niet opgaat is daarom vanzelfsprekend. De sikkels komen hier niet vandaan. Dit wordt bevestigd door het feit dat de A-sikkels technologisch gezien niet onderdoen voor de B-sikkels. Tot slot wijzen ook de afmetingen van de sikkels daarop. Zouden de A-sikkels zijn gemaakt in een gebied met inferieure (Saalien) vuursteen, dan mogen we verwachten dat ze gemiddeld kleiner zijn dan de B-sikkels. Bij het in Duitsland bestudeerde sample blijkt juist het tegendeel. De gemiddelde lengte van de A-sikkels (alleen die van de gevlekte vuursteen) bedraagt 15,8 cm (de in Nederland bestudeerde A-sikkels zijn gemiddeld 15,4 cm lang). De gemiddelde lengte van de B-sikkels is 13,2 cm (de B-sikkel uit Weerdinge is 9,6 cm lang).

5. DE HERKOMST VAN DE A-SIKKELS

Er zijn binnen het verspreidingsgebied van de A-sikkels slechts drie primaire vuursteenvoorkomens te vinden, namelijk Hemmoor, Lüneburg en Helgoland. Het is ondenkbaar dat de sikkels uit Hemmoor afkomstig zijn, omdat de daar voorkomende vuursteen van volstrekt andere aard is dan het voor de sikkels gebruikte materiaal. Bovendien wijst het verspreidingsbeeld van de sikkels (fig. 5) niet op een produktie van deze voorwerpen in de buurt van Hemmoor. Het laatste geldt ook voor Lüneburg. Bovendien liggen vuursteenhoudende lagen daar zo diep dat ze voor de prehistorische mens niet bereikbaar waren. Alleen Helgoland resteert dan als herkomstgebied.

Naast de reeds genoemde rode soort komen op Helgoland ook vier andere vuursteensoorten voor, waarvan vooral twee grijze bruikbaar zijn voor de vervaardiging van werktuigen. Schmid & Spaeth (1981: pp. 36-38) beschrijven deze als volgt:

Dunkelgrauer bis schwarzer Brocken und Knochenflint.
Die nicht seltenen Exemplare dieses Typs stellen den grössten Anteil der Feuersteingerölle Helgolands. Ihre sehr unterschiedliche Grösse schwankt zwischen etwa walnussgrossen und bis gegen 35 cm maximaler Länge erreichenden Stücken. Es handelt sich um gerundete, z.T. knollenförmige Brocken beziehungsweise länglich-knochenförmige Gebilde. Diese Feuersteine bestehen aus einem dunkelgrauen bis schwarzen Kern ohne Rinde. Die Oberfläche ist im allgemeinen glatt.
Plattenflint.
Es handelt sich um flache Konkretionen, die mehrere Zentimeter Dicke erreichen und zwischen 30 und 40 cm Durchmesser zeigen können. Die Farbe des Kerns ist grau bis grauschwarz, teilweise treten auch blassbräunliche Farbtöne auf. Eine Rinde ist nicht ausgebildet, die Oberfläche ist mehr oder weniger glatt, gelegentlich mit wulstigen Auswüchsen.

De door Schmid en Spaeth beschreven *Knochenflint* is door de vorm waarin het voorkomt niet geschikt voor het maken van sikkels, dolken en bijlen. De *Plattenflint* daarentegen heeft een vorm die perfect is voor de produktie van sikkels en dolken. De knollen zijn ook nu nog in grote hoeveelheden op het strand van de oostelijk van het eigenlijke Helgoland gelegen Düne te verzamelen. De kenmerken van de plaatvormige vuursteenknollen van Helgoland komen overeen met de kenmerken van de A-sikkels. De meestal transparante matrix bevat vrijwel zonder uitzondering in wolken voorkomende vlekjes zonder structuur en met een grootte van gemiddeld 0,2 tot 1 mm. Ook komen flarden met dezelfde kleur voor. Hun aantal kan tot enkele tientallen per cm² oplopen (fig. 7). Wolken van kleine spikkels, vaak rond inclusies en met naaldvormige structuren erin komen in de vuursteen voor. Afgezien van de naaldvormige structuren komen in de vuursteen slechts sporadisch bryozoën voor. Grofkorrelige inclusies treden wel vaak op. De kleur van de vuursteen is grijs maar vaak komt een bruin 'zweem' voor (vergelijk ook Schmid & Spaeth, 1981: p. 38). De kleur kan zelfs naar beige-achtig overgaan. Er lijken verschillen te bestaan in de

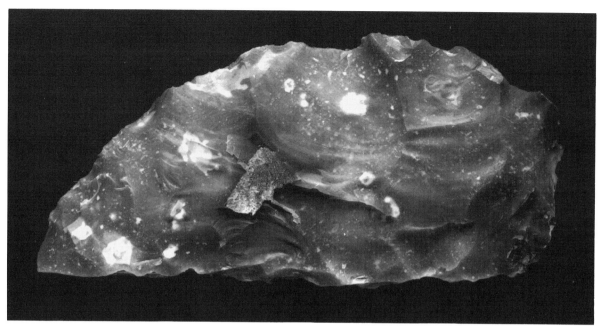

Fig. 7. Recent bewerkt stuk *Plattenflint* van Helgoland (lengte 19,3 cm).

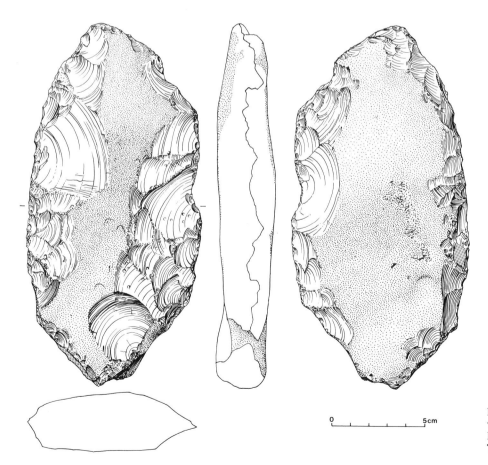

0   5cm

Fig. 8. Halffabrikaat van een
vuurstenen sikkel, gevonden te
Middelstum-Boerdamsterweg.
Tekening H.R. Roelink, B.A.I.

mate van voorkomen van vlekjes in de verschillende kleurvarianten maar voortgezet onderzoek is noodzakelijk om daar meer definitieve uitspraken over te kunnen doen.

De overeenkomst tussen de plaatvormige vuursteen van Helgoland en de A-sikkels is zo evident dat er geen twijfel kan bestaan over de herkomst van de meeste A-sikkels. Zelfs minder voorkomende varianten als de beige-achtige soort zien we terug in het bestand aan sikkels. Interessant is in dit verband de vondst van een halffabrikaat van een sikkel te Middelstum-Boerdamsterweg (fig. 8; Stapert, 1988). Het voorwerp is gemaakt van een afgeplatte vuursteenknol en bezit verder wat betreft de vuursteen alle kenmerken van de A-sikkels. Op grond van dit voorwerp kunnen we veronderstellen dat sikkels soms als halffabrikaten werden geïmporteerd, maar dit lijkt meer uitzondering dan regel te zijn geweest.

Al bij de bestudering van de rode vuursteen van Helgoland is gebleken dat sporen van prehistorische vuursteenwinning en -bewerking op het eiland niet of nauwelijks worden gevonden. Sinds de prehistorie is veel van het eiland aan de zee ten prooi gevallen. Van de kalkkliffen waarin de vuursteenknollen voorkwamen resteert niets meer. Van Helgoland zelf is slechts één sikkelfragment bekend. Helaas is het tijdens bombardementen in de oorlogsjaren in het Nordseemuseum op Helgoland verloren gegaan (Ahrens, 1966: p. 355).

Ook van de rode vuursteen van Helgoland zijn sikkels gemaakt zij het in veel mindere mate dan van de plaatvormige vuursteen.[12] Tot nu toe zijn er slechts zes fragmenten bekend. In Nederland werd bij Velsen een mediaal fragment gevonden. In Duitsland werden sikkelfragmenten gevonden bij Salzderhelden (Beuker, 1988: p. 100), Sievern (Beuker, 1988: p. 98), Appeln[13] (Schön, 1989: p. 19), Westerende (mond. meded. W. Schwarz) en Boberg (Beuker, 1988: p. 98).

6. CONCLUSIE EN SLOTOPMERKINGEN

Zoals gezegd is de plaatvormige vuursteen van Helgoland meestal op het oog goed herkenbaar. Een combinatie van de kenmerken is belangrijk bij de herkenning. Een aantal kenmerken van de plaatvormige vuursteenknollen komt ook wel voor bij sommige vuursteensoorten die bijvoorbeeld aan de oostkust van Sleeswijk-Holstein te vinden zijn. Toch is het karakter daarvan vaak zo afwijkend dat eenduidig kan worden vastgesteld dat we met ander materiaal te maken hebben. Zo komen vlekjes vaak voor in combinatie met een grote hoeveelheid bryozoën. In andere gevallen liggen ze weliswaar in wolken in de matrix (en ontbreken de bryozoën) maar zijn ze gebonden aan fluïdale structuren in de vuursteen, iets wat we bij de vuursteen van Helgoland niet tegenkomen.

Alle, of vrijwel alle A-sikkels uit Nederland zijn van de plaatvormige Helgoland-vuursteen gemaakt. Slechts bij enkele stukken is de determinatie niet geheel zeker.

Bij een sikkel uit Hartwerd is de matrix doorspekt met naaldvormige structuren. In de vergelijkingscollectie in het Drents Museum is deze vuursteen niet als variant van de Helgolander *Plattenflint* aanwezig. Wel kennen we stukken waarvan delen vol zitten met 'naalden'. De grondstof voor twee andere sikkels, te weten een uit Rhee en een uit West-Friesland, is niet in alle opzichten typerend voor Helgoland-vuursteen maar botst ook niet met de kenmerken ervan.[14]

Van de in Sleeswijk-Holstein bestudeerde 13 A-sikkels bleken slechts 2 niet van Helgolandvuursteen gemaakt te zijn. Onder de bestudeerde B-sikkels bevond zich echter geen enkel exemplaar van deze vuursteen. Dit type werd gemaakt van vuursteen van de oostkust. Ook de Nederlandse B-sikkel zou daar vandaan kunnen komen.

Het onderzoek van de sikkels heeft duidelijk gemaakt dat zich ook onder de bijlen exemplaren bevinden die gemaakt zijn van de Helgolander *Plattenflint*. Het aantal bijlen van dit materiaal is echter relatief gering. Vermoedelijk is dit te verklaren door de vorm van de vuursteen. De 'knollen' zijn vaak zo dun dat er moeilijk bijlen van kunnen worden gemaakt. Bij de tot nog toe bekende bijlen gaat het dan ook meest om relatief kleine en platte exemplaren. Dat het aantal bekende bijlen van de rode vuursteen van Helgoland groter is dan die van de hier onderzochte vuursteen kan overigens ook iets met esthetische aspecten te doen hebben. Misschien waren de rode meer geliefd (Beuker, 1988: p. 113 en 1986: p. 134).

Een interessante vondst, zowel wat betreft de rode vuursteen van Helgoland als de plaatvormige vuursteen is het vuursteendepot van Een. Het bestaat uit een halffabrikaat van de rode vuursteen, twee bijlen, twee *Knochenflinten* en twee *Plattenflinten*. Een van de bijlen werd al eerder gedetermineerd als afkomstig van Helgoland. Het onderzoek van de sikkels bevestigt deze eerdere inschatting. De bijl is gemaakt van *Plattenflint*. Het oppervlak van de twee plaatvormige vuursteenknollen uit het depot wordt nog grotendeels door cortex gevormd. De vuursteen kan slechts bij één exemplaar goed bestudeerd worden. Deze op grond van zijn vorm en associatie aan Helgoland toegeschreven knol blijkt alle kenmerken te bezitten van de vuursteen waarvan ook de sikkels zijn gemaakt.

Het aantal dolken dat van deze vuursteen is gemaakt is groter dan het aantal bijlen maar veel geringer dan het aantal sikkels. Toch was de vuursteen ook bij uitstek geschikt om dolken van te maken. Kühn (1979: p. 27) zegt over de produktie van dolken:

> Dagegen sprechen Funde von Vorarbeiten für lanzettförmige Flintdolche in Dithmarschen für die Herstellung der frühen Flintdolche in den Altmoränenlandschaften Westholsteins.

Of er specifieke dolktypen van de *Plattenflint* werden vervaardigd is nog niet verder onderzocht maar het is niet ondenkbeeldig dat we bij de door Kühn bedoelde dolken ook met importen van Helgoland te maken hebben.

Over het ontbreken van klingen van de hier besproken vuursteen kunnen we kort zijn. De plaatvormige knollen waren ongeschikt voor de produktie van klingen. Dit is overigens niet het geval voor de Knochenflinten van Helgoland.

De vuursteen waarvan de A-sikkels zijn gemaakt werd al in het Neolithicum gebruikt. In de loop van de Bronstijd en tijdens de vroege en midden-IJzertijd werd het echter pas op grote schaal tot voorwerpen bewerkt. In deze periode valt namelijk de datering van de vuurstenen sikkels.

De in dit artikel gepresenteerde gegevens zijn te beschouwen als de voorlopige resultaten van het onderzoek. Het ligt in de bedoeling de inventarisaties op grotere schaal uit te voeren waarbij niet alleen naar het Nederlandse, maar vooral ook naar het Duitse materiaal zal worden gekeken. Ook zal worden getracht een nog beter inzicht te krijgen in de specifieke kenmerken van verschillende noordelijke vuursteensoorten. Aan het toepassen van natuurwetenschappelijke methoden van determinatie wordt daarbij prioriteit gegeven. We moeten daarbij overigens wel bedenken dat de herkomst van veel noordelijke vuursteensoorten moeilijk te achterhalen is, omdat dit materiaal niet gebonden is aan een bepaalde plaats maar overal in (bijvoorbeeld) de Weichselienmorenes in Noord-Duitsland kan zijn gewonnen.

7. DANKBETUIGINGEN

Voor het totstandkomen van dit artikel ben ik dank verschuldigd aan de volgende personen: drs. J.W. Boersma van het Groninger Museum en drs. J.N. Lanting van het Biologisch-Archaeologisch Instituut, G. Holtrop te Borger, drs. E. Kramer van het Fries Museum, F. Modderkolk te Assen, R. Trip te Emmen, dr. I. Ulbricht en dr. J. Hoika van het Archäologisches Landesmuseum der Christian-Albrechts-Universität, dr. W. Schwarz van de Ostfriesische Landschaft, drs. T.Y. van de Walle-van der Woude en J. Maas van het Westfries Museum en M. Wetterauw te Groningen.

8. NOTEN

1. Dit onderzoek concentreert zich met name op de herkomst van werktuigen die gemaakt zijn van noordelijke vuursteen.
2. In Noord-Nederland komen sikkels van een ander type nauwelijks voor.
3. Hansen & Madsen (1983: p. 53) geven voor de vervaardiging van dergelijke bijlen van ruwe knol tot bijl een gemiddelde gewichtsvermindering van maar liefst 84%.
4. In de collectie van het Drents Museum maken ze 6% uit van het totale aantal bijlen.
5. Uitgaande van de complete exemplaren. Veel van de fragmenten konden ook typologisch worden toegewezen, echter niet alle stukken. In dit artikel is de laatste categorie als A-sikkel beschouwd.

6. Hetzelfde geldt voor alle sikkelfragmenten.
7. In sommige gevallen komt een minder scherpe begrenzing voor. In enkele gevallen hebben de vlekjes een rafelig uiterlijk.
8. Bij de bijlen, dolken en klingen is bijvoorbeeld het voorkomen van vlekjes vaak gebonden aan het voorkomen van bryozoën.
9. Daarin is wel een aantal groepen te onderscheiden.
10. Daarnaast zijn er veel halffabrikaten van bijlen en dolken gevonden.
11. Op grond van het verspreidingsbeeld van sikkels en dolken in Schonen kwam Malmer tot de conclusie dat sikkels in de omgeving van vuursteenvoorkomens in groter aantal werden vervaardigd dan in vuursteenarme gebieden waar het schaarse materiaal blijkbaar vooral voor het maken van dolken werd gebruikt.
12. De rode vuursteen is zelfs op Helgoland relatief zeldzaam.
13. De vindplaats bij Appeln leverde tot nu toe maar liefst 35 sikkels op.
14. Ook op Helgoland worden minder typerende stukken gevonden.

9. LITERATUUR

AHRENS, C., 1966. *Vorgeschichte des Kreises Pinneberg und der Insel Helgoland*. Neumünster.

BAKKER, J.A., 1979. *The TRB West Group. Studies in the chronology and geography of the makers of hunebeds and Tiefstich pottery*. Amsterdam.

BEUKER, J.R., 1986. De import van Helgoland-vuursteen in Drenthe. *Nieuwe Drentse Volksalmanak* 103, pp. 111-135.

BEUKER, J.R., 1988. Die Verwendung von Helgoländer Flint in der Stein- und Bronzezeit. *Die Kunde* 39, pp. 93-116.

BEUKER, J.R., 1990. The importation of Heligoland-flint in the province of Drenthe (The Netherlands) In: M.R. Séronie-Vivien & M. Lenoir (éd.) *Le silex de sa genèse à l'outil, Actes du V Colloque International sur le silex, Bordeaux, 17 Sept.-2 Oct. 1987* (= Cahiers du Quaternaire 17). Paris, pp. 311-319.

BOERSMA, J.W., 1988. De datering van een vuurstenen sikkel uit Middelstum-Boerdamsterweg. In: M. Bierma et al. (eds), *Terpen en wierden in het Fries-Groningse kustgebied*. Groningen, pp. 31-35.

BRUNSTING, H., 1962. De sikkels van Heilo. *Oudheidkundige Mededelingen van het Rijksmuseum van Oudheden te Leiden* 43, pp. 107-115.

BRUYN, A., 1986. Een vuurstenen sikkel uit Medemblik. *Jaarverslag Rijksdienst voor het Oudheidkundig Bodemonderzoek 1984*, pp. 89-94.

BUTLER, J.J., 1990. Bronze Age metal and amber in the Netherlands (I). *Palaeohistoria* 32, pp. 47-110.

DORST, M.C., 1980. Archeologische documentatie en automatisering. In: M. Chamalaun et al. (eds), *Voltooid verleden tijd*. Amsterdam, pp. 229-243.

GROENMAN VAN WAATERINGE, W. & J.F. VAN REGTEREN ALTENA, 1961. Een vuurstenen sikkel uit de voor-Romeinse IJzertijd te Den Haag. *Helinium* 1, pp. 141-146.

GIJN, A.L. VAN & H.T. WATERBOLK, 1984. Colonization of the salt marshes of Friesland and Groningen. *Palaeohistoria* 26, pp. 101-122.

GIJN, A.L. VAN, 1988. The use of Bronze Age flint sickles in the Netherlands: a preliminary report. In: S. Beyries (ed.), *Industries lithiques, tracéologie et technologie* (= BAR International Series 411). Oxford, pp. 197-186

HANSEN, P.V. & B. MADSEN, 1983. An experimental investigation of a flint axe manufacture site at Hastrup Vaenget, East Zealand. *Journal of Danish Archaeology* 2, pp. 43-59.

KÜHN, H.J., 1979. *Das Spätneolithikum in Schleswig-Holstein*. Neumünster.

LOUWE KOOIJMANS, L.P., 1985. *Sporen in het land. De Nederlandse delta in de prehistorie*. Amsterdam.

SCHMID, F. & C. SPAETH, 1981. Feuersteintypen der Oberkreide Helgolands, ihr stratigraphisches Auftreten und ihr Vergleich mit anderen Vorkommen in N.W. Deutschland. In: F.H.G. Engelen (ed.), *Derde internationale symposium over vuursteen, 24-27 mei 1979, Maastricht* (= Staringia 6). Heerlen, pp. 35-38.

SCHÖN, M.D., 1989. *Feuerstein. Rohstoff der Jahrtausende.* Cuxhaven.

STAPERT, D., 1988. Een sikkel en een halffabrikaat van Middelstum-Boerdamsterweg. In: M. Bierma et al. (eds), *Terpen en wierden in het Fries-Groningse kustgebied.* Groningen, pp. 36-49.

WAALS, J.D. VAN DER, 1972/73. Vondsten van de Uitwedsmee bij Onstwedde. *Groningse Volksalmanak*, pp. 167-182.

EEN MENSELIJK SKELET UIT DE ASCHBROEKEN BIJ WEERDINGE (DRENTHE) RECONSTRUCTIE VAN EEN MISVERSTAND

W.A.B. VAN DER SANDEN
Drents Museum, Assen, Netherlands

C. HAVERKORT & J. PASVEER
Laboratorium voor Anatomie en Embryologie, Groningen, Netherlands

ABSTRACT: In 1992 a human skeleton turned up in the stores of the Biologisch-Archaeologisch Instituut. This skeleton, now in the Drents Museum in Assen, is fairly complete. Unfortunately the skull is no longer present. A note indicates that this skeleton was found in the bog near Weerdinge, municipality of Emmen, province of Drenthe.

In this article it is argued that this skeleton can only be the so-called skeleton of Aschbroeken, discovered in 1931. This means that the skeleton that was published several years ago as the Aschbroeken skeleton must belong to the bog body of Zweeloo, discovered in 1951 (van der Sanden, 1990: pp. 89-90 and 115-117).

The real Aschbroeken skeleton is that of an adult male, who died at an age of 35-45. His stature is estimated at 170.8 ±2.99 cm. Aschbroeken Man had several growth arrests during his childhood and adolescence. He broke his right upper arm. The fracture healed, but not perfectly. His lumbar vertebrae show the beginning of lipping. The ^{14}C date (OxA-3917; 2940±75 BP) indicates that he was deposited in the Bourtanger Moor at the end of the Middle Bronze Age or the beginning of the Late Bronze Age.

The article concludes with a summary of the information now available about the Zweeloo bog body, the bog corpse that now has its skeleton again, i.e. the former Aschbroeken skeleton. Zweeloo Woman died in the Roman period and was deposited in a small bog in late summer or early autumn.

KEYWORDS: Drenthe, bog body, physical anthropology, Middle and Late Bronze Age, pathology, Harris-lines.

1. INLEIDING

In augustus 1992 werd tijdens opruimingswerkzaamheden in het depot van het Biologisch-Archaeologisch Instituut een reeks houten kisten met Drentse bodemvondsten aangetroffen. Deze kisten werden naar het Drents Museum gebracht, waar de inhoud aan een nader onderzoek werd onderworpen. Een van de kisten bevatte een opmerkelijke vondst. Onder een aantal runderhorens lag een grote hoeveelheid donkergekleurde menselijke botten, duidelijk van één individu afkomstig. Meest opvallende afwezige was de schedel. Het kaartje in de kist vermeldde niet meer dan 'Emmen Weerdingerveen'. Nadere informatie over dit veenskelet ontbrak.[1]

Al spoedig drong het besef door dat het bij deze vondst eigenlijk niet anders kon gaan dan om het 'skelet van Aschbroeken', een ontdekking uit 1931. Deze constatering bracht meteen een nieuw probleem met zich mee, want er was al een skelet uit de Aschbroeken bekend. Over dit skelet werd in 1990 gepubliceerd in een studie gewijd aan de Noordnederlandse veenlijken (Van der Sanden, 1990: pp. 89-90). Dat bewuste skelet heeft toebehoord aan een vrouw van ongeveer 35 jaar (Uytterschaut, 1990: pp. 115-117). De nieuwe vondst maakte duidelijk dat er destijds een vergissing in het spel moet zijn geweest. Wij zijn thans de mening toegedaan dat het vermeende skelet van Aschbroeken

hoort bij het veenlijk van Zweeloo.

In het onderstaande zal eerst op het in 1992 'ontdekte' skelet worden ingegaan; daarbij zullen alle beschikbare bronnen – brieven, krantartikelen en mondelinge mededelingen – worden gecombineerd. Aan het slot van het artikel zal worden nagegaan wat een en ander betekent voor het veenlijk van Zweeloo.

2. DE VINDPLAATS

Het skelet van Aschbroeken werd op 3 juni 1931 in de omgeving van Weerdinge, gemeente Emmen, gevonden; de ontdekking vond plaats in veenput 7 (fig. 1 en 2). Al vrij snel daarna werd het skelet gelicht. De heer A. Katuin uit Valthe herinnerde zich nog dat het een dag op een baggelschot heeft gelegen en dat het tegen betaling van een dubbeltje bezichtigd mocht worden.[2] Op 4 juni werd het skelet voor een bedrag van Hfl. 35,- aangekocht door H.F. Buiskool, hoofd der school te Weerdinge.

Enkele dagen later bezocht L. Postema, werkzaam als tekenaar op het B.A.I., de vindplaats. In een op 7 juni gedateerde brief deelde hij aan Van Giffen het volgende mee.[3]

De tocht naar Weerdinge heeft niet aan de verwachtingen voldaan, wij hadden ons gepr. op alles om zoo mogelijk het skelet in zijn geheel op te kunnen nemen. De voornaamste deelen zijn nog

155

aanwezig de schedel en onderkaak zijn prachtig het is een rond-
kop. De diepte van het skelet was 1.80 M op zijn minst was de
geheele hoogte van het veen 3.00 M. Het skelet hebben wij
opgenomen en in de auto medegenomen naar Groningen. Jan kan
het nu in lijmwater koken, de beenderen zijn iets vergaan. Zoals
u zeker zult weten is er 35 gulden voor betaald door den heer
Buiskool. Nogmaals jammer is het dat alles niet meer was als het .
gevonden is ....

Fig. 1. De ligging van Zweeloo en Weerdinge.

Op 8 juni ontving ontdekker A. Lubach van Buiskool
Hfl. 5,-

> voor het bewaken van het skelet ..., voor het nemen van grond-
> monsters en bijbehorende kist.

Op 9 juni schreef Van Giffen – als conservator van het
Provinciaal Museum – aan Buiskool dat hij de pen-
ningmeester opdracht had gegeven om Hfl. 35,- over te
maken. In de brief klinkt enige kritiek door:

> Het is alleen jammer, dat het reeds was opgenomen en daardoor
> toch onvolledig is. Ook zijn enkele kiezen verloren gegaan.

Verder vroeg Van Giffen om meer informatie. Buiskool
schreef hem terug:

> Het skelet lag 1,80 m. boven den beganen grond. De monsternemers
> hebben vergeten op te merken dat hier ±1 M bonkaarde was
> verwijderd. Het skelet lag op de grens van dargveen en sapropheel
> of zwartveen, dat hier overal "pikveen" wordt genoemd.... De
> bedoelde veenput was dit jaar verhuurd aan den heer J. Westerhof,
> die uit een oogpunt van tijd- en geldverlies bezwaar maakte, het
> skelet in den oorspronkelijken toestand te laten liggen.

Op een schets van het profiel, die in het correspondentie-
archief van het B.A.I. te voorschijn kwam (fig. 3), staan
tien (grond)monsterpunten aangegeven, één ter hoogte
van het skelet, twee eronder en zeven erboven. Die
grondmonsters zijn – zo weten we uit een van de brieven
– genomen door A. Lubach. Het kaartje is vermoedelijk
door een medewerker van het B.A.I. vervaardigd; dit
blijkt uit een brief die Van Giffen op 8 februari 1933 aan
de paleobotanicus F. Florschütz schreef. Of Florschütz
de grondmonsters ooit geanalyseerd heeft is niet bekend,
het heeft in ieder geval geen weerslag gevonden in de

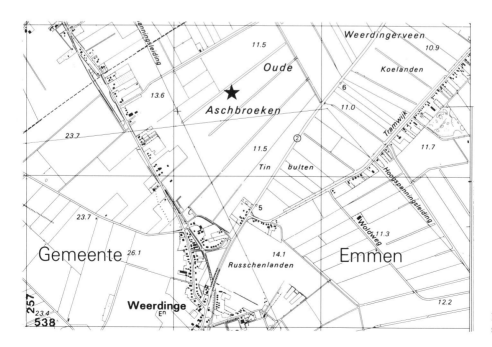

Fig. 2. De vindplaats van het
skelet van Aschbroeken.

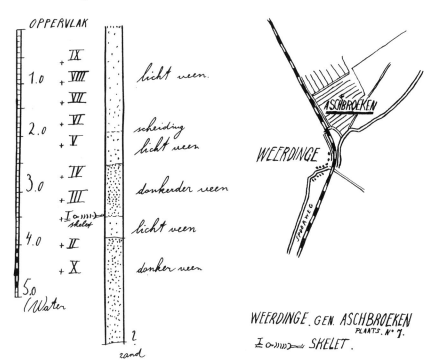

Fig. 3. Schets uit de jaren dertig waarop de vindplaats van het skelet is aangegeven alsmede het profiel met de tien monsterpunten.

wetenschappelijke literatuur (Van der Sanden, 1992: pp. 146-147).

Een volgende bericht dateert uit 1938. Het *Nieuwsblad van het Noorden* vermeldde op 29 oktober van dat jaar in een berichtje over de ontdekking van het veenlijk van Emmer-Erfscheidenveen:

> Het lijk, eenige jaren geleden in de Asschebroeken te Weerdinge gevonden dat in tegenstelling met het thans gevondene een volledig geraamte bezat, was eveneens met bosch bedekt.

Deze informatie verkreeg de verslaggever ongetwijfeld van H.J. Bellen, een legerkapitein met archeologische belangstelling.

Een laatste vermelding stamt uit het jaar 1942. Het is een kranteberichtje van 6 oktober van dat jaar. Naar aanleiding van de vondst van een kano in het Noordveen bij Weerdinge wordt aan het slot van die vondstmelding meegedeeld:

> Op plm. 100 meter van deze zeldzame vondst werd voor een tiental jaren het geraamte gevonden van een mensch, dat bewaard wordt in het Biologisch Archeologisch Instituut te Groningen.

Het skelet en helaas ook het kranteberichtje raakten spoedig in de vergetelheid.

De informatie die thans beschikbaar is, kan als volgt worden samengevat. Het skelet van Aschbroeken was op het moment van de ontdekking waarschijnlijk compleet. Het lag noord-zuid en was (mogelijk) bedekt met rijshout. Het werd op relatief grote diepte aangetroffen, volgens de schets op een diepte van ca. 3,5 m. De schedel van het skelet was opvallend rond van vorm.

Waar deze zich thans bevindt is niet bekend. Een uitgebreide zoektocht heeft geen resultaat gehad. Zoals wel vaker met bijzondere vondsten gebeurt, is een aparte behandeling hem noodlottig geworden.

3. DATERING

Een van de ribben van het skelet is gebruikt voor een ^{14}C-datering (AMS) in Oxford. De uitkomst daarvan is 2940±75 BP (OxA-3917); δ^{13}C is -19,3‰, een normale waarde voor mensen met een terrestrisch dieet. De gecalibreerde ouderdom ligt tussen 1258-1238 of 1216-1022 BC (1 sigma), respectievelijk 1376-1348 of 1316-928 BC (2 sigma's).[4] Het skelet dateert dus uit de overgang van de midden- naar de late Bronstijd.

Uit de brief van Postema van 7 juni 1931 blijkt dat de botten met lijmwater behandeld zouden worden. Het uiterlijk maakt duidelijk dat die behandeling inderdaad heeft plaatsgevonden (zie 4.1). Hoewel de gebruikte lijmsoort niet wordt genoemd, mag aangenomen worden dat het beenderlijm betrof, d.i. een produkt op basis van dierlijk collageen. Deze lijm zal ongetwijfeld beter oplosbaar zijn in water dan het ruwe collageen uit de gedateerde rib, en zal daarom bij de chemische voorbehandeling in Oxford grotendeels verwijderd zijn. Desondanks kan niet uitgesloten worden dat de ^{14}C-datering iets te jong is uitgevallen.

4. HET FYSISCH-ANTROPOLOGISCH ONDERZOEK

4.1. Inleiding

Opvallend aan het skelet uit de Weerdinger Aschbroeken is de zeer donkere, bijna zwarte kleur: een gevolg van het verblijf in het veen. Daarnaast hebben de botten een onnatuurlijke glans. Ongetwijfeld heeft dit te maken met de lijm waarmee ze ruim zestig jaar geleden behandeld zijn. Het skelet verkeert in goede staat van conservering. De botten zijn licht maar stevig. Hoewel het vaak voorkomt dat botten die in het veen hebben gelegen vervormd zijn, is daar bij dit skelet nauwelijks sprake van.

Zoals vermeld ontbreken de schedel en de onderkaak. Het postcraniale skelet is grotendeels compleet. De ontbrekende delen zijn: het sternum, vijf cervicale wervels, drie thoracale wervels, de rechter twaalfde rib, het coccyx, alle metacarpalia en enkele metatarsalia, de rechter calcaneus en een aantal phalangen van beide

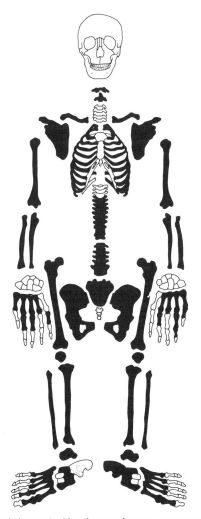

Fig. 4. Het skelet van Aschbroeken, met in zwart aangegeven de delen die gevonden zijn.

handen en voeten (fig. 4). De rechter vierde rib was oorspronkelijk wel aanwezig maar is gebruikt voor de ^{14}C-datering.

Bij het onderzoek van het skelet van Aschbroeken werden verschillende visueel-morfologische en morfometrische methoden toegepast. Daarnaast werden ook röntgenopnamen gebruikt. Bij de bespreking van de resultaten zal kort op de gehanteerde methoden worden ingegaan.

4.2. Geslacht

Om het geslacht van een skelet vast te stellen wordt meestal in eerste instantie gekeken naar de vorm van de pelvis en enkele karakteristieke delen van het cranium. Deze delen van het skelet zijn het meest onderscheidend bij beide geslachten. De pelvis vormt in dezen het meest betrouwbare deel van het skelet, omdat hierin de minste overlap tussen mannen en vrouwen bestaat. Op basis van de pelvis en het cranium, en in mindere mate de lange beenderen, kan een vrij betrouwbare geslachtsdiagnose worden gesteld. Voor de geslachtsbepaling van het skelet van Aschbroeken kan door het ontbreken van het cranium enkel gebruik worden gemaakt van de pelvis. De bouw hiervan is enigszins robuust, en zowel de grootte van de hoek van het schaambeen (angulus pubis), de inkeping aan de achterkant van beide bekkenhelften (incisura ischiadica major), als de vorm van het 'geboortekanaal' (pelvis minor) geven zonder twijfel aan dat het om een mannelijk individu gaat. De resultaten zijn dermate overtuigend dat de afwezigheid van de schedel in dit geval geen probleem vormt.

4.3. Leeftijd

De man van Aschbroeken was in ieder geval een volwassen individu, aangezien alle delen van het skelet volledig ontwikkeld zijn. Voor het bepalen van de leeftijd van een individu aan de hand van het skelet zijn verschillende methoden ontwikkeld. Bij kinderen is de leeftijd vrij nauwkeurig te bepalen omdat de leeftijd waarop verschillende botdelen fuseren en gebitselementen doorbreken min of meer vastligt. Bij volwassenen is de leeftijdsbepaling aanmerkelijk moeilijker. Bij deze leeftijdscategorie kunnen we onder meer gebruik maken van de mate van slijtage van het gebit, degeneratieve verschijnselen aan de gewrichten en de afbraak van de interne botstructuur op specifieke plaatsen.

Om de leeftijd van de man van Aschbroeken vast te stellen is de 'complexe methode' (Acsádi & Nemeskéri, 1970) gebruikt. Bij deze methode worden vier kenmerken aan het skelet (de schedelnaden, de pubissymfyse (schaambeengewricht) en de interne botstructuur in humerus en femur) beoordeeld op hun mate van 'slijtage' of degeneratie. Deze methode is relatief betrouwbaar, mits alle vier kenmerken samen worden gebruikt.

De kenmerken afzonderlijk geven in feite meer de biologische leeftijd dan de kalenderleeftijd aan. Door bepaalde factoren, zoals de leefwijze of bijzondere fysische activiteiten, kunnen sommige gewrichten namelijk sneller of trager degenereren dan 'normaal' is voor een bepaalde leeftijd. Dit levert dan een vertekening van het beeld op. Wanneer alle eerder genoemde kenmerken samen worden beoordeeld, kan een indruk worden verkregen van de kalenderleeftijd. Het is belangrijk de resultaten, indien mogelijk, ook te vergelijken met die van andere methoden, zoals bijvoorbeeld de gebitsslijtage en de microscopische botstructuren (histologisch onderzoek).

Voor het skelet van Aschbroeken was het vanzelfsprekend niet mogelijk de schedelnaden bij de complexe methode te betrekken, zodat de diagnose alleen is gebaseerd op de slijtage van de pubissymfyse en de degeneratie van de interne botstructuur in de kop van de humerus en de femur. De laatste twee konden worden beoordeeld met behulp van röntgenfoto's. De resultaten van de complexe methode geven aan dat de leeftijd ligt tussen 42-56 jaar ±3 jaar.

De interne botstructuur van de clavicula kan ook een indicatie geven voor de leeftijd (Walker & Lovejoy, 1985); ook hierbij gaat het om de mate van afbraak van deze structuur. Op basis van het beeld van röntgenfoto's van de sleutelbeenderen kon een leeftijd worden bepaald van 35-39 jaar. De algemene indruk van de leeftijd op basis van de röntgenfoto's – dat wil zeggen: van de interne botstructuur in alle gefotografeerde botten – is 'jong volwassen'.[5]

Aan leeftijd gerelateerde pathologische verschijnselen aan de botten kunnen een extra ondersteuning van de leeftijdsbepaling geven. Vooral de wervels beginnen na het 30e-35e levensjaar vatbaar te worden voor degeneratieve arthritis. Behalve in samenhang met de leeftijd, kan arthritis ook ontstaan door andere oorzaken, zoals een overbelasting van bepaalde gewrichten. Het is herkenbaar aan typerende veranderingen van het oppervlak van het gewricht, zoals het optreden van porositeit, het uitgroeien van osteofyten aan de randen van het oppervlak ('lipping'), en/of het polijsten van het botoppervlak (eburnatie). Bij het skelet van Aschbroeken is beginnende lipping vastgesteld aan enkele lumbale wervels. Omdat de overige botten geen tekenen vertonen van dergelijke aandoeningen of andere hieraan gerelateerde ziekten, kan worden verondersteld dat de gesignaleerde lipping het gevolg is van het natuurlijke ouderdomsproces. Op basis van dit gegeven is het aannemelijk dat de leeftijd van dit individu boven de 35 jaar ligt.

Samenvattend kan gesteld worden dat de leeftijd van het individu tussen de 35-45 jaar zal hebben gelegen en meer waarschijnlijk in de eerste helft van deze range.

4.4. Lichaamslengte

Het schatten van de lichaamslengte is vooral van belang

Tabel 1. Aschbroeken; lengte van de lange beenderen (in cm).

	Rechts	Links
Femur	44,5	44,6
Tibia	38,2	37,7
Fibula	36,7	36,6
Humerus	31,1	32,3
Radius	25,2	25,0
Ulna	27,1	27,0

bij demografisch onderzoek. De lichaamslengte heeft een genetische basis, maar staat tevens onder invloed van externe omstandigheden, zoals de beschikbaarheid van voedsel en het optreden van ziekten. Wanneer de gemiddelde lichaamslengte bij een bepaalde populatie in een bepaalde periode verandert, kan dit een aanwijzing zijn voor veranderende leefomstandigheden.

Voor het schatten van de lichaamslengte zijn regressie-formules ontwikkeld, waarbij de lengte van de lange beenderen wordt gebruikt. Hierbij moet rekening worden gehouden met het geslacht van het individu en met de populatie waartoe hij of zij behoort: de gemiddelde lengte is voor mannen en vrouwen verschillend en bovendien populatiegebonden. De meest toegepaste formules zijn die van Trotter & Gleser (1958) en die van Trotter (1970), beide gebaseerd op moderne blanke, negroïde en mongoloïde populaties. De resultaten zijn relatief betrouwbaar te noemen, maar hierbij moet men zich wel realiseren dat het niet bekend is in hoeverre prehistorische en recente populaties vergelijkbaar zijn, met andere woorden: in hoeverre deze methoden toepasbaar zijn op de prehistorische mens.

De meest betrouwbare schatting voor de lichaamslengte wordt verkregen op basis van de lengten van de femur en de tibia. Hiervoor geldt de volgende formule: $1.30 \, (\text{fem} + \text{tib}) + 63.29$ (Trotter, 1970). Wanneer we de waarden van de betreffende botten van het skelet van Aschbroeken invullen (tabel 1) levert dat een lengte op van 170,8 ±2,99 cm.

4.5. Trauma en pathologie

De botten van het skelet zijn uitvoerig onderzocht op aanwijzingen voor pathologie en trauma. Dit gebeurt doorgaans in eerste instantie morfologisch, maar kan daarnaast ook met behulp van radiologische technieken (röntgenfoto's) geschieden. Uiteraard kunnen alleen die ziekten of verwondingen worden vastgesteld die het bot aantasten. Het ontbreken van sporen van botpathologie hoeft dus niet te betekenen dat het individu zelden of nooit ziek was. Beide genoemde methoden zijn bij het onderzoek toegepast.

Een direct waarneembare afwijking aan het skelet van Aschbroeken is een geheelde fractuur aan het proximale deel van de rechter humerus (fig. 5). De beide delen zijn niet in de oorspronkelijke stand aan elkaar gegroeid. Het distale deel is enigszins naar buiten

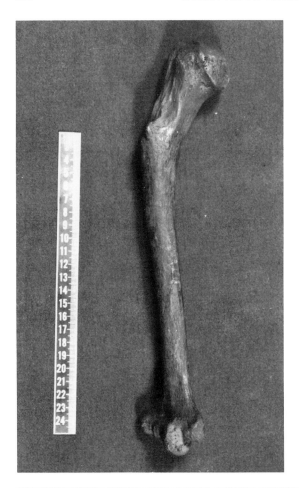

Fig. 5. Aschbroeken: geheelde fractuur aan het proximale deel van de rechter humerus. Duidelijk is te zien hoe de stand van de arm enigszins afwijkt van wat normaal is.

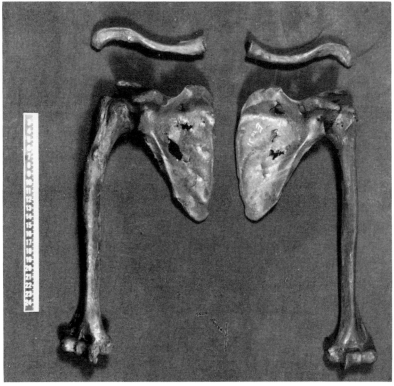

Fig. 6. Aschbroeken: het rechter schouderblad vertoont iets sterker ontwikkelde richels voor de aanhechting van spieren.

gedraaid, waardoor de rechterarm iets naar buiten zal hebben gewezen.

De rechter scapula vertoont iets sterker ontwikkelde richels voor de aanhechting van spieren (fig. 6). Een dergelijke mate van verschil in ontwikkeling tussen linker en rechter schouderblad wordt vaak geassocieerd met normale lateraliteit (links- of rechtshandigheid) (Schulter-Ellis, 1980). Het zou echter ook in verband kunnen staan met de (verkeerd) geheelde breuk, aangezien de rechterarm mogelijk extra werd belast door de ongebruikelijke positie. Omdat de breuk verder geen ernstige afwijkingen van het bot heeft veroorzaakt, mag verondersteld worden dat de arm wel normaal heeft kunnen functioneren.

De lumbale wervels vertonen, zoals vermeld, in zeer geringe mate lipping. Verder is de wervelkolom niet door arthritis aangetast. We kunnen hier dus spreken van een normale leeftijdgebonden degeneratie.

Aan de overige skeletdelen zijn geen verdere bijzonderheden waargenomen. Scheurtjes en enkele gaten in de schouderbladen en de beide bekkenhelften zijn te wijten aan postmortale processen.

Door het radiologische onderzoek kon de interne botstructuur worden bestudeerd. Van de geheelde fractuur is inwendig niets meer te zien, hetgeen inhoudt dat de breuk ruimschoots de tijd heeft gehad te herstellen en ontstaan is lang voordat het individu overleed (fig. 7).

De röntgenfoto's brachten ook Harris-lijnen aan het licht. Een Harris-lijn is een plaatselijke verdichting in het bot die het gevolg is van een groeistop. Door verschillende oorzaken kan de groei onderbroken worden. In de meeste gevallen betreft het een ziekte of voedselgebrek. De ernst van de ziekte of het voedselgebrek – al dan niet gecorreleerd – bepaalt hoe duidelijk de Harris-lijn zich zal manifesteren. Niet elke ziekte laat echter Harris-lijnen na en niet ieder individu zal even ontvankelijk zijn voor de vorming van Harris-lijnen.

De leeftijd waarop zich een groeistop heeft voorgedaan kan worden berekend aan de hand van de plaats van de lijn in het bot. Hier markeert de Harris-lijn de

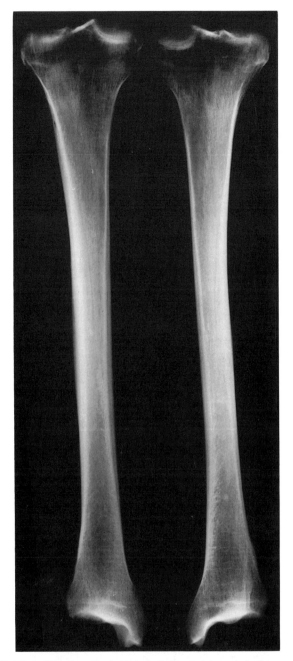

Fig. 8. Aschbroeken: Harris-lijnen in de distale delen van de tibiae.

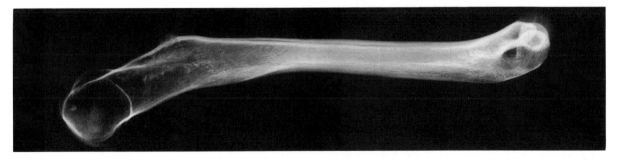

Fig. 7. Aschbroeken: röntgenfoto van de rechter bovenarm. Van de fractuur is niets meer te zien.

locatie waar zich op dat moment de epifyse bevond: het uiteinde van het bot waar de lengtegroei plaatsvindt. Aan de hand van de afstand tussen de Harris-lijn en het zogeheten 'ossificatiecentrum' – daar waar de groei begint – kan de leeftijd worden berekend waarop zich de stoornis heeft voorgedaan (Maat, 1984).

De verdichtingen in de botten van het skelet van Aschbroeken zijn het duidelijkst aanwezig in de distale delen van de tibiae (fig. 8). De positie van de lijntjes in de diafyse van de botten wijst op een aantal groei-stoornissen dat zich op jonge leeftijd heeft voorgedaan. Omdat de lijntjes niet uitgesproken sterk ontwikkeld zijn, zullen zij geen weerspiegeling zijn van extreme stressperioden, of hebben zij in elk geval geen grote invloed gehad op de gezondheid van dit individu. Volgens de methode van Maat (1984) kon berekend worden dat de eerste Harris-lijn is ontstaan op een leeftijd van ca. 9,5 jaar. Daarna zijn er op vrij regelmatige afstand nog meer lijntjes te zien, tot aan het punt waarop uiteindelijk de epifyse is vergroeid. We kunnen hieruit afleiden dat zich vanaf het 9e/10e levensjaar gedurende de gehele verdere groeiperiode met enige regelmaat een lichte groeistop heeft voorgedaan, samenhangend met gezondheidsproblemen.[6]

5. HET VEENLIJK VAN ZWEELOO EN HET MISVERSTAND

De ontdekking van het echte skelet van Aschbroeken heeft, zoals eerder gezegd, er uiteindelijk toe geleid dat het eerder als skelet van Aschbroeken gepubliceerde geraamte nu bij het veenlijk van Zweeloo gevoegd is. Dat veenlijk uit Zweeloo is een ontdekking uit 1951 (Van Zeist, 1953; 1956; Van der Sanden, 1990, pp. 91-92). In het onderstaande zal kort uiteengezet worden hoe de vergissing is ontstaan.

Tijdens het in 1987-1989 uitgevoerde onderzoek van de Nederlandse veenlijken leidde één zin in een verslag van A.E. van Giffen naar de verkeerde toewijzing. Van Giffen schreef[7]:

> Van de in 1931 door het Museum aangewonnen vondsten zijn de voornaamste ...: een skeletvondst in het hoogveen, in plaats 7, in de zoogenaamde Aschbroeken bij Weerdinge-Valthe.

Het inventarisboekje over het jaar 1931 repte weliswaar met geen woord over deze aanwinst, maar daaraan werd door de eerste auteur geen groot gewicht toegekend (zie onder). Omdat er eenvoudigweg geen ander veenskelet in de collectie aanwezig was, kon het gebeuren dat het botmateriaal van Zweeloo tot 'skelet van Aschbroeken' werd bestempeld. Deze keuze was overigens niet zo vreemd. De botten en de huid waren gescheiden opgeborgen en niet in de inventarisboekjes geregistreerd. Daar komt nog bij dat noch in het opgravingsrapport noch in de eerste, summiere publikatie over het veenlijk van Zweeloo over de aanwezigheid van botten wordt gesproken.

De beste aanwijzing voor het bij elkaar horen van huid en botten wordt geleverd door een klein fragmentje bot dat nog vastzit in de huid van de linkerschouder: het lijkt te passen aan het sterk vervormde linker schouderblad van het voormalige skelet van Aschbroeken.[8] Van Zeist vermeldde in zijn publikatie dat het Laboratorium voor Anatomie en Embryologie het veenlijk van Zweeloo determineerde als een jonge, ca. 1,70 m lange vrouw. De geslachtsbepaling zou gebaseerd kunnen zijn op analyse van het skelet, de lichaamslengte is vermoedelijk afgeleid uit de lengte van de resterende huiddelen (Van Zeist, 1953: p. 546 en Van der Sanden, 1990: p. 93, noot 34).

De achterzijde van de huid van Zweeloo is in relatief goede staat, de voorkant daarentegen bestaat uit verschillende losse stukken en is zeker niet compleet te noemen. Verschillende delen van het lichaam – zoals het hoofd, de onderarmen en de voeten – ontbreken. Vermoedelijk is er oorspronkelijk wel meer aanwezig geweest, want Van Zeist (1953: p. 549) schrijft dat hij tussen de tenen van de onbeschadigde voet zijn pollenmonsters heeft genomen.

De huid is in februari 1993 aan een nieuw onderzoek onderworpen. Dit gebeurde onder leiding van dr. M. Voortman van het Laboratorium voor Gerechtelijke Pathologie te Rijswijk. Voortman slaagde erin de huiddelen van de sterk gehavende voorkant van het lichaam op goede wijze met elkaar te verbinden. Hij stelde de aanwezigheid van de vulva en één borst – de linker – vast. Ook de huid wijst er dus op dat we in het geval van het veenlijk van Zweeloo met een persoon van het vrouwelijk geslacht te maken hebben.

Van zowel de huid als het skelet is een ^{14}C-datering beschikbaar: 1835±40 BP (GrN-15458) en 1940±70 BP (OxA-1727). De gemiddelde datering is 1861±35 BP. Na calibratie betekent dat een datering tussen 128-216 AD (1 sigma), respectievelijk 82-238 AD (2 sigma's).

De vrouw van Zweeloo stierf in de eerste helft van de Romeinse tijd, op een leeftijd van ongeveer 35 jaar. Zij was ongeveer 1,70 m lang en had waarschijnlijk bloedgroep O. Ze was (licht?) geïnfecteerd met de parasieten zweepworm en spoelworm. Onderzoek van de ingewanden heeft uitgewezen dat haar laatste maaltijd bestond uit onder andere gierstepap en bramen. Naar alle waarschijnlijkheid werd ze naakt en op haar rug in een kuil in het veen gelegd, haar hoofdhaar had men kortgeknipt. De gebeurtenis vond plaats tussen augustus en oktober. Hoe ze om het leven is gekomen is niet meer vast te stellen.[9]

6. NOTEN

1. Dit skelet heeft thans inventarisnummer 1992/VIII 21.
2. Mondelinge mededeling A. Katuin, december 1987.
3. De brieven die in dit artikel genoemd worden, bevinden zich in het correspondentie-archief van het B.A.I. en/of het Rijksarchief Drenthe (museumarchief).

4. Alle calibraties in dit artikel zijn verricht met het programma Cal 15 (april 1993) van Van der Plicht.
5. Mondelinge mededeling Prof.dr. C.J.P. Thijn.
6. Op de röntgenfoto's is ook langs alle lange beenderen een corticale botafzetting waarneembaar (fig. 7). Dit verschijnsel wordt *hypertrophische osteoarthropathie* genoemd (Thijn, mond. mededeling). Deze botafzetting kan verschillende oorzaken hebben, maar wordt bij hedendaagse patiënten veelal in verband gebracht met chronische longaandoeningen of chronische hart- en vaatziekten. Men zou dit gegeven kunnen projecteren op de gezondheidstoestand van de man van Aschbroeken, hetgeen suggereert dat hij aan een dergelijke aandoening heeft geleden. Wij aarzelen echter bij het trekken van deze conclusie, omdat postmortemprocessen het beeld kunnen hebben beïnvloed.
7. Zie het *Verslag van de Commissie van Bestuur over het Provinciaal Museum ... over 1931*, pp. 7-8.
8. De heer M. Naber van de vakgroep Techniek van de Rechercheschool te Zuthpen was zo vriendelijk om onze observatie te controleren en de breukvlakken onder een microscoop te bekijken. Zijn conclusie is dat de breukvlakken niet sluitend passen. Hij tekent daarbij aan dat dit te maken zou kunnen hebben met de verschillende behandeling (al dan niet met een conserveringsmiddel behandeld) van de beide botdelen. De auteurs zijn hem zeer erkentelijk.
9. De schrijvers willen graag hun dank uitspreken aan verschillende personen: dr. R. Housley, Radiocarbon Accelerator Unit Oxford, voor de snelle behandeling van het monster; drs. J.N. Lanting, B.A.I., voor zijn hulp op vele terreinen; Prof.dr. C.J.P. Thijn, afdeling Radiodiagnostiek van het AZ Groningen, voor de discussie over de röntgenfoto's; de heer G.K. Visser, afdeling Radiodiagnostiek van het AZ Groningen, voor het maken van de röntgenfoto's; dr. M. Voortman, Laboratorium voor Gerechtelijke Pathologie, Rijswijk, voor het onderzoek van de huid van Zweeloo; P. van der Sijde, Fotodienst Centrum voor Histologisch en Elektronenmicroscopisch Onderzoek, Groningen, voor het afdrukken van de röntgenfoto's en het vervaardigen van de figuren 5 en 6; en ten slotte J. Bruggink en M.A. Weijns, repectievelijk Drents Museum en B.A.I., voor het tekenwerk bij dit artikel. De auteurs zijn ook de Drents Prehistorische Vereniging zeer erkentelijk voor de verleende subsidie uit het Wetenschappelijk Fonds, waarmee het fysisch anthropologisch onderzoek mogelijk werd gemaakt.

7. LITERATUUR

ACSADI, G. & J. NEMESKERI, 1970. *History of human life span and mortality*. Budapest

MAAT, G.J.R., 1984. Dating and rating of Harris's lines. *American Journal of Physical Anthropology* 63, pp. 291-299.

SANDEN, W.A.B. VAN DER, 1990. De vondstgeschiedenis. In: W.A.B. van der Sanden (red.), *Mens en moeras; veenlijken in Nederland van de bronstijd tot en met de Romeinse tijd*. Assen, pp. 80-93.

SANDEN, W.A.B. VAN DER, 1992. Mens en moeras: het vervolg. *Nieuwe Drentse Volksalmanak* 109, pp.140-154.

SCHULTER-ELLIS, F.P., L.A. HAYEK & D.J. SCHMIDT, 1985. Determination of sex with a discriminant analysis of new pelvic bone measurements: Part II. *Journal of Forensic Sciences* 30, pp. 178-185.

TROTTER, M., 1970. Estimation of stature from intact long limb bones. In: T.D. Stewart (ed.), *Personal identification in mass disasters*. Washington, pp. 71-84.

TROTTER, M. & G.C. GLESER, 1958. A re-evaluation of estimation of stature based on measurements of stature taken during life and of long bones after death. *American Journal of Physical Anthropology* 16, pp.79-123.

UYTTERSCHAUT, H.T., 1990. De anatomische beschrijving. In: W.A.B. van der Sanden (red.), *Mens en moeras; veenlijken in Nederland van de bronstijd tot en met de Romeinse tijd*. Assen, pp. 104-124.

WALKER, R.A. & C.O. LOVEJOY, 1985. Radiographic changes in the clavicle and proximal femur and their use in determination of skeletal age at death. *American Journal of Physical Anthropology* 68, pp. 67-78.

ZEIST, W. VAN, 1953. Zur Datierung einer Moorleiche. *Acta Botanica Neerlandica* 1, pp. 546-550.

ZEIST, W. VAN, 1956. Palynologisch onderzoek van enkele Drentse veenlijken. *Nieuwe Drentse Volksalmanak* 74, pp. 199-209.

PROJECT PEELO
HET ONDERZOEK IN DE JAREN 1977, 1978 EN 1979 OP DE ES*

P.B. KOOI

Biologisch-Archaeologisch Instituut, Groningen, Netherlands

ABSTRACT: This paper is the first of three about the excavations carried out in 1977-1988 near the hamlet of Peelo, municipality of Assen, in the northern part of the province of Drenthe (fig. 1). In this article the excavations of 1977, 1978 and 1979 in the field area (*es*) west of Peelo are published (fig. 3).

The most important aims of the project were to investigate the development of a settlement in time and the construction of house plans, and to compare these with other earlier excavated settlements.

Result. The oldest traces found belong to the Single Grave culture: a ring ditch (fig. 5: square Y/6) with a north-south orientated findless grave pit and a second grave pit which has an east-west orientation, partly disturbed by a later fence (fig. 5: square Y/10), containing a Protruding Foot beaker (fig. 6), dating from a period between 2600-2450 BC (Drenth & Lanting, 1991).

Settlement traces. The oldest settlement traces date from about 400 BC with house plans of Hijken type (figs 7-8). Some fragments of house plans may date from the Late Iron Age (fig. 9). Between c. 100 BC and 200 AD two farm yards surrounded by fences are concentrated in the highest part of the excavated area, with houseplans of Noord Barge and Wijster A type (figs 10-11). The period 200-300 AD is characterized by house plans of Peelo A type (figs. 7 and 12-13) while the settlement is shifting to the lower, western part of the area. The number of farms increases to three.

During the late Roman era the settlement as a whole is located in the lower western part of the excavation, with house plans of Wijster B and Peelo B types (figs 14-17). During this stage a smithy (fig. 21) is present in the northeastern part, with a number of ovens to prepare iron (fig. 69). From the later periods in this area, only a fragment of the 8th century was excavated with a Odoorn C house plan.

During the different settlement phases several sheds, huts and granaries were built. In the Iron Age we find several small granaries next to the houses and in the field (fig. 56: Nos 111-137). From the beginning of the era on, a more complex set of bigger granaries, sunken huts and wells came into use (figs 21-67).

The majority of the finds consists of handmade pottery. Pottery from the Iron Age (figs 83-84) and the Roman period (figs 85-93) confirm the development of the settlement. Finds of imported terra sigillata and terra nigra, dating from the 3rd and 4th century, are rare.

Discussion. Comparison of the results of Peelo with earlier excavated settlements leads to some changes in the existing models. First of all the dating of house types has changed; the Fochteloo type should be placed in the first century AD (instead 250-100 BC) and Wijster C can be contemporary with Peelo B and therefore be dated between 400 and 500 AD. In general, the lay-out of the settlement and the farm yards during the late Roman period is less compact than in Wijster and the whole is more rapidly shifting. Maybe this is due to the fact that Peelo was a smaller settlement in a different landscape.

KEYWORDS: Grave yard, Single Grave culture, Protruding Foot beaker, settllement traces, Iron Age, Roman period, house plans, sheds granaries, sunken huts, wells, fences, ovens, pits, pottery.

1. INLEIDING

In de jaren 1977 tot en met 1988 werden in en rond het gehucht Peelo (gemeente Assen) grootschalige opgravingen uitgevoerd, voorafgaande aan de realisering van woningbouw in de plangebieden Peelo en Marsdijk (fig. 1). De resultaten en analyses worden in drie delen gepubliceerd. De presentatie van de op-gravingen krijgt in eerste instantie de meeste aandacht, waarbij vergelijkingsmateriaal uit de naaste omgeving wordt betrokken. Een verdergaande synthese betreffende de bewoningsgeschiedenis zal in het laatste deel aan de orde komen, omdat dan een totaaloverzicht kan

* De figuren 5 en 7 bevinden zich in de map achterin dit tijdschrift..

165

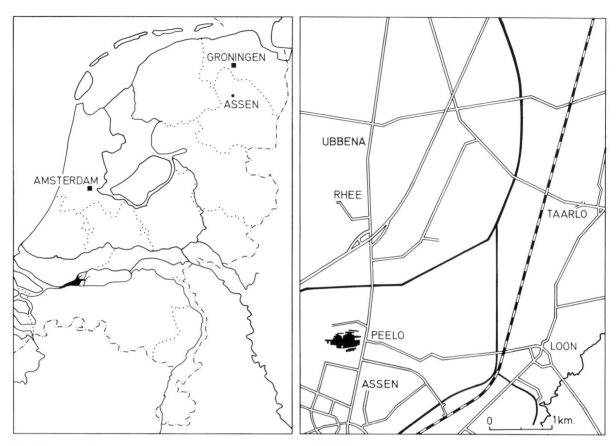

Fig. 1. Topografische kaart met de ligging van het opgravingsterrein in 1977-1979.

worden gepresenteerd. In deze publikatie worden de eerste drie jaren van het onderzoek behandeld, te weten het onderzoek op de es. In deel twee volgt het onderzoek ten noorden van de Marsdijk en tenslotte het Kleuvenveld, het Nijland en de Burcht met een samenvatting van de resultaten.

De eerste contacten tussen het Biologisch-Archaeologisch Instituut (B.A.I.) en de gemeente Assen over de planvorming voor Peelo en Marsdijk vonden plaats in 1968. Vanaf het begin stond de gemeente Assen welwillend tegenover de belangstelling van archeologische zijde. In het voorjaar van 1977 waren de plannen zo ver gevorderd, dat met de opgravingen moest worden begonnen. Regelmatig werden met de dienst Gemeentewerken besprekingen gevoerd. In een sfeer van wederzijds respect werd onder leiding van ir. J. Koolhaas met vertegenwoordigers van de afdelingen Weg- en Waterbouw, Grondzaken en Stedebouw de koers bepaald, waaraan het gemeentebestuur en de directeur Gemeentewerken stilzwijgend hun fiat gaven. Het staat vast dat het goede resultaat van het onderzoek sterk bepaald is door de inzet en de medewerking van alle betrokkenen.

De kosten van het project werden uit verschillende bronnen gefinancierd. Uit het budget van de subfacul-

teit der Prehistorie werd door het B.A.I. jaarlijks een aanzienlijk bedrag beschikbaar gesteld, alsmede de personeelslasten van de technische staf, bestaande uit een tekenaar en een voorgraver. Het tekenwerk in het veld werd vrijwel volledig uitgevoerd door G. Delger, die ook de uitwerking van de kaarten en plattegronden voor zijn rekening heeft genomen en daarmee een substantiële bijdrage aan de publikatie heeft geleverd. Bij de overhaaste aanvang van de opgraving in 1978 zijn de eerste werkputten door J.H. Zwier in tekening vastgelegd. K. Klaassens was bijna voortdurend verantwoordelijk voor de technische leiding en het reinigen en reconstrueren van het vondstmateriaal. Zijn inzet was onmisbaar voor het slagen van het project. Bij de start van de opgraving in 1978 werd zijn plaats voor korte tijd ingenomen door A. Meijer.

In 1977 en 1978 verstrekte het provinciaal bestuur van Drenthe aanzienlijke subsidies om het tekort op de begroting van twee opgravingscampagnes te dekken. Daarmee werd benadrukt, dat men in Drenthe hecht aan het onderzoek van de eigen cultuurhistorie. In 1977 en 1978 werd het project door het Regionaal Coördinatie College voor Openbare werken als E-object aangemerkt, waardoor de lonen van vier arbeiders bijna volledig werden gesubsidieerd. In 1979 werd de opgraving door het directoraat voor de Arbeidsvoorziening opengesteld als object in het kader van de werkgelegenheids-verruimende maatregelen, waardoor via het District Noord van de Aanvullende Civieltechnische werken wederom de lonen van vier arbeiders werden gesubsidieerd.

De voor het onderzoek benodigde percelen waren alle in eigendom van de gemeente Assen en werden om niet beschikbaar gesteld. Met de gebruikers, de landbouwers J. Geerts, G. Geerts, J. ten Brink en Th. Langelo, werd in overleg een passende regeling getroffen, zodat de schade beperkt bleef.

Niet alleen voor het onderzoek van het B.A.I., maar ook voor het onderwijs van de subfaculteit bleek Peelo een dankbaar object. Enkele studenten deden er de verplichte velderuaring op en voorlopige uitwerking van de resultaten werd vastgelegd in een drietal scripties. Met enige nadruk dient hier de scriptie van A.C. Bardet te worden vermeld, die mede de basis heeft gevormd voor de beschrijving van het aardewerk.

De belangstelling tijdens de opgravingen was zeer groot. In 1978 werd de commissaris der koningin, mw. T. Schilthuis en een deputatie van het provinciaal bestuur rondgeleid en over de vorderingen geïnformeerd. Regelmatig werden groepen medewerkers van de dienst Gemeentewerken ontvangen. Stimulerend was de groeiende belangstelling van de inwoners van Peelo, die er mede borg voor heeft gestaan, dat er weinig ongewenst bezoek is geweest.

De discussies over de ontwikkeling van het Drentse boerenhuis met H.T. Waterbolk en C.S.T.J. Huijts waren zeer nuttig. Over een aantal punten zijn de meningen verdeeld en dit wordt bij de betreffende plattegronden besproken.

De voorwerpen zijn grotendeels door J.M. Smit in tekeningen vereeuwigd.

Mw. F.T. Sandmann-Cornelis verzorgde de uitwerking van de tekst.

2. HET ONDERZOEK

2.1. Vraagstelling

Het project Peelo is een onderdeel van het nederzettingsonderzoek in Noord-Nederland, een zwaartepunt in het onderzoekprogramma van het B.A.I. In dat kader werden in de periode vóór 1977 grootschalige opgravingen op diverse locaties in Drenthe uitgevoerd, waarbij agrarische nederzettingen van verschillende ouderdom werden onderzocht, zoals: Elp: midden- en late Bronstijd; Angelslo/Emmerhout: midden-Bronstijd- vroege IJzertijd; Hijken: midden-Bronstijd en IJzertijd; Noord Barge: late Bronstijd-Romeinse tijd; Wijster: IJzertijd-Romeinse tijd; Odoorn: vroege middeleeuwen en Gasselte: middeleeuwen.

Met elkaar leverden deze opgravingen de basis voor het opstellen van een ontwikkelingsreeks van huisplattegronden (Waterbolk, 1980; 1985) en gegevens

over de structuur van nederzettingen, economie en materiële cultuur van de midden-Bronstijd tot en met de middeleeuwen (Waterbolk, 1987; Harsema, 1980). Voorts leverden deelstudies van o.a. *Celtic fields* (Brongers, 1976) en urnenvelden (Kooi, 1979) aanvullende gegevens op over de economie in de IJzertijd en de bevolkingsdichtheid in de late Bronstijd en de vroege IJzertijd.

De bovengenoemde studies en onderzoekingen maakten een aannemelijke reconstructie van de diverse aspecten van de bewoningsgeschiedenis mogelijk. Er bleven echter toch enkele vragen onbeantwoord, namelijk:

Is de ontwikkeling van huisplattegronden en nederzettingsstructuren geldig voor het gehele Drents plateau of waren er lokale variaties?

Stemt het beeld van de ontwikkelingen, samengesteld uit verschillende verspreide locaties overeen met de ontwikkelingen in een klein gebied, bijvoorbeeld een landschappelijk begrensde 'nis'?

Is er sprake van een doorlopende ontwikkeling of zijn er onderbrekingen die wijzen op discontinuïteit in de bewoning?

Deze vragen zouden, naar gehoopt werd door het onderzoek te Peelo geheel of gedeeltelijk kunnen worden beantwoord. Tevens zou op verzoek van de botanici een representatief aantal grondmonsters genomen kunnen worden ten behoeve van het onderzoek van verkoolde zaden.

Om de richting van het onderzoek te kunnen bepalen werd aan de hand van eerdere opgravingen, vondsten en waarnemingen een hypothese voor de bewoningsgeschiedenis opgesteld. De landschappelijke situatie bleek uitermate geschikt te zijn voor het vaststellen van een afgesloten onderzoeksgebied, een landschappelijke 'nis' met grotendeels natuurlijke grenzen. De grenzen van het gebied rond de buurtschap Peelo zijn bepaald aan de hand van de Topografische Kaart van 1896, waarop o.a. zeer gedetailleerde informatie staat over natte en droge gronden en inmiddels verdwenen venen (fig. 2). Aan de oostzijde lag een dalsysteem, het Hamelbroek, met een brede erosiehelling ten westen daarvan. Dwars daarop stonden twee smallere dalsystemen, namelijk aan de noordzijde de Wilde Stroet en het Busebroek die naar de Messen stroomden en aan de zuidzijde de Maarzen, de Landjes en de Zwarte Hullen. Ten westen lagen de uitlopers van het Zeijerveen namelijk het Peelerveld, het Zwarte Water en het Veenland. De oppervlakte van dit gebied lag in dezelfde orde van grootte als ook eens voor nederzettingsterritoria in de late Bronstijd was bepaald (tussen de 270 en 648 ha; Kooi, 1979). (Daarbij is het Stoepveld niet meegerekend, dat door een doorgang in de Landjes vanaf het Kleuvenveld bereikbaar was.)

Bewoningssporen in het beschreven gebied waren uit verschillende perioden bekend. Daar het onderzoek zich richtte op de agrarische component werden de

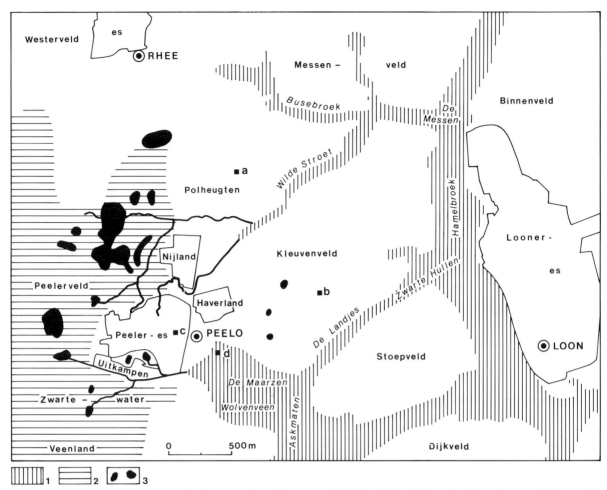

Fig. 2. Vereenvoudigde overzichtskaart van Peelo en omgeving. 1. De oorspronkelijke beekdalen; 2. Reconstructie van de venen; 3. Vennen. Naar de Topografische Kaart van 1896.

vondsten en opgravingen vanaf het Neolithicum bekeken. In het Provinciaal Museum te Assen zijn geregistreerd:

– Een bijl van de Enkelgrafcultuur bij de noordrand van de Peeler es;

– Vuursteenmateriaal uit het Paleolithicum, Mesolithicum en Neolithicum bij het meest noordelijke vennetje op het Peelerveld;

– Grafgiften uit drie graven van de Standvoetbekercultuur op de Peeler es (inv.nr. 1935/X-2, 3, 4, 7, 8, 9);

– Scherven van aardewerk uit de Romeinse tijd, bij Peelo (inv.nr. 1986/IV-35);

– Een kraal van glaspasta uit de Volksverhuizingstijd (inv.nr. 1966/I-1) en scherven uit de middeleeuwen (inv.nr. 1967/III-1) bij Peelo.

In het verleden werden bij ontginningswerkzaamheden en zandafgravingen bewoningssporen ontdekt die in een aantal gevallen tot opgravingen hebben geleid. In 1936 werd een opgraving op de Polheugten uitgevoerd. Naast vuursteenmateriaal uit het laat-Mesolithicum en

het Neolithicum werden een neolithisch graf, resten van een twee-perioden grafheuvel uit de midden-Bronstijd en een urnenveld uit de late Bronstijd – vroege IJzertijd gevonden (fig. 2a) (van Giffen, 1938). Op de Topografische Kaart van 1896 staan hier nog twee heuvels aangegeven. Het jaar daarop werden er op het Kleuvenveld van een groep van 10 brandheuvels uit de IJzertijd twee onderzocht (fig. 2b; Van Giffen, 1939). Op de topografische kaart van 1896 staan hier 8 heuvels aangegeven. In dezelfde omgeving waren op luchtfoto's sporen van een *Celtic field* te zien (Brongers, 1976). In 1925 waren bij een opgraving nederzettingssporen uit de Romeinse tijd onderzocht op de es van Peelo (fig. 2c; Van Giffen, 1926).

Voorts waren er nog een aantal waarnemingen van belang. Op luchtfoto's zijn in de zuidoostelijke hoek van het Stoepveld sporen van een *Celtic field* te zien (Brongers, 1976). Tenslotte zijn er nog twee oude boerderijen te Peelo aanwezig, een met het jaartal 1629 in een balk van de schuur en de ander met het jaartal 1703 boven een deur van het woongedeelte.

De bovengenoemde bewoningssporen waren zowel verspreid over verschillende perioden als ook verspreid over het gebied. Als werkhypothese is daaruit afgeleid dat de gevonden sporen het gevolg kunnen zijn van continue bewoning door de mens vanaf het Neolithicum. Het is bekend dat verplaatsing van akkergronden ten gevolge van uitputting van de bodem periodiek noodzakelijk was en het ligt voor de hand te veronderstellen dat men in de nabijheid van het akkerland bleef en dus ook een nederzetting periodiek werd verplaatst. De verplaatsing van de bewoning zou in het geval Peelo als volgt kunnen zijn:

Van de es (Standvoetbeker-grafveld) naar de Polheugten (neolithisch graf, Bronstijd-graven, vroege IJzertijd). Vandaar naar het Kleuvenveld (brandheuvels en *Celtic fields*), mogelijk uitlopend naar het Stoepveld (*Celtic field*), en verder westwaarts terug naar de es (nederzetting Romeinse tijd) om tenslotte in de middeleeuwen op de plek van het historische Peelo terecht te komen. Door de introductie van bemesting zou de nederzetting daar tot stilstand zijn gekomen.

2.2. Werkwijze (fig. 3 en 4)

Het in 1977-1979 onderzochte terrein lag binnen de Peeloëresweg, de Europaweg-Noord en de Groningerstraat. De eerste opgravingscampagne in 1977 was er op gericht om de aard en omvang van de nederzetting vast te stellen, waarvan de sporen in 1925 waren ontdekt en gedeeltelijk onderzocht (Van Giffen, 1926). Om te beginnen werden een aantal min of meer oost-west georiënteerde werkputten (1-6) uitgezet. In de meest noordelijke, werkput 3, werden sporen uit de IJzertijd aangetroffen en in de werkputten 1 en 2 nederzettingssporen uit de Romeinse tijd. Vervolgens werd het aantal werkputten uitgebreid, waarbij duidelijk werd dat de gezochte nederzetting zich ongeveer tussen de werkputten 5 en 10 moest bevinden. Door het graven van aansluitende werkputten werd zo dit deel van de hoge es onderzocht met uitzondering van een paar gedeelten, namelijk een diep ontzand perceel tussen werkput 1 en 13 en de zandweg langs werkput 5. Voorts werd afgezien van nader onderzoek van het oostelijke deel tussen werkput 1 en 3 en verder noordwaarts, omdat de kwaliteit van de sporen daar zeer slecht was ten gevolge van intensief ploegen en door gedeeltelijke ontzanding. De laatste werkput, nr. 18, die in 1977 werd onderzocht, was bedoeld als testsleuf naar het westelijke, lage deel van de es.

In 1978 was de financiële situatie aanvankelijk zeer ongunstig en was er geen grote campagne voorzien. Toch werden bij het graven van een wegcunet waarnemingen uitgevoerd en toen daarbij meerdere nederzettingssporen werden ontdekt was snel handelen geboden. Na een moeizame start werden aansluitend naast de weg de werkputten 19, 20 en 21 onderzocht. Vervolgens werd ondanks de financiële beperkingen werkput 22 onderzocht, waarbij zodanige vondsten

werden gedaan, dat een verder onderzoek met succes kon worden bepleit. Aldus werd een groot deel van de westelijke es opgegraven, tot en met werkput 31. Na een voorlopige analyse van de resultaten van de opgravingscampagnes in 1977 en 1978 werd besloten om in 1979 nog een aantal aanvullende werkputten te onderzoeken. In eerste instantie werd zo de helling naar een veentje aan de zuidzijde opgegraven met de werkputten 32 tot en met 35. Ten zuiden van de zandweg over de hoger es, werd werkput 24 uit 1978 uitgebreid met de werkputten 36, 40 en 41. De ruimte tussen de werkputten uit 1977 en 1978 werd verder onderzocht door middel van de werkputten 37, 38, 39, 43, 44 en 45. Voorts werd tussen de werkputten 4 en 16 een aansluitende werkput 42 opgegraven en tenslotte werd met werkput 46 de voortzetting van de sporen in werkput 27 onderzocht. Deze laatste werkput leverde het bewijs dat de nederzettingssporen zich nog westwaarts voortzetten, maar door de vergevorderde aanleg van wegen, kabels, leidingen etc. was het inmiddels niet meer mogelijk daar nog verder onderzoek uit te voeren. Een vondst van scherven uit esgreppels in het deelplan De Akkers, aan de westelijke rand van het escomplex, gaf een indicatie van de mogelijke uitgestrektheid (fig. 2).

2.3. De bodem

Het onderzoeksgebied van 1977-1979 lag in de Peeler es, een licht golvend terrein dat grotendeels voor de akkerbouw werd gebruikt. Het hoogste gedeelte, tot ca. 13 m +NAP lag als een brede rug parallel aan de Groningerstraat en liep naar het westen af tot 11,50 m +NAP. Tussen deze rug en de Peeloër esweg lagen nog twee zandkoppen. Van de hoogste kop was in de jaren 1925 en 1935 een gedeelte afgegraven waardoor het bodemarchief ernstig was verstoord, juist in een gedeelte met veel bewoningssporen (werkput 11-15). De ontzanding was in het opgravingsvlak herkenbaar als een systeem van evenwijdige donkere kuilen, gevuld met humeuze grond.

In de diepere ondergrond kwam in het gehele terrein een keileempakket voor in wisselende dikte en samenstelling. Daarboven waren in het Weichseliën dekzandlagen gevormd, deels meegolvend met de andere afzettingen en deels in wisselende lagen, van minstens 50 cm tot meer dan 1,50 m. Door het gebruik als akkergrond vanaf de middeleeuwen was een esdek ontstaan met een variabele dikte van 80 cm op het hoge oostelijke deel van de es, tot nauwelijks de dikte van de huidige bouwvoor van 30 cm in het lagere westelijke deel. Dit verschil kan samenhangen met de natuurlijke vruchtbaarheid van de bodem die ongelijk was of ook met de gebruiksduur. Een verschijnsel dat met deze ontwikkeling samenhangt, was het voorkomen van esgreppels in verschillende percelen. In werkput 8 waren deze greppels in het esdek gegraven en reikten niet tot de vaste ondergrond, in de werkputten 3, 4, 16, 17, 25, 26, 27, 28 en 42 waren ze in het opgravingsvlak

Fig. 3. Kadastrale kaart van het opgravingsterrein (situatie 1977) met de ligging van de opgravingsputten van 1977-1979.

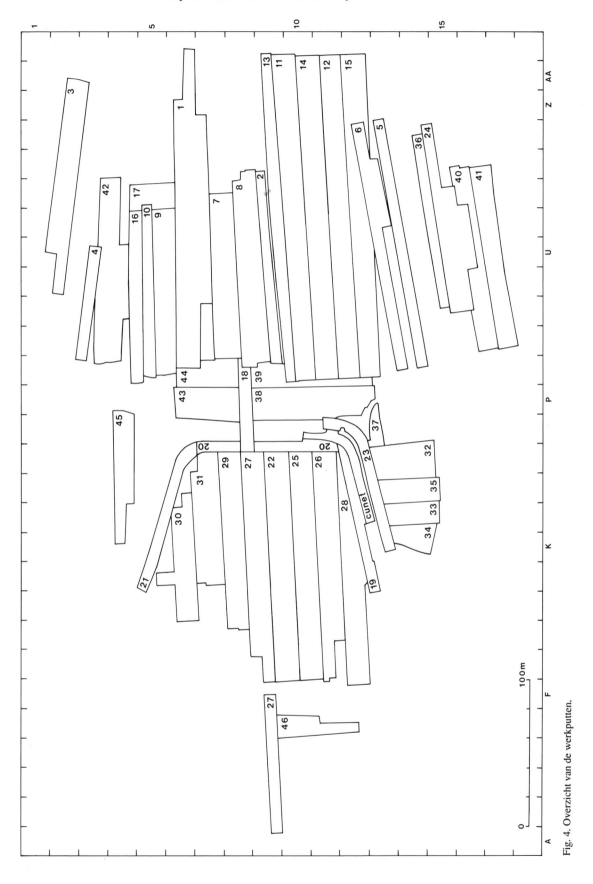

Fig. 4. Overzicht van de werkputten.

172 P.B. Kooi

te zien als series evenwijdige, met humeuze grond gevulde greppels. De kwaliteit van het bodemarchief zal door deze manier van ontginnen aanzienlijk zijn verminderd. Overigens had het esdek bij voldoende dikte een nuttig effect, omdat daardoor de ondergrond voor aantasting door het ploegen in latere tijd werd beschermd.

De afwatering van het gebied vond plaats in oostelijke richting, namelijk aan de noordzijde naar de Wilde Stroet en aan de zuidzijde naar de Landjes.

Uit de kartering van de bewoningssporen zien we een duidelijke correlatie tussen de relatieve hoogteligging en de afwatering. Globaal gezien concentreert de bewoning zich op de hogere delen van het terrein, zoals aangegeven in figuur 79.

3. RESULTATEN

De resultaten van de opgravingen zijn op de volgende wijze bewerkt. Alle grondsporen zijn op een plattegrond in vier delen aangegeven. Langs de rand is een kader met een indeling in vakken aangebracht (fig. 5). Eerst worden de oudste sporen, behorend bij een bekergrafveld, besproken. Daarna volgt de behandeling van de nederzettingssporen, die in acht categorieën zijn verdeeld: huisplattegronden, schuren, spiekers, hutkommen, waterputten, omheiningen, kuilen en ovens. Tenslotte worden de overige sporen volledigheidshalve genoemd. De huisplattegronden, schuren, spiekers,

hutkommen, waterputten en de kuilen en ovens zijn op een schematisch overzicht aangegeven (fig. 7).

3.1. Neolithische graven

Op twee plaatsen zijn sporen van neolithische graven aangetroffen. In werkput 1 (vak Y/6) een standgreppel met een doorsnede van 2,8 m, waarbinnen het laatste restant van het noord-zuid georiënteerde vondstloze graf van 2,2 m lang en 0,7 m breed. In werkput 14 lag een grafkuil van 1,8 m lang en 0,8-1,0 m breed, oost-west georiënteerd met een standvoetbeker (nr. 254), die door latere ingraving van een omheining was verstoord (vak Y/10) (fig. 6). De versiering op de beker is een mengvorm van het type 1b, met visgraat en groeflijnen en het type 1e met rijen schuine streepjes, zonder groeflijnen. De beker dateert daarom uit de periode 2600-2450 v.Chr. (Drenth & Lanting, 1991). In de directe omgeving werd nog een bijltje, gemaakt van gabbro, in het esdek gevonden. De sporen zullen behoren bij een vlakgrafveld van de Enkelgrafcultuur dat op het hoogste punt van de rug heeft gelegen en waarvan bij de eerder gemelde zandafgraving reeds grafgiften te voorschijn waren gekomen (Glasbergen, 1971).

3.2. Boerderijplattegronden

Bij de beschrijving van de plattegronden zijn die configuraties van paalgaten opgenomen, die als hoofdgebouwen zijn te beschouwen. De doorlopende typereeks van

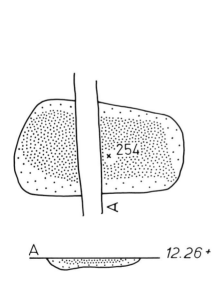

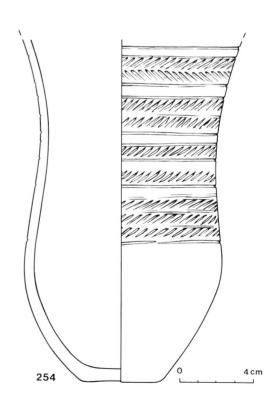

Fig. 6. Grafkuil met standvoetbeker.

de Bronstijd tot en met de middeleeuwen zoals die indertijd door Waterbolk (1980) werd opgesteld en gepubliceerd was met uitzondering van de Peelo typen gereed en dienden als uitgangspunt. Inmiddels is ook de studie van Huijts over de voor-historische boerderijbouw in Drenthe gepubliceerd, waardoor de inzichten op bepaalde punten zijn gewijzigd. Voor zover mogelijk zullen deze in de beschouwingen worden betrokken. De door hem geformuleerde kenmerken zijn per type vermeld.

De plattegronden van de grote gebouwen zijn op schaal 1:200 afgebeeld. Grondsporen, die tot één bouwperiode zijn gerekend worden zwart aangegeven (soms met paalkern). Verbouwingen en reparaties zijn aangegeven door middel van een arcering. Paalgaten waarvan niet zeker is, dat ze bij het gebouw horen zijn open gelaten. Ter verduidelijking zijn ontbrekende paalgaten met een kruisje aangegeven en bij afwezigheid van grotere aantallen paalgaten zijn streepjeslijnen getrokken om de afmetingen globaal aan te kunnen geven. Ingangen zijn voorzien van een zwarte pijl, als ze bij de periode horen. Voor verbouwingen met ingang zijn open pijlen gebruikt. Bij twijfel is bij de pijl een vraagteken geplaatst.

Onder iedere plattegrond zijn de dieptes van de paalgaten aangegeven ten opzichte van een vaste, gekozen hoogte en bijpassend bij de codes van de plattegrond zwart, gestreept of open getekend.

Type Hijken (fig. 8: nrs 3, 17 en 27)
Drieschepig, met een woon- en stalgedeelte, palen buiten de wand als steun voor het dak, één paar brede ingangen tegenover elkaar in de lange wanden.

Nr. 3. Vrijwel complete plattegrond. Opmerkelijk is het verschil tussen het woon- en het stalgedeelte. In het woongedeelte zijn de wandpalen op regelmatige afstanden dichter bij elkaar geplaatst, terwijl in het stalgedeelte de palen per stalscheiding met de staanders corresponderen en dus verder uit elkaar staan. Lengte over de buitenpalen 17,7 m; breedte over de buitenpalen 7,4 m; lengte over de binnenpalen 16,5 m; breedte over de binnenpalen 7,4 m;

Nr. 17. Fragment van een plattegrond. Met hulplijnen is de omtrek van buitenpalen aangegeven. Van de wandpalen en de staanders zijn juist voldoende aanwezig om de plattegrond te kunnen completeren. Lengte over de buitenpalen 17,60 m; breedte over de buitenpalen 5,8 m; lengte over de wandpalen <p/m>16,70 m; breedte over de wandpalen <p/m>5 m;

Nr. 27. Vrijwel complete plattegrond, waarvan het stalgedeelte is verlengd. De plaats van de oorspronkelijke korte stalwand is herkenbaar aan de grotere afstand tussen de staanderparen 7 en 8 van de stal en het onregelmatige verloop van de staanders in het verlengde deel. Bovendien werd daarbij een aantal middenpalen op de overgang geplaatst. Lengte over de buitenpalen 23,0 m; breedte over de buitenpalen 7,2 m; lengte over de wandpalen 22,3 m; breedte over de

wandpalen 6,3 m; lengte van de verlenging 9,0 m.

Overige boerderijplattegronden uit de IJzertijd (fig. 9: nrs 5, 12, 25, 26)
Deze groep bestaat uit fragmenten van plattegronden die door hun aard en ligging in de IJzertijd zijn te dateren. In alle gevallen zijn de brede ingangspartijen aan te wijzen, die als zwaar element bewaard zijn gebleven. Voor een meer nauwkeurige typologische ordening zijn deze plattegronden niet geschikt.

Nr. 5. Twee middenpalen aan de korte zijde geven de totale lengte: 19,1 m. De afstand van de staanders wijst er op dat het stalgedeelte aan de oostzijde ligt. Het woongedeelte is breder dan het stalgedeelte.

Nr. 12. Mogelijk zijn de dubbelpalen aan de korte einden een indicatie voor de totale lengte (18,6 m). De tussenafstand van de staanderparen wijst er op dat het stalgedeelte aan de oostzijde ligt.

Nr. 25. Opvallend van dit fragment is de rij van 5 palen in de middenas. Overigens zijn slechts de ingangspalen en de staanders bewaard gebleven. Het stalgedeelte is herkenbaar aan de oostzijde. De lengte is minimaal 15,4 m geweest bij een breedte van 6,0 m.

Nr. 26. Ook hier zijn waarschijnlijk middenpalen aanwezig. De breedte bij de ingangen is 5,4 m.

De plattegronden uit de IJzertijd vallen in twee groepen uiteen. Dit is grotendeels een gevolg van de verschillen in kwaliteit. Onder het type Hijken zijn drie plattegronden beschreven, waarvan nr. 3 het ideaaltype het dichtst benaderd. Bij nr. 27 is de wandconstructie veel zwaarder en staan de buitenpalen vrij dicht tegen de wand. De restgroep vertoont een grotere verscheidenheid in de plattegronden. Nr. 12 zal wellicht nog tot het type Hijken hebben behoord, maar de wandconstructie is onzeker.

Een duidelijk verschil tussen de beide groepen is de afstand tussen de staanders in het stalgedeelte. In de nrs 3, 17 en 27 is deze respectievelijk 1,6, 1,8 en 1,6 m, in de nrs 5, 12 en 25 is dit 1,4 m. Dit kan een aanwijzing zijn dat de tweede groep in het algemeen jonger gedateerd moet worden.

Type Noord Barge (fig. 10: nrs 19, 22, 23)
Eén-, twee- of drieschepig met een wandgreppel.

Nr. 19. Fragment van een ingegraven wand met buitenpalen en enkele staanders. De ligging en de wandconstructie maken een indeling bij dit type waarschijnlijk (fig. 6: vak W/11).

Nr. 22. Deze plattegrond heeft twee rijen staanders en is dus drieschepig. Van de wandgreppel is alleen het westelijke deel bewaard gebleven. Ingangen zijn niet met zekerheid aan te geven. Onduidelijk is ook de indeling van het gebouw (stal-, woongedeelte). Palen in de wandgreppel of daarbuiten hebben het dak gesteund. Lengte: ca. 16,6 m; breedte 4,8 m.

Nr. 23. Deel van een korte plattegrond, waarvan het zuideinde was verstoord. Ten noorden van de ingangen zijn twee paar palen tegen de wand geplaatst, aan de zuidzijde twee staanders. Dakdragende palen staan buiten tegen de wand. De functie van het gebouw is niet zeker. De grote ruimte suggereert een woonfunctie, de plaats langs de omheining van de nederzetting duidt meer op een stal. Deze plattegrond is bij uitzondering ongeveer noord-zuid georiënteerd. Lengte: 9,9 m; breedte 5,2 m.

Type Wijster A (fig. 11: nr. 21)
Drieschepig, met een stalingang, vaak stalboxen en ingangskuilen.

Nr. 21. Groot deel van een plattegrond met een stalgedeelte in het westen en ingangssporen op de overgang naar het woninggedeelte. De zuidelijke met een ingangskuil. Deels verstoord, waardoor stalingang ook niet meer aanwijsbaar is. Later verlengd en gewijzigd (zie onder type Peelo A, fig. 12). Lengte 24,6 m; breedte 5,8 m.

Dit type is helaas slecht vertegenwoordigd ten gevolge van de ontzanding van de hoogste kop in het terrein. De stalfragmenten van plattegronden nr. 14 en 18 kunnen ook tot dit type behoren. De stalboxen (nr. 14) en de variërende staanderafstanden wijzen in die richting (fig. 19).

Bij dit type is aannemelijk dat de vrij dicht op elkaar geplaatste buitenpalen het dak hebben gedragen (Huijts, 1992: p. 109).

Type Peelo A (fig. 12, 13: nrs 2, 6, 7, 13, 21, 41-a, 45, 46)
Drieschepig, met twee paar ingangen in de lange wanden, driedelig, met stalingang, één paar dubbelpalen bij de overgang naar de stal.

Nr. 2. Goed voorbeeld van het type met zware staanders in het middendeel. Het westelijke einde van de stal is iets versmald. Het woongedeelte eindigt met twee paalgaten in het verlengde van de staanderrijen. De drie paalgaten op één meter daarbinnen zullen bij een indeling horen (zie ook nr. 13). Rond het westelijke staleinde waren greppels gegraven, die kennelijk bedoeld waren om dit laag gelegen deel te ontwateren. Lengte 36,2 m; breedte 5,5 tot 6,0 m.

Nr. 6. Deze zeer lange plattegrond is ontstaan door een verbouwing. De eerste periode is vergelijkbaar met de voorgaande. Voor twee ingangen ligt een ingangskuil. Een aantal stalboksen is nog aanwezig. (Door verlenging met een woongedeelte met drie paar dubbelstijlen is een plattegrond van het type Wijster B ontstaan.) Extra paalgaten aan het begin van de stal en direct ten oosten van de ingangen tussen de stal en het middenhuis suggereren, dat daar tussenwanden zijn geweest. Lengte 31,4 m; breedte 6,0 m.

Nr. 7. Van deze plattegrond is het westelijke deel minder goed bewaard gebleven. De plattegrond is in twee fasen ontstaan, door een verlenging van het woongedeelte in dezelfde trant. Deze eindigt, evenals

bij nr. 2, in twee paalgaten in de korte wand.

De verlenging van 5 meter is constructief uitgevoerd door een extra middenpaal en een paar staanders bij te plaatsen. De stand van de wandpalen suggereert dat op de aansluiting met fase 1 een tweetal dubbelpalen is geplaatst. De verlenging is gepaard gegaan met het aanbrengen van twee nieuwe ingangen. Lengte (fase 1) 30,8 m; breedte 6,0 m.

Nr. 13. Plattegrond met veel sporen van verbouwingen en/of reparaties. Diagonaal door deze plattegrond loopt een smalle strook tussen de werkputten 2 en 13 die niet is opgegraven. In het stalgedeelte ontbreken mede daardoor een aantal staanders.

De eerste fase is driedelig met één ingang aan het begin van de stal in de zuidwand. Deze is later iets verplaatst naar het oosten. In het stalgedeelte zijn extra paren ingangen aan de noord- en zuidzijde gemaakt en de stalingang zelf is ook nog oostwaarts verplaatst. Drie palen aan de oostwand wijzen op een indeling van het woongedeelte, zoals bij nr. 2. Lengte 30,0 m; breedte 5,8 m.

Nr. 21. Ontstaan door verbouwing van het type Wijster A, namelijk door verlenging aan de oostzijde met 14 m, waarin twee paar ingangen in de lange wanden en inkorting van de stal met 4,5 m. De verlenging aan de oostzijde werd door Huijts (p. 110) beschreven als een apart gebouw. De aansluiting bij de oudere fase en de speciale uitbreiding van de omheining rond deze verbouw pleiten daar tegen. Lengte 33,8 m; breedte (verlenging) 5,2 m.

Nr. 41a. Meermalen verbouwde, gerepareerde en vernieuwde plattegrond. Het is niet goed mogelijk om complete plattegronden met wandpalen af te zonderen uit de grote hoeveelheid paalgaten. Het westelijke einde is karakteristiek voor het type Wijster B. Uit de hoeveelheid wandpalen in het midden is af te leiden, dat het om zeker drie fasen gaat. Voorts is de mogelijkheid niet uitgesloten, dat ongeveer in de as van de huisplattegronden een serie spiekers heeft gelegen, zoals dat ook elders in de nederzetting voorkomt. Daarom is gekozen voor een andere benadering. De staanderrijen en ingangspartijen zijn zo goed mogelijk per fase zwart aangegeven. Lengte ca. 22 m; breedte ca. 6 m.

Nr. 45. Restant van een plattegrond, gedeeltelijk verstoord bij de aanleg van een wegcunet. De stalingang en drie van de vier ingangen in de lange wanden zijn aantoonbaar. Een extra paalgat op de overgang naar de stal wijst op een tussenwand. Lengte 30,4 m; breedte 6,4 m.

Nr. 46. Ontstaan uit verbouwing van nr. 45 en gedeeltelijk verstoord door een wegcunet en tevens in het westen door een latere sloot, waardoor daar de ingangen niet met zekerheid zijn aan te geven. De ingangen in de lange wanden zijn bij de stal gewijzigd. Lengte 34,8 m; breedte 6,4 m.

Een van de kenmerken van dit type is de aanwezigheid van dubbelpalen op de overgang naar de stal. Deze dubbelpalen zijn als grondspoor niet altijd even duidelijk. Bij nr. 7 zijn ze als afzonderlijke paalkuilen te zien. Bij de plattegronden nrs 2 en 6 zijn ze aanwezig ten westen van de ingangen, in de vorm van een langgerekt grondspoor waarin twee palen gestaan hebben. Soms is slechts één van beide dubbelpalen bewaard gebleven, zoals bij de nrs 13 en 46, of door verstoringen niet aantoonbaar. Wel is toeschrijving aan dit type aannemelijk door overeenkomsten met de vier overige plattegronden. Het middendeel van de plattegronden omvat constant drie paar staanders en dit patroon zet zich voort in de typen Wijster B en Peelo B.

Een ander belangrijk kenmerk van dit, en de volgende typen, is de mogelijke driedeling in de lengte, die wordt aangegeven door de paren ingangen. Er is dus sprake van een voorhuis, een middenhuis en een achterhuis (Lanting, 1977). Uit de plaatsing van tussenwanden die voornamelijk uit extra paalgaten van deurposten is vast te stellen, kan worden afgeleid dat het voorhuis en het middenhuis als woongedeelte werden gebruikt, terwijl het achterhuis als stal vaak herkenbaar is door de indeling met scheidingen (Huijts, 1992: pp. 117, 129). Een bevestiging van de functie van het voor- en middenhuis blijkt uit de verlengingen, c.q. verbouwingen van de boerderijen nrs 21 en 6, waarbij het voorhuis en het middenhuis samen als verlenging zijn toegevoegd of vernieuwd.

Type Wijster B (fig. 14, 15: nrs 6a, 8, 29, 33, 36, 37, 40, 41b, 41c)
Als Peelo A maar met dubbelpalen in het voorhuis.

Nr. 6a. Ontstaan door verlenging van nr. 6 met 12 meter, namelijk een voorhuis en een middenhuis met twee paar staanders en drie paar dubbelpalen en een paar ingangen. Extra palen bij de aansluiting met nr. 6 wijzen op de aanwezigheid van een tussenwand. Lengte 43,2 m. De breedte neemt naar het oosten af van 6,0 tot 5,0 m.

Nr. 8. Fragment van een plattegrond met nog juist voldoende paalgaten om vast te kunnen stellen dat het om het betreffende type gaat. Lengte 26,5 m; breedte ca. 6 m.

Nr. 29. Regelmatige plattegrond met een paar dubbelpalen in het voorhuis. De noordelijke ingang in het voorhuis lijkt onzeker, omdat de paalgaten op de beoogde plaats niet binnen de lijn van de wandpalen liggen. Lengte 23,8 m; breedte 5,0 m.

Nr. 33. Bijna volledige plattegrond, aan de noordzijde gedeeltelijk verstoord. Bij de ingangen tussen middenhuis en stal zijn restanten van ingangskuilen aangetroffen. Het voorhuis heeft twee paar dubbelpalen, die overigens ontbreken bij het begin van de stal. Een extra paalgat tussen voor- en middenhuis wijst op de aanwezigheid van een tussenwand. Merkwaardig zijn de extra palen bij de ingangen aan de zuidzijde die mogelijk een portaal hebben gevormd. Lengte 22,0 m; breedte 5,8 m.

Nr. 36. Plattegrond, in twee fasen ontstaan door verbouwing van een type Wijster B (tot een Peelo B?). In de eerste fase zijn in het voorhuis een paar dubbelpalen en een extra paar staanders aanwezig. Van de noordelijke dubbelpaal is er één afwezig. De extra staanders zijn hier mogelijk later geplaatst bij de verlenging van het voorhuis. Daarin zijn twee paar dubbelpalen gebruikt waarvan aan de zuidzijde één paar niet is teruggevonden. Daarnaast zijn over de lengte van de eerste fase buitenpalen geplaatst en de stal is ongeveer 1 meter verlengd, waarbij de stalingang waarschijnlijk is vervallen. Eerste fase: lengte 18,5 m; breedte 5,0 m. Tweede fase: lengte 22,4 m; breedte 6,2 tot 6,4 m.

In een eerste, voorlopige publikatie is deze plattegrond ten onrechte gepresenteerd als het type Peelo B (Bardet e.a., 1983). Door de definitieve analyse en de aangescherpte definiëring van de typen is deze plattegrond nu bij het type Wijster B ondergebracht. Helaas is de eerste interpretatie in de latere literatuur overgenomen (Waterbolk, 1980; 1985; Huijts, 1992). Overigens was de plattegrond door de toevoeging van extra palen buiten de wand reeds a-typisch.

Nr. 37. Deze plattegrond is incompleet, maar heeft duidelijke bij het type behorende kenmerken. De noordelijke ingang in het voorhuis is niet meer aantoonbaar. Een extra paal op de overgang van het middenhuis naar het achterhuis wijst op de aanwezigheid van een tussenwand. Lengte 18,4 m; breedte 5,3 m.

Nr. 40. Plattegrond, in twee fasen ontstaan door verbouwing van een type Wijster B (tot een Peelo B). De eerste fase heeft een paar dubbelpalen in het voorhuis en een paar aan het einde van de stal. Het voorhuis is met een deel van ca. 6 m verlengd, waarin drie paar dubbelpalen en een paar nieuwe ingangen. Aan de zuidzijde zijn daarna nog reparaties aangebracht in de vorm van nieuwe dubbelpalen. De stal is ingekort en mogelijk is daarbij de stalingang in de korte wand vervallen, zodat de tweede fase in dat geval een type Peelo B zou zijn. Extra palen wijzen op een dwarswand in de tweede fase, op de overgang tussen voorhuis en middenhuis. Eerste fase: lengte 20,8 m; breedte 5,4 m. Tweede fase: lengte 22,7 m; breedte 5,4 m.

Nr. 41b. Uit de verschillende fasen, c.q. verbouwingen zijn naast de eerder besproken plattegrond van het type Peelo A nog twee van het type Wijster B te construeren, waarbij alleen de essentiële paalkuilen in zwart zijn aangegeven. Nr. 41b heeft aan de zuidzijde twee duidelijke ingangskuilen. Lengte ca. 20 m; breedte ca. 5 m.

Nr. 41c. De jongste fase heeft een zwaar gebouwd nieuw voorhuis met drie paar dubbelpalen, een paar staanders en een paar nieuwe ingangen. Lengte 19,4 m; breedte 5,0 m.

Uit analyse van de boerderijtypen is gebleken dat het type Wijster B in Peelo voorkomt. Dit was bij de eerste analyse niet opgemerkt en komt dus in een eerdere, voorlopige publikatie niet voor (Bardet e.a., 1983).

Het aantal dubbelpalen in het woonhuis varieert van

één tot drie paar. In het algemeen is er een tendens naar een korter achterhuis, zodat een stal in het type Peelo B lijkt te ontbreken. Als overgang naar het type Peelo B zijn de nrs 36 en 37 te beschouwen, waarbij van een stal wellicht nauwelijks nog sprake is. De aanwezigheid van een (stal)ingang in de korte wand is gebruikt om deze nog toe te schrijven aan het type Wijster B, tegenover het ontbreken van een stalingang in het type Peelo B. De plaats van de dubbelpalen varieert in het achterhuis van de traditionele positie op de overgang naar de stal tot dubbelpalen in het achterhuis, zoals bij nr. 40 en mogelijk ook bij nr. 36.

Type Peelo B (fig. 16, 17: nrs 11, 28, 30, 31, 34, 42, 47)
Driedelig als Wijster B, maar zonder stalingang.
Nr. 11. Het oostelijke einde van deze plattegrond is verstoord door een jongere sloot, met uitzondering van een paal in de middenas. Het voorhuis heeft twee paar dubbelpalen met extra palen voor een tussenwand in het westelijke deel. De stal aan de oostzijde is kort met nog twee paar staanders. Lengte 26,0 m; breedte 5,6 m.
Nr. 28. In deze plattegrond ontbreken dubbelpalen op de overgang naar het achterhuis. Het voorhuis heeft waarschijnlijk drie paar dubbelpalen gehad, waarvan drie paar zijn verdwenen; twee aan de noordzijde en één paar aan de zuidzijde. Lengte 18,0 m; breedte 5,3 m.
Nr. 30. Plattegrond waarvan de westelijke wand van het woonhuis ontbreekt. Het is onduidelijk of de extra paal ten westen van het paar dubbelpalen als staander of als restant van de wand moet worden gezien. Ook aan de zuidzijde tussen de ingang en het paar dubbelpalen is een extra paal aanwezig. Het achterhuis is mogelijk door een tussenwand ter hoogte van de dubbelpalen afgescheiden van het voorhuis. Lengte 18,6 m; breedte 5,2 m.
Nr. 31. Zeer korte plattegrond met het voorhuis aan de westzijde. Een extra paal geeft de plaats van een tussenwand aan bij de overgang naar het achterhuis. Lengte 14,4 m; breedte 5,0 m.
Nr. 34. In twee fasen ontstane plattegrond met het voorhuis aan de oostzijde waarin naast de dubbelpalen ook nog één paar staanders voorkomt. Het achterhuis is in de tweede fase verlengd en van een eigen ingang voorzien, zodat typologisch een type Wijster B is ontstaan. Evenals bij nr. 31 geeft hier een extra paal de plaats van de tussenwand tussen midden- en achterhuis aan. Lengte eerste fase 14,2 m; breedte 4,8 m.
Nr. 42. Groot fragment met twee dubbelpalen in het voorhuis aan de oostzijde. Door het ontbreken van de westwand is het onzeker of er wel of niet een (stal)ingang is geweest, maar de beknopte vorm van de plattegrond wijst in de richting van het type Peelo B. Lengte ca. 15 m; breedte 5,2 m.
Nr. 47. Restant van een plattegrond, waarvan de dubbelpalen in het voorhuis, de ingangen in het voorhuis met ingangskuilen en vijf paar staanders als voornaamste bouwelementen zijn teruggevonden. Door het ontbreken van sporen van een stalingang is toewij-

zing aan het type Peelo B mogelijk. Lengte ca. 18 m; breedte 5,6 m.

In een aantal gevallen kan men zich afvragen of de indeling in een woon- en een stalgedeelte nog wel van toepassing kan zijn, omdat het stalgedeelte bijvoorbeeld bij de nrs 31, 34 en 42 zeer kort is en aan de functie als stal kan worden getwijfeld. De mogelijkheid dat het hier om aparte schuren zou gaan, zoals door Waterbolk (1991: *Abb.* 19) wordt gesuggereerd, is in strijd met de differentiatie van de constructie van de gebouwen, waardoor op zijn minst een functionele tweedeling gehandhaafd blijft. Bovendien komen er duidelijk herkenbare schuren voor, die gelijktijdig zijn met dit type (zie onder schuren: nrs 35, 38, 39, 44 en 48).

Type Odoorn C (fig. 18)
Driedeling met dakdragende dubbelpalen, waarvan de binnenste in de wand en de buitenste op enige afstand buiten de wand. Twee paar ingangen in de lange wanden en géén stalingang.
Nr. 50. Fragment van een plattegrond. Kenmerkend is het ontbreken van grondsporen van de korte wanden, die kennelijk minder diep waren ingegraven. Het is uit de plattegrond niet duidelijk op te maken waar het voorhuis of de stal zich heeft bevonden. Lengte 30,5 m; breedte 6,8 m over de buitenpalen en 5,0 m over de wand.

Fragmenten van boerderijplattegronden (fig. 19)
In deze categorie zijn een zestal plattegronden ondergebracht die door het ontbreken van essentiële paalkuilen niet eenduidig aan een type zijn toe te schrijven.
Nr. 9. Fragment, waarvan nog drie paar staanders en een deel van de lange wanden met ingangen zijn gevonden. Door de aanwezigheid van een aantal dubbelpalen in de wand zou de plattegrond kunnen behoren tot de reeks Wijster B, Peelo B, Wijster C.
Nr. 14. Stalgedeelte met boksen, gezien de ligging behorend tot het type Wijster A of Peelo A.
Nr. 15. Bij staanders in werkput 13 en een wandfragment (met ingangskuil?) in werkput 2 van een niet nader te identificeren huisplattegrond (niet afgebeeld).
Nr. 16. Wandfragment van een schuur- of huisplattegrond (niet afgebeeld).
Nr. 18. Stalgedeelte van een vrij lange plattegrond met eigen ingang, mogelijk behorend tot het type Wijster A of Peelo A.
Nr. 19. Diep ingegraven wandfragment en enkele staanders van een schuur- of huisplattegrond (niet afgebeeld) mogelijk behorend tot het type Noord Barge of Fochteloo.
Nr. 20. Wandfragment en enkele staanders van een niet nader te identificeren huisplattegrond (niet afgebeeld).
Nr. 32. Centraal blok van 3 paar staanders en twee delen van de lange wanden.
Nr. 43. Vijf staanders en twee wandfragmenten.

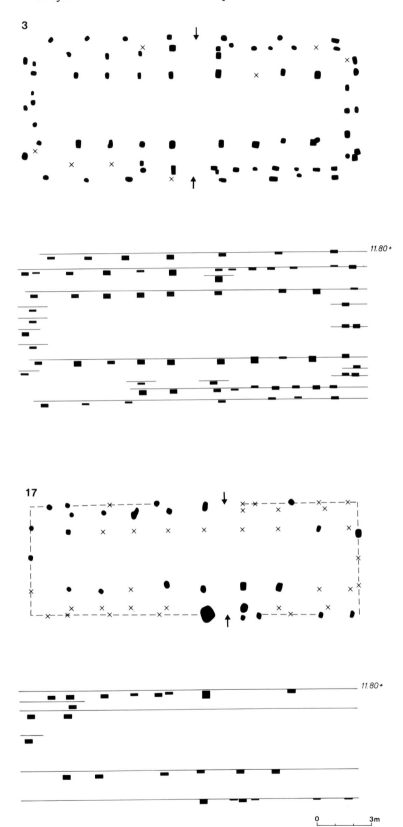

Fig. 8. Boerderijplattegronden van het type Hijken (de nummers corresponderen met de nummering op fig. 7)

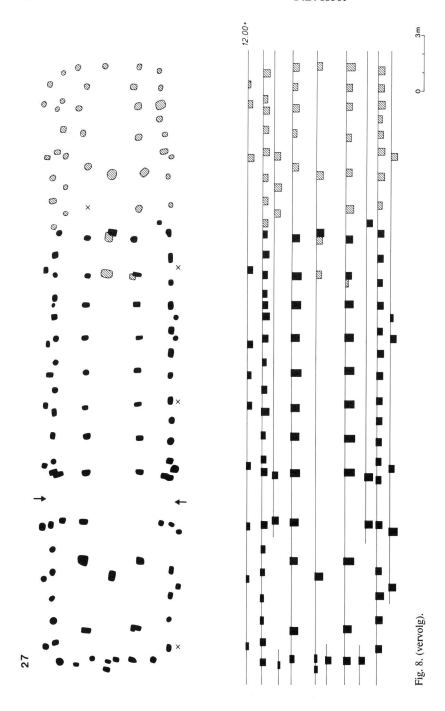

Fig. 8. (vervolg).

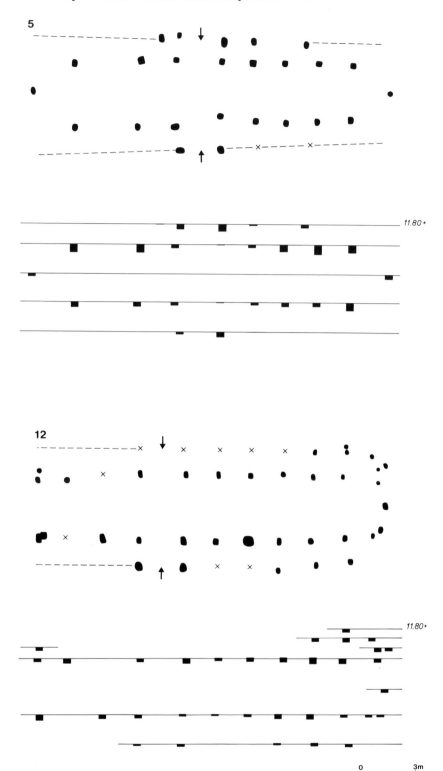

Fig. 9. Overige boerderijplattegronden uit de IJzertijd.

25

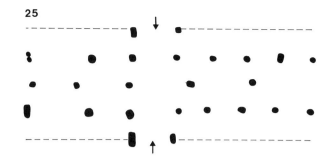

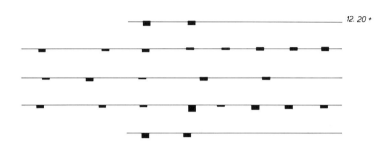

26

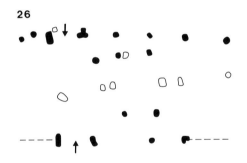

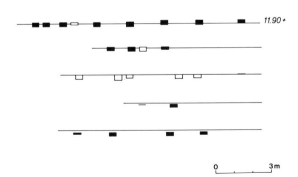

Fig. 9. (vervolg).

0 ___ 3 m

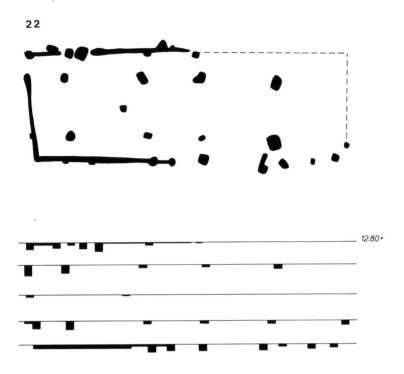

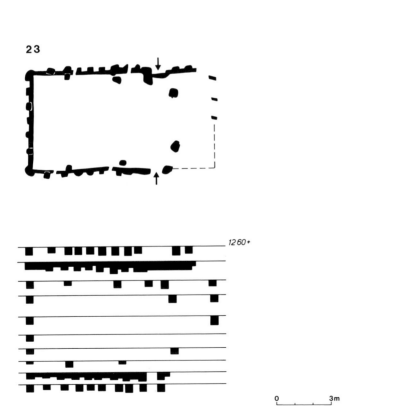

Fig. 10. Boerderijplattegronden van het type Noord Barge.

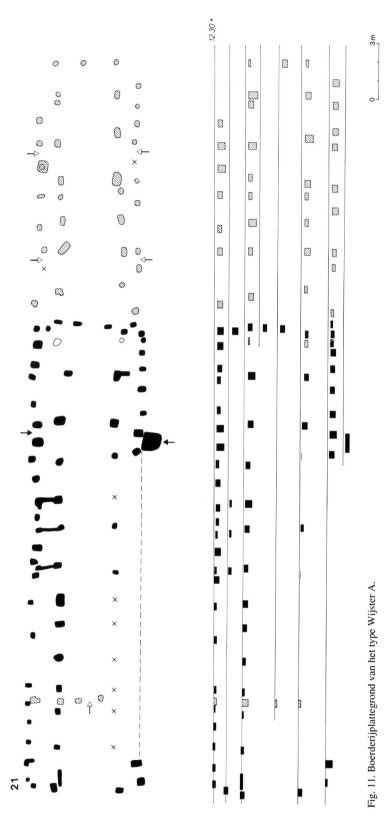

Fig. 11. Boerderijplattegrond van het type Wijster A.

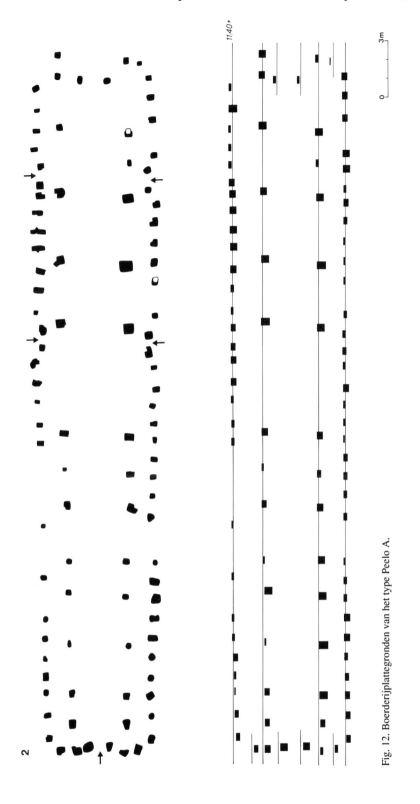

Fig. 12. Boerderijplattegronden van het type Peelo A.

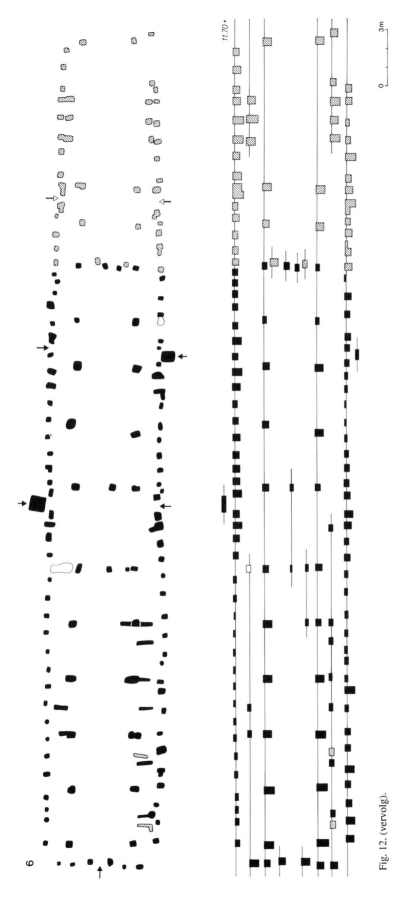

Fig. 12. (vervolg).

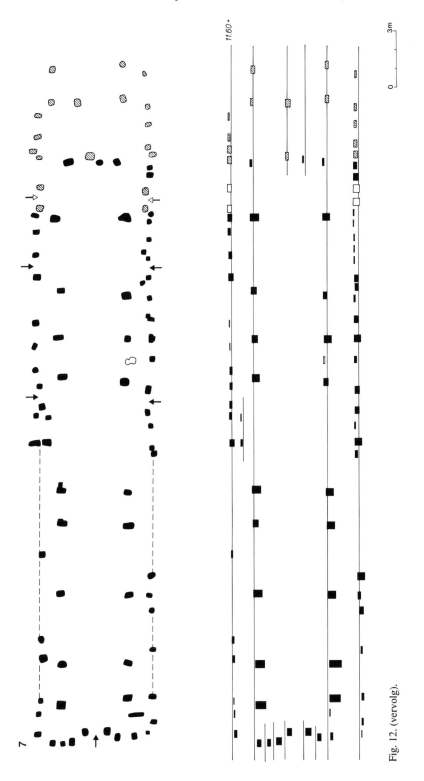

Fig. 12. (vervolg).

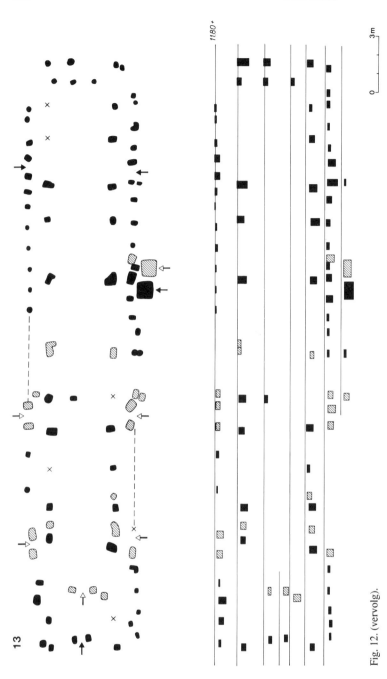

Fig. 12. (vervolg).

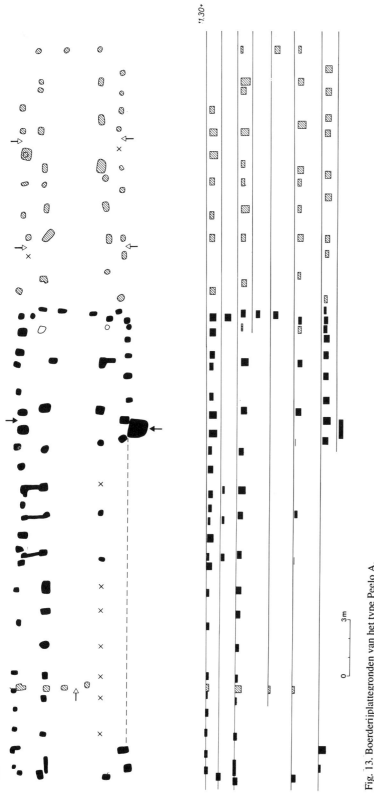

Fig. 13. Boerderijplattegronden van het type Peelo A.

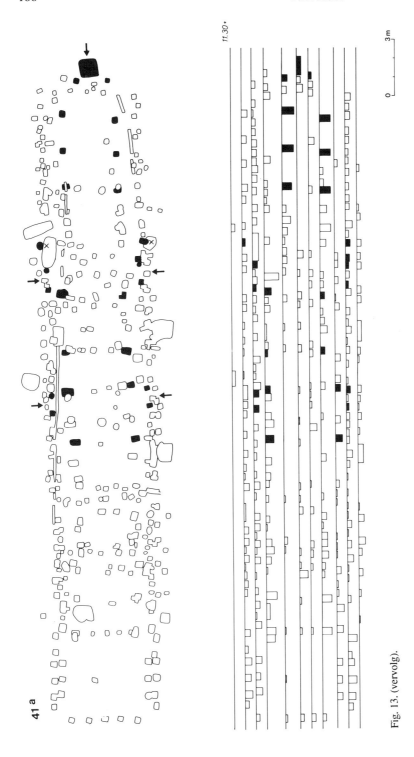

41 a

11.30+

3m

Fig. 13. (vervolg).

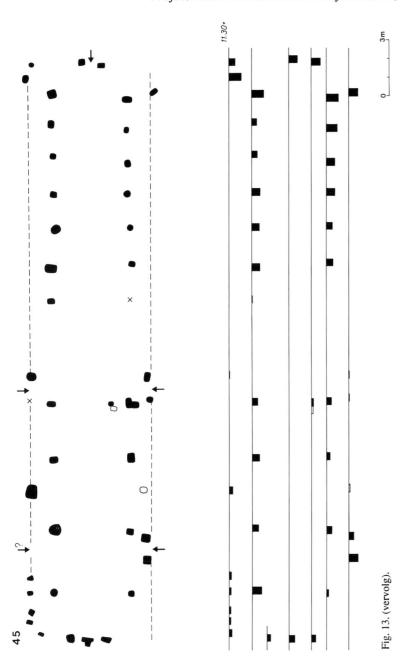

Fig. 13. (vervolg).

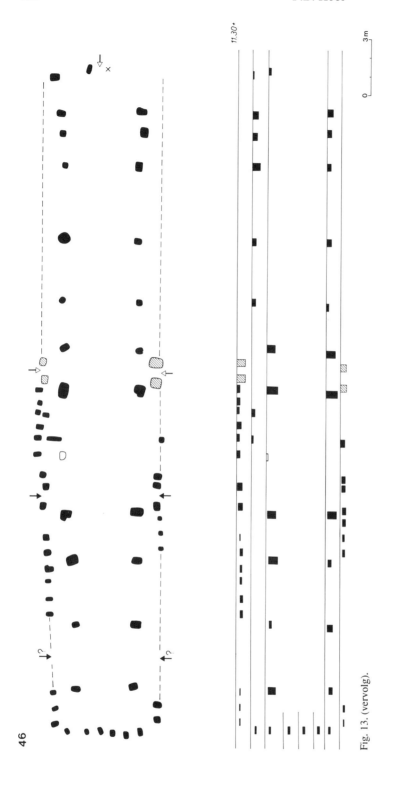

Fig. 13. (vervolg).

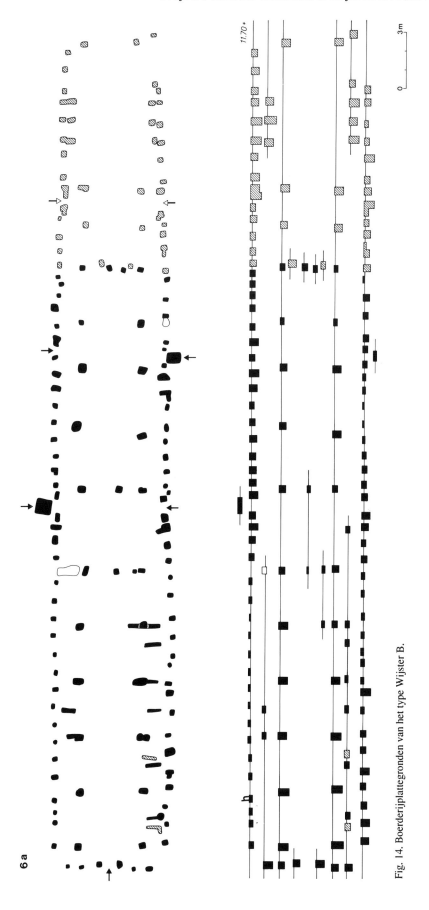

Fig. 14. Boerderijplattegronden van het type Wijster B.

8

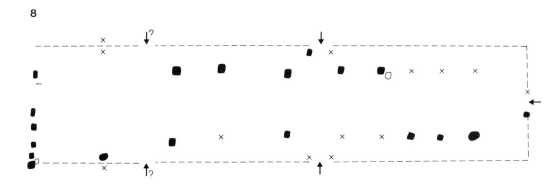

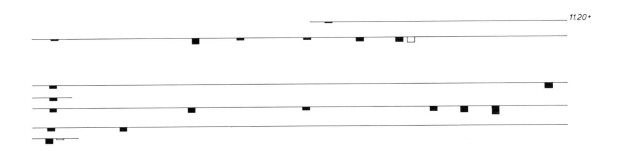

11.20+

29

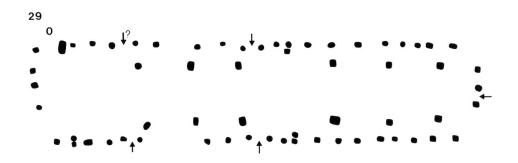

11.40+

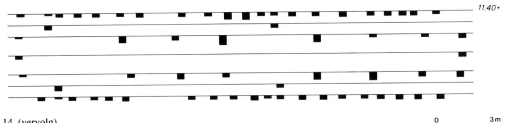

Fig. 14. (vervolg).

0 _____ 3 m

33

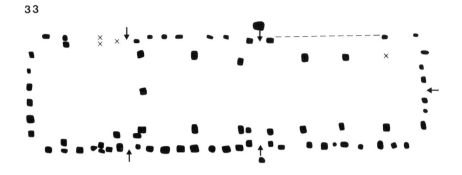

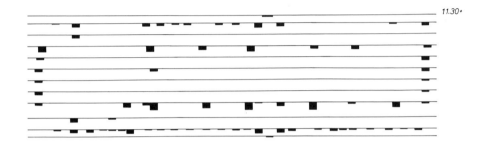

11.30⁺

36

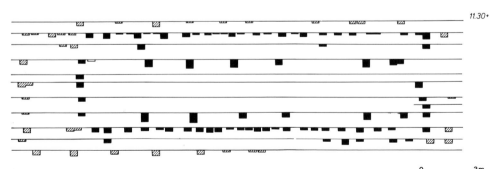

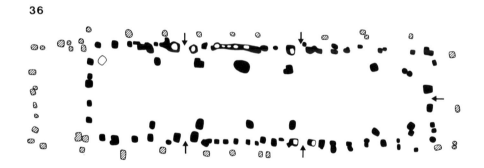

11.30⁺

Fig. 14. (vervolg).

0 _____ 3m

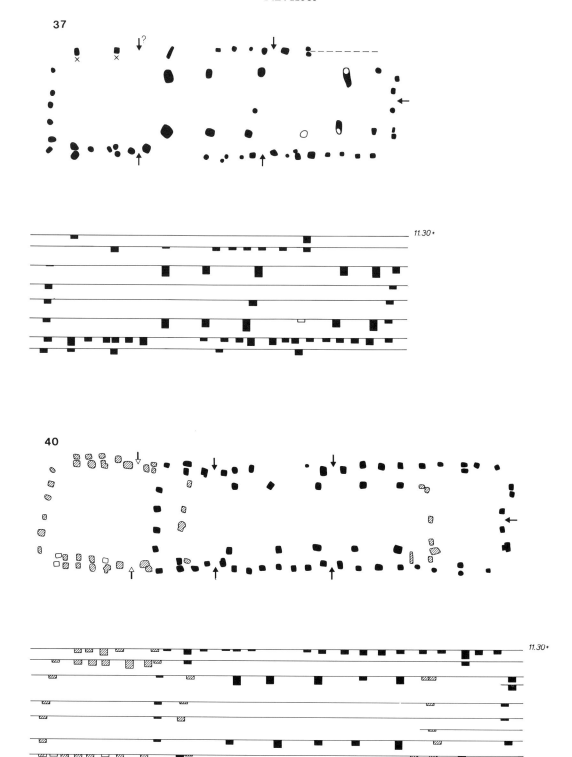

Fig. 15. Boerderijplattegronden van het type Wijster B.

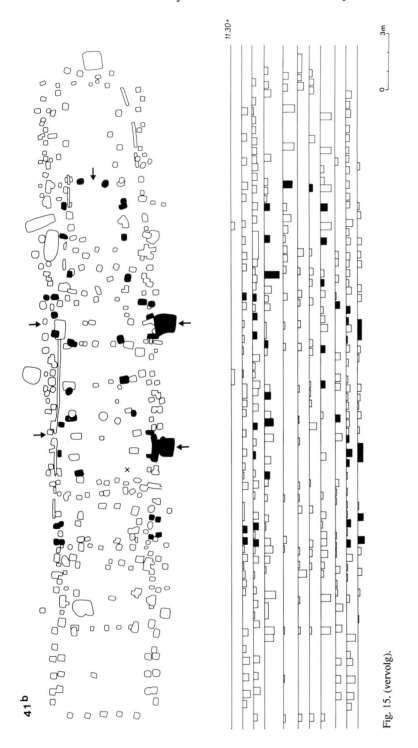

Fig. 15. (vervolg).

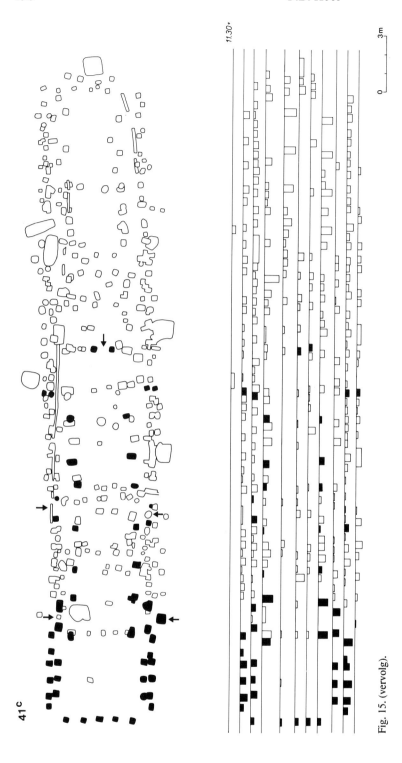

Fig. 15. (vervolg).

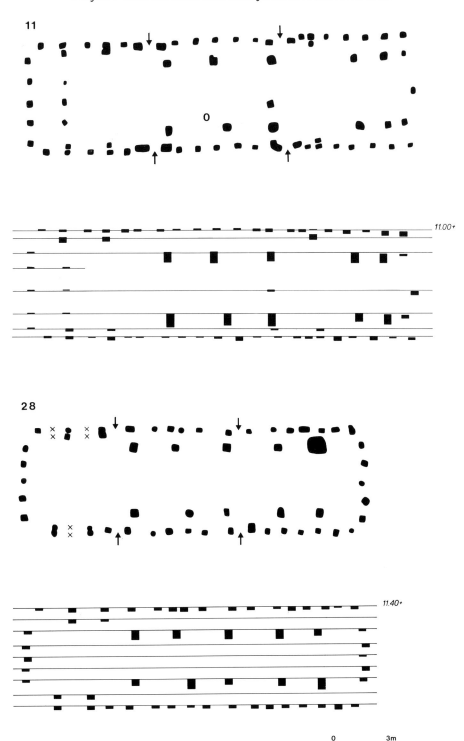

Fig. 16. Boerderijplattegronden van het type Peelo B.

30

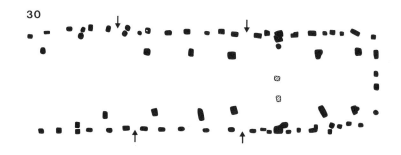

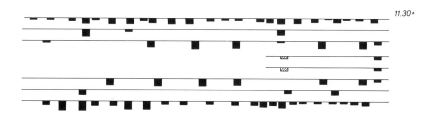

31

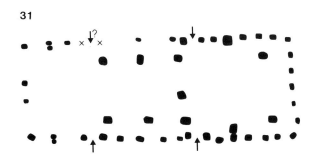

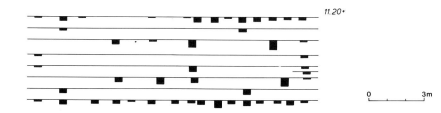

Fig. 16. (vervolg).

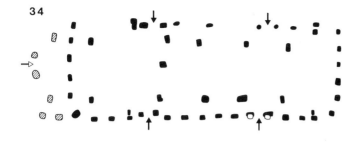

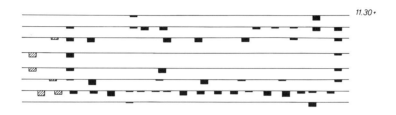

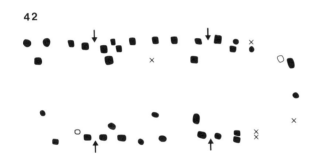

Fig. 16. (vervolg).

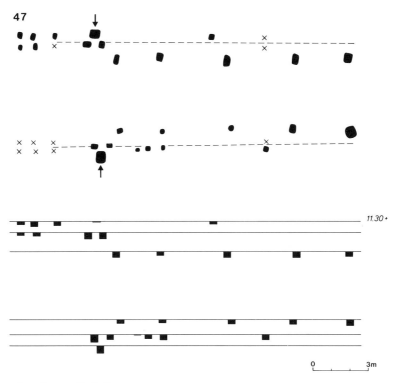

Fig. 17. Boerderijplattegrond van het type Peelo B.

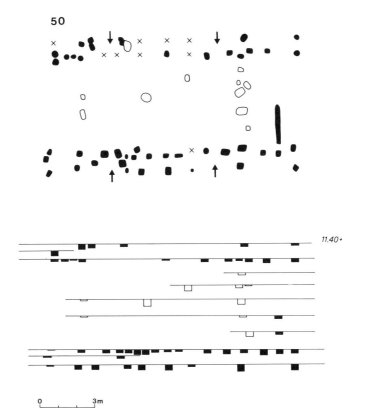

Fig. 18. Boerderijplattegrond van het type Odoorn C.

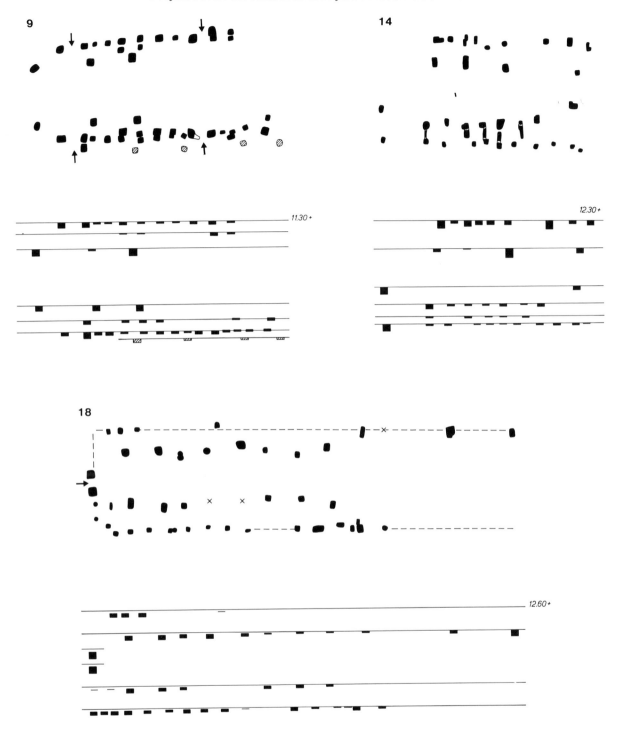

Fig. 19. Fragmenten van diverse boerderijplattegronden.

P.B. Kooi

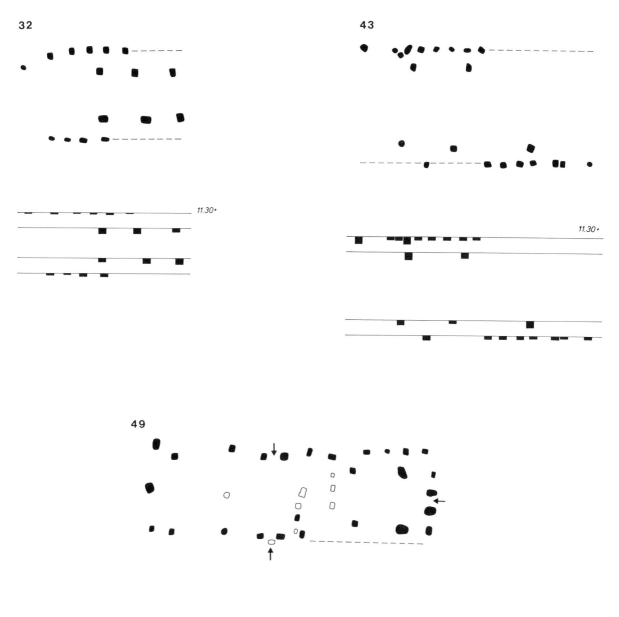

Fig. 19. (vervolg).

Nr. 49. Fragment van een plattegrond met een stalgedeelte aan de oostzijde, met een eigen ingang. Op de overgang naar de stal zijn een paar ingangen aanwezig, maar over de constructie c.q. plattegrond van het voorhuis bestaat geen zekerheid. Het is mogelijk een gereduceerde Peelo B of een Wijster C plattegrond.

De ontwikkeling van het Drentse boerenhuis
Het totale aantal beschreven boerderijtypen in Peelo is nu te vergelijken met hetgeen voor de aanvang van het onderzoek bekend was. Naast een bevestiging van wat ook elders reeds was gevonden, kan ook een aantal afwijkingen worden geconstateerd.

Als duidelijke representant van de bewoning in de IJzertijd is het type Hijken aanwezig. De resterende plattegronden uit die periode zijn helaas slechts fragmentarisch bewaard gebleven en daardoor typologisch niet goed in te delen. Zoals eerder opgemerkt wijst de breedte van de stallen op een datering in de late IJzertijd. Daar staat tegenover dat juist de slechtere kwaliteit van de plattegronden ten opzichte van het type Hijken een datering in de vroege IJzertijd suggereert. Toeschrijving van deze fragmenten aan een type stuit nog op een ander probleem. Bij opstelling van de typelijst door Huijts en Waterbolk zijn een aantal restgroepen geformeerd tot één type. In dit geval hebben we te maken met de variant Elp en de variant Hijken. Datering op basis van archeologische gegevens wordt door hen niet mogelijk geacht. Volgens het gepubliceerde overzichtsschema wordt de variant Elp in de periode ca. 900-800 v.Chr. gedateerd en de variant Hijken ca. 300-250 v.Chr.

Bij deze twee varianten zijn drie plattegronden opgenomen, die opvallende overeenkomsten vertonen, namelijk de plattegronden Emmerhout X 1 en 2 en Peelo 2 (Kleuvenveld). Alle drie zijn gevonden in relatie met een *Celtic field*, waarvan in Emmerhout de staketselrijen zijn aangetroffen en in Peelo de wallen als lichtere banen op luchtfoto's. Kenmerkende overeenkomst van de plattegronden is de versmalling van het woongedeelte naar het einde van de stal en de middenpalen, zowel in het woongedeelte als in de overgang naar de stal. Enerzijds zijn er overeenkomsten aan te wijzen met het type Hijken waarbij in een aantal gevallen het stalgedeelte ook aanzienlijk smaller wordt, anderzijds lijken de dubbele paalgaten in het woongedeelte veel op het overgangstype Hijken. De variatie in de plaatsing van de dakdragende palen zou echter ook duiden op een type dat voorafgaat aan het type Noord Barge. Er zullen meer plattegronden met dateerbaar materiaal gevonden moeten worden om dit probleem op te kunnen lossen. Opvallend lijkt de afwezigheid in Peelo van het type Fochteloo. Daarvoor zijn verschillende verklaringen mogelijk. In de eerste plaats is er maar een zeer beperkt aantal exemplaren van dit type bekend (Huijts: 5x), en deze bevinden zich alle in het zuidelijk deel van het Drents plateau. Het type heeft dus een beperkte verspreiding.

Een andere mogelijkheid is, dat met de afgraving van het hoogste deel van de es plattegronden van dit type zijn verdwenen. In dat geval zou het fragment nr. 19 een laatste restant kunnen zijn. Er is echter nog een belangrijker aspect aan de orde, namelijk de datering van het type. Bij de beschrijving van de nederzettingssporen op het Haverland te Peelo (Kooi, 1987) is reeds gewezen op de overeenkomst tussen de daar gevonden plattegrond uit de 2e-3e eeuw en de plattegrond Fochteloo 1 (Huijts, 1992: afb. 128), die grote overeenkomst vertoont met het type Peelo A (zie Kooi, 1987: fig. 7). Bouwtechnisch gezien is er bovendien nauwelijks verschil tussen het type Fochteloo en de plattegrond Fochteloo 1 (Huijts, 1992: p 131). Bij de publicatie van de opgravingen te Fochteloo door Van Giffen, geeft deze een datering van beide nederzettingscomplexen in de Romeinse tijd, op basis van het gevonden aardewerk (Van Giffen, 1954). Het is duidelijk, dat het type Fochteloo niet algemeen voorkomt en dat het mede op grond van bouwtechnische kenmerken in de ontwikkeling van de huistypen waarschijnlijk geplaatst moet worden naast het type Noord Barge.

De Romeinse tijd leverde in Peelo de nieuwe typen Peelo A en B op. Bij revisie en herinterpretatie van het onderzoek te Wijster, blijkt het type Peelo A ook daar aanwezig te zijn. In tegenstelling tot wat Huijts (en Waterbolk) beweert is het type Peelo B in Wijster niet met zekerheid aan te tonen. Daarvoor zijn de aangevoerde voorbeelden te onvolledig. Er staat tegenover, dat het type Wijster C in Peelo ontbreekt. Wanneer we de typen Peelo B en Wijster C met elkaar vergelijken valt er een overeenkomst en een verschil op te merken. De overeenkomst is, dat de totale lengte van de boerderijen ten opzichte van de voorgaande typen aanzienlijk is bekort. Het verschil is, dat dit bij het type Peelo B het gevolg is van de reductie of het verdwijnen van de stalruimte, terwijl bij het type Wijster C het middenhuis verdwijnt en de stal juist wel blijft bestaan. Een dergelijk verschil zal dus te correleren zijn met een verschil in de landbouweconomie tussen beide nederzettingen: in Wijster bleef de veehouderij relatief belangrijk, terwijl in Peelo de akkerbouw mogelijk de overhand had.

De conclusie hieruit is dat beide typen gelijktijdig kunnen voorkomen. Het lokale verschil wordt veroorzaakt door economische verschillen.

Het resultaat van de discussie ziet er als volgt uit:

Waterbolk + Huijts 1992		Kooi
type	datering	gewijzigde chronologie
Elp	1200-800 v.Chr.	X
variant Elp	ca. 900-800	?
overgangstype		
Hijken	800-400	X
Hijken	400-250	X
variant Hijken	ca. 300-250	? ca. 250-100
Fochteloo	250-100	
Noord Barge	100 v.Chr.-100 na Chr.	X
		Fochteloo 0-100 na Chr.

Waterbolk + Huijts 1992		Kooi	
type	datering	gewijzigde chronologie	
Wijster A	100-250	X	100-250
Peelo A	200-400	X	200-350
Wijster B	200-400	X	300-450
Wijster C	300-400	X	400-550
Peelo B	400-550	X	400-550

Een ander aspect van de boerderijbouw is de bouwkundige ontwikkeling zoals door Huijts is beschreven. Hoewel de gekozen oplossingen gedegen zijn beredeneerd, kan vanuit een andere (archeologische) invalshoek naar alternatieve oplossingen worden gekeken. Een van de eerste duidelijke veranderingen in de grondsporen is de overgang van het type Hijken naar het type Noord Barge en Fochteloo. De zware constructie van de wand in deze typen zou niet alleen kunnen betekenen dat het een dakdragende wand is geweest, maar bovendien zou de wandhoogte al ongeveer op stahoogte geweest moeten zijn. De plaats van de ingangen in de wand geeft dat aan (Huijts, 1992: p. 95). Aansluitend bij deze constatering zou een overspanning van wand tot wand als mogelijkheid moeten worden gezien, waardoor de onregelmatige binnenconstructie van het type Noord Barge beter verklaarbaar is. De voortzetting van deze constructie lijkt in het type Wijster A weer afwezig te zijn (Huijts, 1992: p. 107). Toch zou

de grote dichtheid van de wandpalen bij dit en volgende typen daarvoor een aanwijzing kunnen zijn. Dat de wandpalen niet overal in een rechte lijn liggen sluit niet uit, dat ze aan de bovenzijde door een rechte ligger verbonden zijn geweest. De ligger zou juist daar een regulerend effect kunnen hebben. Ook hoeft dit geen beletsel te zijn om de flexibele vlechtwerkwand aan deze wandpalen te bevestigen. Het verstevigt bovendien het onderlinge verband. Illustratief is in dit geval de plattegrond van Fochteloo 2. Ondanks het zeer strakke wandspoor blijken de wandpalen zeker aan de zuidzijde niet op één lijn te liggen. Ook ten opzichte van de ingangen geeft deze plattegrond duidelijk twee mogelijkheden voor de aansluiting van de wand bij de ingangsstijlen. De wand is bij de ingangen in de lange wanden ingeklemd tussen de ingangsstijl en een wandstijl. Bij de stalingang eindigt de wand met een extra stijl, die in lijn staat met de ingangsstijl. Bij de eerstgenoemde constructie staan de ingangsstijlen binnen de lijn van de wand en dus vrij ver naar binnen. Tot slot kan worden opgemerkt dat het ontbreken van een wandstijl in de hoeken hier samen gaat met rechte hoeken. Als we deze bevindingen nog eens vergelijken met de plattegronden uit de Romeinse tijd, dan lijkt de suggestie van een vrijstaande wand zoals Huijts suggereert niet aannemelijk. Het naar binnen liggen van de ingangsstijlen is daarvoor géén overtuigend argument, noch de plaatsing van de wandpalen. De aanname van Huijts (1992: p. 109) dat de Wijster A plattegronden verge-

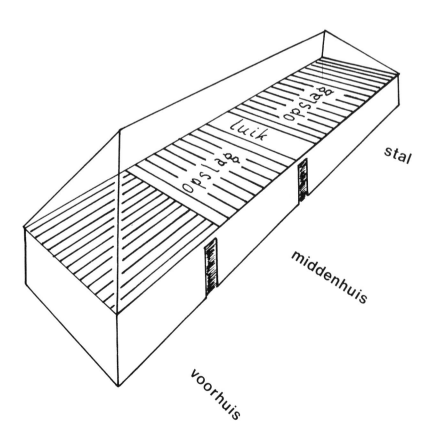

Fig. 20. Mogelijke constructie van een Peelo A huis met berging in de kap.

lijkbaar zijn met het type Fochteloo is in tegenspraak met de eerder door hem gebruikte argumenten, maar bevestigt wel de gewijzigde datering van het type Fochteloo.

In de ontwikkeling van de typen zien we met het type Peelo A ook door het optreden van dubbelpalen een aanwijzing dat een koppeling van wand tot wand werd toegepast. Een koppeling van de wanden zou ook voor het gebruik voordelen gehad kunnen hebben, door op deze dwarsbalken een vloer aan te brengen. De bovenliggende ruimte zou dan benut kunnen worden voor opslag, terwijl de temperatuur in de verschillende woon- en stalgedeelten minder afhankelijk was van de buitentemperatuur. De vraag is natuurlijk in hoeverre de constructie een extra belasting van opslag kan dragen. Om extra steun en stevigheid te verkrijgen, zouden de dwarsverbindingen van wand tot wand met een keep of halfhouts gekoppeld kunnen worden aan de staanders. Op de plaatsen waar dubbelstijlen voorkomen is deze koppeling niet mogelijk en de consequentie is dan dat daar geen extra belasting aanwezig is. De dubbelpalen op de overgang van het middenhuis naar het achterhuis zouden dan verband kunnen houden met de plaats in het huis waar een open verbinding was van de begane grond met de opslagruimte onder de kap (fig. 20).

3.3. Bijgebouwen

Eenschepige schuren (fig. 21: nrs 4, 35 en 51)
Nr. 4. Plattegrond met één paar ingangen in de lange wanden en twee paar dubbelpalen ter weerszijden van de ingangen. Aan de noordzijde zijn de dubbelpalen incompleet. Voor de ingang aan de zuidzijde bevindt zich een ingangskuil. Twee van de vier wandpalen aan de westzijde zijn extra zwaar. Lengte 7,0 m; breedte 4,4 m.

Nr. 35. Plattegrond met vijf paar dubbelpalen, waarvan aan de noordzijde één ontbreekt. Eén paar ingangen, waarvan de zuidelijke is gewijzigd. Lengte 7,8 m; breedte 4,4 m.

Nr. 51. Deze plattegrond heeft evenals nr. 35 vijf paar dubbelpalen en één paar ingangen. Lengte 7,6 m; breedte 3,6 m.

Dit type schuren is gebouwd volgens hetzelfde principe als het voorhuis van de huistypen Wijster B, C en Peelo B, zij het met een geringere breedte.

De smederij (fig. 21: nr. 10)
Nr. 10. Vierkante plattegrond van 5,0x5,0 m, waarvan alle vier wanden met dubbelpalen. Iets uit het centrum liggen twee grote haardplekken. De ingang zal zich bevonden hebben aan de westzijde, waar een ietwat afwijkende constructie van de wand voorkomt. Opvallend is de aanwezigheid van twee extra palen buiten de wand in het midden van de noord- en zuidzijde. Beide hoeken aan de zuidzijde zijn gerepareerd. Daar

deze plattegrond in de nabijheid ligt van een aantal ijzeroventjes is een functie in verband met ijzerbewerking waarschijnlijk.

Schuur met dubbelpalen en staanders (fig. 21: nr. 44)
Deze plattegrond heeft ten oosten van het paar ingangen twee paar dubbelpalen, terwijl het oostelijke deel bestaat uit vier staanders zonder wandpalen. In aanleg is deze plattegrond vergelijkbaar met een schuur op het Haverland, die weliswaar alleen staanders heeft maar ook daar een tweedeling te zien geeft (Kooi, 1984). Lengte ca. 9 m; breedte 4,1 m.

Tweeschepige schuren (fig. 22: nrs 1, 38 en 39)
Nr. 1. Plattegrond, bestaande uit twee rijen palen op regelmatige tussenafstand, en een standgreppel aan de noordzijde. De afstand tussen de twee palenrijen is groter dan tussen de standgreppel en de aangrenzende rij palen. Lengte 21,6 m; breedte 5,1 m.

Nr. 38. Plattegrond, bestaande uit twee rijen paalgaten en een wandgreppel aan de noord- en westzijde. De plattegrond is onregelmatiger dan de voorgaande, mogelijk ten gevolge van reparaties of verbouwingen, zoals ook uit de deels dubbele greppels blijkt. Lengte ca. 16 m; breedte 4,4 m.

Nr. 39. Fragment van een plattegrond, bestaande uit een standgreppel, drie palen van de middenrij en één paal van de buitenste rij. Lengte (greppel) 18,4 m; breedte 5,0 m.

De tweeschepige schuren liggen alle drie langs de rand van een erf. Bij nr. 38 sluit het omheiningsspoor van de erfbegrenzing aan bij het wandspoor van de schuur. De plattegronden suggereren, dat het kapschuren zijn geweest, waarvan de kap op twee rijen palen heeft gerust en die aan drie zijden open is. De hoogte zal in verband met deze constructie beperkt zijn geweest.

Tweeschepige schuur met verdiepte vloer (fig. 22: nr. 24)
Nr. 24. Deze plattegrond bestaat uit een grote rechthoekige kuil waarbinnen een rij middenpalen en palen langs de wand. De ingangen waren in het vlak te zien als twee donkere kuilen die aan de noord- en zuidzijde uitstulpten. Aan de noordzijde ligt de rand van de verdieping van de schuur op 30-50 cm parallel aan een omheining. In eerste instantie ging de noordelijke ingang door de omheining, maar werd in een later stadium afgesloten. Gezien de positie ten opzichte van de omheining en de verdiepte vloer is het waarschijnlijk, dat het dak rondom op een wand van zoden heeft gerust. De palen langs de rand van de kuil hebben mogelijk gediend om de binnenzijde van de kuil en de zodenwand (misschien met vlechtwerk) te steunen. Lengte (van de kuil) 10,7 m; breedte 4,6 m.

Stal (fig. 23: nr. 184)
Nr. 184. Plattegrond met forse paalgaten en een

drempelsleuf. Deze structuur komt slechts een keer voor. De situering van dit schuurtje bij het einde van de stal van plattegrond 40 wijst op een aparte stal. Lengte 5,4 m; breedte 5,4 m.

Komhutten (fig. 24-31)
In deze categorie zijn 53 plattegronden opgenomen, die onderling aanzienlijk verschillen.
Op grond van deze verschillen is een onderverdeling te maken namelijk:
a. Komhutten met twee palen, in het midden van de korte zijde; nrs 33, 48, 77, 89 en 104;
b. Komhutten met palen in het midden van de korte zijden en op de hoeken; nrs 7, 8, 9, 19, 20, 22-28, 41, 49, 51, 52, 73-79, 88, 109, 139-147, 149, 162, 167, 169, 170, 172, 173, 174, 185, 191, 193. Bij deze groep zijn de nrs 74 en 75 uitzonderlijk groot, terwijl in nr. 75 nog sporen van planken of balken in een deel van de vloer werden aangetroffen. Opmerkelijk is ook dat deze betrekkelijk eenvoudige bouwsels werden herbouwd of grondig gerepareerd, zoals de nrs 20, 22, 24, 76, 143, 145 en 174;
c. Bijzondere vormen; nrs 10 en 11, twee plattegronden met een duidelijke verdiept gedeelte, zijn omgeven door een onregelmatige paalzetting. Mogelijk is van deze hutten slechts een deel van de vloer verdiept geweest.

Onduidelijk is de toewijzing van de nrs 44 en 194, waar naast de palen midden in de korte zijden slechts twee palen aan één lange zijde zijn gevonden. Deze indeling wordt echter mede bepaald door variaties in het niveau waarop de sporen bewaard zijn gebleven en met de bouwwijze. Rond een rechthoekige verdiepte vloer werd een wand van opgestapelde zoden gemaakt, waarvan geen sporen zijn aangetroffen. Het tentdak werd gedragen door twee palen in het midden van de korte zijde en door de zodenwand. Palen op de hoeken of extra palen langs de lange zijden hebben wanden met vlechtwerk of planken gestut en dienden eventueel als extra steun voor het dak. De ingang bevond zich aan een van de korte zijden. Een redelijk complete plattegrond is bijvoorbeeld nr. 7, waarin veel constructieve elementen voorkomen, mede door de gunstige diepte waarop de grondsporen bewaard zijn gebleven. Uit de schematische weergave van de diepte der verschillende sporen is te zien wat het resultaat is van een opgravingsvlak op dieper niveau, waarbij alleen de diepere paalgaten overblijven, zoals het geval is bij nr. 9. Het is zelfs denkbaar, dat de palen op de hoeken van de kuil nauwelijks zijn ingegraven, zodat ze geen waarneembaar spoor hebben achtergelaten.
Naast de typologische indeling is ook een groepering mogelijk naar de plaats of fase binnen de nederzetting. We krijgen dan de volgende clusters:
Nrs 7-11 bij de boerderijplattegronden nrs 2, 6 en 7;
Nrs 19-28 bij de boerderijen nrs 8 en 9;
Nrs 33, 48 en 49 bij de boerderij nr. 13;

Nrs 41-75 bij de boerderijen nrs 14, 15 en 16;
Nrs 76-109 bij de boerderijen nrs 20 en 21;
Nrs 139-149 bij de boerderijen nrs 28-34, 36 en 37;
Nrs 162-170 in het uiterste westen;
Nrs 162, 171-194 bij de boerderijen nrs 40, 41, 42, 43, 45, 46 en 47.

Spiekers, mijten en roedenbergen (fig. 32-56)
In totaal zijn onder deze categorie 136 plattegronden ingedeeld waarvan wordt verondersteld dat ze hebben gediend voor de oogstberging. Daarin zijn op basis van de plattegronden een aantal verschillende typen te onderscheiden:
a. Met vier palen, min of meer regelmatig in een vierkant of rechthoek geplaatst. De afmetingen variëren van 1,4x1,4 m tot 3,6x3,6 m. Soms komen extra palen voor, die als reparaties zijn op te vatten, zoals bijvoorbeeld nrs 4, 16, 32, 50 en 69. Voorts is opmerkelijk dat een aantal twee- à driedubbele palen heeft, zoals bijvoorbeeld nrs 2, 47, 96, 154 en 165;
b. Met zes palen in een vierkant of rechthoek geplaatst. Te onderscheiden zijn vierkante paalzettingen met twee palen extra in twee tegenoverliggende zijden, zoals bijvoorbeeld nrs 5, 97, 101, 151 en 153, rechthoeken die als het ware een verdubbeling of uitbreiding van een vierkant vormen, zoals bijvoorbeeld nrs 14, 60, 83, 86, 90, 94 en 105;
c. Met palen min of meer in een vijfhoek geplaatst. In totaal gaat het om zes exemplaren nrs 55, 156, 157, 161, 163 en 195, waarvan de kleinste (nr. 53) een diameter heeft van ca. 3 m en de rest een diameter van ca. 4 m;
d. Overige vormen (nrs 46, 148 en 177):
Nr. 46. Plattegrond van 4,0x5,5 m, met een serie van vier palen in de middenas en zes palen aan de lange zijden, gegroepeerd in twee keer drie palen, waarvan de middelste geringer in diameter is geweest. Een dergelijk zware bouw wijst op een extra belasting (van de vloer?) en duidt mogelijk op de aanwezigheid van een verdieping.
Nr. 148. Fragment van een plattegrond van 5,0x7,5 m, met vier palen aan de lange zijden en een rij middenpalen. De bouw is minder zwaar dan van nr. 46.
Nr. 177. Rechthoekige plattegrond van 4,5x5,0 m, met negen palen in een rechthoek geplaatst en een aanbouw met drie palen.

Over de bouwwijze is weinig concreets bekend; deze wordt in het algemeen afgeleid en beredeneerd aan de hand van oogstberging uit recente en historische tijd (Zimmermann, 1991; 1992a). In de terminologie worden spiekers, mijten en roedenbergen onderscheiden. Kortheidshalve zal in deze publikatie het woord spieker als verzamelnaam worden gebruikt.
Een aantal aspecten van de spiekers verdient nadere aandacht. Om inzicht te krijgen in de soorten spiekers en de mogelijke ontwikkeling, zijn ze in groepen verdeeld. Een aantal groepen spiekers is toe te schrijven

Fig. 21. Plattegronden van eenschepige schuren met dubbelpalen.

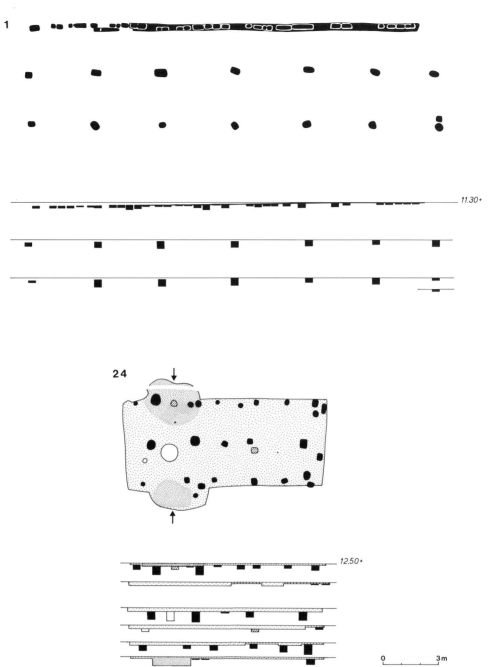

Fig. 22. Plattegronden van tweeschepige schuren, waarvan nr. 24 met verdiepte vloer.

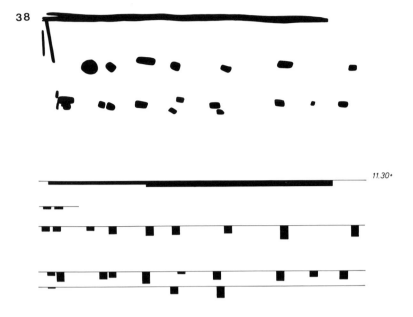

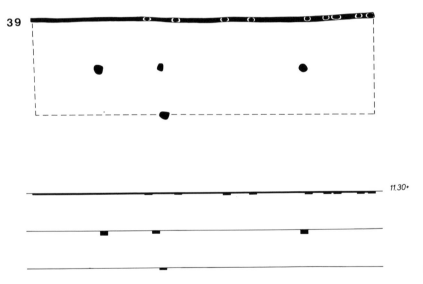

Fig. 22. (vervolg).

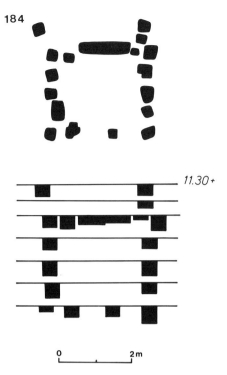

Fig. 23. Plattegrond van een kleine stal.

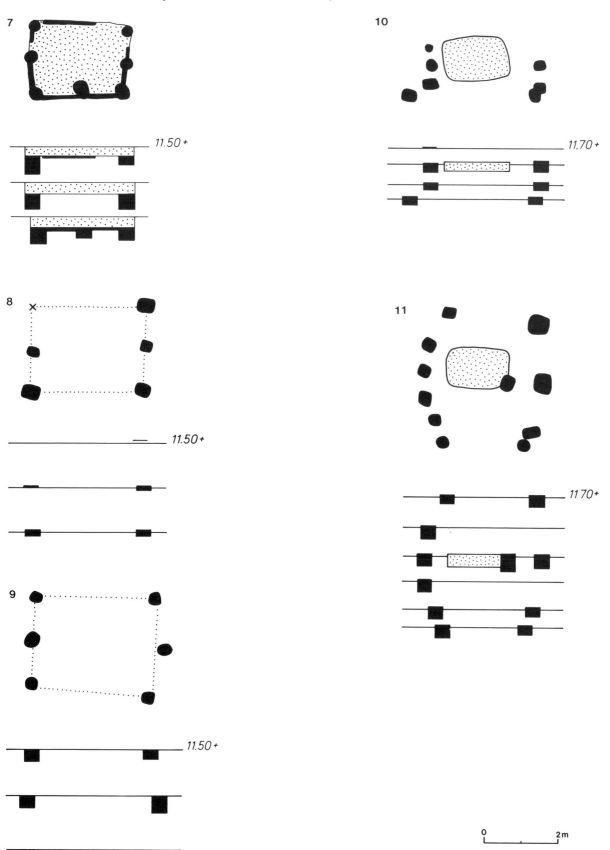

Fig. 24. Plattegronden van komhutten.

P.B. Kooi

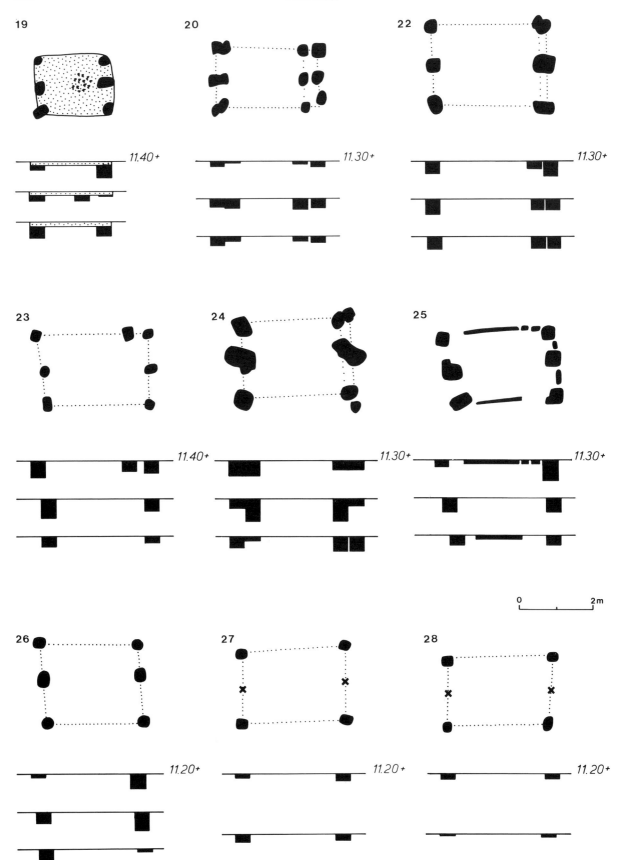

Fig. 25. Plattegronden van komhutten.

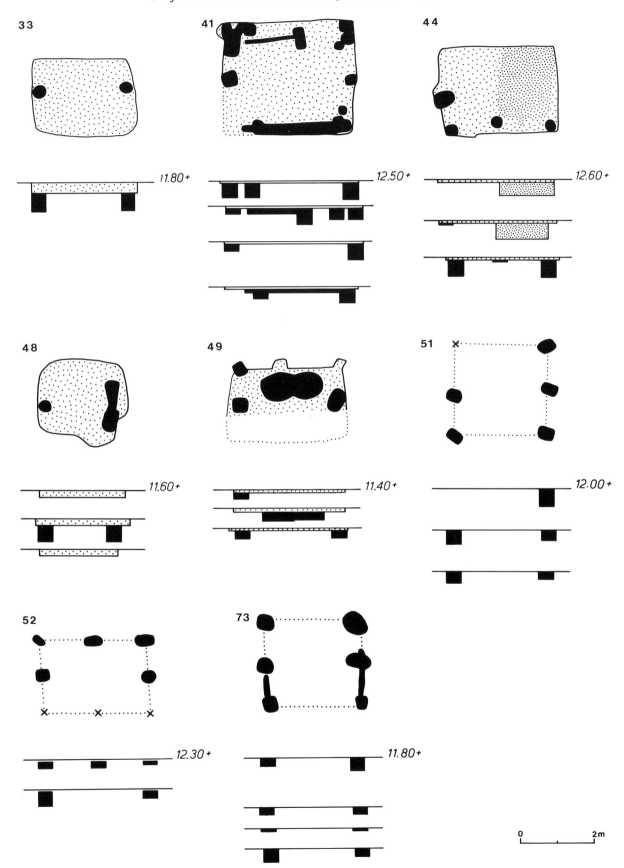

Fig. 26. Plattegronden van komhutten.

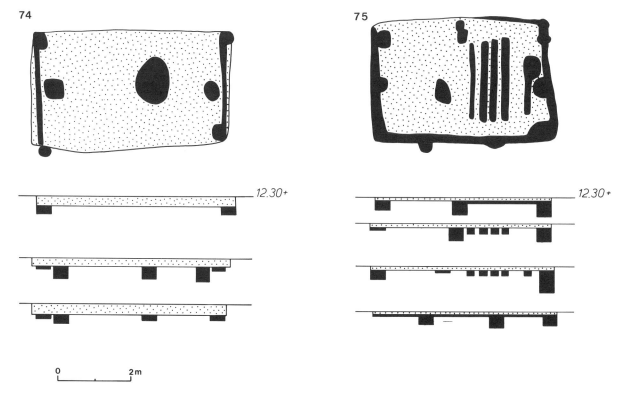

Fig. 27. Plattegronden van komhutten.

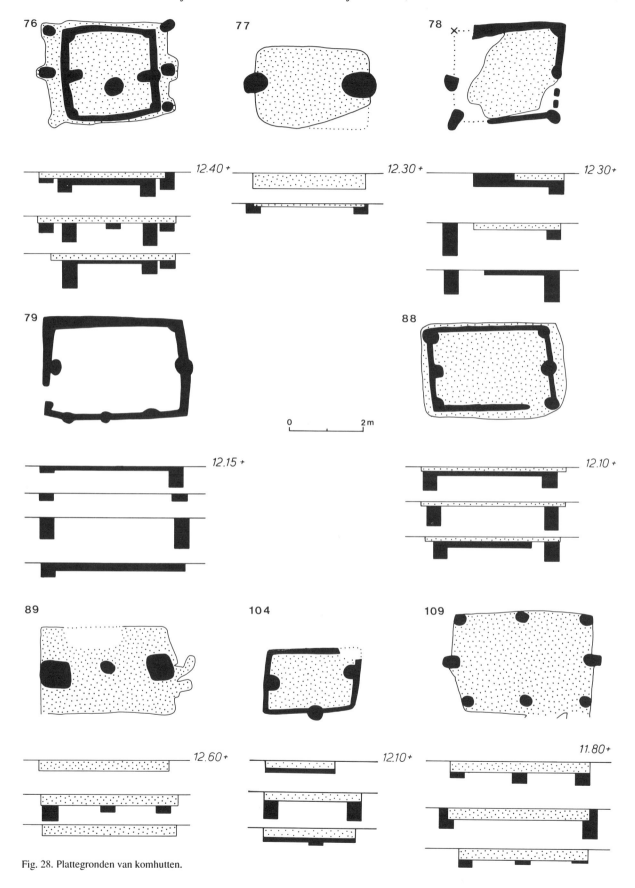

Fig. 28. Plattegronden van komhutten.

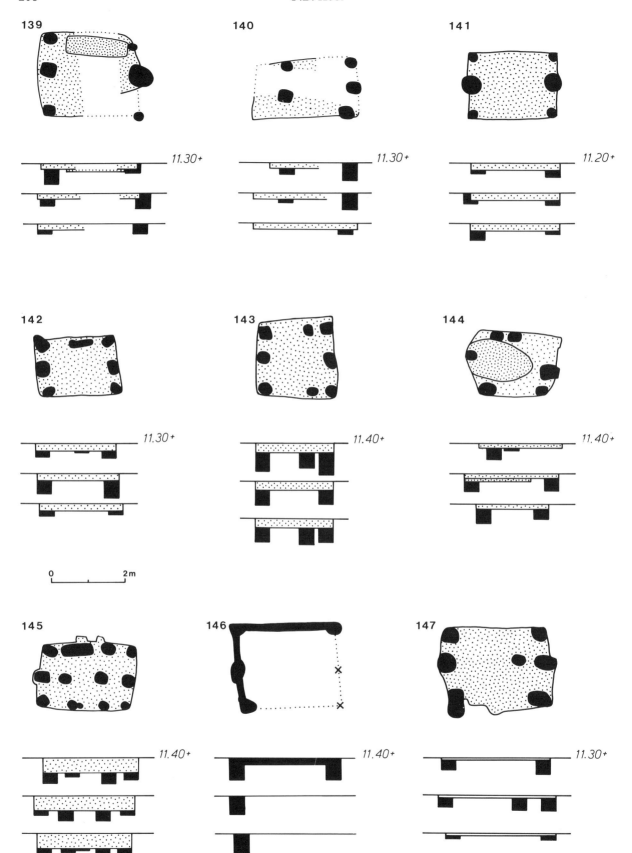

Fig. 29. Plattegronden van komhutten.

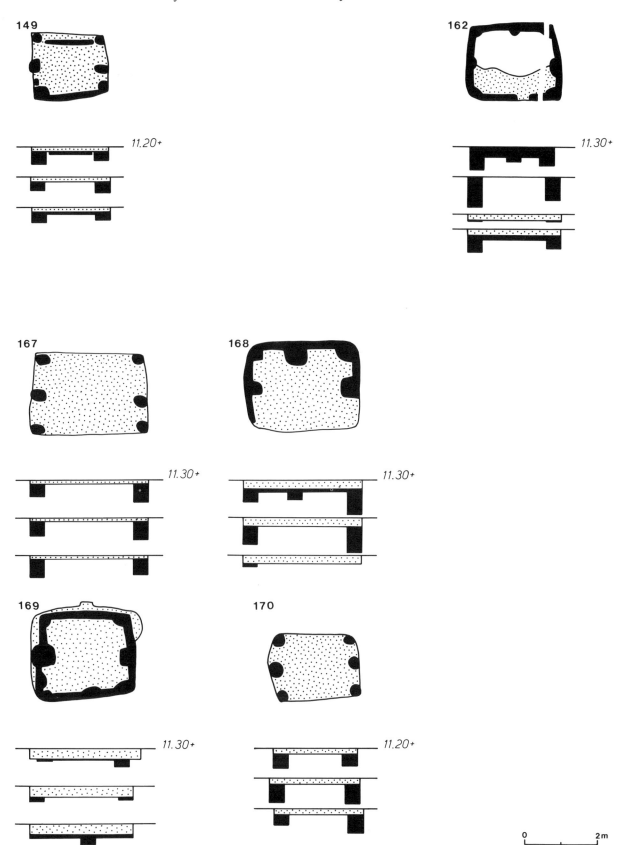

Fig. 30. Plattegronden van komhutten.

218   P.B. Kooi

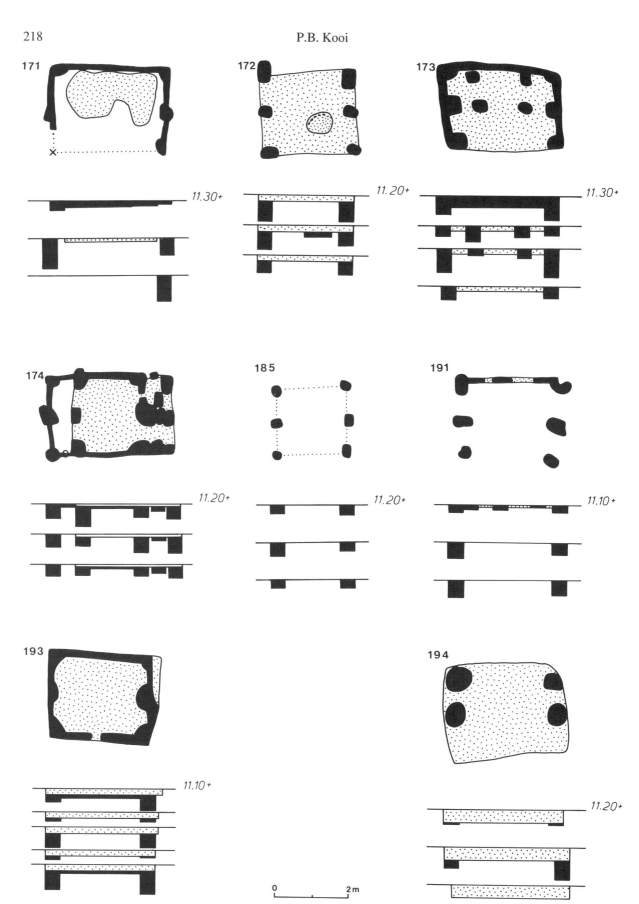

Fig. 31. Plattegronden van komhutten.

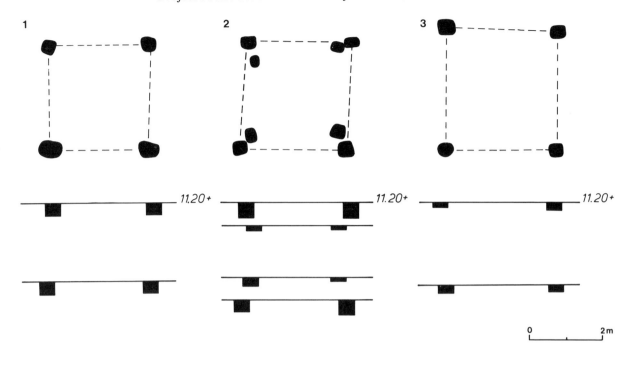

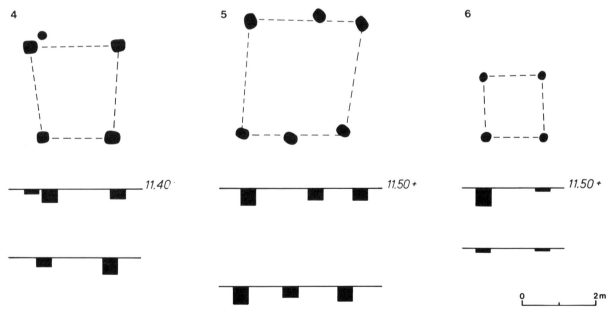

Fig. 32. Plattegronden van spiekers, mijten en roedenbergen.

P.B. Kooi

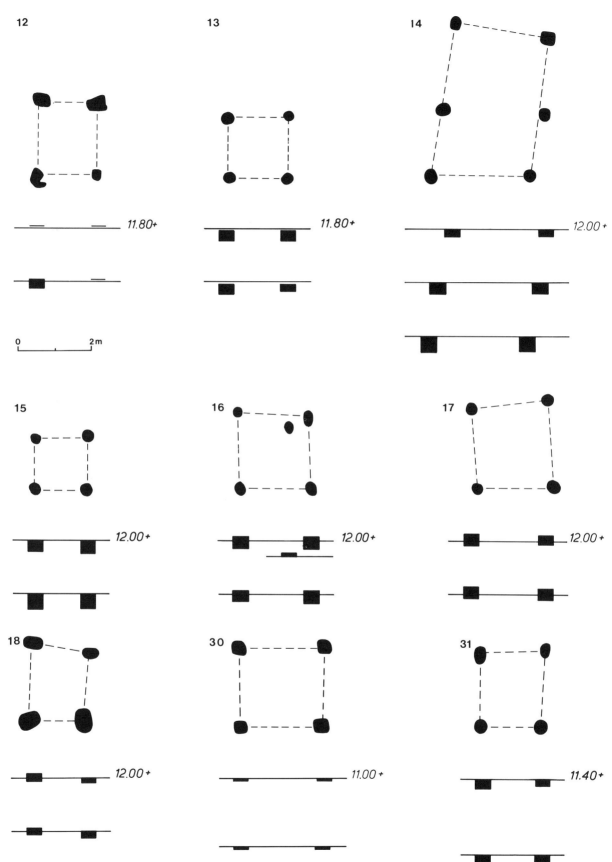

Fig. 33. Plattegronden van spiekers, mijten en roedenbergen.

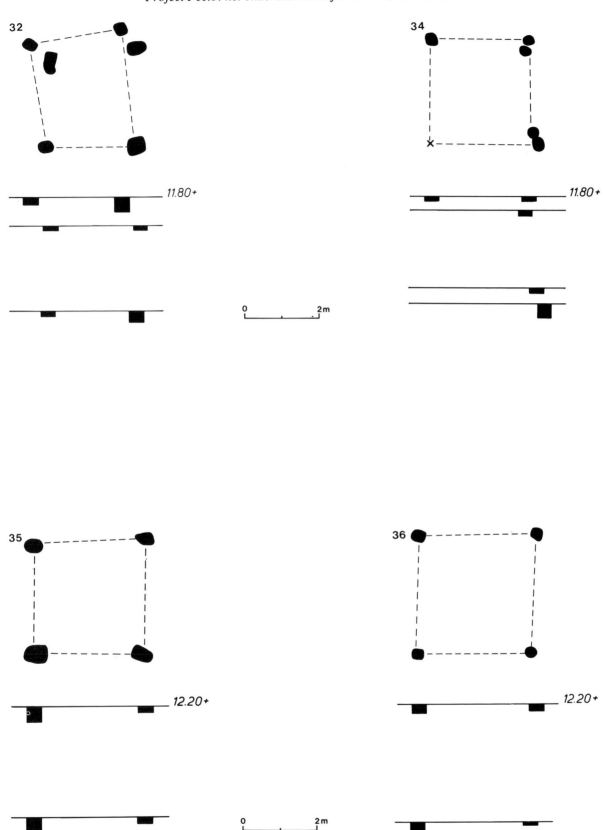

Fig. 34. Plattegronden van spiekers, mijten en roedenbergen.

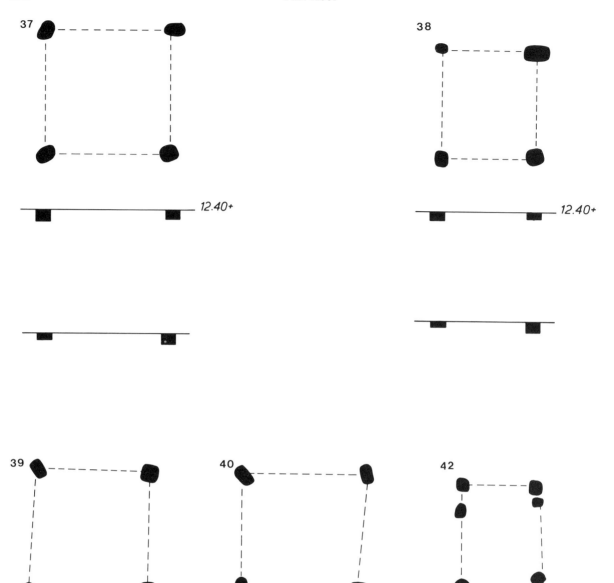

Fig. 35. Plattegronden van spiekers, mijten en roedenbergen.

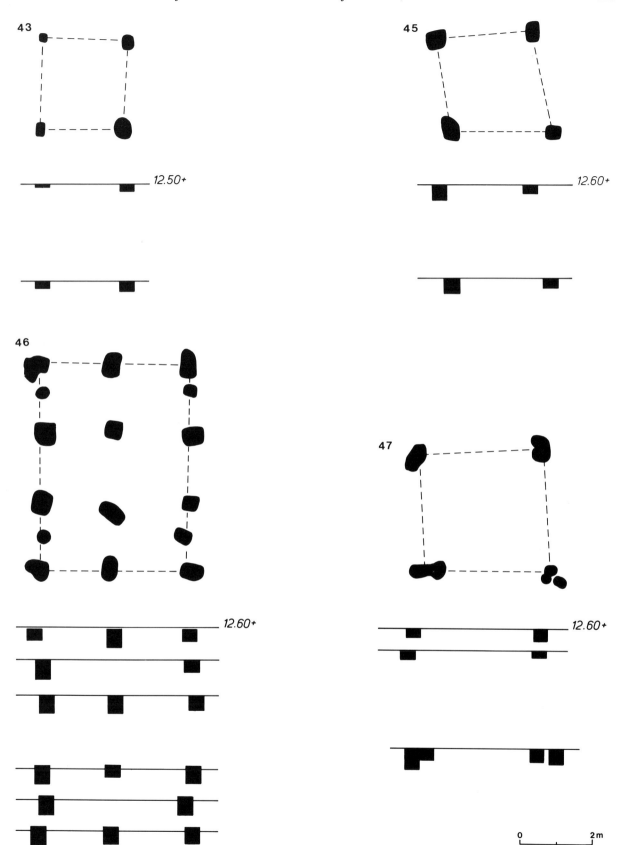

Fig. 36. Plattegronden van spiekers, mijten en roedenbergen.

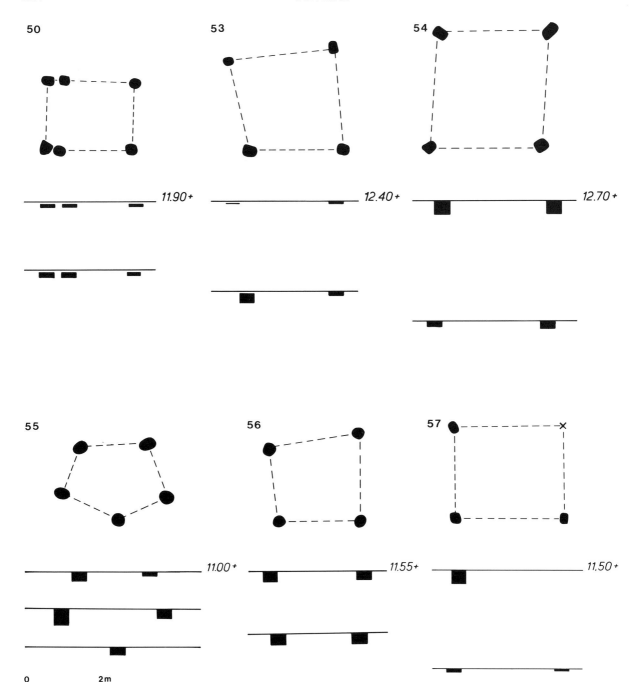

Fig. 37. Plattegronden van spiekers, mijten en roedenbergen.

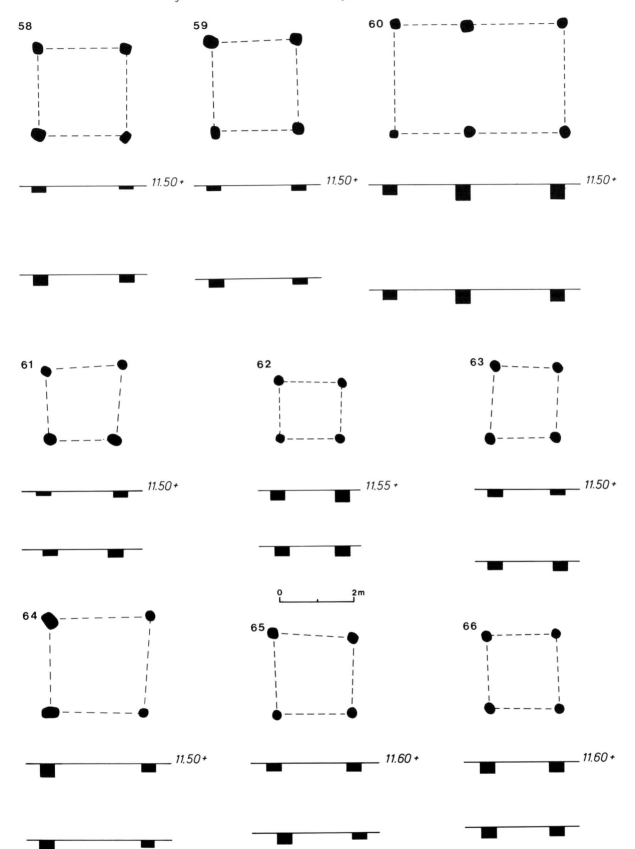

Fig. 38. Plattegronden van spiekers, mijten en roedenbergen.

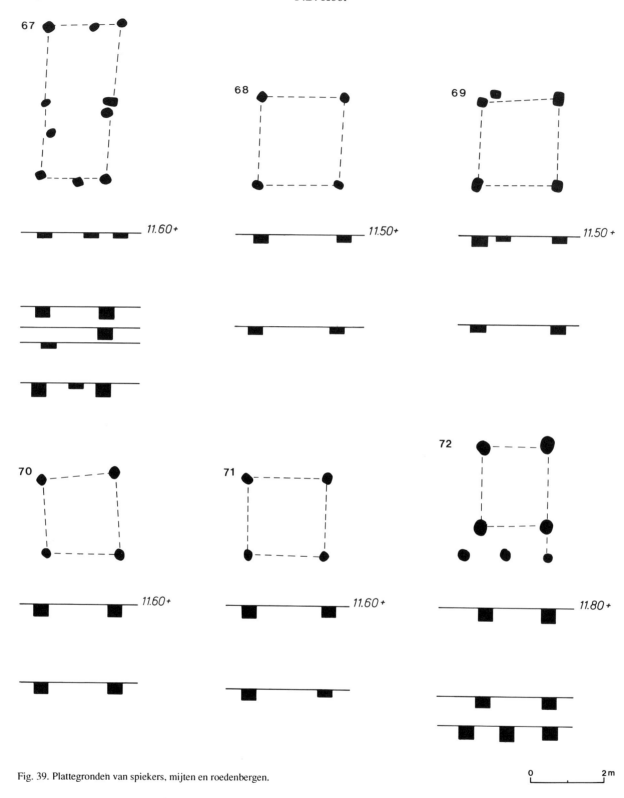

Fig. 39. Plattegronden van spiekers, mijten en roedenbergen.

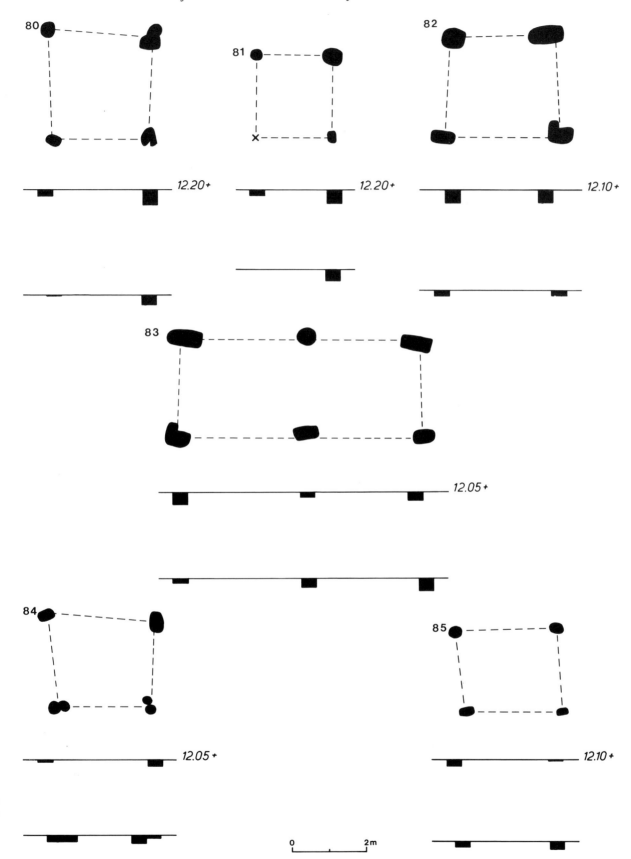

Fig. 40. Plattegronden van spiekers, mijten en roedenbergen.

P.B. Kooi

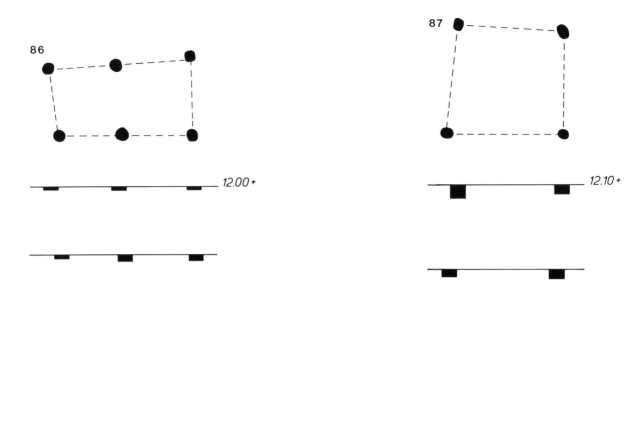

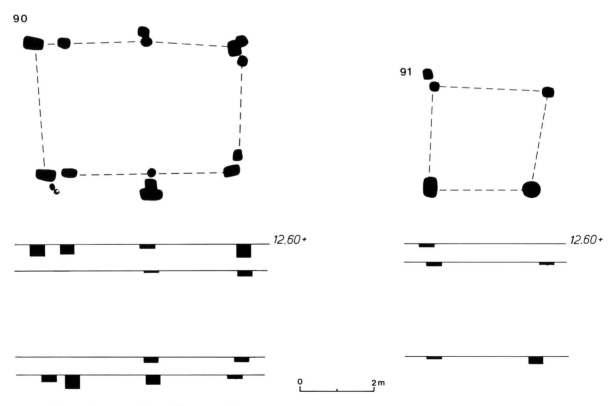

Fig. 41. Plattegronden van spiekers, mijten en roedenbergen.

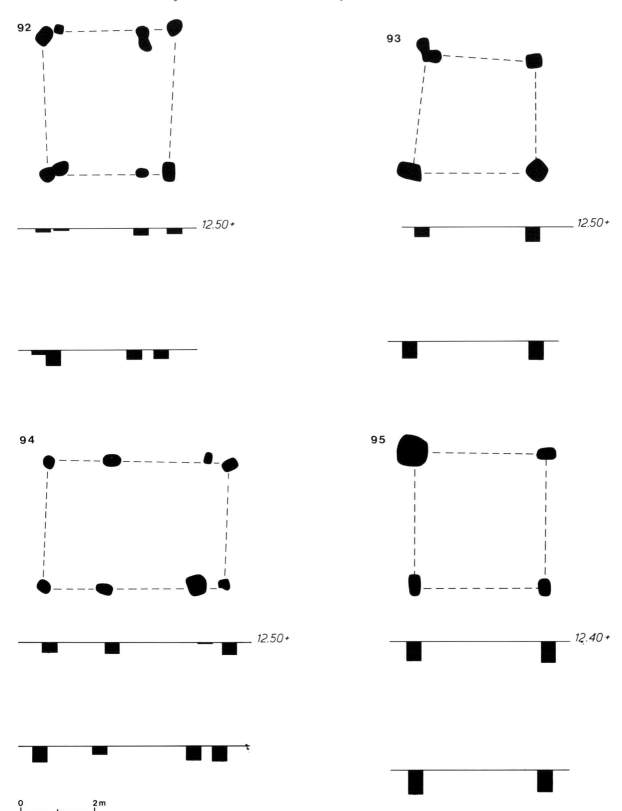

Fig. 42. Plattegronden van spiekers, mijten en roedenbergen.

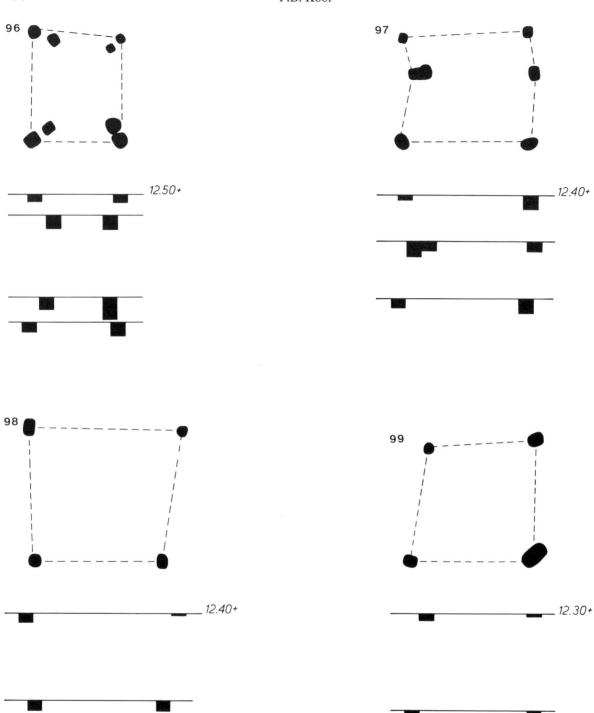

Fig. 43. Plattegronden van spiekers, mijten en roedenbergen.

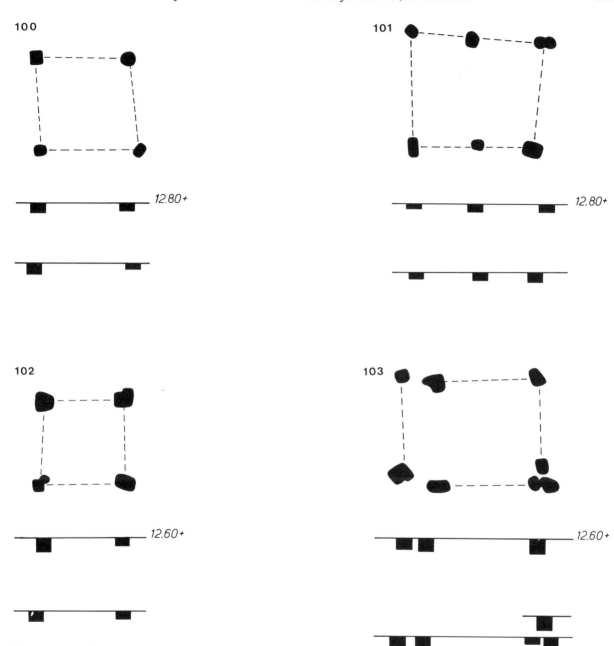

Fig. 44. Plattegronden van spiekers, mijten en roedenbergen.

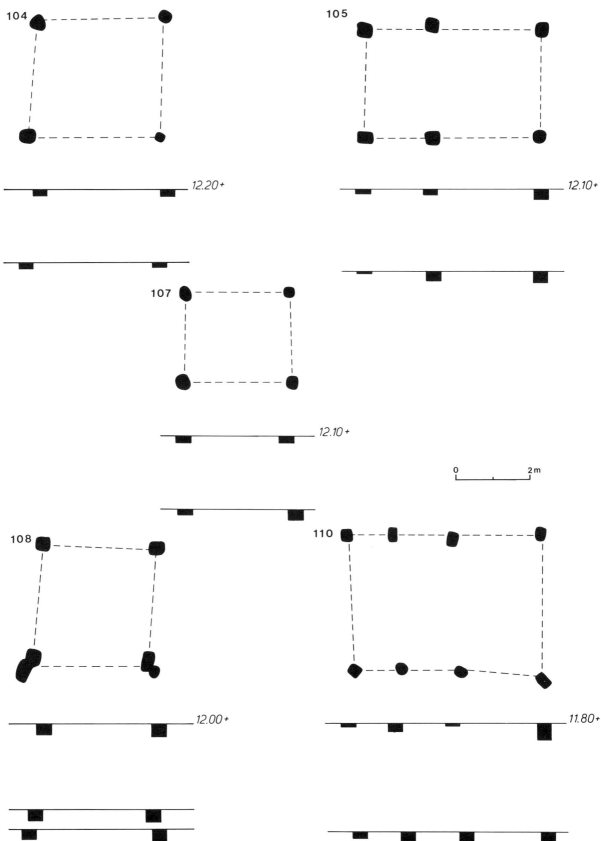

Fig. 45. Plattegronden van spiekers, mijten en roedenbergen.

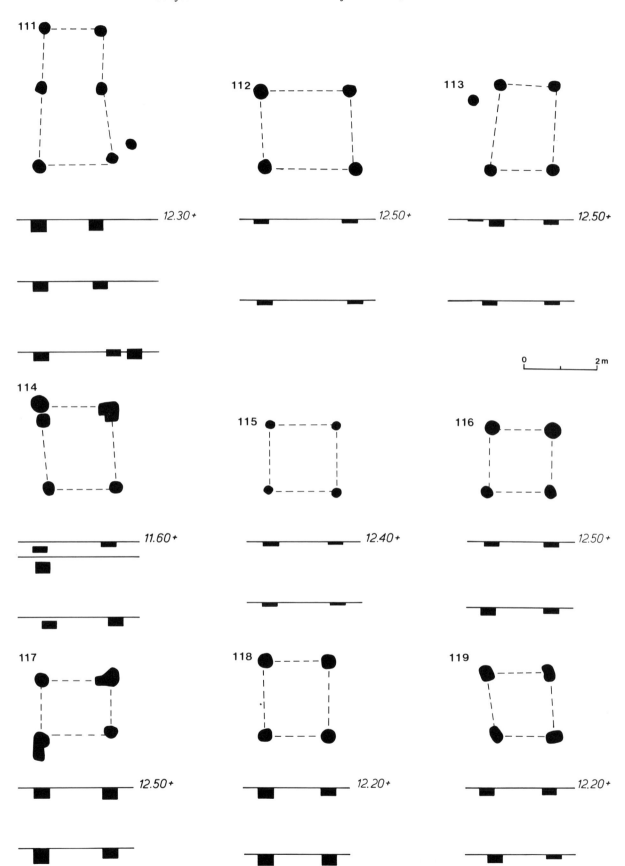

Fig. 46. Plattegronden van spiekers, mijten en roedenbergen.

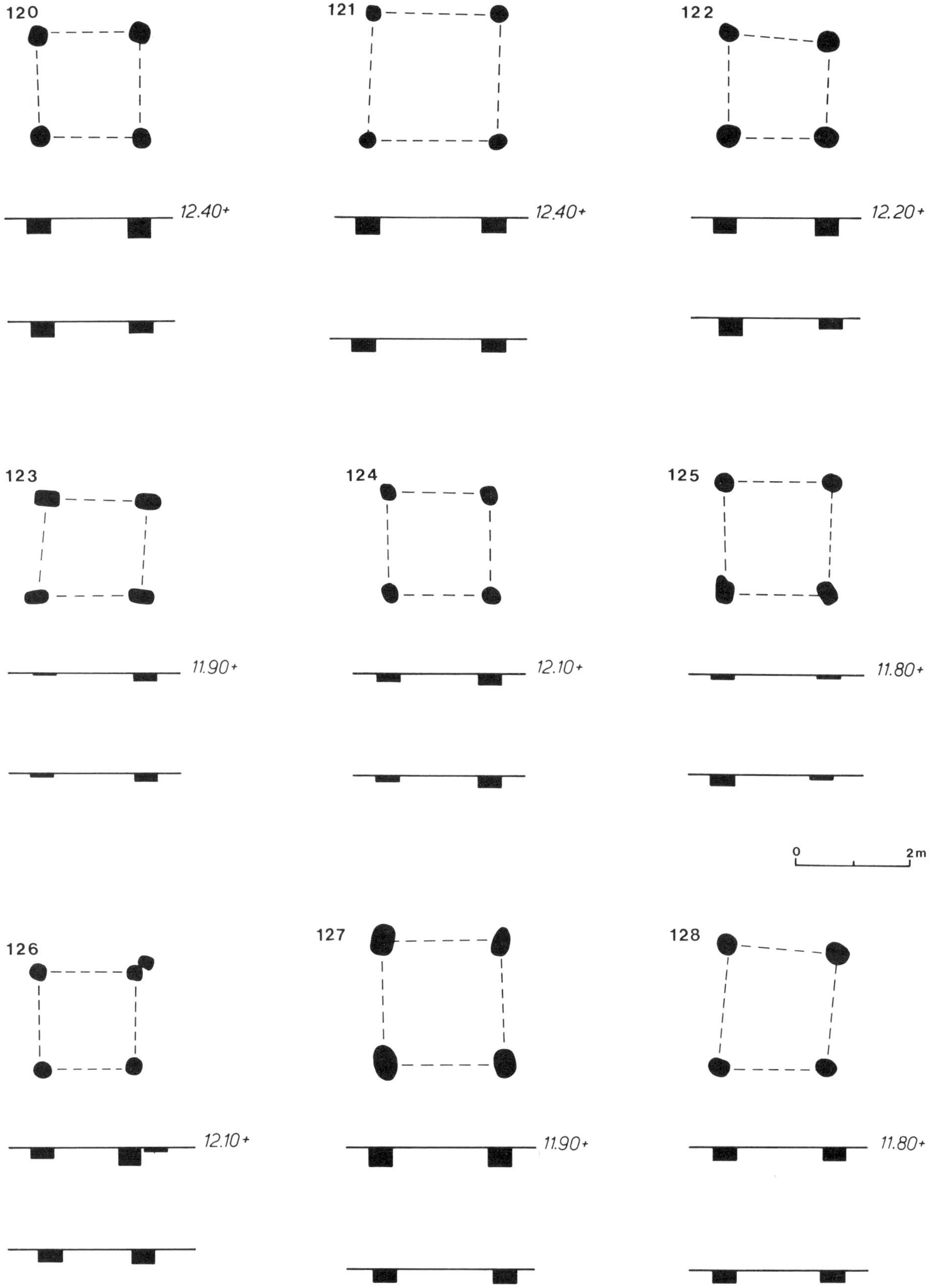

Fig. 47. Plattegronden van spiekers, mijten en roedenbergen.

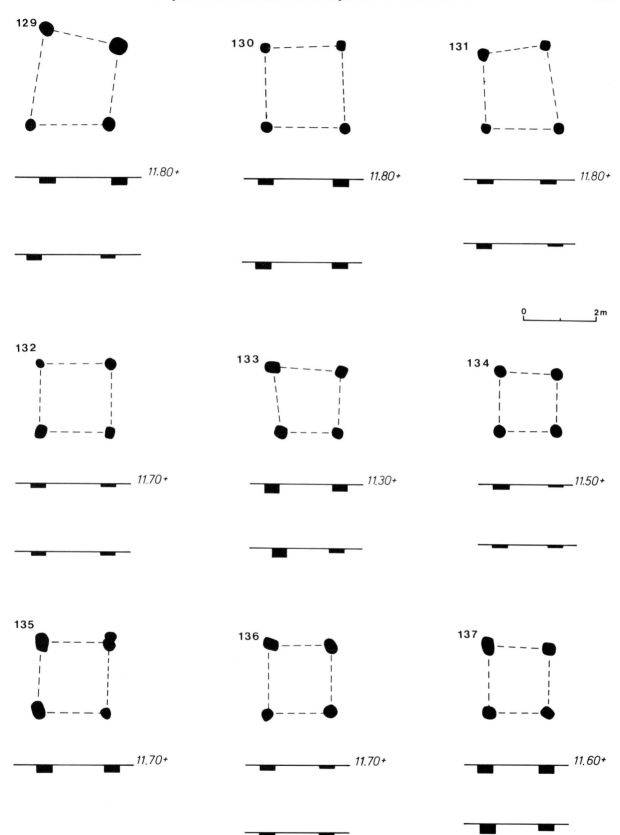

Fig. 48. Plattegronden van spiekers, mijten en roedenbergen.

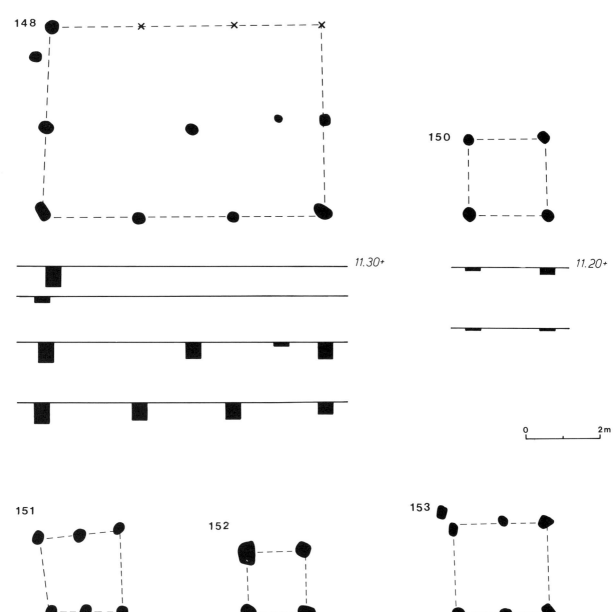

Fig. 49. Plattegronden van spiekers, mijten en roedenbergen.

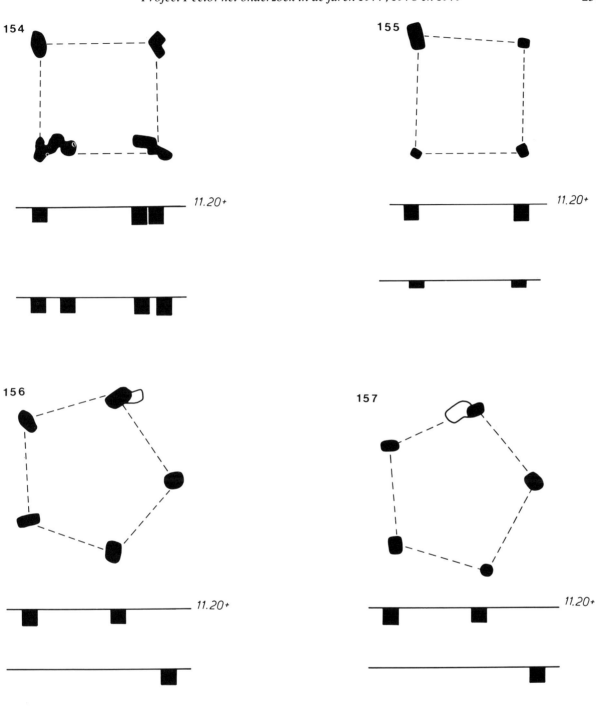

Fig. 50. Plattegronden van spiekers, mijten en roedenbergen.

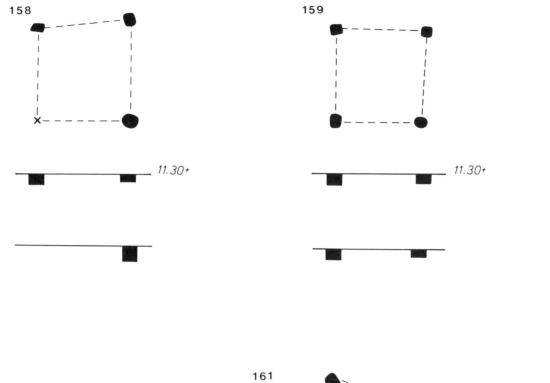

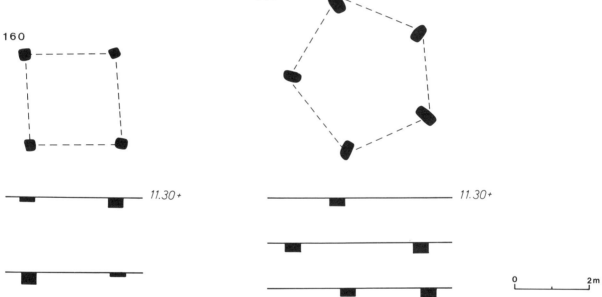

Fig. 51. Plattegronden van spiekers, mijten en roedenbergen.

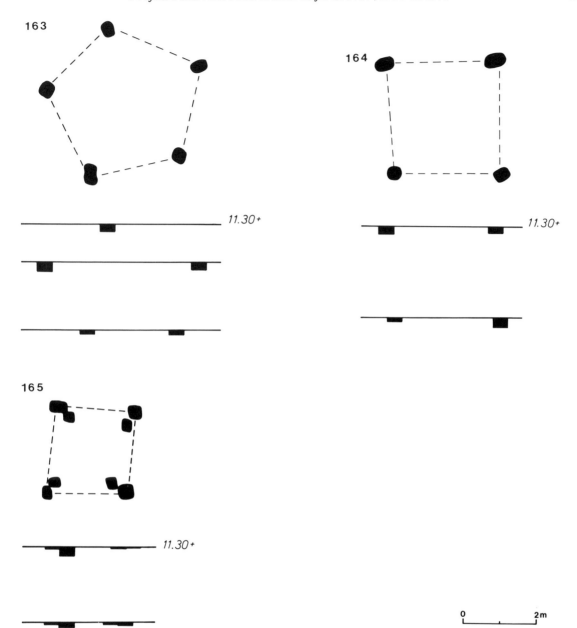

Fig. 52. Plattegronden van spiekers, mijten en roedenbergen.

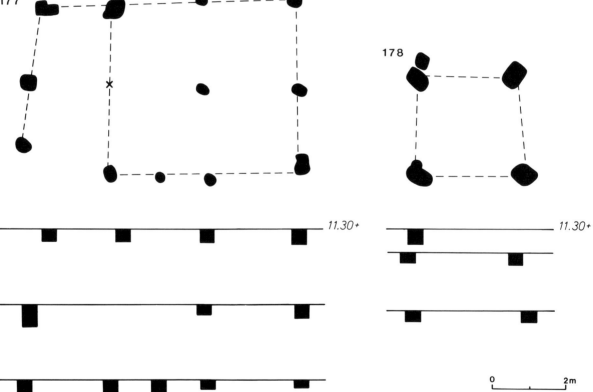

Fig. 53. Plattegronden van spiekers, mijten en roedenbergen.

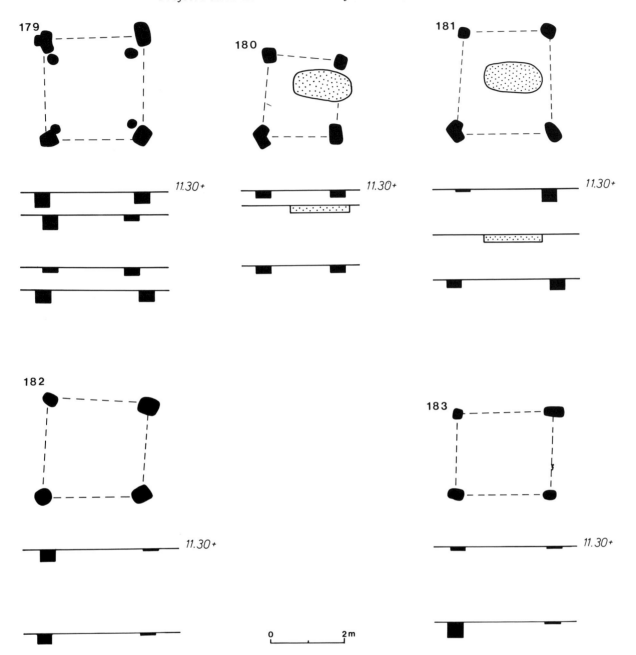

Fig. 54. Plattegronden van spiekers, mijten en roedenbergen.

P.B. Kooi

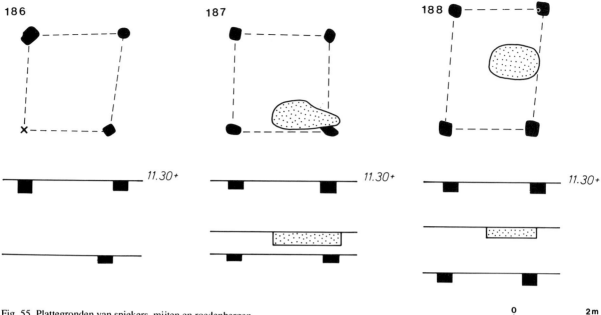

Fig. 55. Plattegronden van spiekers, mijten en roedenbergen.

0   2m

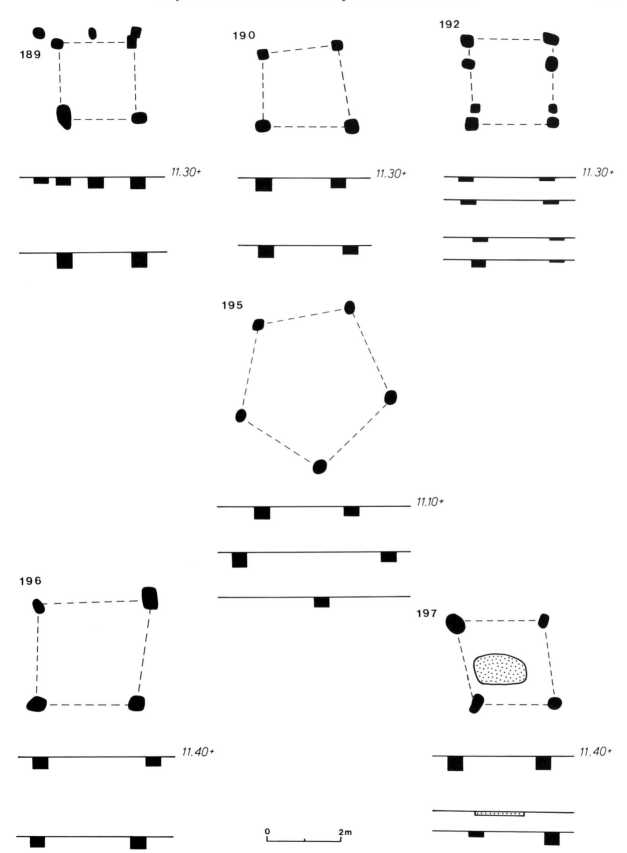

Fig. 56. Plattegronden van spiekers, mijten en roedenbergen.

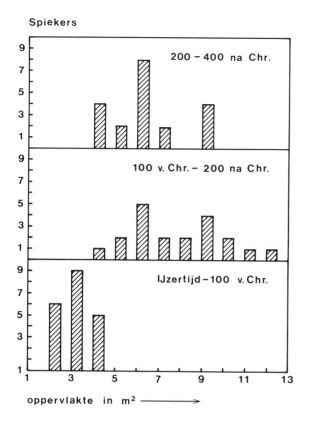

Spiekers

Fig. 57. Blokdiagram van de oppervlakteverhoudingen van de oogst-opslag, uit de IJzertijd, de periode van ca. 100 v.Chr.-200 na Chr. en de periode van ca. 200-400 na Chr. (er zijn steeds 20 exemplaren per periode gemeten).

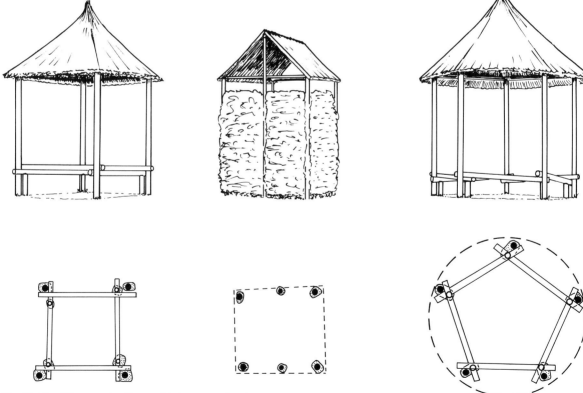

Fig. 58. Mogelijke reconstructies van drie typen oogstberging.

aan een bepaalde periode of boerderijtype:

Nrs 1-3: behorend bij Peelo A – Wijster B;

Nrs 35-47, 80-85, 104-110: Wijster A – Peelo A;

Nrs 90-99: Noord Barge – Peelo A;

Nrs 111-137: IJzertijd;

Nrs 154-157, 158-161, 163-165: Peelo A – Wijster B;

Nrs 175-183, 186-188: Peelo A – Peelo B.

De bovenstaande groepering is ook bij de afbeeldingen aangehouden.

In de eerste plaats valt het op dat de spiekers uit de IJzertijd veel geringere afmetingen hebben dan die uit de latere perioden. Bovendien liggen ze meer gespreid, namelijk zowel in de directe nabijheid van de boerderijen, als ook op grotere afstand. De spiekers uit de perioden daarna zijn groter en liggen in rijen evenwijdig aan de boerderijen. Indien deze kenmerken uit verschillende perioden dateren, kunnen in het algemeen de kleinere spiekers aan de bewoning uit de IJzertijd worden toegeschreven, zoals nrs 6, 12, 13, 15, 16, 17, 18, 31, 61, 62, 63, 65, 66, 81. De opslag van de oogst in de IJzertijd lijkt, gezien de spreiding zowel bij de boerderij als bij de akkers te hebben plaats gevonden in gelijksoortige spiekers. De mogelijkheid is echter niet uitgesloten, dat in de Romeinse tijd, naast een bepaalde opslagcapaciteit bij de boerderij in grotere spiekers, er tevens nog spiekers op de akkers in gebruik waren (fig. 57).

De constructie van de spiekers zal zijn aangepast aan de wijze van oogstopslag, waarbij mag worden aangenomen dat deze bescherming boden tegen bederf en vraat. Reconstructies van spiekers uit de IJzertijd zijn voorzien van een verhoogde vloer en dak, soms ook nog van een wand van vlechtwerk. Belangrijk is vanzelfsprekend ook of alleen de aren werden opgeslagen of dat tevens een groot deel van de stengel met aar werd opgestapeld. In de plattegronden zijn aanwijzingen te vinden voor diverse constructies. Ten eerste vinden we de eerder genoemde dubbele palen (zie nr. 2), hetgeen er op wijst, dat afzonderlijke palen zijn geplaatst om de vloer te dragen (de binnenste) en ter ondersteuning van het dak (fig. 58a). Ook de langgerekte paalgaten kunnen een aanwijzing zijn voor een dergelijke constructie, waarbij de dakdragende en de vloerdragende palen dicht naast elkaar in één kuil hebben gestaan. Daarnaast kunnen er ronde bergen of mijten met zowel vier als vijf enkelvoudige paalzettingen zijn geweest met een eveneens rond dak. Voor de plattegronden met vijf palen lijkt een dergelijke vorm vanzelfsprekend, waarbij tevens de mogelijkheid van dubbele palen in langwerpige paalgaten mogelijk is, zoals bij de nrs 156, 157 en 161 (fig. 58c). Dubbele palen kwamen ook voor bij grotere opslag, zoals de nrs 38 en 90. Een andere constructie is denkbaar bij de eerder genoemde zes palen-configuraties, zoals nr. 5, waarbij de extra palen in het midden van twee zijden staan. Dit suggereert een overkapping met een tentdak, die naast de steun van de hoekpunten

ook een noksteun heeft gehad (fig. 58b). De algemene conclusie uit een aantal constructies kan dus zijn dat er bij dubbelpalen en extra palen zeker een overkapping is geweest, terwijl bij de enkelvoudige constructies ook sprake van een verhoogde vloer kan zijn, waarbij een losse afdekking met stro een tijdelijke bescherming vormde. Of de kappen ook in hoogte verstelbaar waren is uiteraard niet uit de sporen af te leiden. Kuilen binnen de paalzettingen van nrs 180, 181, 187, 188 en 197 duiden op gebruik van de ruimte onder de verhoogde vloer voor andere doeleinden, zoals ook in later tijd gebruikelijk is (Goutbeek e.a., 1988; Zimmermann, 1991: p. 87).

3.4. Waterputten (fig. 59-67)

In het geheel zijn 22 restanten van waterputten aangetroffen. In de meeste gevallen was het onderste deel van een houten bekisting bewaard gebleven, voor zover deze zich onder het grondwaterniveau bevond. Sinds de tijd waarin de nederzetting heeft bestaan is dit niveau gedaald, maar metingen van de gemiddelde hoogste stand in deze tijd geven toch een indruk van de situatie (fig. 5). Daaruit valt onder andere af te lezen, dat het niveau met het reliëf fluctueert. De ontzanding op het hoogste punt van de es heeft daar de bodemstructuur ingrijpend gewijzigd en waarschijnlijk een plaatselijke daling van de grondwaterspiegel veroorzaakt (zie fig. 5: tussen de 500 en 600 m vanaf de Peeloër esweg). In het algemeen bestaan de waterputten uit een trechtervormige kuil, waarin een bekisting is geplaatst, waarna de ruimte rond de bekisting weer met grond is opgevuld. In de profielen is soms een trapvormige insteek te zien in de ingraving (zie nrs 2, 4, 17).

Voor de bekistingen zijn vier verschillende systemen gebruikt:

– Een uitgeholde boomstam (nrs 2 en 21). Nr. 21 was tot een hoogte van 0,9 m bewaard gebleven en bestond uit een in drie stukken gekloofde en uitgeholde eiken stam. De naden waren met planken dichtgezet;

– In het rond geplaatste paaltjes met vlechtwerk (nr. 6);

– Een vierkant van verticaal geplaatste paaltjes met een raamwerk van horizontale balkjes (nrs 7, 9, 10, 11, 13, 14, 15(?) en 17;

– Een vierkant van gestapelde planken of balkjes aan de binnenzijde op de hoeken gestut door verticale palen.

Van een aantal diepe kuilen kan worden aangenomen dat ze eveneens als put zijn gebruikt, maar dat het hout van de bekisting volledig is vergaan (nrs 5, 8, 12). Twee kuilen (nrs 3 en 4) zijn mogelijk te beschouwen als een mislukte poging om een put aan te leggen. De kuilen zijn voldoende diep, maar de vulling is breed gelaagd. In nr. 4 werd een deel van een ladder gevonden met stijlen van essehout en sporten van els gemaakt (fig. 60). De trapsgewijze ingravingen en de ladder zullen dienst hebben gedaan bij het graafwerk, dat vanwege

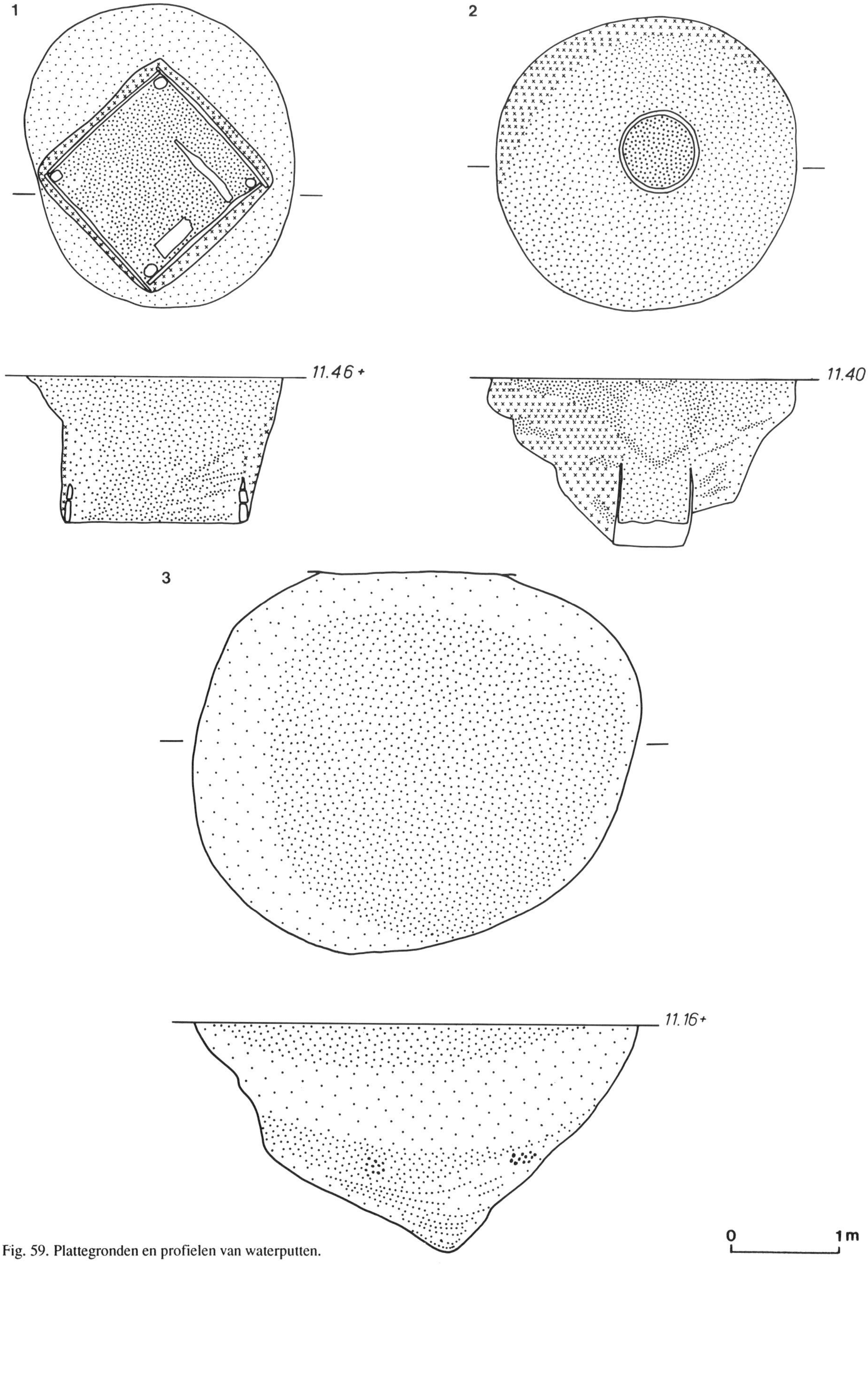

Fig. 59. Plattegronden en profielen van waterputten.

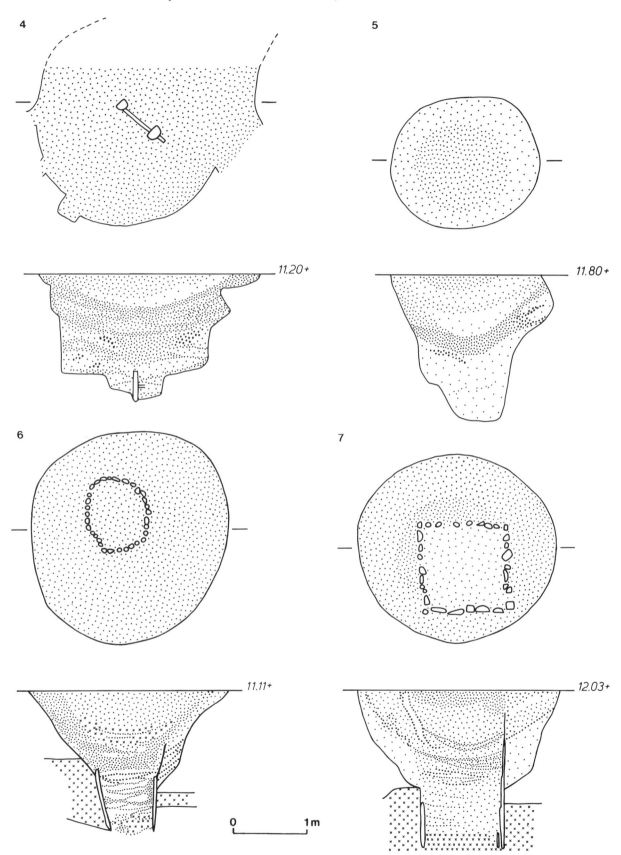

Fig. 60. Plattegronden en profielen van waterputten.

P.B. Kooi

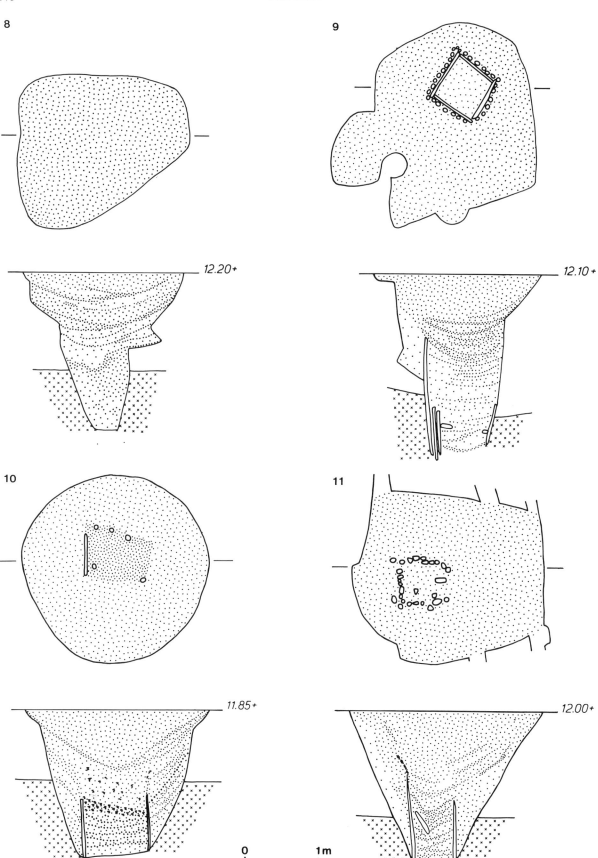

Fig. 61. Plattegronden en profielen van waterputten.

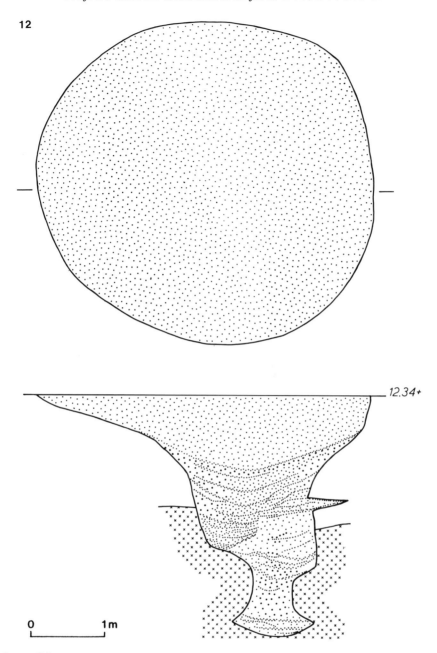

Fig. 62. Plattegrond en profiel van een waterput.

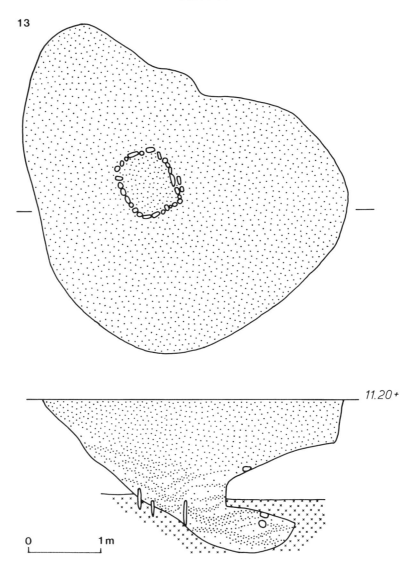

Fig. 63. Plattegrond en profiel van een waterput.

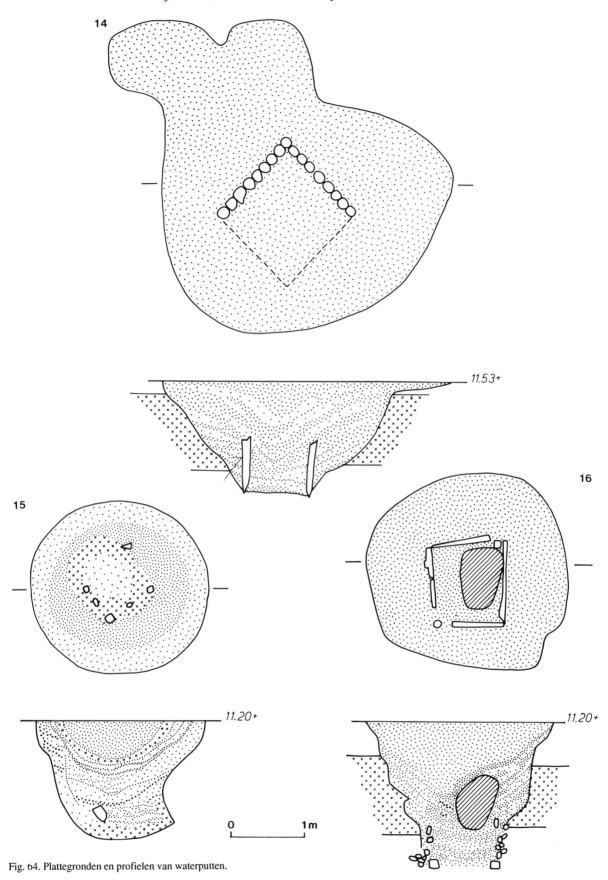

Fig. 64. Plattegronden en profielen van waterputten.

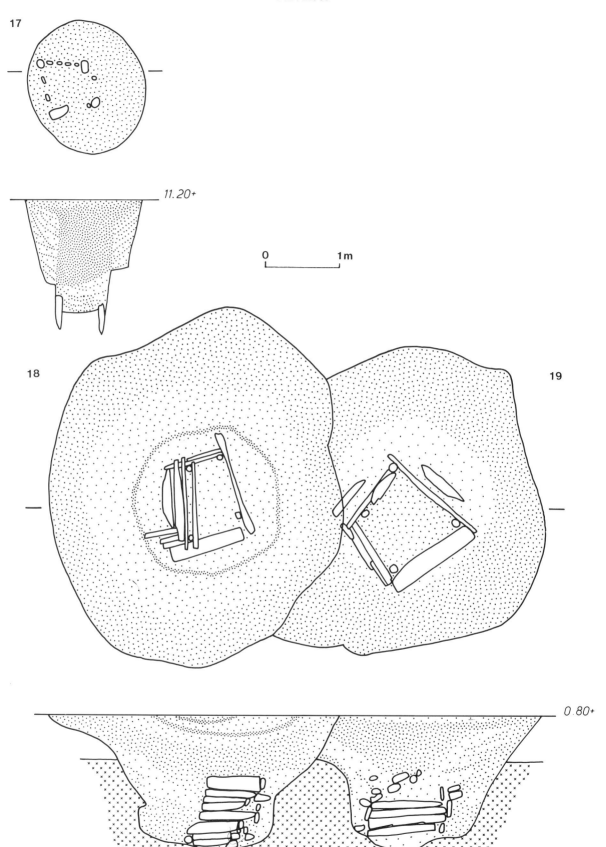

Fig. 65. Plattegronden en profielen van waterputten.

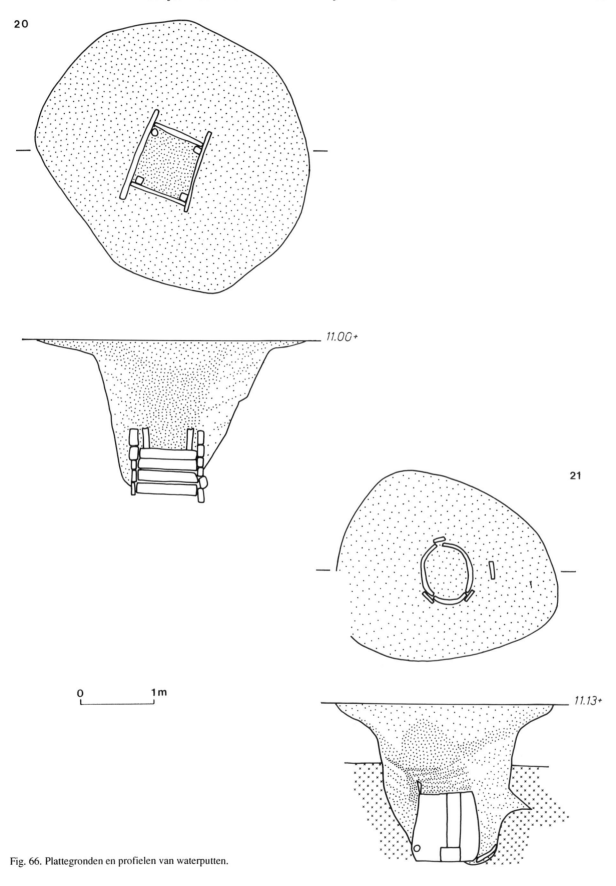

Fig. 66. Plattegronden en profielen van waterputten.

22

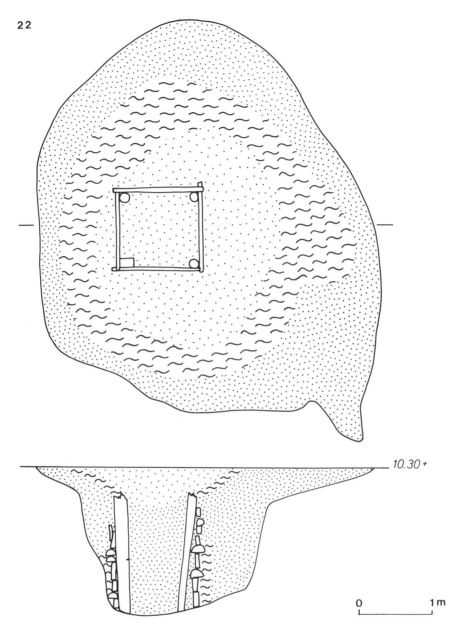

10.30 +

0   1 m

Fig. 67. Plattegrond en profiel van een waterput.

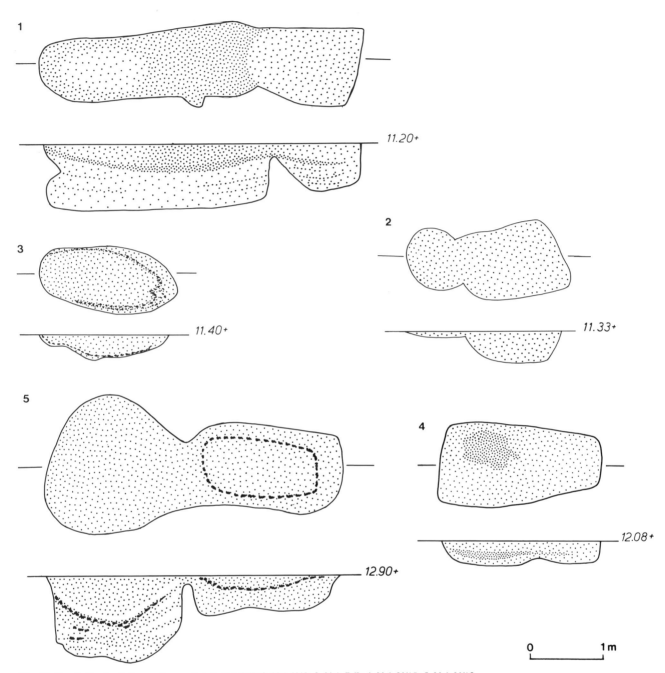

Fig. 68. Ovenkuilen in plattegrond en profiel. 1. Vak R/7; 2. Vak K/8; 3. Vak D/9; 4. Vak Y/10; 5. Vak V/13.

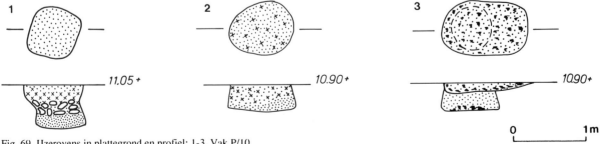

Fig. 69. IJzerovens in plattegrond en profiel: 1-3. Vak P/10.

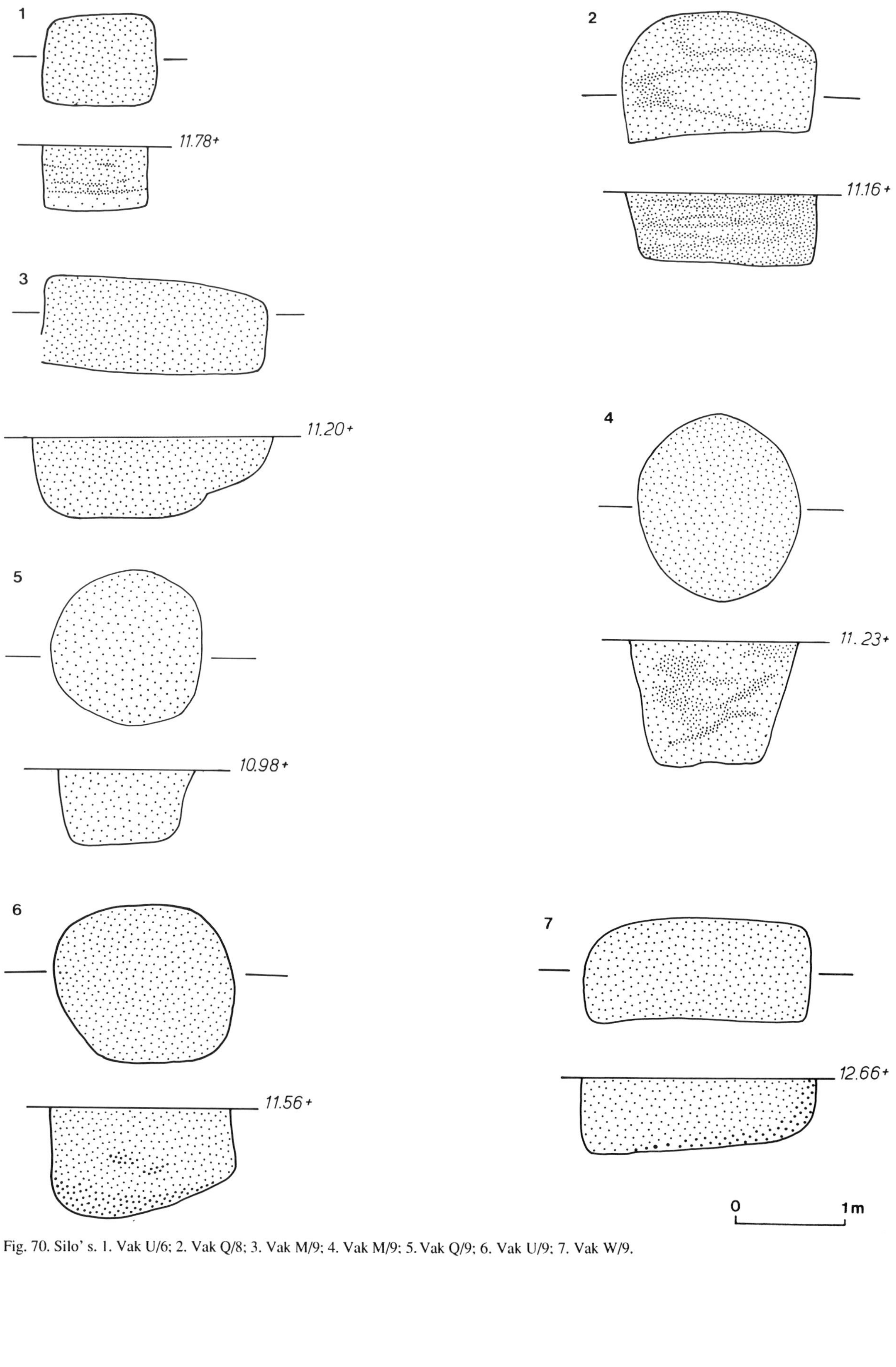

P.B. Kooi

Fig. 70. Silo' s. 1. Vak U/6; 2. Vak Q/8; 3. Vak M/9; 4. Vak M/9; 5. Vak Q/9; 6. Vak U/9; 7. Vak W/9.

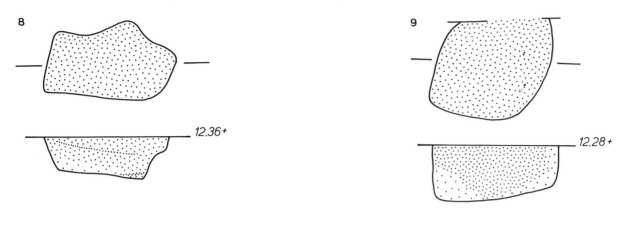

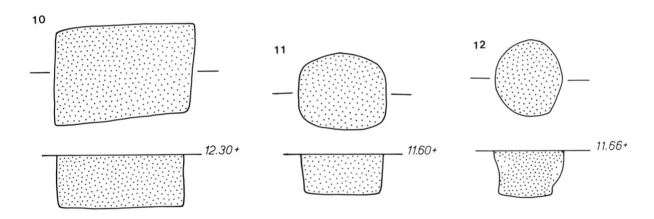

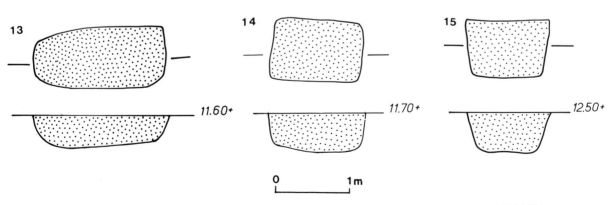

Fig. 71. Silo' s. 8. Vak Y/9; 9. Vak V/19; 10. Vak V/10; 11. Vak Y/10; 12. Vak I/11; 13. Vak S/11; 14. Vak T/11; 15. Vak X/11.

P.B. Kooi

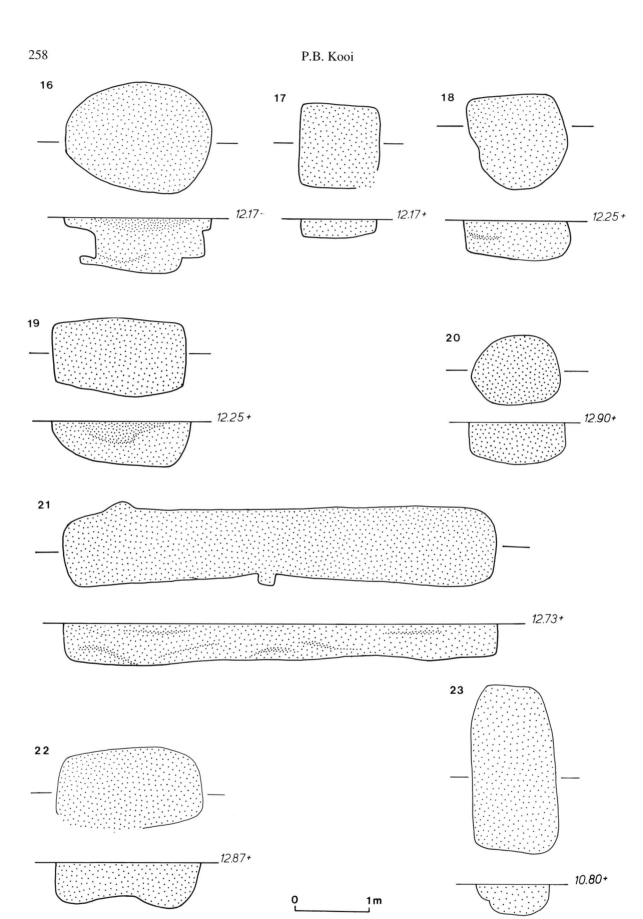

Fig. 72. Silo's. 16. Vak Y/11; 17. Vak Y/11; 18. Vak Y/11; 19. Vak Y/11; 20. Vak V/12; 21. Vak W/12; 22. Vak W/12; 23. Vak N/14.

het gevaar van instorting en kwel in een hoog tempo moet zijn uitgevoerd.

Uit de waterputten zijn 132 houtmonsters genomen en door W.A. Casparie geanalyseerd. Er werden 7 verschillende houtsoorten gedetermineerd:

Quercus (eik)	46	35%
Fraxinus (es)	8	6%
Alnus (els)	48	36%
Betula (berk)	22	17%
Salix (wilg)	3	2%
Populus (populier)	4	3%
Sorbus	1	< 1%

De belangrijkste constructieve elementen, te weten de hoekpalen van vierkante bekistingen, waren uit vergegraven om toevoer van brandstof of het uitruimen mogelijk te maken.

IJzeroventjes (fig. 69)

Drie naar boven taps toelopende kuilen, gevuld met houtskool, verbrande leem en ijzerslakken. Uit de vorm en inhoud van de kuilen is af te leiden, dat het restanten zijn van ijzeroventjes met een flesvormige mantel van leem, waarin de ijzerhoudende grondstof werd verhit. Door aanjagen van het vuur ontstond een aaneenklitting van ijzer, maar de temperatuur werd onvoldoende hoog om ijzer vloeibaar te maken. De oventjes werden na eenmalig gebruik afgebroken en het bruikbare deel van de inhoud werd verder uitgesmeed en bewerkt (Pleiner, 1964). Voorwerpen van ijzer zijn in nederzettingscontext niet aangetroffen. Wel zijn er buiten de ijzeroventjes bij de smederij een aantal ijzerslakken gevonden, die als afvalmateriaal van ijzerbereiding kunnen worden be-

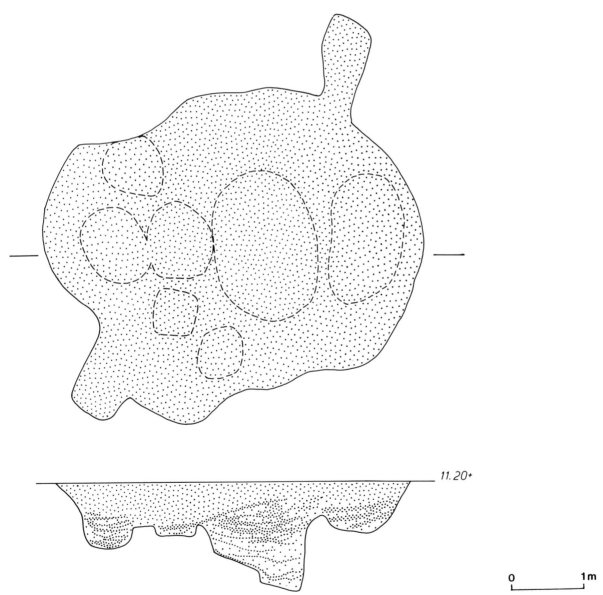

11.20+

0   1m

Fig. 73. Kuilencomplex (vak K/10).

schouwd. Van de twintig vondstnummers met ijzerslakken bevinden zich vijftien op het oostelijke deel van de es, twee in de omgeving van de smederij en slechts schillende houtsoorten gemaakt: *Quercus* 8x, *Alnus* 8x, *Betula* 4x, *Fraxinus* 3x en *Salix* 2x.

3.5. Kuilen

Onder deze noemer zijn ovenkuilen, silo' s en afvalkuilen samengevat.

Ovenkuilen (fig. 68)
Een viertal kuilen bestaat uit twee delen en de vijfde is een kuil met houtskool. De kuilen zijn waarschijnlijk het laatste restant van ovens die voor verschillende doeleinden zijn gebruikt. De tweedeling is ontstaan doordat voor de eigenlijke vuurmond een kuil werd drie in het westelijke deel. Met andere woorden, over de verschillende perioden bezien is de verdeling globaal als volgt: vijftien vondsten uit de periode 2 tot en met 3 en vier vondsten aan het einde van periode 5. Dit impliceert dat er een wijziging in de ijzerbewerking of verwerking heeft plaatsgevonden in de nederzettingsfase met de boerderijplattegronden van het type Wijster B en een groot deel van de fase met het type Peelo B. Waarschijnlijk is de verwerking van ijzer in de perioden 2 en 3 meer gespreid per erf uitgevoerd en heeft zich daarna door specialisatie op één erf met een smederij geconcentreerd.

Silo' s (fig. 70, 71 en 72)
Uit de totale hoeveelheid kuilen is een selectie van 23 stuks afgebeeld, die regelmatig van vorm zijn en vrij steile of vertikale wanden hebben. De mogelijkheid bestaat dat deze als ondergrondse opslag (silo) zijn gebruikt.

Kuilencomplex (fig. 73)
Een min of meer ovale kuil van maximaal 5 m diameter met ogenschijnlijk willekeurig gelegen diepere gedeelten, waarvan het ontstaan of de functie onbekend is.

3.6. Omheiningen

Een aantal sporen zijn afkomstig van omheiningen. In het algemeen betreft het smalle greppels, maar daarnaast komen ook paalzettingen voor. Over de constructie van de omheiningen valt in detail weinig te zeggen. In het algemeen zal het om vlechtwerk zijn gegaan, waarbij soms 'geprefabriceerde' elementen werden geplaatst, die op regelmatige afstanden werden gestut (vak L-N/ 8-9 en vak R-U/9-10). Soms is het vlechtwerk ter plaatse gemaakt met in de grond geplaatste palen (vak W-Z/10-13). Naar functie zijn een aantal toepassingen te onderscheiden:
– Veekraal (fig. 74). Een ovaal van acht palen naast boerderij nr. 27, waardoor een datering in de IJzertijd waarschijnlijk is (vak V/16);

– Erfbegrenzing. Dit is de meest voorkomende vorm van omheining in de nederzetting vanaf de fase met huizen van het type Noord Barge. De omheiningen worden regelmatig vernieuwd, verplaatst en aangepast. Voor de verlenging van boerderij nr. 20 is bijvoorbeeld de omheining rond het erf aangepast (vak AA/10). In een aantal gevallen zijn kleinere delen van een erf apart omheind. Het kan bijvoorbeeld bedoeld zijn om akkertjes af te schermen (vak S-U/11-12) of als omheining van een groep spiekers (vak S-U/10-11);
– Veesluis/schutplaats. In de vakken V-Z/11-13 is een combinatie van erfomheining en veekraal aangetroffen, waarbij het vee kon worden opgevangen tussen twee parallelle omheiningen, die in twee perioden heeft bestaan. Aan de oostzijde is een brede ingang met centraal een spoor van doortrapte grond (fig. 75a). Waarschijnlijk is hier een afsluiting geweest met een dubbel hek. Aan de westzijde sloot de dubbele omhei-

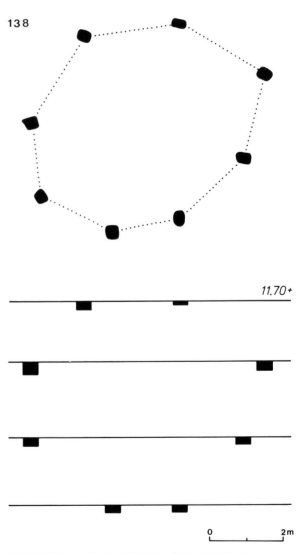

Fig. 74. Kleine veekraal (zie fig. 7, nr. 138) uit de IJzertijd.

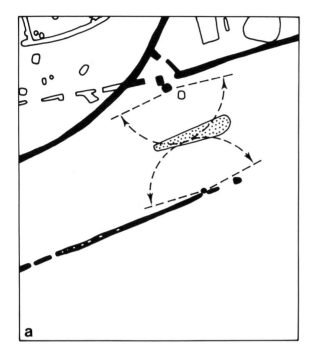

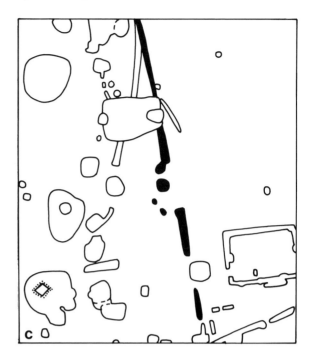

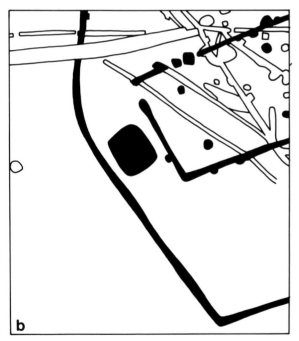

Fig. 75. a. Ingang van de schutplaats voor het vee met een spoor van doortrapte grond (vak Y/12); b. Veerooster tussen boerderij 22 en de omheining (vak V/12); c. Doorgangen met paalzetting (vak Y/9).

ning in de eerste fase aan bij boerderij nr. 22. Tussen de buitenste omheining en de korte wand van het huis lag een grote kuil, die waarschijnlijk als veerooster heeft gediend om de doorgang voor het vee onmogelijk te maken (fig. 75b). Doorgangen in de omheiningen kunnen

bestaan uit een simpele onderbreking, maar er zijn ook plaatsen waar dit gepaard gaat met extra palen, waarschijnlijk bedoeld om de doorgang te kunnen blokkeren (fig. 75c). Voorts zijn er mogelijk, naast veeroosters, vernuftige constructies bedacht die het vee tegenhouden en de bewoners doorlaten. Een dergelijke oplossing schijnt aanwezig te zijn bij het brede hek (fig. 75a).

Overigens moeten we er rekening mee houden dat een aantal vormen van erf- en akkerbegrenzing geen sporen heeft nagelaten. In de IJzertijd kunnen walsystemen met begroeiing als scheidingen hebben gediend, die na egalisatie niet zijn terug te vinden. De gegroepeerde spreiding van de spiekers in die tijd geeft daarvoor een indicatie. Daarnaast kunnen heggen en doornige struiken voor een goede begrenzing van erf en akkers hebben gezorgd.

3.7. Ploegsporen en ontginningsgreppels

Op verschillende plaatsen zijn sporen van bewerking van de grond voor de akkerbouw gevonden.

De oudste sporen van grondbewerking zijn afkomstig van een eergetou. Ze zijn bewaard gebleven in een komvormige laagte naast boerderij nr. 41 (vak K-L/11) die later is opgevuld met afval en humeuze grond. De sporen lopen in twee, iets verschillende richtingen ongeveer oost-west en zijn op doorsnede min of meer symmetrisch V-vormig (fig. 76). Ze dateren waarschijnlijk uit de IJzertijd, of de vroeg-Romeinse tijd.

Veel jonger zijn de sporen van een keerploeg (vak K-L/8-9) waarvan één kant van de vore recht is met een iets donkere vulling en de andere kant van de vore iets

Fig. 76. Foto van boerderij 41, uit het oosten gezien. Aan de rechterzijde zijn de ploegsporen van een eergetou te zien. Op voorgrond zijn de diepste verstoringen van esgreppels zichtbaar als horizontale rijen schopsteken.

Fig. 77. Foto van werkput 27 met boerderijplattegronden, hutkommen en aan de linkerzijde sporen van een keerploeg.

P.B. Kooi

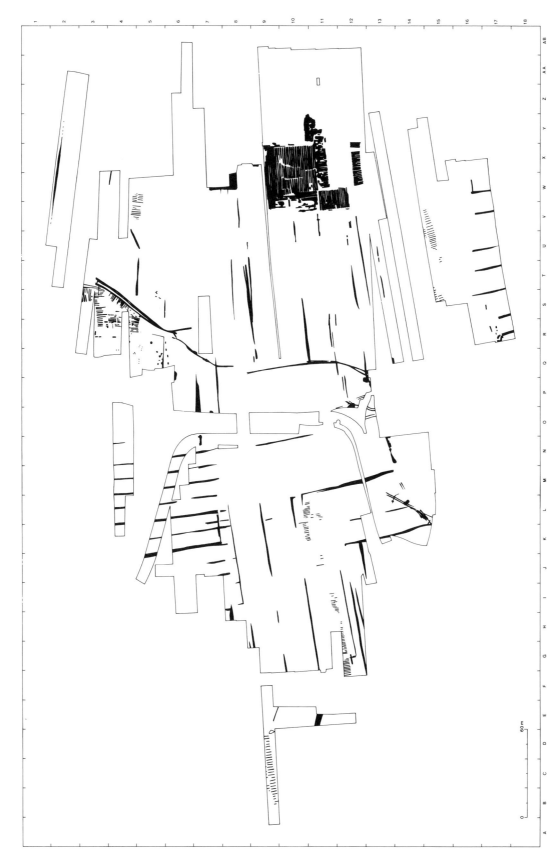

Fig. 78. Overzichtskaart van alle sporen, die vanaf de middeleeuwen tot in recente tijd zijn ontstaan.

rafelig en lichter van vulling. De sporen lopen ongeveer oost-west en noord-zuid en zijn jonger dan de nederzettingssporen (fig. 77). Ze kunnen uit de middeleeuwen dateren.

Een storend verschijnsel zijn de esgreppels (fig. 78). In verschillende delen van het opgravingsvlak zijn zogenaamde esgreppels waargenomen. Deze methode van greppelen was bedoeld als grondverbetering en is hier perceelsgewijs uitgevoerd, dat wil zeggen met de greppels dwars op de lengterichting van het perceel. Deze greppelsystemen zijn respectievelijk te zien in de werkputten 4, 42, 16, 17 in het noorden, 36 en 24 in het zuiden en 25 en 28 in het westen. Ook in het esdek in de werkputten 8, 2 zijn esgreppels waargenomen, die merkwaardigerwijs niet tot in de vaste ondergrond reikten. Het esdek was daar ongeveer 80 cm dik.

Perceelscheidingen en akkerbegrenzingen in de vorm van greppels zijn het gevolg van de percelering op de es. Veelal corresponderen ze met de indeling volgens de kadastrale minuut.

Karresporen komen vooral in het centrale drassige deel van het terrein voor (vakken L-M/13-14 en O-P/12-13; fig. 6 en 78) en bestaan meestal uit evenwijdige humeuze banen, waarin spoellaagjes voorkomen.

Ten behoeve van zandwinning werd in 1935 het hoogste deel van de es afgegraven (vakken W-AB/7-9) tot een niveau ver onder het opgravingsvlak in de aangrenzende putten, waardoor nader onderzoek daar niet zinvol was. Wel onderzocht werd een gedeelte in de vakken V-Y/9-12 waar in 1925 ook een onderzoek van Van Giffen werd uitgevoerd. De plaats van de destijds gevonden waterput kon worden getraceerd.

De jongste grondsporen waren sleuven, die ter voorbereiding van de woningbouw voor de aanleg van leidingen en kabels werden gegraven.

4. DE ONTWIKKELING VAN DE BEWONING

Voor een fasering of periodisering van de nederzetting kunnen verschillende gegevens worden gebruikt. Uit oversnijding van grondsporen, kan de volgorde van bepaalde constructies en bouwsels worden bepaald. Vondsten kunnen een relatieve datering opleveren, maar ook de gevonden boerderijtypen zijn over het algemeen redelijk goed gedateerd. Daarom zou een overzicht van de ligging van deze typen een indicatie kunnen geven van de ontwikkelingen die zich in de loop der tijd hebben voorgedaan. Vervolgens kunnen de landschappelijke situatie en omheiningssporen bij een

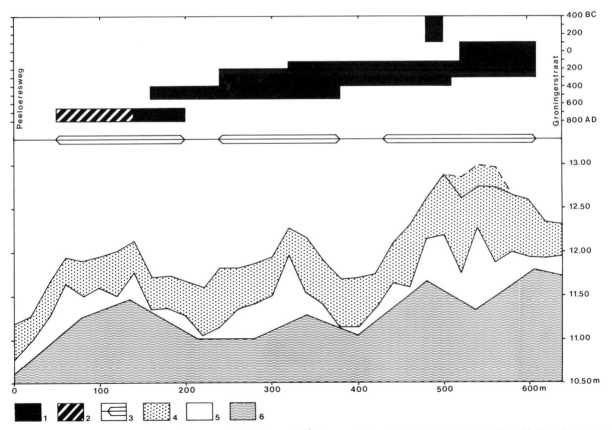

Fig. 79. Schematische doorsnede door de es, van de Peeloëresweg (links) naar de Groningerstraat (rechts). Legenda: 1. Nederzettingssporen aangetoond; 2. Nederzettingssporen vermoedelijk aanwezig; 3. Grootste dichtheid van de nederzettingssporen; 4. Humuslaag; 5. Ongestoorde ondergrond; 6. Hoogste gemiddelde grondwaterstand.

P.B. Kooi

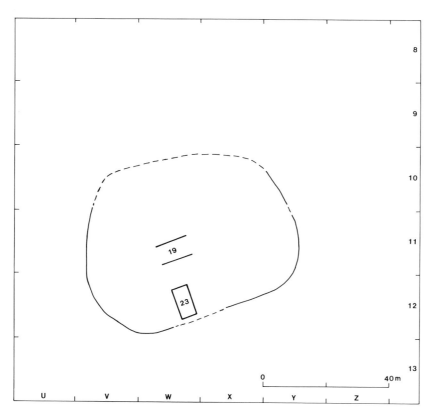

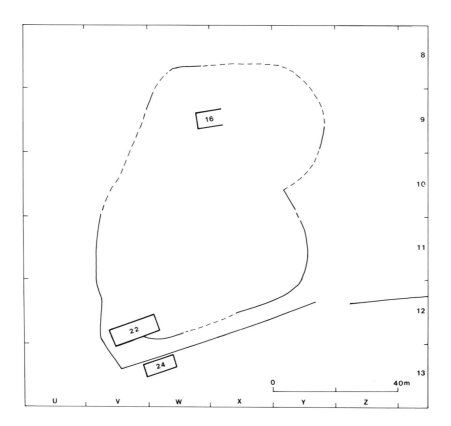

Fig. 80. Globaal overzicht van de ontwikkeling van de nederzetting gedurende periode 2 in vier fasen.

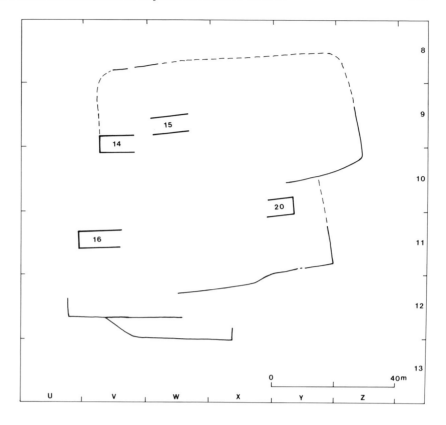

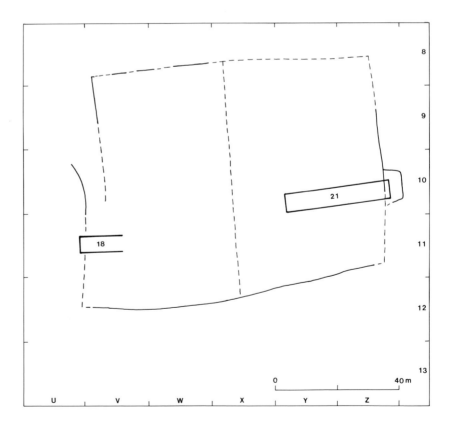

Fig. 80.(vervolg).

fasering van belang zijn. Tenslotte kan het ontstane model getoetst worden door middel van daterende vondsten.

Periode 1, 400-100 voor Chr.
De boerderijplattegronden uit de IJzertijd (nrs 3, 5, 12, 17, 25, 26 en 27) liggen in een ruime boog aan de westflank van het hoogste deel van het terrein. De groepering van deze plattegronden wijst erop, dat er mogelijk drie bedrijven gelijktijdig hebben bestaan. Dit wordt bevestigd door het feit, dat de plattegronden 3, 17 en 27 typologisch gelijk zijn (type Hijken) en op ruime afstand van elkaar liggen. De andere plattegronden kunnen eveneens bij de drie bedrijven horen, bijvoorbeeld respectievelijk 5 bij 3, 12 bij 17, en 25 en 26 bij 27. Duidelijk begrensde erven zijn in deze fase niet waargenomen. Vergelijking met nederzettingssporen uit dezelfde periode geeft aan, dat de erven in deze periode onderdeel zijn geweest van een *Celtic field*-systeem en dus vallen binnen een door wallen omgeven vak. Op het Kleuvenveld ging een dergelijke situatie samen met een concentratie van graanschuurtjes in verscheidene akkers. Sporen van walletjes van een *Celtic field* zijn op de es niet waargenomen. Zij zullen in de Middeleeuwen zijn opgeruimd toen het terrein opnieuw voor de akkerbouw werd ingericht, echter volgens een opstrekkend verkavelingssysteem. Perceelscheidingen in de vorm van staketsels en omheiningen, zoals op het Hijkerveld in het *Celtic field* zijn gevonden werden in Peelo niet aangetroffen. De sporen uit de IJzertijd in de vakken R-W/13-17 zijn vrij van latere nederzettingssporen en bieden de beste vergelijkingsmogelijkheid. Wanneer we uitgaan van de veronderstelling dat we rond de plattegronden 25, 26 en 27 met één erf te maken hebben, dan tekent zich hier een cluster van schuurtjes af (nrs 118-121 en 123-137) die bij verschillende stadia van de ontwikkeling van het erf kunnen behoren. Daarbij ligt tevens een veekraal (nr. 138) en een afvalkuil in vak R/17. Nemen we deze laatste twee sporen als uiterste grens, dan is de afmeting van het bijbehorende erf in oost-west-richting ca. 40 meter, terwijl de afmeting noord-zuid minimaal 20 m bedraagt van spieker 118 naar 134. Dergelijke afmetingen vallen binnen de schaal van de *Celtic fields* (Brongers, 1976). Door deze constatering wordt ook de situatie bij de andere boerderijplattegronden duidelijker. Spiekers als de nrs 6, 12, 13, 15 en 16 kunnen tot het erf van plattegrond 5 respectievelijke 3 worden gerekend. De groep spiekers tussen de boerderijen 12 en 17 kunnen voor zover het de kleine betreft (o.a. nrs 50, 61 en 63) op het bijbehorende erf hebben gelegen, dat een afmeting van minimaal 24 m van noord naar zuid heeft gehad. Een indicatie van de grootte in oost-west-richting is moeilijk te geven.

Periode 2, 100 voor Chr.-200 na Chr.
De tweede periode van de bewoning wordt gekarakteriseerd door een cluster bewoningssporen op het hoogste deel van het terrein, beginnende met plattegronden van het type Noord Barge en eindigend met het type Peelo A. Door de omvangrijke verstoringen in dit deel van de nederzetting is het overgeleverde beeld fragmentarisch, maar uitgaande van de ontwikkeling van de huistypen, in combinatie met de omheiningen en clusters van bijgebouwen is een globale indeling in vier fasen te maken (fig. 80).

2-a. De eerste fase omvat fragmenten van twee boerderijen van het type Noord Barge, namelijk nrs 19 en 23. Door hun ligging en oriëntatie zijn ze te combineren met een ovale omheining. De afwijkende noord-zuid-oriëntatie van nr. 23 maakt het waarschijnlijk dat het een bijgebouw (stal?) is, terwijl de centrale ligging van nr. 19 erop wijst, dat het om een boerderij gaat. Het is aannemelijk, dat het om de restanten van één erf gaat.

2-b. Vrij spoedig heeft zich een tweede erf aan de noordzijde bij de eerste gevoegd. Mogelijk is nog een wandfragment van de boerderij afkomstig (nr. 16). Tegelijkertijd is aan de rand van het zuidelijk erf een nieuwe boerderij gebouwd (nr. 22). De omheining is daartoe aan de zuidzijde iets uitgelegd en door de aanleg van een tweede omheining evenwijdig aan de eerste ontstond een schutplaats voor het vee, waarlangs aan de buitenzijde een verdiepte stal (nr. 24) werd gebouwd. Deze verving mogelijk het bijgebouw nr. 23.

2-c. In de derde fase worden de erven in omtrek hoekiger. Zowel het noordelijke als het zuidelijke erf bevatten tenminste twee fragmenten van boerderijen. De dubbele omheiningssporen aan de zuidzijde wijzen op het handhaven van de schutplaats voor het vee in deze fase. Langs de randen van de erven liggen rijen spiekers en hutkommen.

2-d. In deze fase vindt er een herverdeling van de erven plaats in combinatie met een vergroting (samen ca. 90 are). In plaats van een noordelijk en een zuidelijk erf wordt er nu een oostelijk en een westelijk erf gevormd. In beide erven is het restant van één boerderij aan deze fase toe te schrijven.

Periode 3, 200-300 AD
In tegenstelling tot de voorgaande periode is de ontwikkeling van de nederzetting in periode 3 veel minder verstoord. Deze periode is vooral bepaald door het voorkomen van boerderijplattegronden van het type Peelo A, namelijk nrs 2, 6, 7, 13, 41a, 45 en 46. Twee daarvan zijn verbouwd tot boerderijen van het type Wijster B. Vanwege de verdere ontwikkeling van de nederzetting is periode 3 in twee fasen.

3-a. Vanuit de bewoning op het hoogste deel van de es verschuift de bewoning naar het westen, waarbij de nederzettingssporen voornamelijk liggen op de westflank van de hoge es en de oostkant van de lager gelegen esgronden. Deze worden gescheiden door een enigszins drassige zone, die minder geschikt was voor het plaatsen van gebouwen. In totaal zijn in deze periode drie erfsituaties onderscheiden, namelijk een noordelijk erf met o.a. de boerderijplattegronden 2, 6 en 7, een zuidelijk

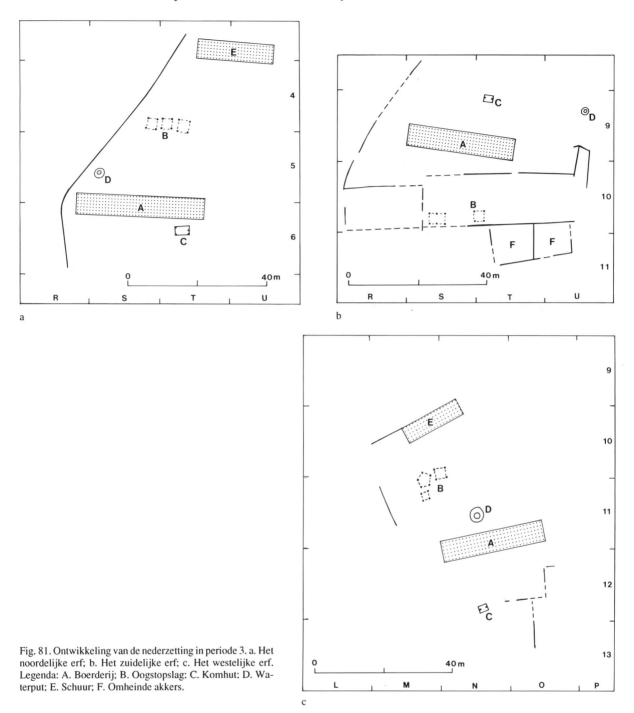

Fig. 81. Ontwikkeling van de nederzetting in periode 3. a. Het noordelijke erf; b. Het zuidelijke erf; c. Het westelijke erf. Legenda: A. Boerderij; B. Oogstopslag; C. Komhut; D. Waterput; E. Schuur; F. Omheinde akkers.

erf met plattegrond 13 en een westelijk erf met de plattegronden 45 en 46. Naast overeenkomsten in de bouwwijze zijn er tevens individuele verschillen te zien in de indeling en samenstelling van de bijgebouwen. Het noordelijk erf strekt zich naar het noorden uit tot en met de schuur nr. 1, die waarschijnlijk evenals de schuren 38 en 39 in het westelijk erf langs een omheining heeft gelegen. Aan de westzijde ligt een drassige laagte. De grens aan de zuidzijde zal tussen de platte-

gronden 7 en 13 gelegen hebben en de oostgrens waarschijnlijk tot het hoekpunt van de omheining van periode 2 naar het noorden. Het hoofdgebouw is twee keer herbouwd, waarbij de meest waarschijnlijke volgorde nrs 2, 7, 6 zal zijn geweest. Rond het staleinde van nr. 2 zijn greppels gegraven die duiden op wateroverlast. Nr. 7 ligt iets hoger en nr. 6 is door verbouwing een tussenfase naar een volgende periode van de ontwikkeling. Als bijgebouwen zijn aan te wijzen de al genoemde schuur nr. 1, de spiekers nrs 1, 2, 3, 4 en 14, de komhutten nrs 7, 8, 9, 10 en 11 (schuur nr. 4 behoort door zijn bouwwijze met dubbelpalen in de volgende periode). Voor de watervoorziening zijn tenminste twee putten aanwezig geweest. Een mogelijke momentopname is gegeven in figuur 81a.

Het zuidelijke erf met boerderij nr. 13 is aan de zuidzijde begrensd door een stelsel van omheiningen. Aan de oostzijde liggen de omheiningen van de vorige periode als grens en aan de westzijde een omheining, die vrij ver doorloopt naar het noorden, zodat mag worden aangenomen dat de grens met het voorgaande erf op korte afstand van plattegrond nr. 7 heeft gelegen. De boerderij nr. 13 vertoont sporen van een ingrijpende verbouwing. Onder de bijgebouwen is geen schuur aan te wijzen, zoals aan de noordzijde van het eerste erf is gevonden, maar wel de komhutten 48 en 49, de spiekers 32, 34 en de grotere spiekers van de cluster 57-72 (de kleinere horen bij de bewoning uit de IJzertijd) en een tweetal waterputten. Een mogelijke momentopname is gegeven in figuur 81b. De begrenzing van het erf is aan de noord- en westzijde vrij nauwkeurig te bepalen. Aan de noordzijde is het erf stapsgewijs uitgebreid. Dit is af te leiden uit de positie van de bijgebouwen, die een aantal keren zijn verplaatst, zoals de schuur nr. 39 en de spiekers 163-165 (zie fig. 81c). De spiekers zijn verplaatst naar de hoek van het erf (nrs 158-161) en tenslotte is de schuur nr. 38 gebouwd met de spiekers 154-157, die tezamen tot een volgende fase behoren. De westelijke begrenzing is bepaald door een aantal parallelle omheiningssporen, die met de verplaatsing van de boerderijen nrs 45-46 en de spiekers iets naar het westen is verschoven. Aan de oostzijde is voor deze periode geen duidelijke begrenzing aan te geven. Wellicht heeft deze gelegen in de niet opgegraven strook tussen de werkputten 20 en 38. Aan de zuidzijde ligt een aantal fragmenten van omheiningssporen, waarvan echter gezien de grotere afstand niet duidelijk is of ze bij deze periode behoren.

3-b. In de volgende fase treedt een geringe wijziging in het nederzettingspatroon op. Het zuidelijk erf verdwijnt en aan de westzijde verschijnt een nieuw erf met boerderij nr. 41a, zodat het totaal aantal van drie erven gehandhaafd blijft. De boerderijplattegronden veranderen in deze fase door verbouwing in het type Wijster B. Een schuurtje als nr. 4, met dubbelpalen is kenmerkend voor deze ontwikkeling.

Periode 4, 300-550 AD
Tijdens deze periode blijft het bewoonde oppervlak van de nederzetting stabiel en ligt volledig op het lage gedeelte van de es. De hoofdgebouwen zijn van het type Wijster B (nrs 8, 9, 36, 37, 40, 41) en Peelo B (nrs 11, 28, 30, 31, 34 en 42). Bij een poging tot indeling in erven en fasering doet zich een aantal problemen voor. In de eerste plaats ontbreken samenhangende omheiningssporen van een aantal erven. Voorts zijn niet overal de grenzen van de nederzetting voldoende bekend en

tenslotte is van een aantal gebouwen de aard of het type niet meer te bepalen vanwege het fragmentaire karakter (nrs 9, 32, 43 en 49). Dit lijkt het gevolg te zijn van het dunne esdek, waardoor de kwaliteit is aangetast ten gevolge van dieper ploegen in recente tijd. Een prominente plaats binnen de nederzetting wordt in deze periode ingenomen door een omheind gebied, groot ca. 65 are, waarbinnen een drietal boerderijplattegronden met schuren, spiekers, komhutten, waterputten, de smederij en ijzeroventjes ligt. Het westelijke deel is een voortzetting van het erf met de boerderijen 45 en 46 uit periode 3, het oostelijke gedeelte is toegevoegd. Hoewel het blijkbaar om een dubbel erf gaat zijn er onvoldoende sporen om een scheiding aan te geven. Deze kan gelegen hebben in de niet onderzochte strook tussen de werkputten 20 en 38. In het westelijke deel liggen mogelijk twee fragmenten van boerderijplattegronden uit deze periode, namelijk 47 en 49. Plattegrond nr. 47 kan zowel het restant zijn van een boerderij van het type Wijster B als van het type Peelo B, terwijl plattegrond 49 zelfs lijkt op het type Wijster C. Tegen de omheining aan de noordzijde liggen de spiekers 154-157 en de schuur nr. 38. De schuur nr. 48 en de spieker 195 met plattegrond 49 liggen langs de omheining aan de zuidzijde. Daarbuiten ligt nog een waterput op de helling naar het veentje (vak N/14). Uit bovenstaande is af te leiden, dat de bewoning van dit erf in de loop van deze periode geleidelijk iets lager, naar het zuiden opschuift.

De rest van het omheinde terrein grenst in het oosten aan het erf van de boerderij 13 uit de voorafgaande periode. Het middendeel is komvormig en tamelijk drassig geweest. Daar zijn ook geen sporen van gebouwen aangetroffen. In het noorden ligt de boerderijplattegrond nr. 11 met daarnaast de smederij (nr. 10), een waterput, spieker en ijzeroventjes (fig. 82). Binnen de nederzetting neemt dit erf door de sporen van ijzerbewerking een bijzondere plaats in.

Het erf bij de boerderij 41 blijft in deze periode staan, waarbij het hoofdgebouw wordt opgevolgd door de nrs 40, 42 en 43, die telkens op een andere plaats zijn opgebouwd. De locatie van de spiekers uit de verschillende fasen blijft daarbij vrijwel gelijk, de serie nrs 186-188 is wellicht ouder dan de serie 175-183. De schuur nr. 44 is qua functie waarschijnlijk vergelijkbaar met nr. 48 van het naburige erf. Uitzonderlijk is het bijgebouwtje 184, mogelijk een zwaar uitgevoerde aparte stal. Een systeem van kleine omheinde stukken grond liggend naast en door boerderijplattegrond 41 is vergelijkbaar met de indeling, die we in periode 3 bij het erf van boerderij 13 hebben beschouwd als omheinde akkers. In dit geval kunnen ze niet gelijktijdig met plattegrond 41 zijn, maar zullen bij de jongere fasen horen. Hoewel de begrenzing van dit erf aan de oostzijde bepaald wordt door meerdere omheiningssporen uit verschillende fasen, die als scheiding met het naburige erf hebben gediend, is er voor het overige geen spoor van erfomheiningen gevonden. De analyse van de resterende bewoningssporen ten noorden van de reeds besproken

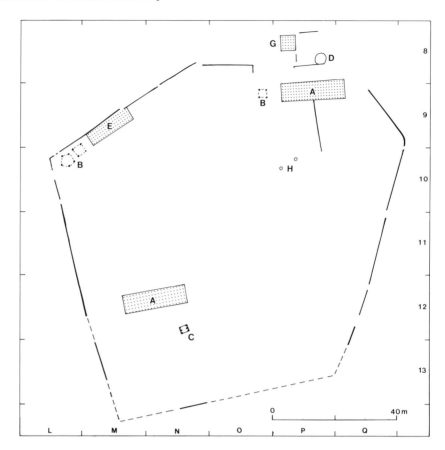

Fig. 82. Dubbel erf uit periode 4.
Legenda: A. Boerderij; B. Oogst-
opslag; C. Komhut; D. Waterput; E.
Schuur; G. Smederij; H. IJzeroven.

erven is zo mogelijk nog problematischer. Door de smalle percelering uit latere tijd, gescheiden door greppels is in combinatie met het gebruik als akkergrond sinds de middeleeuwen een deel van de minder diepe grondsporen verploegd. In totaal zijn er zeker vijf boerderijplattegronden van het type Wijster B (nrs 8, 29, 33, 36, 37) en vier van het type Peelo B (nrs 28, 30, 31, 34) gevonden, met nog twee onduidelijke fragmenten (nrs 9 en 32). Ondanks het kwaliteitsverlies kan het volgende worden geconstateerd. In de eerste plaats valt het op dat er zeven komhutten bij de boerderijen 8 en 9 (nrs 19-25) liggen en slechts negen bij het grotere aantal overige boerderijplattegronden. In de tweede plaats zijn in dit gedeelte van de nederzetting geen spiekers gevonden. Dat deze selectief door verstoring zouden zijn verdwenen is onwaarschijnlijk. De schuur nr. 148 lijkt het enige bouwsel dat nog voor oogstberging in aanmerking komt. Mogelijk wordt dit verschijnsel veroorzaakt door een verandering in de economie. Ook een andere oogstberging zou hiervoor een verklaring kunnen zijn, maar is minder waarschijnlijk.

Periode 5, 8e eeuw
De verschuiving van de nederzetting naar het westen gaat voort, maar is daar slechts zeer ten dele onderzocht. Na een vrijwel lege noord-zuid-verlopende zone (vakken G-H/9-13) zijn er meer westelijk weer nederzettings-

sporen aanwezig in de vorm van paalkuilen, afvalkuilen, komhutten (nrs 167-170), twee waterputten in de werkputten 22, 25, 26 en 28 en een schuurtje met dubbelpalen in werkput 46. In de werkputten 27 en 46 ligt een huisplattegrond van het type Odoorn en twee spiekers. Het ligt voor de hand te veronderstellen, dat in het niet onderzochte deel van het terrein tussen de werkputten meer nederzettingssporen uit de vroege middeleeuwen hebben gelegen. Voor het overige blijft het gissen; de meest westelijke aanwijzing voor bewoning, bestaande uit aardewerkscherven is gevonden in het cunet van een fietspad liggende tussen het verste punt van werkput 27 en de Peeloëresweg.

5. VONDSTEN

De hoeveelheid vondsten is gering in verhouding tot de opgegraven oppervlakte en in vergelijking met vondst- complexen uit dezelfde periode, zoals Wijster. Dit is verklaarbaar als we rekening houden met de spreiding van de nederzettingssporen of de intensiteit van de bewoning per opgegraven oppervlak. Immers, het afval van de bewoning uit een bepaalde periode komt maar ten dele in diepere grondsporen uit diezelfde periode terecht. Soms zijn afvalkuilen gegraven, maar b.v scherven zullen voor een groot deel aan de oppervlakte

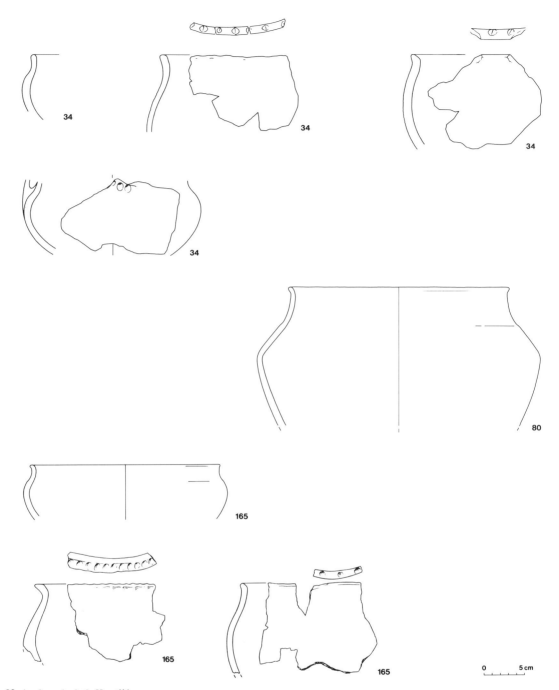

Fig. 83. Aardewerk uit de IJzertijd.

blijven liggen. Ze belanden dan bij toeval in grondsporen die in een volgende periode zijn gegraven. De plaatsen die het meeste materiaal opleveren zijn hutkommen en waterputten die in onbruik zijn geraakt en als afvalkuilen zijn gebruikt. Vondstmateriaal uit de IJzertijd is schaarser, omdat toen geen waterputten of komhutten werden gebruikt. Omgekeerd heeft een enkele scherf uit een paalgat dus ook een betrekkelijk geringe daterende waarde. Het gevonden nederzettingsmateriaal is

in vier groepen gesplitst, namelijk aardewerk, steen, sierplaatjes en glas.

Aardewerk

Bij de beschrijving van het aardewerk is gebruik gemaakt van de bestaande typologieën opgesteld voor Wijster, Paddepoel en Zeijen (versterkte nederzettingen; Van Es, 1967; 1970; Waterbolk, 1977). Een eerste onderverdeling is te maken tussen het aardewerk uit de

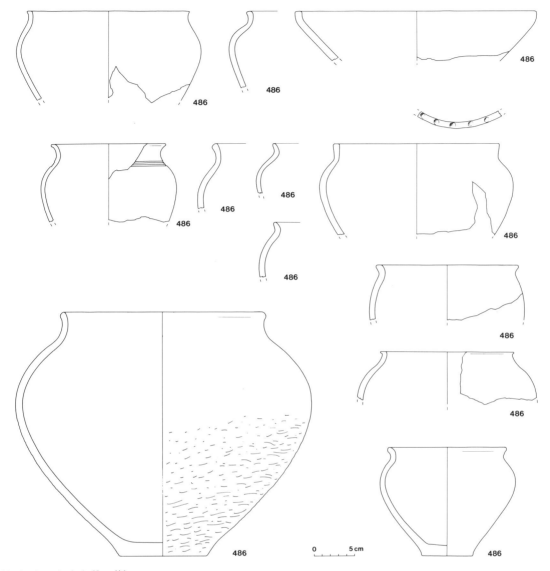

Fig. 84. Aardewerk uit de IJzertijd.

vóór-Romeinse IJzertijd en de Romeinse tijd. De IJzertijd manifesteert zich het Ruinen-Wommels-aardewerk (RW) in acht vondstnummers, die grotendeels afkomstig zijn uit de zone met de boerderijplattegronden uit periode 1, en bovendien voorkomt in het oosten van werkput 3. Dit betekent dat de daar voorkomende grondsporen, die niet tot één plattegrond zijn samen te brengen in de IJzertijd te dateren zijn, en dat de bewoning zich in die richting heeft uitgestrekt. In het vondstmateriaal komen in deze periode de typen RW I en II niet voor. Het oudst dateerbare materiaal behoort tot het type RW III, waarvan de vroegste datering ca. 300 v.Chr. is. Een relatief vroege vorm is nr. 80 (fig. 83) uit een kuil nabij boerderij 12. In de overige afgebeelde vondstcomplexen nrs 34, 165 en 486 komt RW III-aardewerk voor, met dien verstande dat daarvan nr. 34 de meest karakteristieke samenstelling heeft, zonder

RW IV-kenmerken, zodat er een datering tussen ca. 300 en 200 v.Chr. aan te geven is. De beide overige complexen vertonen vormen die verwant zijn aan het aardewerk uit Paddepoel (PP IV), dat omstreeks het begin van de jaartelling wordt gedateerd en gekenmerkt wordt door een kortere hals en zwaardere uitstaande randen, die bij het type PP IV A aan de binnenzijde gefaceteerd zijn. Een ander verschijnsel waarin het aardewerk van de voor-Romeinse IJzertijd zich onderscheidt is het voorkomen van indrukken op de rand van de pot. Deze indrukken zijn aangebracht met de vingertoppen of met een spatel of stokje, waardoor de rand min of meer gekarteld is. In de Romeinse tijd komen de indrukken overwegend aan de buitenzijde van de randen voor. Kartering van de twee verschillende kenmerken levert een verrassend resultaat op. De randen met indrukken op de rand liggen verspreid over de zone met bewoning

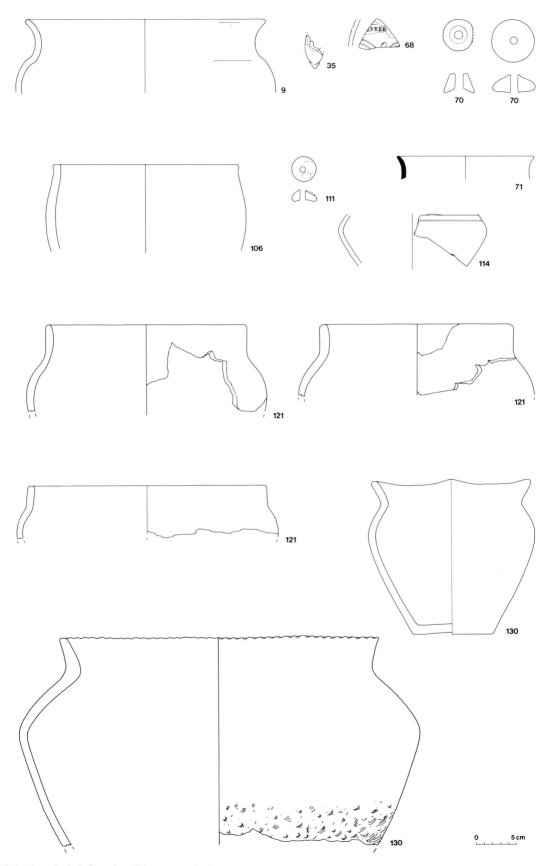

Fig. 85. Aardewerk uit de Romeinse tijd tot en met de 6e eeuw.

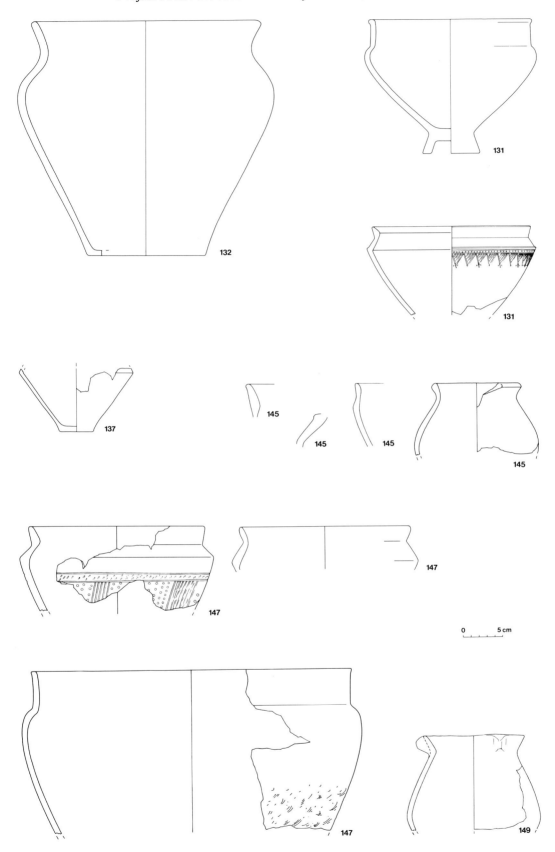

Fig. 86. Aardewerk uit de Romeinse tijd tot en met de 6e eeuw.

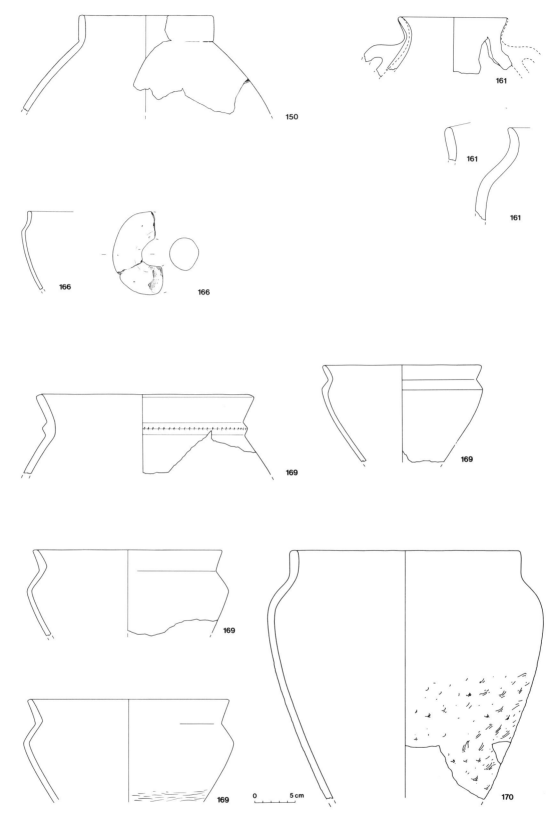

Fig. 87. Aardewerk uit de Romeinse tijd tot en met de 6e eeuw.

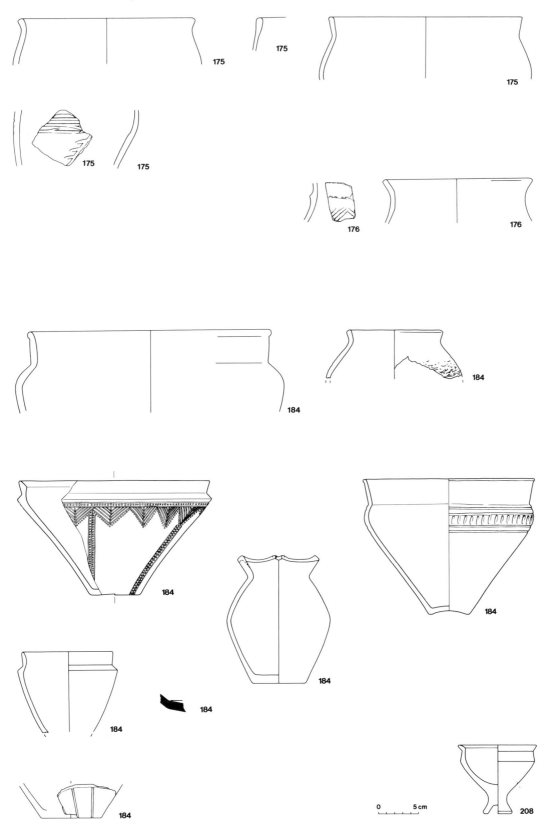

Fig. 88. Aardewerk uit de Romeinse tijd tot en met de 6e eeuw.

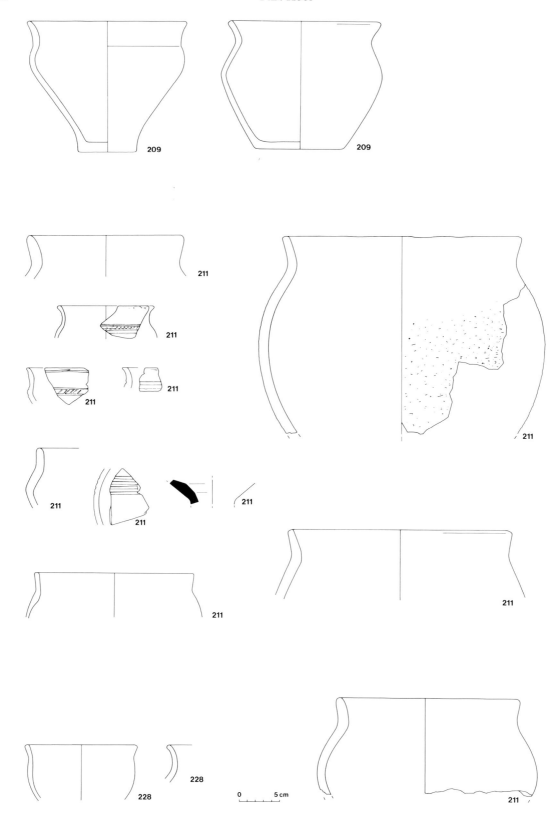

Fig. 89. Aardewerk uit de Romeinse tijd tot en met de 6e eeuw.

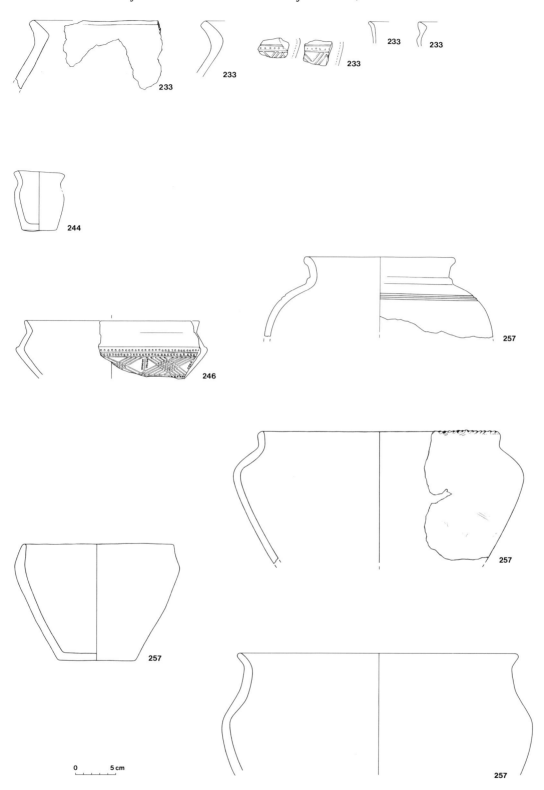

Fig. 90. Aardewerk uit de Romeinse tijd tot en met de 6e eeuw.

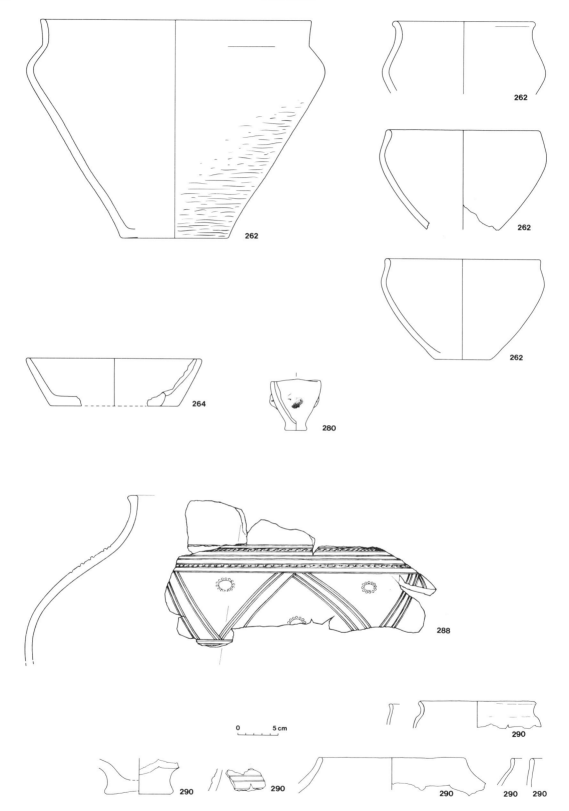

Fig. 91. Aardewerk uit de Romeinse tijd tot en met de 6e eeuw.

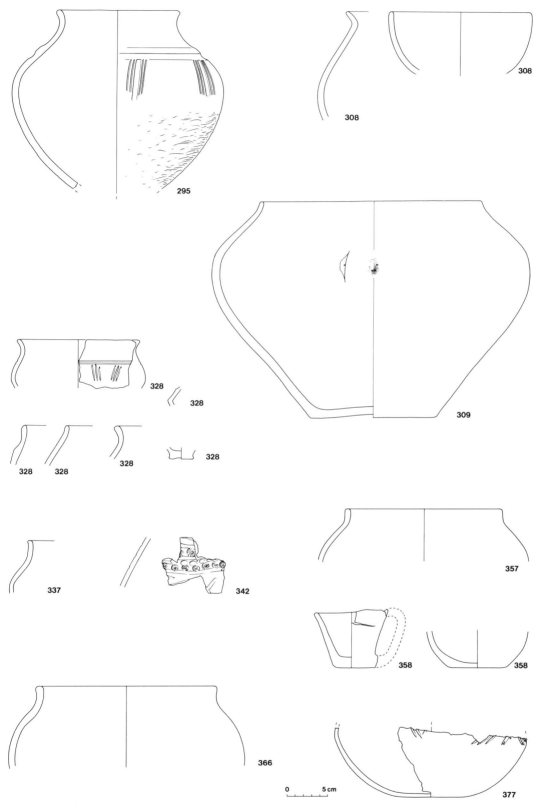

Fig. 92. Aardewerk uit de Romeinse tijd tot en met de 6e eeuw.

P.B. Kooi

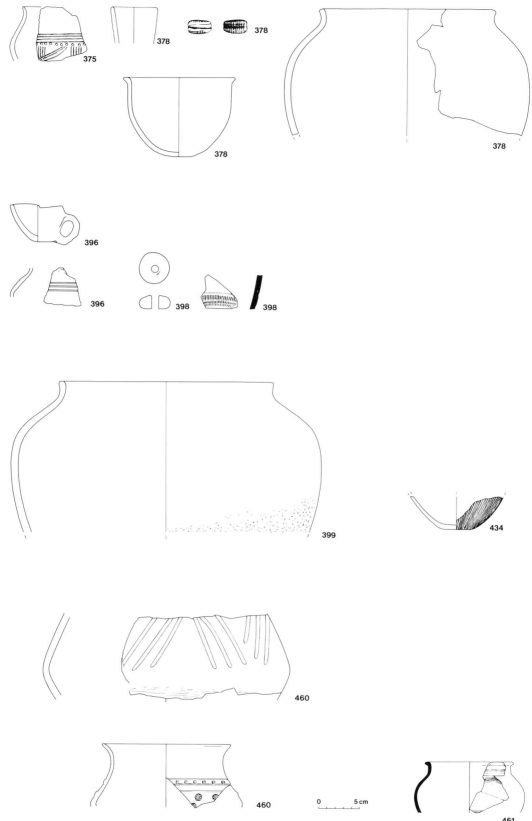

Fig. 93. Aardewerk uit de Romeinse tijd tot en met de 6e eeuw.

uit de IJzertijd, de randen met indrukken op de buitenzijde zijn gerelateerd aan de omheinde erven op het hoogste deel van de es met bewoning uit periode 2. Dit kenmerk is dus een veel sterker onderscheid dan op basis van de toch geleidelijke overgang verwacht mag worden. Overigens ontbreken de typen die met enige zekerheid in de eeuwen voor en na de jaartelling kunnen worden gedateerd, hoewel dit op basis van de boerderijtypen wel verwacht mag worden.

Aardewerk uit de 2e-3e eeuw is goed vertegenwoordigd, waarbij de typen IA, IB, IIB en IIIA als gidsvormen kunnen worden beschouwd. Bij kartering van deze typen blijkt dat de verspreiding zich beperkt tot het gebied binnen de omheinde erven uit periode 2 en 3a, terwijl een grotere verspreiding naar de erven met Peelo A-boerderijen te verwachten zou zijn. Vondstcomplexen die met de periode 2 gecorreleerd kunnen worden, zijn de nrs 121, 130, 131, 145, 147, 161, 169, 175, 184, 209, 233, 257 en 262. De oudste van deze reeks is nr. 184 (vak Y/9) op grond van de combinatie van de typen IA, IB2, IIIA en Von Uslar-type III te dateren in de tweede helft van de 2e eeuw. In de tweede helft van de tweede eeuw tot en met de eerste helft van de derde eeuw vallen de complexen 121 (vak Y/9) o.a. met het type IIB2, 130 (vak V-10) met de typen IIB2 en IIIA1, 131 (vak Y/11) met de typen IB2, Von Uslar-type II en *terra sigillata*-Chenet 320, 147 (vak Y/11) met de typen IB1 en 3, 161 (vak X/11) met de typen IIIA1, IVA en Von Uslar-type II, 175 (vak T/9) met de typen IIB3, IIIA1 en IVF, 209 (vak Z-10) met de typen IIB en 3 en tenslotte 233 (vak V/12) met de typen IB2 en IIB2. Deze reeks bevestigt globaal de periodisering van de erven. Een drietal complexen is nog iets nauwkeuriger te dateren, namelijk in de eerste helft van de 3e eeuw. Het zijn de nrs 145 (vak Y/9) met de typen IB3 en IIIB, 169 (vak Y/9) met de typen IB, IIB, en IIIB en 257 (vak Y/11) met de typen IIB, IIIA1 en VIIB, die in verspreiding aansluiten bij de voorgaande. In het vormspectrum komt één detail voor dat enige aandacht verdient, namelijk het aardewerk met een niet verdikte uitstaande randlip. Deze komt in Wijster slechts weinig voor, maar is in Peelo in ruime mate vertegenwoordigd bij de typen IB2 en 3, IIB3 en Von Uslar-*Form* II. Het gaat hierbij om kwalitatief beter aardewerk dat ook een aantal keren met versiering voorkomt zoals bijvoorbeeld in nr. 184 met twee tot vijf horizontale groeflijnen waarvan één zone met verticale indrukken. De vrij steile rand met iets uitgebogen randlip is ook toegepast op geheel andere vormen en typen zoals in nrs 147 en 184 te zien is. Op grond van vondsten te Einswarden en Jemgumerkloster (Schmid, 1965) wordt dit kenmerk in het aangrenzende Noordduitse gebied gedateerd in het einde van de tweede eeuw tot het einde van de derde eeuw. Het laatste vondstcomplex in deze reeks is nr. 262 (vak Y/10) uit de tweede helft van de derde eeuw. Dit illustreert de laatste fase van de bewoning in dit deel van de nederzetting, zoals nr. 175 het begin van de verschuiving naar het westen bevestigt. Wat tevens opvalt is, dat de bewoning

met de type Peelo A-boerderijen aan de noord- en westzijde nauwelijks te dateren aardewerk heeft opgeleverd. Een complex uit het einde van de derde eeuw en het begin van de vierde eeuw is nr. 211 (vak V/10-11) met de typen IB, IIIB, IVB en *terra sigillata*. Belangrijke vormen die in de 4e eeuw voor het eerst voorkomen zijn die welke in het algemeen als Angelsaksisch worden beschouwd. Complexen uit deze periode zijn de nrs 328 (vak K/11) met de typen ID/VIIIA, 396 (vak K/8) met de typen VIA en VIIIA en 460 (vak N/14) met het type VIIIB. Daarnaast zijn er een aantal scherven zoals nr. 228 met plastische ribbels, chevrons en rozetten, nrs 375 en 377. In de complexen gaan deze typen samen met wijdmondige, eenvoudig geprofileerde kommen met uitstaande randlip en grof gevormde kopjes met knobbeloortjes. Deze aardewerkvormen lopen door in de 5e eeuw waarbij de versiering op het type VIII wordt uitgebreid met facetten op de buikknik en ingestempelde motieven. Vier vondstnummers met de typen ID en VIII liggen nog in het oostelijke deel van het terrein bij de boerderijtypen uit de periode 2-3, maar het merendeel is gevonden in het oostelijke, lage deel van de es met de boerderijtypen Wijster B en Peelo B. Een duidelijke verschuiving van de nederzetting wordt daarmee bevestigd. Het geïmporteerde, op de draaischijf vervaardigde aardewerk is, als groep voorlopig nog buiten beschouwing gebleven, omdat het zeer gering van omvang is. Er zijn in totaal fragmenten gevonden van vier stuks *terra sigillata* en acht stuks *terra nigra*-achtige potten. Van deze laatste is één stuk zwart gevernist en een andere (nr. 461) is van het type Chenet 342. Onder het grovere materiaal bevindt zich het type Chenet 320. Toch geeft ook de verspreiding van dit schaarse importmateriaal een bevestiging van de verschuiving van de nederzetting naar het westen die zich einde 3e, begin 4e eeuw voltrekt.

Onder de vorige aardewerkvondsten zijn enkele spinklosjes (o.a. nrs 70 en 111) en weefgewichten (o.a. nr. 166), waarvan één met een driehoekige vorm en drie gaten, zoals vanaf de Latène-tijd tot in de vroeg-Romeinse tijd voorkomen (Wilhelmi, 1977). Bij het vondstnr. 35 behoort een halve keramische slingerkogel die meer zuidelijk regelmatig voorkomen (Verwers, 1972). Uiterst merkwaardig is een klein napje (nr. 16) van gesinterd aardewerk met fragmenten van drie 'pootjes' . De aard van dit fragment is niet nader te bepalen.

Voorwerpen van steen (fig. 94)
In een paalgat van spieker nr. 72 werd het stompe einde van een strijdhamer van het type Muntendam (Mu1) gevonden. De gebruikte steensoort is diabaas. De spieker is door zijn afmetingen en ligging te dateren in de midden-IJzertijd. Door Achterop en Brongers (1979) zijn deze strijdhamers in verband gebracht met ijzerbewerking. Dit is echter niet af te leiden uit de verspreiding van de vindplaatsen, die meer duidt op offers

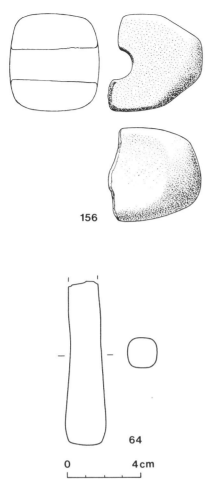

Fig. 94. Fragment van een strijdhamer van het type Muntendam (a) en een halve strekel (b).

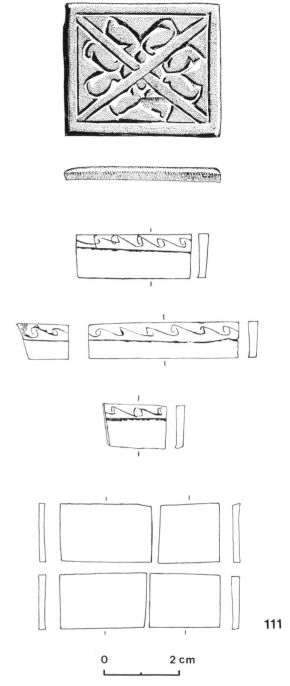

Fig. 95. Sierplaatjes van pek.

in een nat milieu (Groenendijk, 1993).

Fragmenten van maalstenen gemaakt van basaltlava werden op diverse plaatsen gevonden, maar deze waren zo gering van afmeting, dat reconstructie of toewijzing aan een bepaald type onmogelijk is.

Verspreid over het opgravingsterrein zijn enkele fragmenten van strekels gevonden (fig. 94). Deze voorwerpen van kwartsiet of *Tonschiefer* zijn waarschijnlijk vanaf de prehistorie tot in het begin van deze eeuw gebruikt om zeis of sikkel te scherpen (Van der Zweerde, 1979). Meestal zijn ze door veelvuldig gebruik sterk gesleten en daardoor gebroken.

Sierplaatjes (fig. 95; nr. 111)
In de vulling van hutkom 33 werden negen vlakke stukjes zwart materiaal aangetroffen waarvan vijf versierd zijn. Aanvankelijk werd verondersteld, dat de plaatjes van git was gemaakt, maar bij nadere bestudering lijkt het pek te zijn. Op verschillende plaatsen in het materiaal zijn namelijk kleine luchtbellen te zien. Waarschijnlijk betreft het hier een inheems produkt mogelijk

vervaardigd uit berketeer, dat na verhitting in vloeibare toestand in een vorm is gegoten en vervolgens op maat gemaakt. Het grootste versierde stuk van 3,5x4,2 cm is 3,5 mm dik. De versiering bestaat uit een rand, waarbinnen een diagonaal kruis. In de vier driehoeken tussen de armen van het kruis zijn twee aan twee S-vormige figuurtjes geplaatst. Het motief lijkt er in gekerfd, maar dit is waarschijnlijk een gevolg van het feit dat de gietmal is gemaakt naar een voorwerp van hout of been.

Geheel anders is de versiering op een vijftal smalle repen, bestaande uit een ingekrast motief van golfjes, afgezet door een rechte groeflijn. Twee van deze repen zijn compleet, de andere twee zijn afgebroken. Waarschijnlijk zijn er oorspronkelijk twee van 4 cm lang bij 0,9 cm breed en twee van 3 cm lang bij 1,1 cm breed geweest. Tenslotte zijn er nog vier onversierde plaatjes waarvan de oppervlakte tezamen ongeveer overeenkomt met die van de versierde plaat. De vondst is uniek in zijn soort en dat maakt dat de betekenis onduidelijk is. Wellicht waren de plaatjes oorspronkelijk om een kern van vergankelijk materiaal bevestigd, bijvoorbeeld een doosje of hanger. De hutkom waarin de plaatjes zijn gevonden kan in de 3e eeuw worden gedateerd.

Glas
Uit de opgravingen zijn zeer weinig glasvondsten te voorschijn gekomen. Een paar scherfjes van geel-groen glas zijn niet aan een bepaalde vorm toe te schrijven. Het enige determineerbare stuk is een halve, oorspronkelijk vier-lobbige kraal van witte en violette glaspasta (vondstnr. 378, fig. 93). De vondstomstandigheden wijzen op een datering in de 5e eeuw.

6. NAWOORD

Deze publicatie kwam tot stand dank zij een financiële bijdrage van de gemeente Assen.

7. LITERATUUR

ACHTEROP, S.H. & J.A. BRONGERS, 1979. Stone cold chisels with handle (Schlägel) in the Netherlands). *Berichten van de Rijksdienst voor het Oudheidkundig Bodemonderzoek* 29, pp. 255-356.

BARDET, A.C., P.B. KOOI, H.T. WATERBOLK & J. WIERINGA, 1983. *Peelo, historisch-geografisch en archeologisch onderzoek naar de ouderdom van een Drents dorp. Mededelingen van de K.N.A.W.* (= Nieuwe Reeks, deel 46, nr. 1). Amsterdam.

BRONGERS, J.A., 1976. *Air photography and Celtic field research in the Netherlands* (= Nederlandse Oudheden 6). Amersfoort.

DRENTH, E. & A.E. LANTING, 1991. De chronologie van de Enkelgrafcultuur in Nederland: enige voorlopige opmerkingen. *Paleo-aktueel* 2, pp. 42-46.

ES, W.A. VAN, 1967. Wijster. A native village beyond the imperial frontier, 150-425 A.D. Diss. Groningen.

GIFFEN, A.E. VAN, 1926. De proefgravingen te Peelo. *Jaarverslagen van de Vereniging voor Terpenonderzoek* 9-10, pp. 32-35.

GIFFEN, A.E. VAN, 1938. Het urnenveld en de vergraven palissadeheuvel te Peeloo, Gem. Assen. *Nieuwe Drentse Volksalmanak* 56, pp. 110-114.

GIFFEN, A.E. VAN, 1939. De brandheuvels 5 en 6 bij Peeloo, gem. Assen. *Nieuwe Drentse Volksalmanak* 57, pp. 128-129.

GIFFEN, A.E. VAN, 1954. Praehistorische huisvormen op de zandgronden. *Nederlands Kunsthistorisch Jaarboek* 5, pp. 11-41.

GLASBERGEN, W., 1971. *Graves containing beakers with protruding foot* (= Inventaria Archeologica. The Netherlands NL5). Bonn.

GOUTBEEK, A. & E. JANS, 1988. Hooibergen in Oost-Nederland. Zutphen.

GROENENDIJK, H.A., 1933. Landschapsontwikkeling en bewoning in het Herinrichtingsgebied Oost-Groningen, 8000 BC-1000 AD. Diss. Groningen, p. 104.

HAARNAGEL, W. & P. SCHMID, 1984. Siedlungen. *Archäologische und naturwissenschaftliche Untersuchungen an ländlichen und frühstädtischen Siedlungen im deutschen Küstengebiet vom 5. Jahrhundert v.Chr. bis zum 11. Jahrhundert n.Chr.* Band 1. Weinheim, pp. 167-244.

HARSEMA, O.H., 1980. *Drents boerenleven van de bronstijd tot de middeleeuwen* (= Museumfonds publicatienr. 6). Assen.

HILLS, C.M., 1978a. Sächsische und angelsächsische Keramik. In: *Sachsen und Angelsachsen* (= Veröff. des Helms Museums Hamburg nr. 32). Harburg, pp. 135-142.

HILLS, C.M., 1978b. Gestempelte Keramik. In: *Sachsen und Angelsachsen* (= Veröff. des Helms Museums Hamburg 32). Harburg, pp. 143-152.

HUIJTS, C.S.T.J., 1992. De voor-historische boerderijbouw in Drenthe. Diss. Arnhem.

KLUNGEL, A.E., 1963. De sleufakkers van de Westerwoldse essen. *Boor en Spade* 13, pp. 27-39.

LANTING, J.N., 1977. Bewoningssporen uit de IJzertijd en de vroege middeleeuwen nabij Eursinge, gem. Ruinen. *Nieuwe Drentse Volksalmanak* 59, pp. 213-249.

PLEINER, R., 1964. Die Eisenverhüttung in der 'Germania Magna' zur römischen Kaiserzeit. *Bericht der Römisch-Germanische Kommission* 45, pp. 12-86.

VERWERS, G.J., 1972. Das Kamps Veld in Haps in Neolithikum, Bronzezeit und Eisenzeit. Diss., Leiden.

WATERBOLK, H.T., 1973. Odoorn im frühen Mittelalter. Bericht der Grabung 1966. *Neue Ausgrabungen und Forschungen in Niedersachsen* 8, pp. 25-89.

WATERBOLK, H.T., 1977. Walled enclosures of the Iron Age in the north of the Netherlands. *Palaeohistoria* 19, pp. 97-172.

WATERBOLK, H.T., 1979. Siedlungskontinuität im Küstengebiet zwischen Rhein und Elbe. *Probleme der Küstenforschung im südlichen Nordseegebiet* 13, pp. 1-23.

WATERBOLK, H.T., 1980. Hoe oud zijn de Drentse dorpen? *Westerheem* 29, pp. 198-206.

WATERBOLK, H.T., 1985. Archeologie. In: J. Heringa et al., *Geschiedenis van Drenthe*. Meppel, pp. 15-90.

WATERBOLK, H.T., 1987. Terug naar Elp. In: *De historie herzien*. Vijfde bundel 'Historische avonden' uitgegeven door het Historisch Genootschap te Groningen ter gelegenheid van zijn honderjarig bestaan. Hilversum, pp. 183-215.

WATERBOLK, H.T., 1991. Das mittelalterliche Siedlungswesen in Drenthe. Versuch einer Synthese aus archäologischer Sicht. In: H.W. Böhme (Hrsg.), *Siedlungen und Landesausbau zur Salierziet*. I. *In den nördlichen Landschaften des Reiches*. Sigmaringen, pp. 47-108.

WILHELMI, K., 1977. Zur Funktion und Verbreitung dreieckiger Tongewichte der Eisenzeit. *Germania* 55, pp. 179-184.

ZIMMERMANN, W.H., 1976. Die eisenzeitlichen Ackerfluren Type 'Celtic field' von Flögeln-Haselhörn (Kr. Wesermünde). *Probleme der Küstenforschung im südlichen Nordseegebiet* 11, pp. 79-90.

ZIMMERMANN, W.H., 1978. Die Siedlung Flögeln bei Cuxhaven. In: *Sachsen und Angelsachsen* (= Veröff. des Helms Museum Hamburg 32). Harburg, pp. 363-386.

ZIMMERMANN, W.H., 1988. Regelhafte Inngliederung prähistorischer Langhäuser in den Nordseeanrainerstaaten. *Germania* 66, pp. 465-488.

ZIMMERMANN, W.H., 1991. Erntebergung in Rutenberg und Diemen aus archäologischer und volkskundlicher Sicht. *Néprajzi Értesítö* 71-73, pp. 71-104.

ZIMMERMANN, W.H., 1992a. The 'helm' in England, Wales, Scandinavia and North America. *Vernacular Architecture* 23, pp. 34-43.

ZIMMERMANN, W.H., 1992b. Die Siedlungen des 1. bis 6. Jahrhunderts nach Christus von Flögeln-Eekhöltjen, Niedersachsen: Die Bauformen und ihre Funktionen. *Probleme der Küstenforschung im südlichen Nordseegebiet* 19, pp. 8-360.

ZWEERDE, H. VAN DER, 1979. 'k Mag sterven als' t niet waar is. p. 98.

ROMAN IRON AGE PLANT HUSBANDRY AT PEELO, THE NETHERLANDS

W. VAN ZEIST & R.M. PALFENIER-VEGTER

Biologisch-Archaeologisch Instituut, Groningen, Netherlands

ABSTRACT: Rye (*Secale cereale*) and hulled barley (*Hordeum vulgare*) were major crop plants at Roman Iron Age Peelo. Very likely also common oat (*Avena sativa*) was grown. Broomcorn millet (*Panicum miliaceum*) and flax (*Linum usitatissimum*) were of minor importance. No pulse-crop seeds were found. It is not clear whether weed seeds, such as those of *Chenopodium album*, served for human consumption. As usual, weedy vegetations (of cornfields and ruderal habitats) are well represented in the archaeological seed record. Although a fair number of grassland species has been demonstrated, the grassland acreage in the Peelo area was probably of limited extent.

KEYWORDS: Rye, barley, wild food plants, vegetation types, feeding livestock.

1. INTRODUCTION

In this paper the results will be presented of the examination of plant remains recovered from the Late Iron Age/Roman Iron Age settlement on the terrain called 'de Es' at Peelo (for location, see fig. 1). The final archaeological report on the excavations carried out in 1977-1979 has been published by Kooi (1991) in this volume. The publication of the archaeological and botanical examination of other sections of the Peelo settlement site, among which the medieval occupation, is scheduled for the next few years.

The soil samples taken by the excavators for botanical analysis are listed in table 1. As usual the plant remains were retrieved from the samples by means of manual water flotation. The flotation residues were completely or partly sorted for seeds and fruits. Almost all samples are from the fill of post-holes (posts, uprights) and various kinds of pits. The conditions in these features, viz. well-aerated, sandy soil above the groundwater level, allow the preservation of only carbonized plant remains. Non-charred seeds and fruits have been discarded because they must have been due to modern intrusions, e.g. by burrowing animals. In dry, sandy areas such as at Peelo, non-carbonized plant remains are preserved only in the waterlogged fill of wells and such-like. From the medieval settlement several well samples with excellently preserved waterlogged plant remains have been retrieved. The only well sample from 'de Es' (No. 209) is from a thick layer of charcoal that had been dumped in the well after this had fallen into disuse. This sample yielded only carbonized remains, but its composition differs from that of the other samples included in this study (see table 2).

As it appears from table 1, only a few samples date from the Late Iron Age, from the fourth to first century BC. Almost all samples are from Roman Iron Age (first five centuries AD) occupation features. Although for chronological reasons it would have been obvious to treat the Late Iron Age samples first, this is not done here because very little can be said about these samples. Thus, the below discussion is almost wholly devoted to the Roman Iron Age plant remains (as also indicated in the title of this paper).

Table 1 mentions for a number of samples 'no seeds'. Some of these did yield sclerotia of the fungus species *Cenococcum geophilum*. In many samples these durable sclerotia were present, sometimes in considerable quantities, but no attention will be paid here to these remains, the significance of which in an archaeological context is still obscure. From various samples charred stem fragments, presumably of *Calluna vulgaris* (heather), were retrieved.

In addition to the 'no seed' samples, a considerable number of samples turned out to be poor in seeds (the term 'seeds' used here includes cereal grains and other morphologically-defined fruits). In fact, relatively few samples yielded more than twenty seeds. Presentation of analyses of all samples in one or more tables consisting predominantly of zero scores was not considered meaningful. For that reason only the samples relatively rich in plant remains are shown here (table 2). In fact, it is these samples which provide most of the information on the plant husbandry of the site and on which the discussion is largely based. The samples are arranged in four groups. In the samples of the first group cereal grains are most numerous, while in the second group total weed-seed numbers are higher than those of cereals. The samples of the third group are fully dominated by weed seeds. The last group includes a sample of threshing remains and the well sample mentioned above.

Table 3 lists all plant taxa attested for the Roman

Iron Age occupation of 'de Es'. The total numbers of seeds are those actually found; for samples examined in part the numbers of seeds have not been converted for the whole of the sample.

The vegetable remains in the fill of pits and post-holes were there in secondary position. They must have been present in the soil which was shoveled into the pit or post-hole. It is likely that many samples are of mixed origin, that is to say, that they consist of plant remains of more than one provenance. It goes without saying that this is a complicating factor in interpreting the vegetable remains in terms of ancient plant husbandry practices. A few samples are definitely not of mixed origin, e.g. sample 80 (table 2) which represents the remains of a processed rye supply.

2. CULTIVATED PLANTS

The archaeobotanical data (tables 2 and 3) suggest that rye (*Secale cereale*) and hulled barley (*Hordeum vulgare*) were the most important crop plants at first to fifth century AD Peelo. It is interesting to note that rye played such a prominent part. The intentional cultivation of this corn crop could be attested also for occupation

phase IV at Noordbarge (for location, see fig. 1), dated to 100 BC–AD 100 (van Zeist, 1981(1983)). There can be no doubt that in the present-day province of Drenthe, rye cultivation had started already in the first centuries AD and not only in medieval times as was formerly assumed. It is perhaps too far-fetched to see a connection with the Roman presence in continental western Europe. One could, for instance, speculate that in regions outside the occupied territory corn was grown for the provisioning of the Roman army and that, for one reason or another, rye was required. However, transportation of substantial corn supplies from Drenthe to the Roman-occupied territory must have met with serious difficulties, unless a network of overland routes and waterways already existed at that time. Be this as it may, we can only ascertain that in the first centuries AD rye cultivation had been firmly established in Drenthe.

In addition to barley and rye, very probably common oat (*Avena sativa*) was grown by the Peelo farmers. The oat grains themselves provide no clue as to whether the domestic or the wild species (*Avena fatua*) is concerned. *Avena fatua* has been a corn-field weed since Neolithic times and could have been found in the fields of the Peelo farmers together with other field weeds (see below). No oat flower bases, which would have permitted

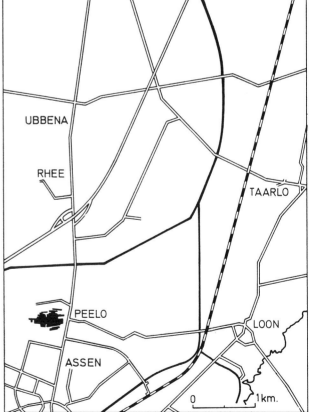

Fig. 1. Map showing the location of Peelo.

Table 1. Peelo, 'de Es' (excavations 1977, 1978, 1979). Samples taken for botanical examination. Dates in centuries AD, unless otherwise stated (Late Iron Age samples).

No.	Date	Context of sample	Remarks
1	3-4	Pit	
9	3-4	Pit	Table 2
16	2-4	Pit	
24	3-4	Pit	No seeds
35	2-3	Oven pit	Table 2
37	1-2	Pit	
41	1-2	Pit in sheep pen	Table 2
43	2-3	Pit	
46	3-4	Wall post	Samples 46-50 from same house
47	3-4	Wall post	
48	3-4	Entrance pit	Table 2
49	3-4	Pit	
50	3-4	Pit	No seeds
56	3-4	Post	Table 2, samples 56-67 from same house
62	3-4	Upright	
66	3-4	Upright	Table 2
67	3-4	Upright	
77	4-5	Pit	
80	3-4	Pit	Table 2
81	3-4	Pit	
82	3-4	Sunken hut	
84	3-4	Upright	Samples 84, 99 and 101 from same house
87	3-2 BC	Upright	Samples 87-92 from same house
88	3-2 BC	Upright	
89	3-2 BC	Wall post	
92	3-2 BC	Wall post	
99	3-4	Upright	
101	3-4	Upright	
111	3-4	Sunken hut	
115	2-4	Post, granary	
116	2-4	Upright	
131	2-3	Pit	No seeds
147	1-2	Pit	
169	2-3	Pit	No seeds
184	3-4	Post	
194	2-3	Pit	No seeds
195	2-4	Pit	
203	1-3	Post, granary?	
204	1-3	Post, granary?	
205	1-3	Post, granary	
209	2-4	Fill of well	Table 2, carbonized remains
228	3-4	Pit	Table 2
248	1-2	Post	Samples 248-250 from same house
249	1-2	Post	No seeds
250	1-2	Wall post	No seeds
257	3-4	Oven pit	Table 2, samples 257, 261 and 268-270 from same house
261	3-4	Upright	
267	3-4	Post	
268	3-4	Post	
269	3-4	Post	
270	3-4	Post	
277	4	Pit	GrN-10694: 1680±35 BP
279	5	Pit with iron slag	
288	3-4	Pit	No seeds
289	3-4	Sunken hut	Table 2
293	3-4	Pit	
295	5	Pit in granary	Table 2
306	3-4	Pit	Table 2
309	5	Pit in granary	
316	2-1 BC	Upright	Samples 316 and 317 from same house
317	2-1 BC	Upright	
328	3-4	Entrance pit	Samples 328-338 and 345-351 from same house, 2-3 periods
332	3-4	Entrance pit	
333	4-5	Wall post	
334	3-5	Upright	
338	3-5	Upright	
339	4-5	Pit	
345	3-5	Upright	
346	3-4	Entrance pit	
351	3-4	Entrance pit	
352	3-4	Pit	Table 2
356	4-5	Wall post	No seeds, samples 356 and 357 from same house
357	4-5	Entrance pit	
361	5	Oven pit	Table 2
364	4-5	Entrance pit	No seeds
366	5	Pit	
383	5	Pit	Table 2
384	5	Wall post	
385	4-5	Upright	Table 2, samples 385 and 386 from same house
386	4-5	Wall post	Table 2
389	5	Upright	
393	4-5	Upright	
399	5	Oven (?) pit	Table 2
435	5	Upright	
470	4-5	Wall post	Samples 470 and 472 from same house
472	4-5	Upright	Table 2
477	3-5	Post	
480	3-5	Pit	No seeds
481	3-5	Pit	No seeds
482	5	Upright	Samples 482 and 518 from same house
488	3-4	Post, barn	
491	3-4	Post, enclosure	No seeds
499	3-4	Upright	
509	3-4	Post	
511	3-4	Post	
512	3-5	Ditch	Table 2
513	5	Wall post	Samples 513-517 are from a 'smithy'
514	5	Wall post	No seeds
515	5	Wall post	
517	5	Fire place	No seeds
518	5	Upright	Table 2

a species determination, have been secured. As in a few samples the proportion of oat among the cereal grains is quite large, it is assumed here that *Avena sativa* formed part of the crop-plant assortment. Admittedly, there is no absolute proof for this assumption.

The minimal representation of emmer wheat (*Triticum dicoccum*) and bread wheat (*Triticum aestivum*) (see table 3) suggests that wheat was not intentionally grown here during the Roman Iron Age. Although emmer wheat was no longer a crop of major importance as it had been during the Neolithic and Bronze Age, it was still widely cultivated in continental western and northwestern Europe in the first half of the first millennium AD. It looks if at Peelo the cultivation of emmer wheat had been abandoned in favour of that

W. van Zeist & R.M. Palfenier-Vegter

Table 2. Numbers of seeds and fruits in selected Roman Iron Age samples from Peelo ('de Es'). Of *Calluna* stems only the presence is indicated. I. Predominantly cereal grains; II. More weed seeds than cereal grains; III. Weed seeds by far predominant; IV. Rest group. Plant nomenclature in accordance with *Heukels' Flora van Nederland* (van der Meijden, 1990).

	I								II			
Sample number	9	80	352	383	385	386	512	518	35	228	257	295
Part examined	1/1	1/2	1/2	1/2	1/1	1/1	1/1	1/2	1/1	1/1	1/1	1/2
Avena (*sativa*)	.	.	19	1	.	.	1	.	1	.	9	3
Hordeum vulgare	12	.	50	.	1	1	10	.	62	7	14	31
Hordeum rachis internodes	.	.	1	.	.	.	.	.	.	.	.	.
Secale cereale	120	28	21	32	430	46	14	23	11	20	.	6
Secale rachis internodes	.	.	.	.	2	.	.	.	.	.	.	.
Cereal grain fragments (in grams)	0.14	.	0.35	0.23	0.92	0.08	0.03	+	0.06	.	0.06	0.05
Panicum miliaceum	.	.	.	.	.	.	.	.	3	1	.	.
Linum usitatissimum	.	.	.	.	.	.	.	.	2	.	.	.
Corylus avellana	.	.	.	.	.	.	.	.	.	+	.	.
Rubus spec.	.	.	.	.	.	.	.	.	1	.	.	.
Vaccinium myrtillus	.	.	.	.	.	.	.	2	.	.	.	.
Chenopodium album	1	.	.	1	5	.	.	.	200	40	5	28
Polygonum convolvulus	32	.	.	.	.	.	.	.	1	.	1	1
Polygonum lapathifolium	18	.	2	1	7	.	13	2	11	2	33	66
Rumex acetosella	.	.	10	3	48	7	.	5	.	.	12	.
Spergula arvensis	.	.	6	3	5	1	.	1	.	1	18	100
Vicia spec.	34	.	16	1	10	.	.	.	.	20	5	1
Agrostis spec.	.	.	.	.	.	.	.	.	.	.	.	1
Apera spica-venti	.	.	.	.	.	.	.	.	.	.	.	.
Bromus secalinus	.	.	1	.	.	.	.	.	.	.	.	.
Carex cuprina	.	.	.	.	.	.	.	.	.	.	.	.
Carex flacca	.	.	.	.	.	.	.	.	.	.	.	.
Carex nigra type	.	.	.	.	7	.	.	.	.	.	.	.
Carex oederi	.	.	.	.	1	.	.	.	.	.	.	.
Carex panicea	.	.	.	1	.	1	.	.	.	.	.	.
Carex rostrata/vesicaria	.	.	.	.	.	.	.	.	.	.	1	.
Carex spec.	.	.	.	.	1	.	.	.	.	.	2	.
Claviceps spec.	.	.	.	.	.	.	.	.	.	.	.	.
Conium maculatum	.	.	.	.	.	.	.	.	.	.	.	.
Cuscuta spec.	.	.	.	.	.	.	.	.	.	.	.	.
Echinochloa crus-galli	.	.	.	.	.	.	.	.	.	.	.	.
Eleocharis multicaulis	.	.	.	.	.	.	.	.	.	.	.	.
Eleocharis palustris	.	.	.	.	.	.	.	.	.	.	.	.
Euphrasia type	.	.	.	.	.	.	.	.	.	.	.	.
Festuca pratensis	.	.	.	.	.	.	.	.	.	.	.	.
Galeopsis tetrahit/speciosa	.	.	.	.	.	.	.	.	.	.	.	.
Galium palustre	.	.	.	.	.	.	.	.	.	.	.	.
Galium spec.	.	.	.	.	.	.	.	.	1	.	.	.
Gramineae indet.	.	.	5	.	.	.	.	.	.	.	.	.
Knautia arvensis	1	.	.	.	.	.	.	.	.	.	.	.
Matricaria maritima	.	.	.	.	.	.	.	.	.	.	.	.
Malva spec.	.	.	.	.	.	.	.	.	.	.	.	.
Phleum pratense	.	.	.	.	.	.	.	.	.	.	.	.
Plantago lanceolata	.	.	.	.	.	.	.	.	1	.	.	.
Poa annua	.	.	.	.	1	.	.	.	.	.	.	.
Poa pratensis/trivialis	.	.	.	.	.	.	.	.	.	.	.	.
Polygonum aviculare	.	.	.	.	2	.	.	.	.	.	.	.
Polygonum hydropiper	.	.	.	.	.	.	.	.	.	.	.	.
Polygonum persicaria	1	.	.	.	2	1	.	.	.	.	.	.
Potentilla erecta	.	.	.	.	1	.	.	.	.	.	.	.
Prunella vulgaris	.	.	.	.	.	.	.	.	.	.	.	.
Ranunculus flammula	.	.	.	.	.	.	.	.	.	.	.	.
Ranunculus repens	.	.	.	.	.	.	.	.	.	.	.	.
Ranunculus sardous	.	.	.	.	.	.	.	.	.	.	.	.
Raphanus raphanistrum	.	.	.	.	.	.	.	.	.	.	2	.
Rhinanthus spec.	.	.	.	.	.	.	.	.	.	.	.	.
Rumex crispus	.	.	.	.	.	.	.	.	.	.	.	.
Rumex obtusifolius	1	.	.	.	.	.	.	.	1	.	.	.
Scleranthus annuus	.	.	.	.	.	.	.	.	.	.	1	.

II			III					IV		
306	361	399	48	56	66	289	472	41	209	Sample number
1/1	1/3	1/10	1/1	1/1	1/1	1/3	1/1	1/1	1/1	Part examined
6	1	10	.	.	.	3	2	.	1	*Avena (sativa)*
4	6	19	.	1	4	11	3	1	5	*Hordeum vulgare*
.	.	.	.	.	.	4	.	47	.	*Hordeum* rachis internodes
.	40	32	5	1	1	.	.	2	1	*Secale cereale*
.	.	.	.	.	.	.	.	.	.	*Secale* rachis internodes
+	0.09	0.31	.	.	0.02	0.05	0.02	.	.	Cereal grain fragments (in grams)
.	.	.	.	.	.	.	.	.	.	*Panicum miliaceum*
1	.	.	.	.	.	.	.	.	.	*Linum usitatissimum*
.	+	.	.	.	.	.	.	.	.	*Corylus avellana*
.	.	.	.	.	.	.	.	.	.	*Rubus* spec.
.	.	.	.	.	.	.	.	.	.	*Vaccinium myrtillus*
10	9	10	28	15	2	560	16	.	40	*Chenopodium album*
.	.	1	.	.	.	.	.	.	.	*Polygonum convolvulus*
4	15	40	72	22	20	70	50	1	12	*Polygonum lapathifolium*
24	40	15	.	65	18	70	1	.	10	*Rumex acetosella*
3	15	9	1	64	21	15	2	.	12	*Spergula arvensis*
.	1	8	.	1	.	.	.	.	.	*Vicia* spec.
.	.	190	.	.	.	.	.	.	.	*Agrostis* spec.
.	1	14	.	.	.	1	.	.	.	*Apera spica-venti*
.	.	.	.	.	.	.	.	.	.	*Bromus secalinus*
.	.	.	.	.	.	.	.	.	13	*Carex cuprina*
.	.	.	.	.	.	.	.	.	3	*Carex flacca*
.	.	.	.	1	.	.	1	.	5	*Carex nigra* type
.	.	.	.	.	.	.	.	.	.	*Carex oederi*
.	.	.	.	.	1	.	.	.	1	*Carex panicea*
.	.	.	.	.	.	.	1	.	2	*Carex rostrata/vesicaria*
.	.	.	.	.	.	.	.	.	4	*Carex* spec.
.	1	.	.	.	.	.	.	.	.	*Claviceps* spec.
.	.	.	.	.	.	.	.	.	1	*Conium maculatum*
2	.	.	.	.	.	.	.	.	.	*Cuscuta* spec.
.	.	.	.	.	.	2	.	.	1	*Echinochloa crus-galli*
.	.	.	.	.	.	.	.	.	2	*Eleocharis multicaulis*
.	.	.	.	.	.	.	.	.	4	*Eleocharis palustris*
.	.	.	.	.	.	.	1	.	.	*Euphrasia* type
.	.	.	.	.	.	.	.	.	1	*Festuca pratensis*
1	1	.	.	1	.	.	.	.	.	*Galeopsis tetrahit/speciosa*
.	.	.	.	.	.	.	.	.	1	*Galium palustre*
.	.	.	.	.	.	.	.	.	.	*Galium* spec.
2	5	.	.	.	.	.	.	.	5	Gramineae indet.
.	1	.	.	.	.	.	.	.	.	*Knautia arvensis*
.	.	5	.	.	.	.	.	.	.	*Matricaria maritima*
.	.	.	.	.	.	.	.	.	1	*Malva* spec.
.	3	.	.	.	.	.	.	.	2	*Phleum pratense*
.	.	.	.	.	.	.	.	.	2	*Plantago lanceolata*
.	.	.	.	.	.	.	.	.	.	*Poa annua*
.	2	40	.	.	.	.	.	.	.	*Poa pratensis/trivialis*
.	.	.	.	1	3	.	1	.	11	*Polygonum aviculare*
.	.	.	.	.	2	.	.	.	.	*Polygonum hydropiper*
.	.	.	.	.	.	.	1	.	.	*Polygonum persicaria*
.	.	.	.	.	.	1	.	.	5	*Potentilla erecta*
.	.	.	.	.	.	.	.	.	1	*Prunella vulgaris*
.	.	.	.	.	.	.	.	.	1	*Ranunculus flammula*
.	.	.	.	.	.	.	.	.	6	*Ranunculus repens*
.	.	.	.	.	1	.	.	.	.	*Ranunculus sardous*
.	.	.	.	.	.	.	.	.	.	*Raphanus raphanistrum*
.	.	.	.	.	.	.	.	.	1	*Rhinanthus* spec.
.	.	.	.	.	.	.	.	.	.	*Rumex crispus*
.	1	.	.	.	.	.	.	.	8	*Rumex obtusifolius*
.	.	.	.	.	.	.	.	.	.	*Scleranthus annuus*

Table 2 (Continued).

			I								II			
Senecio cf. *aquaticus*	.	.	1	.	.	.	.	.		.	.	.	.
Solanum nigrum	.	.	.	.	.	.	.	.		.	.	.	.
Trifolium spec.	.	.	.	.	.	.	.	.		.	.	.	.
Umbelliferae indet.	.	.	.	.	.	.	.	.		.	.	.	.
stems (*Calluna*)	.	.	.	.	+	.	.	.		.	.	.	.

of rye. Broomcorn millet (*Panicum miliaceum*) is conspicuously scarce, suggesting that it was at most occasionally grown, this in contrast to Noordbarge, where millet must have been a common crop plant in the period of 100 BC to AD 100 (van Zeist, 1981(1983)).

Although for the oleaginous linseeds chances of becoming preserved in a carbonized condition are much smaller than for cereal grains, the poor representation of flax (*Linum usitatissimum*) in the seed record suggests that at Peelo this species was at most a crop of minor importance. It may be remembered here that in contemporary settlements in the coastal marshes in the north of the Netherlands there is convincing evidence of flax cultivation (van Zeist, 1974).

Pulse-crop seeds were not found, although one should consider the possibility that Celtic bean (*Vicia faba* var. *minor*) was grown by the Peelo farmers. This species is reported for pre-Roman and Roman Iron Age sites in continental western Europe (cf. Bakels, 1991).

3. (POTENTIAL) WILD FOOD PLANTS

Remains of wild fruits and nuts are extremely scarce. This could in part be due to the fact that in general nuts and fruits are poorly represented in a carbonized condition. In this respect waterlogged archaeological deposits offer better prospects. However, the collecting of wild fruits and nuts may have been of only little economic importance at Roman Iron Age Peelo; early-medieval waterlogged well deposits from this site did not yield evidence of intensive wild fruit collecting, either.

In table 2, typical weeds (cf. table 3 and discussion in section 4), which in one or more samples are represented by comparatively great numbers of seeds, are grouped together (*Chenopodium album* up to *Vicia* spec.). How should one interpret these field-weed seeds? In the samples of group one, with predominantly cereal grains, the weed seeds could be considered contaminants of the corn crop that had not yet been cleaned of impurities. In the samples of groups two and three the numbers of weed seeds are higher and very much higher, respectively, than those of cereal grains. Here it looks as if the weed seeds, after they had been removed from the crop, had been set apart to be used for some purpose. One could think of crop-cleaning waste destined

to be thrown away or fed to the animals. However, typical threshing remains, such as rachis internode fragments, are absent from these samples. The only sample with predominantly rachis internode remains (No. 41) yielded just one weed seed. Apparently the weed seeds had been separated from other crop processing waste products. One wonders whether the starch-rich weed seeds could have served for human consumption, as famine food after crop failure. In this connection reference is made to the finds of weed-seed supplies in Iron Age sites in Denmark (Helbæk, 1951, 1960). Almost pure supplies of *Chenopodium album*, *Spergula arvensis* and *Polygonum lapathifolium* are reported. It is suggested that the weed seeds were harvested particularly on fallow fields. Another explanation of relatively large numbers of *Chenopodium album* seeds is that not the seeds but the green parts of the plants were used (as vegetable). Prior to food preparation, the seeds were removed (e.g. by beating or lashing), which would account for the many *Chenopodium album* seeds in settlement sites (Knörzer, 1973).

It remains a matter of speculation whether the relatively high weed seed proportions at Peelo are evidence of the exploitation of these wild food resources for human consumption. Admittedly, a more than occasional necessity to supplement the corn crop with wild weed seeds is difficult to reconcile with the suggestion that the Peelo farmers may have produced grain for a Roman market.

4. SPECULATIONS ON FARMING PRACTICES

One of the questions to which archaeobotanical research is expected to give an answer is that of the season of sowing. Depending upon the time of sowing (and tillage of the fields), so-called winter (autum-sown) and summer (spring-sown) cereals are distinguished. As for the main cereal species attested for Roman Iron Age Peelo, rye is at present a predominantly winter cereal, but spring-sown varieties do occur. Oats, on the other hand, is a typical summer cereal. Of barley, both autumn-sown and spring-sown varieties occur. Broomcorn millet and linseed (flax) are spring-sown.

From the above one may conclude that at least some of the fields were sown in the spring. However, were rye

Table 2 (Continued).

II										
			III				IV			
.	.	.	.	.	.	.	.	.	.	*Senecio* cf. *aquaticus*
.	.	.	.	.	.	.	.	.	2	*Solanum nigrum*
.	.	.	.	.	.	.	.	.	28	*Trifolium* spec.
.	.	.	.	.	.	.	.	.	1	Umbelliferae indet.
.	.	.	+	.	.	.	.	.	+	stems (*Calluna*)

Table 3. Ecological affinity of the plant taxa attested for Roman Iron Age Peelo ('de Es'). In the left column the total numbers of seeds actually recovered are shown. An asterisk indicates that more than half of the seeds were found in one sample. 1. Cultivated plants; 2. Wild fruits; 3. Weeds of winter cereals; 4. Weeds of summer cereals and root crops; 5. Weeds of ruderal habitats; 6. Grassland species; 7. Plants of marshy habitats; 8. Plants of heathlands.

Number of seeds etc.		1	2	3	4	5	6	7	8
63	*Avena* (*sativa*)	x	.	.	.	.	.	.	.
323	*Hordeum vulgare*	x	.	.	.	.	.	.	.
59*	*Hordeum* internodes	x	.	.	.	.	.	.	.
878	*Secale cereale*	x	.	.	.	.	.	.	.
4	*Secale* internodes	x	.	.	.	.	.	.	.
1	*Triticum aestivum*	x	.	.	.	.	.	.	.
2	*Triticum dicoccum*	x	.	.	.	.	.	.	.
5	*Panicum miliaceum*	x	.	.	.	.	.	.	.
4	*Linum usitatissimum*	x	.	.	.	.	.	.	.
+	*Corylus avellana*	.	x	.	.	.	.	.	.
1	*Rubus* spec.	.	x	.	.	.	.	.	.
2	*Vaccinium myrtillus*	.	x	.	.	.	.	.	.
37*	*Polygonum convolvulus*	.	.	x	.	.	.	.	.
2	*Scleranthus annuus*	.	.	x	.	.	.	.	.
16	*Apera spica-venti*	.	.	x	.	.	.	.	.
2	*Arnoseris minima*	.	.	x	.	.	.	.	.
1	*Bromus secalinus*	.	.	x	.	.	.	.	.
105	*Vicia* spec.	.	.	x	.	.	.	.	.
372	*Rumex acetosella*	.	.	x	x	.	.	.	.
340	*Spergula arvensis*	.	.	x	x	.	.	.	.
4	*Raphanus raphanistrum*	.	.	x	x	.	.	.	.
4	*Galeopsis tetrahit/speciosa*	.	.	x	.	x	.	.	.
5	*Echinochloa crus-galli*	.	.	.	x	.	.	.	.
8	*Polygonum persicaria*	.	.	.	x	.	.	.	.
1	*Setaria viridis*	.	.	.	x	.	.	.	.
1	*Anagallis arvensis*	.	.	.	x	.	.	.	.
1006*	*Chenopodium album*	.	.	.	x	x	.	.	.
547	*Polygonum lapathifolium*	.	.	.	x	x	.	.	.
5	*Matricaria maritima*	.	.	.	x	x	.	.	.
4	*Solanum nigrum*	.	.	.	x	x	.	.	.
1	*Stellaria media*	.	.	.	x	x	.	.	.
7	*Polygonum hydropiper*	.	.	.	x	.	.	x	.
1	*Conium maculatum*	.	.	.	.	x	.	.	.
3	*Poa annua*	.	.	.	.	x	.	.	.
23	*Polygonum aviculare*	.	.	.	.	x	.	.	.
16	*Rumex obtusifolius*	.	.	.	.	x	.	.	.
2	*Ranunculus sardous*	.	.	.	.	?	.	.	.
1	*Rumex crispus*	.	.	.	.	x	x	.	.
193*	*Agrostis* spec.	.	.	.	.	.	x	.	.
1	*Senecio* cf. *aquaticus*	.	.	.	.	.	x	.	.
1	*Euphrasia* type	.	.	.	.	.	x	.	.
2	*Knautia arvensis*	.	.	.	.	.	x	.	.
1	*Malva* spec.	.	.	.	.	.	x	.	.
5	*Phleum pratense*	.	.	.	.	.	x	.	.
3	*Plantago lanceolata*	.	.	.	.	.	x	.	.
42*	*Poa pratensis/trivialis*	.	.	.	.	.	x	.	.
1	*Festuca pratensis*	.	.	.	.	.	x	.	.

Table 3 (Continued).

Number of seeds etc.		1	2	3	4	5	6	7	8
1	*Prunella vulgaris*	.	.	.	.	.	x	.	.
6	*Ranunculus repens*	.	.	.	.	.	x	.	.
1	*Rhinanthus* spec.	.	.	.	.	.	x	.	.
28*	*Trifolium* spec.	.	.	.	.	.	x	.	.
3	*Carex flacca*	.	.	.	.	.	x	.	.
16	*Carex nigra* type	.	.	.	.	.	x	x	.
5	*Carex panicea*	.	.	.	.	.	x	.	x
8	*Potentilla erecta*	.	.	.	.	.	x	.	x
7	*Carex rostrata/vesicaria*	.	.	.	.	.	.	x	.
13	*Carex cuprina*	.	.	.	.	.	.	x	.
1	*Carex oederi*	.	.	.	.	.	.	x	.
1	*Galium palustre*	.	.	.	.	.	.	x	.
1	*Ranunculus flammula*	.	.	.	.	.	.	x	.
5	*Eleocharis palustris*	.	.	.	.	.	.	x	.
2	*Eleocharis multicaulis*	.	.	.	.	.	.	.	x
+	*Calluna vulgaris*	.	.	.	.	.	.	.	x
7	*Carex* spec.								
1	Chenopodiaceae indet.								
1	*Claviceps* spec.								
2	*Cuscuta* spec.								
1	*Galium* spec.								
19	Gramineae indet.								
1	Umbelliferae indet.								

Table 4. Total numbers of seeds etc. recovered from six Late Iron Age samples (Nos 87, 88, 89, 82, 316, 317).

Hordeum vulgare	2
Triticum dicoccum glume base	1
Panicum miliaceum	1
cf. *Brassica*	1
Carex oederi	1
Carex pseudocyperus	1
Carex rostrata/vesicaria	1
Chenopodium album	10
Chenopodiaceae indet.	1
Echinochloa crus-galli	3
Poa pratensis/trivialis	1
Polygonum aviculare	1
Polygonum convolvulus	3
Polygonum hydropiper	3
Polygonum lapathifolium	132
Polygonum persicaria	2
Rumex acetosella	2
Solanum nigrum	7
Spergula arvensis	16
Stellaria media	7
stems (*Calluna*)	+

and perhaps part of the barley sown in the autumn? In principle, arable weeds should be informative in this respect. Among these weeds, those of winter-corn and summer-corn fields are distinguished. As already appears from table 3, this differentiation is not particularly clear-cut. Quite a lot of weeds grow in various habitats. Although for Peelo only a rather small number of weeds

typical of autumn-sown cornfields have been attested, one may assume that winter cereals were indeed cultivated.

Rachis internode remains are almost absent in the samples with comparatively large numbers of cereal grains (table 2), suggesting that the grains are of corn supplies that had been threshed and cleaned of threshing waste. Only sample 47 (table 2) represents threshing waste, namely of barley. It is tempting to assume that the corn crop had been stored in a threshed condition.

The great numbers of caryopses ('seeds') of the perennial grasses *Agrostis* and *Poa pratensis/trivialis* in sample 399 (table 2) make one wonder whether this sample includes the remains of a hay supply. However, other grassland species are not represented in this sample. For that reason one could speculate that both grasses occurred as weeds of arable. The comparatively slight disturbance of the soil by ard ploughing, as was practised at the time, did not eliminate perennial weeds as happened with mouldboard ploughing. The other wild grass rather well represented in sample 399 is *Apera spica-venti*; this annual grass is characteristic of winter-corn fields.

The above brings us to the question of the feeding of the livestock which, judging from the size of the farmhouses, must have been quite considerable. We have no information as to how much pasture land (grassland, probably of poor quality, and heathland) may have been available, but one may safely assume that tree foliage was a major animal fodder. The gathering of leaf fodder, i.e. the cutting and drying of the branches of deciduous trees, must have been an important economic activity. The potential importance of *Calluna* heath in the feeding of livestock should not be underestimated. From medieval times until a hundred years ago and even more recently heather played a key role in the agricultural economic system on poor acid soils of Atlantic West and Northwest Europe. If properly managed it provides a suitable fodder for cattle and sheep (Kaland, 1986). One could speculate that the charred *Calluna* stems recovered from the Peelo samples were of heather that had been stored as winter fodder. However, heather could have been used for other purposes, too. It is generally assumed that at the time, in the first centuries AD, heaths were still of limited extent and consequently they could at most have been of secondary importance as grazing land.

5. SOME REMARKS ON THE VEGETATION

In the previous section mention has already been made of differences between the weed communities in fields of winter and summer cereals. Another group of weedy vegetations includes those of ruderal habitats, such as waste places in farmyards, muck-heaps and roadsides. As appears from table 3, various species of ruderal habitats are also found in fields of summer cereals and root crops, among which two species that are particularly well represented in the Peelo seed record, viz. *Chenopodium album* and *Polygonum lapathifolium*. Curiously, the *Atriplex prostrata/patula* seed type, which includes two species common in fields and other disturbed habitats, has not been attested for Roman Iron Age Peelo. It is not clear whether orache (*Atriplex*) was not present at or close to the site or whether its absence in the seed record is more a matter of lack of preservation in a carbonized condition. Fairly characteristic of, but not confined to trodden places are *Polygonum aviculare* and *Poa annua*. The poisonous species *Conium maculatum* is better represented in late-prehistoric sites than one would expect on the basis of its present occurrence.

Various species attested for Peelo are nowadays predominantly found in grasslands. It has already been discussed that *Agrostis* and *Poa pratensis/trivialis* may (also) have occurred in cornfields. *Plantago lanceolata* and *Knautia arvensis* may formerly have been elements of the arable weed flora (Groenman-van Waateringe, 1986; Pals, 1987). It is not the ecology of the species that had changed but the ecological conditions in the fields. Except for coastal areas, where a kind of natural grassland vegetation occurs on the high salt marshes, in the greater part of Europe meadows and pastures are anthropogenic, that is to say, their origin and continued existence are due to the activity of man. Grasslands have replaced forest vegetations. In prehistoric Europe, synanthropic grasslands were at most of limited extent and consequently could have played only a minor part

in the economy (for a review, see van Zeist, 1991). On the other hand, in Roman-occupied Germany, and this applies probably also to the part of the Netherlands under Roman authority, the grassland acreage had expanded considerably, which is associated with an increased emphasis on stock-breeding (provisioning of the Roman army with meat) and with the procurement of grazing land and hay for the Roman cavalry horses.

In the Peelo area, only the wet to moist stream valleys naturally covered by alder and willow carr could have been converted to rather extensive grasslands (hay meadows) of not too poor a quality. However, it is not likely that large-scale cutting of the stream-valley forest had started already in the first centuries AD. This probably did not happen until medieval times. Grassland vegetation in the vicinity of Roman Iron Age Peelo must have been of limited extent, consisting predominantly of rather dry, poor pasture land, only suitable for grazing. Around bog sites (see below) patches of a moister type of 'grassland', with sedge (*Carex*) species, may have occurred.

A modest number of species characteristic of marshy habitats could be demonstrated for Roman Iron Age Peelo. Marsh vegetation was found in small peat-bogs in and around the settlement terrain (cf. Bardet et al., 1983: fig. 3). It is striking that almost all species listed here as marsh plants are represented in sample 209 from the fill of a well (table 2). Although the vegetable remains in this sample are carbonized, the composition compares with that of waterlogged (i.e. non-carbonized) well samples, viz. a variety of species from divergent habitats.

A few species of dry heath (*Calluna vulgaris*, *Potentilla erecta*) and damp heath (*Carex panicea*) are represented, while *Eleocharis multicaulis* is found in oligotrophic pools such as occur in heathlands. Various samples yielded charred stem remains of *Calluna*, sometimes in considerable numbers. As has been mentioned above, no estimate of the heathland acreage can be given.

In addition to the open vegetations discussed above, oak-dominated woodland (open forest) must have been present in the Peelo area. Oak was the main building timber. Hazel (*Corylus avellana*), bramble (*Rubus* spec.) and bilberry (*Vaccinium myrtillus*) are woodland and forest-edge species.

6. THE LATE IRON AGE PLANT REMAINS

The number of Late Iron Age (fourth to first century BC) samples retrieved from 'de Es' is too small to allow a satisfactory evaluation. The total numbers of seeds recovered from six Late Iron Age samples are shown in table 4. *Calluna* stems are fairly numerous, suggesting that heather was exploited for some purpose or other. *Secale* (rye) has not been recorded, which may indicate that this corn crop was not introduced until the beginning of the Christian era. Comparatively great numbers of seeds of some weedy species make one again wonder whether the seeds concerned were collected on purpose to serve for human consumption (cf. section 3).

The co-operation of Mrs. G. Entjes-Nieborg, Mrs. S.M. van Gelder-Ottway and Dr. P.B. Kooi in the preparation of the publication is gratefully acknowledged.

7. REFERENCES

BAKELS, C.C., 1991. Western Continental Europe. In: W. van Zeist, K. Wasylikowa & K.-E. Behre (eds), *Progress in Old World palaeoethnobotany*. Rotterdam etc., pp. 279-298.

BARDET, A.C., P.B. KOOI, H.T. WATERBOLK & J. WIERINGA, 1983. *Peelo, historisch-geographisch en archeologisch onderzoek naar de ouderdom van een Drents dorp* (= Mededelingen der Koninklijke Nederlandse Akademie van Wetenschappen, afd. Letterkunde, Nieuwe Reeks 46, No. 1). Amsterdam.

GROENMAN-VAN WAATERINGE, W., 1986. Grazing possibilities in the Netherlands based on palynological data. In: K.-E. Behre (ed.), *Anthropogenic indicators in pollen diagrams*. Rotterdam etc., pp. 187-202.

HELBÆK, H., 1951. Ukrudsfrø som naeringsmiddel i førromersk Jernalder. *Kuml* 1951, pp. 65-74.

HELBÆK, H., 1960. Comment on *Chenopodium album* as a food plant in prehistory. *Ber. Geobotan. Forschungsinst. Rübel, Zürich* 31, pp. 16-19.

KALAND, P.E., 1986. The origin and management of Norwegian coastal heaths as reflected by pollen analysis. In: K.-E. Behre (ed.), *Anthropogenic indicators in pollen diagrams*. Rotterdam etc., pp. 19-36.

KNÖRZER, K.-H., 1973. Pflanzliche Grossreste. In: J.-P. Farruggia, R. Kuper, J. Lüning & P. Stehli (eds), *Der Bandkeramische Siedlungsplatz Langweiler 2* (= Rheinische Ausgrabungen 13). Bonn, pp. 139-152.

KOOI, P.B., 1991. Project Peelo. Het onderzoek in de jaren 1977, 1978 en 1979 op de es. *Palaeohistoria* 33/34.

MEIJDEN, R. VAN DER, 1990. *Heukels' Flora van Nederland*. 21e druk. Groningen.

PALS, J.P., 1987. Reconstruction of landscape and plant husbandry. In: W. Groenman-van Waateringe & L.H. van Wijngaarden-Bakker (eds), *Farm life in a Carolingian village*. Assen etc., pp. 52-96.

ZEIST, W. VAN, 1974. Palaeobotanical studies of settlement sites in the coastal area of the Netherlands. *Palaeohistoria* 16, pp. 223-371.

ZEIST, W. VAN, 1981(1983). Plant remains from Iron Age Noordbarge, province of Drenthe, the Netherlands. *Palaeohistoria* 23, pp. 169-193.

ZEIST, W. VAN, 1991. Economic aspects. In: W. van Zeist, K. Wasylikowa & K.-E. Behre (eds), *Progress in Old World palaeoethnobotany*. Rotterdam etc., pp. 109-130.

APPENDIX. English and Dutch names of wild plant taxa attested for Iron Age Peelo.

Agrostis spec.	Bent-grass	Struisgras
Anagallis arvensis	Scarlet pimpernel	Gewoon guichelheil
Apera spica-venti	Loose silky-bent	Windhalm
Arnoseris minima	Lamb's succory	Korensla
Bromus secalinus	Chess	Dreps
cf. *Brassica*	Cabbage/mustard	Kool/mosterd
Calluna vulgaris	Heather	Struikheide
Carex cuprina	False fox-sedge	Valse voszegge
Carex flacca	Glaucous sedge	Zeegroene zegge
Carex nigra type	Common sedge	Gewone zegge
Carex oederi (*C. demissa/serotina*)	Yellow sedge (*Carex flava* agg.)	Geelgroene zegge/dwergzegge
Carex panicea	Carnation sedge	Blauwe zegge
Carex pseudocyperus	Cyperus sedge	Cyperzegge
Carex rostrata/vesicaria	Bottle sedge/bladder sedge	Snavel-/blaaszegge
Carex spec.	Sedge	Zegge
Chenopodiaceae indet.	Goosefoot family	Ganzevoetfamilie
Chenopodium album	Fat hen	Melganzevoet
Claviceps spec.	Ergot	Moederkoren
Conium maculatum	Hemlock	Gevlekte scheerling
Corylus avellana	Hazel	Hazelaar
Cuscuta spec.	Dodder	Warkruid
Echinochloa crus-galli	Cockspur grass	Hanepoot
Eleocharis multicaulis	Many-stemmed spike-rush	Veelstengelige waterbies
Eleocharis palustris	Common spike-rush	Gewone waterbies
Euphrasia type	Eyebright	Ogentroost
Festuca pratensis	Meadow fescue	Beemdlangbloem
Galeopsis tetrahit/speciosa	Common/large-flowered hempnettle	Bleekgele hennepnetel/dauwnetel
Galium palustre	Common marsh-bedstraw	Moeraswalstro
Galium spec.	Bedstraw	Walstro
Gramineae indet.	Grass family	Grassenfamilie
Knautia arvensis	Field scabious	Beemdkroon
Matricaria maritima	Scentless mayweed	Reukeloze kamille
Malva spec.	Mallow	Kaasjeskruid
Phleum pratense	Timothy grass	Timoteegras
Plantago lanceolata	Ribwort plantain	Smalle weegbree
Poa annua	Annual meadow-grass	Straatgras
Poa pratensis/trivialis	Meadow grass/rough meadow-grass	Veldbeemdgras/ruw beemdgras

Polygonum aviculare	Knotgrass	Varkensgras
Polygonum convolvulus	Black bindweed	Zwaluwtong
Polygonum hydropiper	Water-pepper	Waterpeper
Polygonum lapathifolium	Pale persicaria	Knopige/viltige duizendknoop
Polygonum persicaria	Persicaria	Perzikkruid
Potentilla erecta	Common tormentil	Tormentil
Prunella vulgaris	Self-heal	Brunel
Ranunculus flammula	Lesser spearwort	Egelboterbloem
Ranunculus repens	Creeping buttercup	Kruipende boterbloem
Ranunculus sardous	Hairy buttercup	Behaarde boterbloem
Raphanus raphanistrum	Wild radish	Knopherik
Rhinanthus spec.	Yellow-rattle	Ratelaar
Rubus spec.	Bramble	Braam
Rumex acetosella	Sheep's sorrel	Schapezuring
Rumex crispus	Curled dock	Krulzuring
Rumex obtusifolius	Broad-leaved dock	Ridderzuring
Scleranthus annuus	Annual knawel	Eenjarige hardbloem
Senecio cf. *aquaticus*	Marsh ragwort	Waterkruiskruid
Setaria viridis	Green bristle-grass	Groene naaldaar
Solanum nigrum	Black nightshade	Zwarte nachtschade
Spergula arvensis	Corn spurrey	Gewone spurrie
Stellaria media	Chickweed	Vogelmuur
Trifolium spec.	Clover	Klaver
Vaccinium myrtillus	Bilberry	Bosbes
Vicia spec.	Vetch	Wikke
Umbelliferae indet.	Carrot family	Schermbloemenfamilie

NIEUWE GEGEVENS BETREFFENDE DE MUNTVONDST VAN MIDLUM VAN 1925

A. UFKES

Biologisch-Archaeologisch Instituut, Groningen, Netherlands

ABSTRACT: In 1925 a small hoard of 5th/6th century Byzantine solidi was found near Midlum in Friesland. This hoard has been published several times (Boeles, 1927; 1951; Zadoks-Josephus Jitta, 1960; Van der Vin, 1992). Recently it was discovered that the archives of the B.A.I. and of Mr. P.C.J.A. Boeles contain additional information regarding the number of coins, the find circumstances and the way in which the majority of these coins were acquired by the Fries Museum. In particular, the role of Dr. A.E. van Giffen, director of the B.A.I., was most peculiar.

The hoard was found in the *terp* Middelstein, west of Midlum, during waterpipe construction works. The coins were sold by the finder, partly through an intermediate, to the Fries Museum (2), to Dr. van Giffen (12) and to a jeweller in Harlingen (1). It is possible, however, that coins were sold to other people, as well. Apart from that it is likely that the intermediate got one or more coins for his services. Van Giffen sold 11 of his coins to the Fries Museum, on the condition that 2 'duplicates' were returned to him. What happened to the three coins in Van Giffen's possession, is unknown. The Fries Museum also acquired the coin bought by the jeweller. This means that the hoard consisted of at least 15, but probably more, coins of which 12 are in the possession of the Fries Museum.

An interview with the son of the jeweller revealed that his father had bought a small ceramic vessel together with the solidus. This pot was still owned by the family, but presented to the author on behalf of the Fries Museum. It is likely that this was in fact the container in which the solidi were hidden during the 6th century.

KEYWORDS: Netherlands, Friesland, Early Middle Ages, *terp*, hoard, Byzantine solidi.

1. INLEIDING

Bij het doorbladeren van het brievenarchief van het Biologisch-Archaeologisch Instituut te Groningen in het voorjaar van 1990, op zoek naar gegevens over vondsten uit het kanaal Buinen-Schoonoord, viel mijn oog op een brief uit 1926 van P. Helfrich te Wijnaldum over gouden munten. Vanwege de connectie met Wijnaldum, waar later dat jaar opgravingen zouden plaatsvinden, was de belangstelling direct gewekt. Bij verder speuren in het archief bleek het echter te gaan om de bekende vondst van 5e/6e-eeuwse solidi uit Midlum die in 1925 werd gedaan en die in de literatuur al eerder beschreven werd (Boeles, 1927 en 1951; Zadoks-Josephus Jitta, 1960; Van der Vin, 1992). De brieven bleken tot nu toe onbekende gegevens te bevatten over het aantal munten, de wijze van verwerving door het Fries Museum van het grootste gedeelte van de vondst, en over de curieuze rol die Van Giffen hierbij speelde.

Aanvullende informatie over de wijze van verwerven was niet alleen te vinden in de inventarisboeken van het Munt- en Penningkabinet van het Fries Museum en in het archief van Mr. P.C.J.A. Boeles (Rijksarchief Leeuwarden) maar werd ook verkregen door gesprekken met P. en K. Helfrich, zonen van de vinder van de munten, J. en G. Feikema, zonen van de arts uit Harlingen die destijds bij de verkoop van de munten bemiddelde

en S. Leydesdorff, zoon van de juwelier die in 1934 één van de munten aan het Fries Museum overdroeg. Leydesdorff bleek bovendien in het bezit van een klein potje van aardewerk, dat gelijktijdig met de munt was gekocht, en dat mogelijk het potje is waarin de munten waren opgeborgen.

Een fragment uit de beschrijving van de terp Middelstein in het Kadaster te Leeuwarden (appendix 1), de aantekeningen uit de inventarisboeken van het Munt- en Penningkabinet van het Fries Museum (appendix 2), de betreffende brieven (appendix 3) en een persoonlijke noot van Van Giffen (appendix 4) zijn als bijlagen bij dit artikel gevoegd, en spreken voor zich. De belangrijkste gegevens kunnen als volgt worden samengevat.

2. VINDPLAATS EN VONDSTOMSTANDIG-HEDEN

Tot dusver is de vindplaats van de muntschat in de literatuur niet nader aangeduid. Zowel Boeles (1927, 1951), A.N. Zadoks-Josephus Jitta (1960), als J.P.A. van der Vin (1992) volstaan met de vermelding 'Midlum' als vindplaats. De kaartcoördinaten die Van der Vin (1992: p. 104) opgeeft zijn zelfs die van de dorpsterp van Midlum. Bij Midlum liggen echter vijf

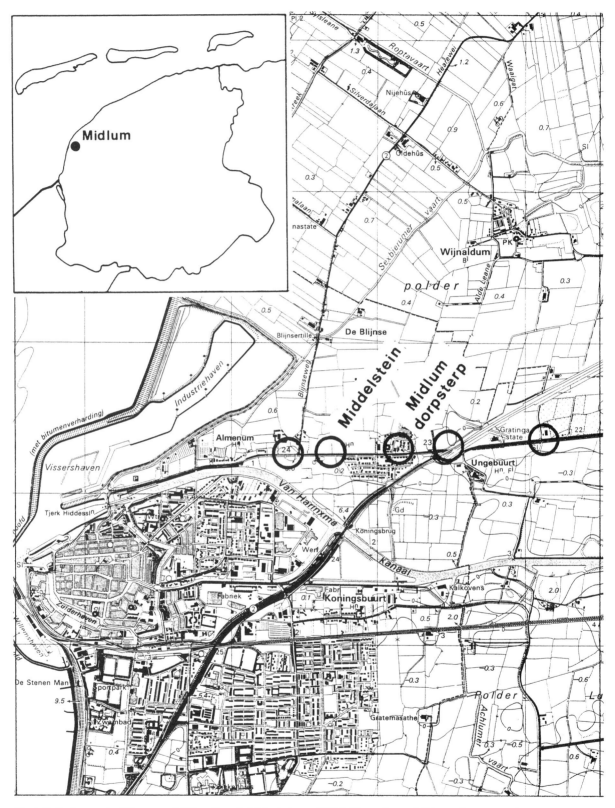

Fig. 1. De ligging van de vijf terpen van Midlum. De waterleidingbuis van 1925 ligt langs de noordzijde van de weg over de terpen.

terpen, inclusief de dorpsterp (fig. 1). Het 100ste *Verslag van het Fries Genootschap* geeft de aanwijzing om welke van de vijf terpen het moet gaan. Onder de inventarisnummers 126 38-42 staan aardewerk en beenderen ingeschreven welke uit dezelfde terp afkomstig zijn als de munten. Het getal 126 is het nummer van de terp die bekend is als Middelstein. Boeles (1951: p. 486) vermeldt deze terp als Midlum-West. Deze terp werd tussen 1905 en 1928 commercieel afgegraven, onder leiding van terpbaas F. Helfrich (fig. 2-3). Zijn zoon P. Helfrich werkte op de afgravingen mee, en was de vinder van de muntvondst. Zijn naam staat ook vermeld in het inventarisboek waarin de aanwinst van de munten in 1927 werd beschreven.

Toch werden de munten niet tijdens de commerciële afgraving van terpaarde gevonden. Volgens een brief van P. Helfrich d.d. 2-2-1927 werden de munten gevonden bij het graven van een waterleiding tegen de glooiing van de terp, op een diepte van 1,70 meter. Bij navraag bij Waterleiding Friesland bleek dat in 1925 inderdaad een waterleidingbuis werd gelegd tussen Harlingen en Herbayum.[1] Thans liggen waterleidingen aan de noord- en zuidkant van de rijksweg. De noordelijke buis is van gietijzer vervaardigd en de zuidelijke van asbestcement. Aangezien asbestcement pas vanaf 1935 wordt gebruikt (Efdée, 1988), moet de vondst gedaan zijn bij het leggen van de noordelijke leiding. Deze buis werd gelegd door het terprestant, dat door F. Helfrich werd geëxploiteerd. Het curieuze feit doet zich voor, dat Mr. Boeles, mogelijk met andere leden van het Fries Genootschap en in gezelschap van P. Helfrich, de vindplaats heeft bezocht. Hij vond het echter niet de moeite waard om de vindplaats nauwkeurig te beschrijven.

3. DE EIGENDOMSVERHOUDINGEN OP DE VINDPLAATS

Het feit dat P. Helfrich betrokken was bij de vondst, doet vermoeden dat de waterleidingbuis gelegd werd door de percelen die commercieel werden afgegraven door terpbaas F. Helfrich. Deze kavels waren, volgens gegevens die bewaard worden bij de Dienst van het Kadaster in Leeuwarden[2], eigendom van de N.V. Kunstmesthandel voorheen Hulshof en Co. te Utrecht. Een zekere J. Visscher te Nijehaske trad op als 'lasthebber' voor deze kunstmesthandel. Pas in 1933 werd de berm, waarin de waterleiding werd gelegd, verkocht aan de staat, ten behoeve van wegverbreding. In 1925 was de kunstmesthandel wettelijk eigenaar van de terp, inclusief de berm.

F. Helfrich was als terpbaas verantwoordelijk voor de afgraving. In het algemeen was een terpbaas niet de eigenaar van de grond, maar een aannemer die personeel aanstelde en zorgde dat afgraving, opslag en vervoer van de terpaarde goed verliep. Vaak werden er afspraken gemaakt tussen eigenaar en terpbaas aan wie

TERPAARDE.

De groote TERP, Middelstein o/d. Midlum, is weder geopend. Gemakkelijk te laden.

De terpbaas HELFRICH.

Tevens te bekomen ongeveer 12 duizend ton ONDERGROND, zeer geschikt voor het verzwaren van polderdijken. Te bevragen bij JOHs. KALMA, Stiens en bij HELFRICH te Midlum.

Fig. 2. Advertentie uit de *Leeuwarder Courant*, 1-4-1910.

eventuele oudheden toe kwamen. Het kon voorkomen dat de eigenaar van de terp niet geïnteresseerd was in oudheden, en alles aan de terpbaas overliet (Arjaans, 1987). In het geval van de terp Middelstein was er niets over dergelijke afspraken te achterhalen.

Ook in 1925 was al wettelijk vastgelegd, dat de eigenaar van de grond recht had op de helft van de waarde van een schatvondst, en de vinder op de andere helft. Het is de vraag wie in dit specifieke geval als vinder moet worden aangeduid: Waterleiding Friesland of P. Helfrich. Het is niet duidelijk op welke manier P. Helfrich bij de graafwerkzaamheden van Waterleiding Friesland betrokken was. Mogelijk hield hij toezicht als zoon van de terpbaas. Het is echter waarschijnlijk dat noch Waterleiding Friesland, noch N.V. Kunstmesthandel voorheen Hulshof en Co., respectievelijk lasthebber J. Visscher ooit van de vondst van de gouden munten heeft geweten.

4. DE OMVANG VAN DE MUNTVONDST (fig. 4)

In oktober 1925 koopt het Fries Genootschap 2 solidi van respectievelijk keizer Marcianus en keizer Leo I. In 1926 ontvangt Van Giffen 12 solidi, die hij na lang aarzelen uiteindelijk alle koopt. Van deze 12 biedt hij er 11 aan het Fries Genootschap aan. Hij bedingt echter dat hij van deze 11 uiteindelijk 2 'doubletten' terug ontvangt, en wel een Leo I en een Anastasius. In de rekeningen van het Biologisch-Archaeologisch Instituut over 1926-1927, die ook in het archief aanwezig zijn, wordt deze aankoop niet genoemd. Kennelijk handelde Van Giffen als privé-persoon. In het rekeningenboek van het Fries Genootschap wordt overigens evenmin de aanschaf van de munten vermeld.

In de periode 1925-1927 is er dus sprake van 14 munten, waarvan Boeles 13 stuks onder ogen heeft gehad. Van deze 14 komen 11 stuks in Leeuwarden terecht, waarvan er 2 rechtstreeks zijn aangekocht en 9 via Van Giffen zijn binnengekomen. Van Giffen heeft 3 munten in zijn bezit, waarvan 1 exemplaar dat Boeles niet kent. De solidi die Boeles in 1927 heeft gezien zijn:

2 solidi van Marcianus (450-457 AD);
8 solidi van Leo I (457-474 AD);
2 solidi van Anastasius (491-518 AD);

302   A. Ufkes

Opgravingen in de terp Middenstein te Midlum (Fr.). — Tot de merkwaardige vondsten van overoude tijden, behooren de hierboven in beeld gebrachte *holpotten*, welke als emmers gebezigd werden, verder een been, een hoorn van een rund en twee ringen, die aan vischnetten bevestigd geweest moeten zijn; alle voorwerpen zijn van steen. — *Boven:* De terp met het dorp Midlum op den achtergrond. *Rechts, onder:* De opgegraven voorwerpen.

Fig. 3. De commerciële afgraving van de terp Middelstein (foto *De Prins*, ergens tussen 1905 en 1910).

1 solidus van Justinianus (527-565; geslagen voor 539).[3]

In 1934 krijgt het Fries Genootschap een solidus van keizer Leo I van juwelier Leydesdorff te Harlingen. Kennelijk stond op grond van de beschikbare informatie vast, dat deze munt tot de schatvondst van 1925 behoorde. Op het kartonnen kaartje bij deze solidus staat geschreven dat deze munt zeker bij de muntvondst hoort. Boeles lijkt zonder meer te hebben aangenomen, dat deze munt dezelfde was als de solidus van Leo I die in 1927 aan Van Giffen werd geschonken. In feite is over de huidige verblijfplaats van de 3 solidi van Van Giffen niets bekend: ze zijn niet in de collectie van het Biologisch-Archaeologisch Instituut terechtgekomen, en waren ook niet in zijn nalatenschap aanwezig. Ongetwijfeld zijn ze ooit te gelde gemaakt. Maar het is niet voor de hand liggend dat Van Giffen daarvoor naar een juwelier in Harlingen zou zijn gereisd. S. Leydesdorff herinnert zich, dat zijn vader de munt al in zijn bezit had vóór 1930, het jaar waarin hij zelf in de zaak kwam. Het feit dat zijn vader bovendien bij diezelfde gelegenheid een klein aardewerk potje had gekregen of gekocht, doet vermoeden dat hij rechtstreeks contact heeft gehad met de vinder of de tussenpersoon. Deze tussenpersoon was

H. Feikema uit Harlingen. Het lijkt inderdaad voor de hand liggend dat Helfrich en Feikema ook munten hebben verkocht aan anderen dan Van Giffen en het Fries Genootschap. Leydesdorff, in zijn positie als juwelier, is een voor de hand liggende kandidaat als koper van een gouden munt. Wellicht is zelfs eerst Leydesdorff benaderd om er zeker van te zijn dat het om gouden munten ging.

H. Feikema was huisarts in Harlingen van 1925 tot 1930. Volgens zijn zonen had hij grote belangstelling voor oudheden en verzamelde hij actief. Helaas is van zijn collectie, die in 1944 als gevolg van oorlogshandelingen in Nijmegen verloren ging, geen beschrijving bewaard gebleven. Het is m.i. echter verre van onwaarschijnlijk, dat Feikema's rol als tussenpersoon niet helemaal vrijblijvend was, en dat hij voor zijn inspanningen één of meerdere munten heeft gekregen.

Problematischer is de toewijzing aan de schatvondst van 1925 van een solidus van Justinus I (518-527 AD) die in 1928 door het Fries Genootschap werd aangekocht van Th. Schaafsma, schoolhoofd te Ried (gem. Franekeradeel). In het 101ste *Verslag van het Fries Genootschap* waarin de aankoop wordt vermeld staat dat de munt 'wellicht uit Midlum' afkomstig is. Hij wordt daar beschreven als een 'bleek' gouden munt, een

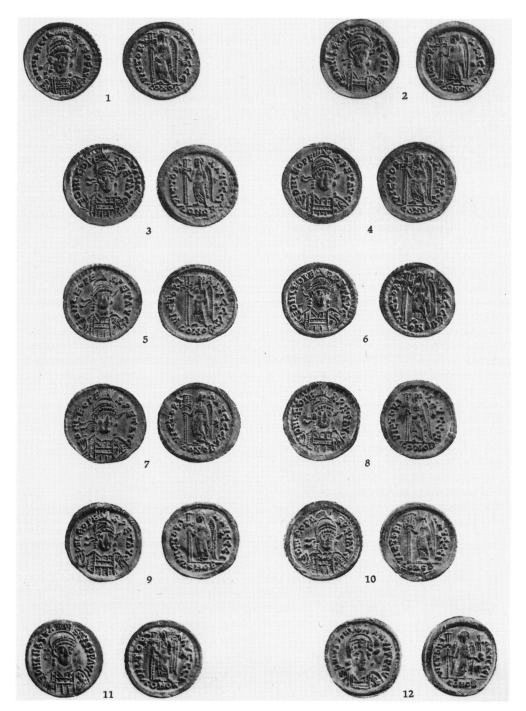

Fig. 4. De 12 solidi van de muntvondst die in het Fries Museum te Leeuwarden worden bewaard (naar Zadoks-Josephus Jitta, 1960).

Germaanse naslag van de Byzantijnse keizer Justinus I. Van der Vin (1992: p. 146 sub 45) spreekt van een 'Merovingische imitatie in bleek goud'. Dat maakt het zeer onwaarschijnlijk, dat deze munt tot de muntvondst van 1925 behoorde. Zadoks-Josephus Jitta (1960) heeft immers al gewezen op het bijzondere karakter van de muntvondst. Van 14 munten staat vast, dat ze in

Constantinopel werden geslagen. Ook het grote aantal munten van Leo I is opvallend, aangezien deze keizer in Westeuropese vondsten zelden voorkomt. (Overigens is uit Blija, gem. Ferwerderadeel, een in Constantinopel geslagen solidus van Leo I bekend, zie Van der Vin, 1992: p. 42). Zij veronderstelt dat de munten via Scandinavië in Midlum terecht zijn gekomen. Oost-romeinse solidi komen in Scandinavië overvloedig voor, en met name Leo I is goed vertegenwoordigd.

Samenvattend kan gesteld worden dat de schatvondst zeker uit 15 munten heeft bestaan, waarvan er 14 gedetermineerd zijn. Van drie van deze munten is de huidige verblijfplaats niet bekend. Het is echter waarschijnlijk, dat de vondst oorspronkelijk een groter aantal munten omvatte.

5. HET POTJE (fig. 5)

Tijdens het eerste gesprek met S. Leydesdorff bleek, dat zijn vader destijds niet alleen een gouden solidus had gekocht, maar dat hij bij dezelfde gelegenheid ook een klein aardewerk potje had gekregen of gekocht. Dit potje was in het bezit van S. Leydesdorff, maar deze bleek bereid het af te staan aan het B.A.I., of eventueel aan het Fries Museum. Overigens bleek dat het potje al rond 1950 aan het Fries Museum was aangeboden, maar geweigerd!

Het betreft een klein potje dat met enige goede wil als buidelvormig kan worden omschreven. Het aardewerk is hard gebakken en ondanks het feit dat er een scherfje uit de rand ontbreekt, is de magering niet goed zichtbaar. De maximale hoogte is 6,2 cm, de maximale diameter is 6,5 cm en de binnenranddiameter is 3,5 cm. Het potje is versierd met verticale groeflijnen.

Hoewel exacte tegenhangers niet bekend zijn, en een datering van dergelijke kleine potjes een hachelijke zaak is, is op grond van vorm en versiering een 6e-eeuwse datering zeker niet uit te sluiten. Uit Beetgum-Besseburen komt een exemplaar dat weliswaar 12 cm hoog is en een grootste diameter van 12,4 cm heeft, maar dat qua vorm en versiering toch wel overeenkomt (Holwerda, 1907: Pl. VIII.14; Knol, 1993: fig. 75.4).

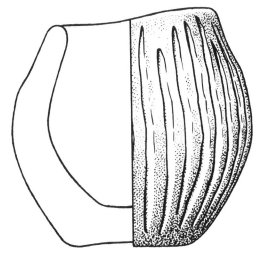

Fig. 5. Het potje uit de collectie Leydesdorff, dat tezamen met één van de solidi werd aangekocht (tek. B.A.I., J.M. Smit). Schaal 1:1.

Uit het feit dat het potje tegelijk met een solidus werd aangeboden aan Leydesdorff, mag misschien worden afgeleid dat potje en munt bij elkaar hoorden, of met andere woorden: dat de 15 (of meer) solidi in het potje werden gevonden. Bij gebrek aan enige vermelding in de brieven van 1925-1927 van dit potje, moet het echter noodgedwongen bij dit vermoeden blijven.

6. SLOTOPMERKINGEN

Het is voor archeologen altijd weer verbazingwekkend, dat numismaten zo weinig geïnteresseerd zijn in het vastleggen van exacte vondstomstandigheden, en het bij elkaar houden van de munten. Ook vandaag de dag nog worden muntvondsten, na beschreven te zijn, zonder veel problemen naar de veiling gebracht, waarna de munten naar alle windstreken verdwijnen.

In dit licht gezien kan de handelswijze van Mr. Boeles, die in de eerste plaats een numismaat was, dan ook niet als vreemd of afwijkend worden gezien. Wel verrassend is het optreden van Van Giffen, van wie als archeoloog een andere houding verwacht had mogen worden.

Verrassend was ten slotte ook nog, dat meer dan 65 jaar na het vinden van de munten, nog nieuwe gegevens boven water konden worden gebracht, hoewel geen van de rechtstreeks betrokkenen van destijds meer in leven is.[4]

7. NOTEN

1. Waterleiding Friesland, 1983. Leidingbeheerkaart 1:1000.
2. Dienst van het Kadaster en de Openbare Registers te Leeuwarden, Leeuwarden 1620-19, dagregister deel 138, nummer 25, 1918 Leeuwarden 1673-35, dagregister deel 143, nummer 55, 1919 Leeuwarden 1921-98, dagregister deel 171, nummer 1106, 1927 Leeuwarden 1930-50, dagregister deel 172, nummer 499, 1928.
3. Zadoks-Josephus Jitta (1960) vermeldt ten onrechte dat de solidus van Justinianus behoort tot een type dat na 538 in zwang kwam. Boeles (1951) had al op een datering voor 538 gewezen.
4. Met dank aan J. Arjaans, J. en G. Feikema, P. en K. Helfrich, S. Leydesdorff, E. Knol, T.B. Volkers, J. Zijlstra, en speciaal aan J.N. Lanting voor zijn kritiek en suggesties.

8. LITERATUUR

ARJAANS, J., 1987. Terpafgravingen in Friesland. Doctoraalscriptie Sociale Geografie, Vrije Universiteit Amsterdam.
BOELES, P.C.J.A., 1927. *Friesland tot de 11e eeuw.* 's-Gravenhage.
BOELES, P.C.J.A., 1951. *Friesland tot de 11e eeuw.* 2e druk, 's-Gravenhage.
EFDÉE, R., 1988. *Putten uit het verleden. Geschiedenis van de drinkwatervoorziening in Friesland.* Leeuwarden.
HOLWERDA, J.H., 1907. *Nederland's vroegste beschaving. Proeve van een archaeologisch systeem.* Leiden.
KNOL, E., 1993. De Noordnederlandse kustlanden in de vroege middeleeuwen. Dissertatie Vrije Universiteit Amsterdam.
Verslag van het Friesch Genootschap van Geschied- Oudheid- en Taalkunde te Leeuwarden 100, 1927-1928. Dokkum.

Verslag van het Friesch Genootschap van Geschied- Oudheid- en Taalkunde te Leeuwarden 101, 1928-1929. Dokkum.

Verslag over het boekjaar 1926 van de N.V. Intercommunicale Waterleiding, gebied Leeuwarden. Leeuwarden.

VIN, J.P.A. VAN DER, 1992. *Die Fundmünze der römischen Zeit in den Niederlanden. Abteilung I, Provinz Friesland.* Berlin.

ZADOKS-JOSEPHUS JITTA, A.N., 1960. Midlum. *Jaarboek voor Munt- en Penningkunde* 47, pp. 94-96.

APPENDIX 1: Fragment uit de beschrijving van de terp Middelstein in het Kadaster te Leeuwarden.

Op 10 mei 1918 worden de betreffende kavels gekocht door de Naamlooze Vennootschap Kunstmesthandel voorheen Hulshof en Co., gevestigd te Utrecht van Johannes Karsjens Kalma, koopman, en Pieter Ytsens van der Werff, veehouder en wethouder der gemeente Leeuwarderadeel, beide wonende te Stiens. Jan Visscher, een handelaar wonende te Nijehaske (Heerenveen) handelt als 'lasthebber' voor de kunstmesthandel. In 1927 wordt een poging gedaan om bij een openbare veiling deze percelen te verkopen. In het dagregister van het Kadaster te Leeuwarden, 1927, wordt over de terp Middelstein het volgende vermeld:

Overgenomen uit: Dienst van het Kadaster en de Openbare Registers te Leeuwarden.

"Nota betreffende de afgraving.
Ten behoeve van de afgraving had den verkoopster van dit perceel ten laste van de perceelen nummers 1189, 1100, 2289 en 2293 recht tot vervoer en opslag van terpaarde met vrije vaart naar de Hopmansvaart, welk recht echter eindigt op een April negentienhonderd achtentwintig.
De kooper treedt vanaf de toewijzing in alle rechten en verplichtingen van de verkoopster betrekkelijk het hebben en onderhouden van een strookje met rails en toebehooren op den rijksweg waarvoor jaarlijks ene recognitie ad vijftig cent verschuldigd is.
De kooper zal zelf op eigen kosten voor eventueele verlenging, vernieuwing of beëindiging dier vergunning moeten zorgdragen. Desgewenscht kan de kooper van de verkoopster overnemen het hok, vijftien kipkarren, circa zevenhonderdvijfenzeventig meter rails met vier wissels, tien kruisplanken, twee breekijzers, een bankschroef, twee kruiwagens, de stelling met barte en toebehooren aan de opvaart op het perceel nummer 2289 en het paard met twee tuigen, alles thans voor de exploitatie in gebruik, samen voor tweeduizend vierhonderd gulden.
Al deze goederen zijn mitsdien buiten den koop, ook voorzoover zich bevindende op dit perceel."

De Naamlooze Vennootschap Kunstmesthandel voorheen Hulshof en Co. heeft de terp waarschijnlijk willen verkopen vanwege het feit dat op 1 april 1928 het recht van vervoer en opslag van de terpaarde zou vervallen en omdat de terp inmiddels grotendeels was afgegraven. Bij de veiling in 1927 wordt de terp echter niet verkocht. Op 23 maart 1928 koopt Jacobus Fontein, cargadoor en expediteur te Harlingen alsnog de kavels 66a, 1188, 1191, 1454 en 1455 inclusief kipkarren, de rails met wissels, de kruiplanken en verdere goederen voor exploitatie voor *f* 4500,-. Het paard wordt bij deze transactie niet meer vermeld.

APPENDIX 2: Fragment van de aantekeningen uit het inventarisboek van het Munt- en Penningkabinet van het Fries Museum.

"A 1678-1688. Muntvondst Terp Midlum 1925. Van de 13 alstoen

aldaar bij elkaar gevonden gouden Byzantijnsche solidi, kreeg het Fr. Genootschap elf stuks en wel eerst A 1678 1 Marcianus (450-457 n.C.) en A 1979 1 Leo I (457-474 n.C.). Dr. van Giffen kocht de rest voor f120.- en deed die voor hetzelfde geld in juni 1927 aan het Fr.Gen. over op voorwaarde dat hij 2 der dubbelen kreeg voor het Biol.Arch.Instituut te Groningen. De 9 die er toen bij kwamen zijn één Marcianus als voren zes Leo I alsvoren één Anastasius (491-518 n.C.) en één Justinianus I (527-565 n.C.). Dr. van Giffen kreeg een Leo I en een Anastasius voor het Instituut.
Over de terp loopt de rijksweg naar Harlingen. Bij onderzoek ter plaatse in 1925 nadat het Fr.Gen. in October 1925 van Dr. Feikema te Harlingen de beide eerst gemelde solidi ter beoordeeling had ontvangen, bleek dat de zoon van de terpbaas te Midlum wat eigenaardig met de vondst was omgesprongen, zoodat het Fr. Museum deze vondst niet rechtstreeks van den terpbaas of eigenaren van de terp heeft kunnen aankopen.

Nos 1678/9 zijn Oct. 1925 ingekomen. De rest in 1927.

A 1679 is afgebeeld door Boeles Friesland t.d. 11e eeuw pl.35.2 Zie aldaar ook blz. 155 en 254.

A 1868 Gouden Solidus van keizer Leo (457-474). Ingek. Maart 1934.
Vz. buste 3/4 en face met lans over linker schouder
  D H L E O P E B P E T A V C
Kz. V I C R O R I A A V C C C I C O N O B
  Victoria staande n. l. met kruisstaf.
Afkomstig uit de terp te Midlum. Behoort vrij zeker bij de vondst Boeles Friesl. tot 11e eeuw p. 284 (35.2)
Gekocht van den goudsmid Leydesdorff te Harlingen voor f12."

Op het kartonnen kaartje behorende bij deze munt staat:
"Keizer Leo 457-474 n.C. M19 A 1868 gouden solidus terp Midlum (behoort zeker bij muntvondst)"

APPENDIX 3: Binnengekomen en uitgaande brieven van de jaren 1925 t/m 1927, afkomstig uit het brievenarchief van het B.A.I., behalve de briefnummers 187, 216 en 341, die afkomstig zijn uit het brievenarchief van P.C.J.A. Boeles, Rijksarchief Leeuwarden.

(aan)   Harlingen 16 - 9 - 25
Weled. Zeergel. Heer v Giffen
te Groningen

Bij dezen wend ik mij tot u met een eigenaardig verzoek. Ik kan u namelijk te koop bieden 2 oude gouden munten uit de 8ste eeuw. Om ze te redden van de smeltkroes heb ik mijn bemiddeling aangeboden, temeer daar ze behooren aan een arme oude man en ik anders bang ben dat hij in verkeerde handen zal vallen. Ik bied ze aan aan de meestbiedende liefhebber. Ik heb enkele aangeschreven. Hiernevens gaat een afdruk. De eigenaar wenscht vooralsnog onbekend te blijven.
  Inmiddels met de meeste hoogachting
  Uw.Dw.Dnr.
  H. Feikema, arts.

Weledl zeergel Heer!   Harlingen 2 - 6 - 26

Hierbij zend ik U bona fida 11 munten, gevonden in deze buurt jaren geleden, waarover ik u destijds schreef. Gaarne wist ik de prijs die U er voor denkt te kunnen besteden.
  Met de meeste hoogachting
  Uw dw dnr
  H Feikema

Den WelEdelGeboren Heer H. Feikema 8 Juli '26
Harlingen.

WelEdele Heer!

Uw schrijven d.d. 2-VII.l. kwam in goede orde in mijn bezit, zoomede de munten, welke U mij bona fide daarbij zond, en dank ik U voor het in mij gestelde vertrouwen.

Intusschen merk ik op, dat U zich bij de telling blijkbaar vergist hebt, er zijn nl. niet 11, zooals U schrijft, doch 12 gouden munten. Over een en ander ben ik nog in correspondentie en schrijf ik daarover alvast even voorloopig. Ik hoop dat U nog een paar weken geduld wilt doen, opdat ik U volledig kan inlichten en een wel overwogen voorstel kan doen.

Inmiddels teeken ik mij met alle hoogachting,
Uw dw. dn.
(Van Giffen)

WelEd. Heer! Wijnaldum, Juli '26

Zoudt U de gouden munten per ommegaande aan Dr. H. Feikema te Harlingen willen terug zenden?
Met de meeste hoogachting
De eigenaar
P. Helfrich.

Den WelEdele Heer P. Helfrich 17 Juli '26
Wynaldum

WelEdele Heer!

Aan het verzoek van Uw schrijven d.d. Juli l.l. hetwelk ik zoo juist ontving, kan ik onmogelijk voldoen, aangezien ik met mijne informaties nog niet gereed ben. Overigens verwondert mij Uw schrijven in zooverre wel eenigszins en begrijp ik er niet veel van. U zoudt mij derhalve verplichten mij wel te willen melden, of en eventueel welke redenen er voor U mogen zijn, mij zoo "bondig", om dit woord maar te bezigen, te schrijven.

Inmiddels teeken ik mij hoogachtend
(Van Giffen)

Weled. Heer! Wijnaldum, Juli '26

In antwoord op uw geëerd schrijven van j.l. het volgende. Kijk Mijnheer, de zaak zit zoo. Ik was in geld verlegenheid en nu dacht ik als ik nu de munten verkoop, dan kan ik me er wel door helpen. Ik moest einde Juni of begin Juli het geld beslist gebruiken en ik zag anders geen uitweg. Nu was hier iemand die bood mij f 10 per stuk, en ziet U, met dat geld zou ik voldoende geholpen zijn. En daarom schreef ik U dat briefje, dat ik graag de munten terug wou hebben. Denkt U nu ook niet dat dit de beste oplossing zal zijn, om me uit den nood te helpen? Had ik nu het geld niet hoognoodig, dan had ik er natuurlijk nooit om geschreven, maar ik was in een moeilijke situatie. Hopende dat het U nu eenigszins duidelijk is, teeken ik mij
Met de meeste hoogachting
P. Helfrich

Den WelEdelGeb. Heer P. Helfrich 13 Aug. '26
Wynaldum

Geachte Heer Helfrich!

Uw schrijven d.d. Juli '26 kwam in mijn bezit. De munten zijn bijna alle afkomstig van Byzantynsche keizers. Ik zal ze U thans gaarne retourneeren, doch verneem nog graag, alvorens dit te doen of ik zelf er nog een paar mag uithouden waarvoor ik wil vergoeden f12.-. Voorts zou ik U nogmaals in overweging geven ze alvorens weg te doen, nog aan te bieden aan het Friesch Genootschap.

Gaarne Uw antwoord tegemoet ziende hoogachtend

P.S. Bewaart U nog beenderen en koppen voor mij en hebt U reeds iets. De muntvondst is immers ook uit de Midlumer terp? Wilt U dit laatste onbekend houden?
(Van Giffen)

Den WelEdelGeb. Heer 31 Aug. '26
Den Heer P. Helfrich
Wijnaldum

Geachte Heer!
Gaarne had ik even antwoord inzake het van U nog steeds bij mij berustende in verband met mijn laatste brief.
Hoogachtend
Van Giffen

WelEdel Geb. Heer! Wijnaldum, 7.9.26.

In antwoord op uw geëerd schrijven van l.l. Dinsdag deel ik u mede dat U er maar een paar van moet uithouden. Mijn schrijven komt wat laat, maar ik ben tegenwoordig gauw eens uit te kaatsen, waar ik veel aan doe. Nu Mijnheer ik schrijf U later nog wel iets omtrent de terp.
Hoogachtend
P. Helfrich

Weled. Heer! Wijnaldum, Nov. 1926

We ontvingen tot nu toe geen bericht van U aangaande de munten. Op uw laatste schrijven toch heb ik bericht gestuurd, dat u kan er wel een paar van afhouden, maar er komt niets geen bericht terug. Zou u zoo goed willen zijn en stuur ze per ommegaande of als U de munten kan plaatsen, dan het geld? 't Is voor mij weer 12 Nov. geweest en de winter is er zoo meteen ook weer.
Hoogachtend
P. Helfrich

Weled. Heer! Wijnaldum, 10.12.'26

Tot mijn verwondering ontving ik tot heden nog geen bericht van U aangaande de munten, die U nog onder Uwe berusting heeft. Mag ik zoo vrij zijn, u hierover te vragen? Ik heb ze deze zomer al reeds kunnen verkoopen aan 't Friesch Museum, maar zoolang ik de munten niet heb kan ik ze ook niet verkoopen. Zou U nu zoo goed willen zijn en zend ze mij zoo spoedig mogelijk? Dan kan ik den heer Draaisma even berichten nietwaar of als U er beter raad mee weet is het mij ook best, maar gaarne had ik deze zaak eens afgedaan. Gaarne Uw antwoord tegemoet ziend
Hoogachtend
P. Helfrich.

Den WelEdelGeb. Heer P. Helfrich   1 Febr. '27
Wijnaldum

Geachte Heer!

Doordat ik sedert eerste helft December ziek was en nog steeds overspannen ben, is allerlei blijven liggen. Daaronder ook Uw Brief. Gaarne retourneer ik U de munten. U vindt dus goed dat ik er een paar van houdt, voor de afgesproken prijs en wilt U mij dan nog bovendien 5 laten voor f10 per stuk. Deze zijn alle gelijk. Ik kan U dan het geld, tegelijk met de overige munten retourneeren. Het Friesch Museum zal de rest wel willen nemen en anders ben ik bereid - omdat ik U heb laten wachten en U in geen geval zou willen dupeeren - ze voor 10 gulden per stuk af te nemen. Wel zou ik gaarne precies vernemen, waarin en wanneer de munten gevonden zijn.

Ik hoop zeer, dat U mij dit jaar zult willen helpen met de beenderen uit Uwe terp. Ik zal er U goed voor betalen, doch zou dan ook gaarne op Uwe medewerking rekenen.

Vertrouwend dat U de vertraging van mijn antwoord zult willen excuseeren, teeken ik mij hoogachtend

(Van Giffen)

Weled. Heer!   Wijnaldum, 2.2.'27

Wegens ziekte kan ik eerst heden uw brief beantwoorden. Met blijdschap vernam ik dat u bereid is de munten van mij over te nemen. Gaarne wilde ik, dat U ze allemaal maar hield, dan behoef ik ook niet meer naar Draaisma de Vries te schrijven. Ik heb de munten gevonden op ongeveer 1 meter 70. Bij 't graven van de waterleiding tegen de glooiing van de terp.

Ik hoop, dat ik U deze zomer met raad en daad terzijde mag staan inzake Uw onderzoekingen in de terp. U kunt ten allen tijde op mij rekenen. Bij voorbaat dank voor het plaatsen van de munten.

Hoogachtend
P. Helfrich

Aan den WelEdelenHeer   5 Mrt '27
P. Helfrich.
te Wijnaldum.

Geachte Helfrich!

Per slot had ik het nog bijna vergeten, doch gelukkig schoot mij juist ons gesprek in gedachte nu ik op het punt sta te vertrekken. Ik haast mij dus hierbij f.100.00 te zenden. U krijgt dan nog f10.00 is het niet?

Toezegging is gaarne Uw antwoord tegemoetziende, na vriendelijke groeten,

Uw. dw.
(Van Giffen)

5 Mrt'27
Ingesloten A.H.060850————f.100.00

Aan den Directeur v/h Friesch Genootschap   Groningen, 13 April
1927   187
van Geschied- Oudheid- en Taalkunde
LEEUWARDEN.

Zeer Geachte Heer!

Voor een paar weken kwam ik door koop in het bezit van eenige gouden munten, gevonden in de terp. *

Ik sprak U daarover reeds bij gelegenheid van de vergadering van de Oudh. bnd. te Zwolle. De zaak heeft lang geduurd, doch ik heb gemeend om allerlei redenen den koop te moeten sluiten, overtuigd dat dit in het algemeen belang was en de munten anders reeds verloren gegaan waren. Intusschen ben ik bereid de bewuste collectie bestaande uit 11 gouden dinarii van Leo I (10x) 457-473 en Justinus I (1x) 518-527 tegen kostprijs geheel of ten deele ter beschikking van het Genootschap te stellen, op voorwaarde dat zij niet verkocht of voor ruil gebruikt worden.

Door velerlei bezigheden gaf ik niet direct gehoor aan mijn bovenstaand voornemen.

Inmiddels teeken ik mij met beleefde groeten,
Hoogachtend,
Directeur Biologisch Instit.
A.E. van Giffen

commentaar van Boeles bij deze brief:

N.B. de beschrijving der munten is niet correct. Zie Friesland tot de 11e eeuw waarin ze alle beschreven zijn.

Van hem overgenomen en 2 dubbelen teruggegeven voor het Biol. Arch. Instituut. B.
* (te Midlum.)

[brievenarchief P.C.J.A. Boeles]

Friesch Genootschap   Leeuwarden "Baens-Ein" 22/4 '27
van Geschied-, Oudheid-, en Taalkunde
te Leeuwarden.

Amice,

Dank voor je kaart. Ik houd me voor nader bericht aanbevolen. Alle dagen van de volgende week zouden mij passen voor een conferentie. Vanmiddag kreeg ik toevallig je brief van 13 April l.l. in handen. Daar, zooals je wellicht weet, de oude muntenafdeeling door mij behartigd wordt, haast ik mij te melden dat de bereidwilligheid om de gouden solidi aan ons tegen kostprijs over te doen zeer gewaardeerd wordt. Om de zaak voor beslissing door de vergadering te prepareeren dien ik natuurlijk te weten wat gij ze betaald en waar ze gevonden zijn. Ik heb omtrent de vindplaats wel een stellig vermoeden, doch zag dit gaarne bevestigd. Als ik in Groningen kom voor boven bedoelde correspondentie zou ik de solidi wel even mee kunnen nemen om ze aan het bestuur te laten zien. Verleden jaar heb ik met Ottema voor je uitgezocht, volgens belofte,
/ twee oorpot met ingezonken hals
/ dito gesmoord met uitgetrokken ooren.
Vandaag bleek mij dat die potten nog niet verzonden zijn, naar ik hoor omdat O. dacht dat er nog meer bij uitgezocht zou worden. Mij was dat onbekend n.l. 't met verzenden.

Met vr. groeten
Steeds gaarne
Boeles

Den Hoogedelgestrengen Heer   Groningen, 4 Mei 1927.
Den Heer Mr P.C.J.A. Boeles   216
Raadsheer
LEEUWARDEN.

Amice!

Met nog een berg correspondentie voor mij, kom ik eindelijk tot eene beantwoording van schrijven 10 April, waarover ik u reeds per kaart schreef.
    Ge zijt dus zoo te zeggen weer in Uw knollentuin teruggekeerd.
    Thans de verlanglijst:
1). Gaarne.
2). Deze teekening is niet te geven,aangezien het veld slechts zeer onvolkomen onderzocht is.Wel kan ik onvolledige teekening en foto verschaffen.
3). Gaarne.(zie terpenverslag,hetwelk dezer dagen verschijnt).
4). ''
5). ''
6). ''
7). ''
8). ''
9). ''
    Gij ziet dus dat ik ouder gewoonte handel.
    Over het runenhoutje zou ik vast wenschen te houden aan de Uwerzijds gedane beloften en de gemaakte afspraak. Het is een speciaal Groningsche vondst en nu alles zoo anders is geloopen dan ik voor jaren had gedacht, doch vooral omdat ik in gebreke ben met de voorloopige publicatie en er bovendien met Prof. Friesen over in verbinding sta, zou ik wenschen het houtje voorloopig nog dood te zwijgen.
    Wat aangaat de munten zoo dit nog noodig mocht zijn, zie ik in eene opname geene bezwaren. Ik zou wel gelegenheid willen geven eens eene proeve te nemen. Mocht ge dus eens willen komen zoo zijt ge zeer welkom.
    Wat de laatste zinsnede betreft, geloof mij,dat ik eene goede samenwerking altijd op prijs gesteld en er ook naar gestreefd heb.
    Zoo gij dus mocht kunnen komen, zal ik gaarne Maandag of Dinsdag e.k. hier zijn. Alleen zag ik gaarne nog van te voren even bericht.
    Geloof mij inmiddels na beste groeten, steeds gaarne,
    t.à.t.,
    A.E. van Giffen

[brievenarchief P.C.J.A. Boeles]

Den HoogEdelgestrengen Heer   Groningen, 5 Mei 1927
Den Heer Mr.P.C.J.A.Boeles
Raadshaar (Harlinger Str.weg 75
LEEUWARDEN

Amice!

Bij het verder afwerken der correspondentie kom ik tot je brief d.d. 22 VI l.l., waaruit ik zag dat gij den brief aan den Heer Nieweg te behartigen hebt. Het lijkt mij het beste dat wij die aangelegenheid en annexen bij je bezoek hier maar afdoen. Ik betaalde 100 gulden doch ben nog schuldig 10 of 20 g., dat weet ik niet precies. De vondstomstandigheden zal ik even weer opzoeken, opdat ge geheel au fait zijt. Het antwoord op je laatste schrijven van 3 Mei l.l. is, zooals ik zie, reeds gegeven in mijn brief van gisteren. Tot spoedig ziens dus.

    Met vriendelijke groeten, steeds gaarne
    t.à.t.,
    (Van Giffen)

Friesch Genootschap   5 Mei '27, Leeuwarden
van Geschied-, Oudheid-, en Taalkunde   Harlinger Straatweg 75
te Leeuwarden.

Amice,

Het was mij hoogst aangenaam heden je brief te ontvangen. De daarin vervatte toezeggingen waren geheel naar wensch. Ik ben nu voornemens Dinsdag 10 Mei a.s. om 9.10 v.m. met den sneltrein te Groningen te komen en mij dan naar het Instituut te begeven. De munten en nog een paar dingen zal ik meebrengen en zou het zeer waardeeren wanneer je fotograaf dien dag van de munten, die op twee plankjes formaat 13×18 cM bevestigd zijn een foto zou kunnen maken, zoowel van de voor als van de keerzijde, dan is het afgieten in gips wellicht niet bepaald nodig. Indien gij de mij toegezegde foto`s en teekeningen alvast klaar kunt laten leggen of afdrukken dan zou dit mijn werk zeer kunnen vergemakkelijken, want ik zou ze graag meenemen. De munten van Leo I enz. kan ik dan zeker ook wel eens zien.
    Inmiddels met de beste groeten
    Boeles

Friesch Genootschap   Harlinger Straatweg 75
van Geschied-, Oudheid-, en Taalkunde   19 Juni '27
te Leeuwarden.

Amice,

Op de vergadering van 1 Juni l.l. heeft het Bestuur overeenkomstig je voorstel besloten om de elf gouden solidi uit de terp te Midlum aan te koopen voor f 120.-. Hierbij heb ik het genoegen een chècque voor dat bedrag in te sluiten. Zoodra ik bericht van goede ontvangst van de chècque heb, zal ik die betaalbaar stellen, ongeveer 3 dagen na ontvangst van dat bericht. Liefst ontvang ik een kwitantie ten name van het Friesch Genootschap. De twee dubbelen krijgt gij dan bij gelegenheid gratis op het voorstel. Daar ik mijn kuitspier v/h rechterbeen wat beschadigde kon ik zelf de vergadering niet bijwoonen. Ik ben nu weer aardig ter been.
    Met beste groeten
    Boeles

Den WelEdelen Zeergeleerden Heer   Groningen, 23 Juni 1927
Den Heer Dr.P.C.Boeles   341
Harlinger Straatweg 75
LEEUWARDEN.

Amice,

Je brief d.d. 19 l.l. heb ik in goede orde ontvangen en is deze aangelegenheid dus in zooverre van beide kanten naar genoegen geregeld. De muntjes ontving ik dan wel gaarne bij gelegenheid toegezonden.
    Ik hoop intusschen, dat ge weer geheel hersteld zijt, want dergl. beschadigingen zijn meestal pijnlijk en overigens lastig.
    Mijn antwoord ondervond eenige vertraging, doordat ik voor een paar dagen buitenslands geweest en pas teruggekeerd ben.
    Na beste groeten steeds gaarne,
    t.à.t.,
    A.E. van Giffen

[brievenarchief P.C.J.A. Boeles]

Weled. Heer!   Wijnaldum, Dec. '27

Toen U deze zomer in Midlum was, hebt U mij beloofd, dat U met Uwe secretaris er over zou spreken, dat ik nog f 24 hebben moest van de munten. U heeft mij f 100 gestuurd en U schreef me, dat U er 2 hield voor f 12 per stuk en de andere 10 voor f 10 per stuk. Dus nu moet ik feitelijk nog f 24 hebben. Zou U het nu ook gelegen komen, dat U ze mij nu stuurde. Ik kan ze nu heel goed gebruiken omreden het winter is en ik nu niets kan uitvoeren met de werkzaamheden en er dus ook geen verdiensten zijn.

     Vertrouwende, dat U het mij niet kwalijk zult nemen, teeken ik mij
U.dw.dn. P. Helfrich

     Van harte gefeliciteerd met Uwe onderscheiding
'k Las het in de Leeuwarder Courant

Den WelEdele Heer P. Helfrich   19 Dec. '27
Wijnaldum

Geachte Helfrich

Bijgaande doe ik U de 24 gulden toekomen met verzoek mij daarvoor ingesloten kwitantie na onderteekening, te willen retourneeren. Mag ik er voorts mederekenen, dat gij komend jaar beenderen, koppen enz. zoo mogelijk ook enkele oudheden voor mij wilt verzamelen?
     Voorts dank ik U voor Uwe felicitatie.
     Na vriendelijke groeten
     hoogachtend
     (Van Giffen)

APPENDIX 4: Een persoonlijke noot van Van Giffen betreffende de transactie.

De muntvondst werd door P.C.J.A. Boeles gepubliceerd in het 100ste verslag van het Fries Genootschap, 1927-28:

     A. 1678-1688. MUNTVONDST MIDLUM 1925. In de terp te MIDLUM (zie rubriek B. 126, 38-42), werden in 1925 samen gevonden dertien fraai geconserveerde gouden Byzantynsche solidi. Het Friesch Museum *) kocht de geheele vondst en stond daarvan twee doubletten, n.l. één Leo en één Anastasius af aan het Biol. Archaeologisch Instituut te Groningen. Zie over deze vondst Boeles, Friesland t. d. 11e eeuw, p. 155, 254 en pl. 35.2. De elf die het Fr. Museum behield zijn:
     1678, MARCIANUS (450-457 n. C.). Sabatier, Monnaies Byzantines I, p. 124 no. 4. Geen letter achter Auggg; 1686, ALS VOREN, met bijletter; 1679-1685, zeven exemplaren van L e o I (457-474 n. C.), Sabatier I, p. 131 no. 4, met verschillende bijletters achter Auggg.; 1687, ANASTASIUS (491-518 n. C.), Sabatier, I, p. 152 no. 2, doch met Dn. Anastasius P.P (niet P. F.) Aug.; 1688, JUSTINIANUS I (527-565 n. C.), Sabatier I, p. 177 no. 2.

In het exemplaar dat in het B.A.I. aanwezig was (nu in de UB Groningen) leverde A.E. van Giffen het volgende commentaar:

*) NB precies verkeerd voorgesteld: ik zelf kocht de vondst aan en stelde het Fr. Gen. via dhr. Boeles in de gelegenheid haar zonder meer over te nemen. Zie de gevoerde correspondentie BAI.

VAN DE MIEDEN, EGESTE EN BROKE
DE MIDDELEEUWSE NEDERZETTINGSGESCHIEDENIS VAN HET ZUIDWESTELIJK WOLD-OLDAMBT IN KORT BESTEK

J. MOLEMA

Biologisch-Archaeologisch Instituut, Groningen, Netherlands

KURZFASSUNG: Im Spätmittelalter sind im westlichen Teil des Wold-Oldambts die folgenden Siedlungsentwicklungen erkennbar. Anfangs gab es ein Gebiet namens Broke oder Broek. Es war begrenzt durch die Flüße Siepsloot und Ae im Osten, der Fluß Siepsloot im Norden und das Moor im Westen und Süden. Anscheinend gab es in Broek eine Primär-Siedlung mit allerdings demselben Namen. Es entstand eine Zweiteilung des Gebiets durch die Gründung einer zweiten Siedlung, Suthabroke oder Zuidbroek genannt. Zum Unterschied bekam die Primär-Siedlung den Namen Nortbroke oder Noordbroek. Die folgende Entwicklung war die Gründung einer dritten Siedlung, Mieden oder Meeden genannt, die ebenfalls vom Gebiet Zuidbroek getrennt wurde. Die drei Siedlungen im Gebiet Broek sind während des 13. und 14. Jahrhunderts durch Nässe infolge Moorschwundes, verschoben worden. So auch Meeden das ebenfalls durch eine sehr langsame Verschiebung eine Zeitlang zwei Siedlungskonzentrationen hatte, Mieden und Avermieden, wobei Avermieden die neueste Siedlung war. Dieses Dorf wird in münsterischen Dekanatsregister auf lateinisch Extengamieden und Exterigamedum genannt. Exteriga- und Extenga- soll man lesen als Extra- und Extra- ist identisch mit Aver-. Der Name Avermieden oder Extramieden ist zu erklären: das Dorf, das außerhalb des Dorfes Mieden liegt. Bisher hatte man unglückigerweise, zuerst Stratingh & Venema (1855), der Namen Exterigamieden oder Extegamedum irrtümlicherweise mit Eextameeden übersetzt und deshalb das Dorf Eexta (Gem. Scheemda) als Mutterdorf von Meeden angemerkt. Eine kirchliche Beziehung zwischen Meeden und Zuidbroek läßt sich aus einer Widmungstext in der Kirche von Zuidbroek konkludieren. Darüber hinaus umfassten die Dörfer Meeden, Zuidbroek und Noordbroek ein gesämtliches Rechtsgebiet. Die Parochiegrenze zwischen Meeden und Eexta wird durch die Flüße Siepsloot und Ae gebildet. Die Grenze zwischen Meeden und Zuidbroek ist größtenteils fabriziert und wird durch eine aufstreckende Parzelle gebildet. Die Relation zwischen Meeden und Zuidbroek ist klar und es gibt keine Beziehungen mit Eexta. Das Gebiet Broek habe ich als eine Primär-Parochie betrachtet. Auf diese Gedanke basiert ist die (hypothetische) Vierteilung des Wold-Oldambts und Reiderlandes. Vier Primär-Parochien die durch natürliche Grenzen geschieden sind. Primär-Siedlungen lassen sich möglicherweise durch frühe Kirchenbauten nachweisen. Die Gebiete liegen momentan weitere Untersuchungen vor.

STICHWORTE: Groningen, Wold-Oldambt, Reiderland, Spätmittelalter, Primär-Parochie, Siedlungsentwicklung.

1. INLEIDING

Het Wold-Oldambt en het woldgebied van het Reiderland lenen zich goed voor onderzoek dat is gericht op de reconstructie van het middeleeuwse cultuurlandschap. Beide gebieden zijn voormalige veengebieden die door de inbraken van de Dollard met een kleipakket bedekt raakten. Hierdoor is het middeleeuwse veenlandschap met bewoningssporen goed bewaard gebleven. Momenteel richt onderzoek zich op de vragen wie het gebied ontgon of liet ontginnen (sociale structuren) en wat de reden tot ontginning was (economische aantrekkelijkheid); wanneer en van waaruit de ontginningen begonnen werden en hoe uiteindelijk het landschap reageerde op de ingrepen van de mens. Kennis over structuur en karakter van de nederzettingen en de nederzettingsgebieden (primaire kern of pioniernederzetting, latere nederzetting, buurschap) zijn van belang om fasering binnen de ontgin-

ning en ontwikkeling van het cultuurlandschap te herleiden. Toponymisch en topografisch (kadastraal) onderzoek en bestudering van middeleeuwse oorkonden en registers kunnen dit beeld verscherpen. Van een onder de klei verdwenen nederzetting is de kerk vanwege haar bouwmassa het meest gemakkelijk te traceren.

Kerken spelen een belangrijke rol omdat ze een aanwijzing geven over de datering van nederzettingen. De stichting van een nederzetting gaat gewoonlijk aan een kerkstichting vooraf. De oudste sporen die tot nu toe in de woldgebieden van het Oldambt en het Reiderland aangetroffen zijn, dateren uit de 12e eeuw. De verdwenen kerk van Midwolda, gebouwd in Romaanse stijl, is hiervan een voorbeeld. De herkomst van de ontginners van de woldgebieden moeten we waarschijnlijk noordwaarts zoeken, in een gordel met nederzettingen die tussen het iets verder noordwaarts gelegen terpengebied en het woldgebied ingeklemd ligt. Blokverkaveling en een Karolingische muntschat in Wagenborgen en kogelpotaardewerk met schelp-gruismagering in Meedhuizen-Wilderhof zijn aanwij-zingen voor bewoning vanaf de negende eeuw in deze gordel. Ten zuiden van de woldgebieden, op de plei-stocene ruggen die aan de woldgebieden grenzen als ook het meer zuidelijk gelegen Westerwolde, zijn geen vroeg-middeleeuwse bewoningssporen bekend.

Het middeleeuwse cultuurlandschap van de zuid-westrand van het Wold-Oldambt wordt in dit artikel belicht. De aandacht is met name gericht op de dorpen Meeden en Zuidbroek. Sociale structuren, dorps-structuren en nederzettingspatronen komen niet aan de orde, de landschapsgenese en economische aantrekke-lijkheid slechts in kort bestek.

Voordat ruilverkavelingen werden uitgevoerd bezat de zuidwestrand van het Wold-Oldambt nog een groot deel van zijn middeleeuws wegennet. Deze wegen waren bewaard gebleven ondanks het feit dat de Dollard in de eerste helft van de 16e eeuw het gebied blank zette. Deze inundatie was in dit gedeelte van het Oldambt slechts van korte duur en daarom niet destructief; be-woning van het veen was evenwel niet meer mogelijk. Het door de Dollard afgezette kleipakket wigt uit tegen de langgerekte pleistocene ruggen waarop de huidige bewoning geconcentreerd is. In de marge is het klei-pakket slechts van geringe dikte; de middeleeuwse veenbewoning schemerde er als het ware door heen. Het moet voor de dorpsbewoners dan ook niet moeilijk geweest zijn om na het terugtrekken en terugdringen van de Dollard het oude wegennet weer in gebruik te nemen. In 1391 en 1420 wordt in akten betreffende waterstaatsaangelegenheden een gedetailleerde be-schrijving gegeven van het zuidelijk Wold-Oldambt en het westelijk deel van Reiderland (Westerlee, Heiligerlee). Deze akten zijn van groot belang voor de kennis van het gebied. Doordat de middeleeuwse in-frastructuur zo goed bewaard was gebleven en tot ca. 1970 ook intact bleef, kunnen we veel van de in de akten genoemde plaatsaanduidingen nasporen (Casparie & Molema, 1990).

2. HET MIDDELEEUWSE LANDSCHAP IN GEDRANG

Waarschijnlijk in de 12e eeuw raakten het Wold-Oldambt en het woldgebied van het Reiderland bewoond. Het in cultuur brengen van deze drassige veengebieden vereiste hydrologische ingrepen. Op welke schaal die ingrepen plaatsvonden is archeologisch nog niet, of niet meer tastbaar. Goede systematiek zoals we die kennen uit de zeventiende-eeuwse Groningse veenontginningen is de middeleeuwer in ieder geval vreemd geweest. Het resultaat was dat door drainage het veen ging krimpen en oxyderen. Agrarisch gebruik versterkte de oxydatie van het veenoppervlak. Het gevolg was dat het maaiveld ging dalen, waardoor een relatieve stijging van de grondwaterspiegel optrad. In 1391, zo'n kleine 300 jaar

na de eerste ontginningen, was het reeds nodig de regionale waterafvoer door uitvoerige en harde afspraken te regelen. Het leek echter wel dweilen met de kraan open. Door de continue daling van het maaiveld was een groot, laaggelegen bekken ontstaan, waarin veenwater uit het ten zuiden van Reiderland en het Wold-Oldambt gelegen Bourtanger Moor en water uit de met afvoer-problemen kampende rivier de Eems gemakkelijk kon-den binnendringen (Casparie & Molema, 1990). De bewoners poogden door het toeslaan van natuurlijke en kunstmatige waterlopen en het leggen van dammen in rivierdalen (Oude Ae, Siepsloot) het veenwater perio-diek tegen te houden. Men kon zodoende toch nog het land bewerken en oogsten. 's Winters liet men het ge-stuwde water de vrije loop (Ramaer, 1909; Achterop et al., 1969; Casparie & Molema, 1990). Het Eemswater en zeewater dat de noordoostrand van het gebied be-dreigde, poogde men door bedijkingen te temmen. Ondanks alle maatregelen mislukte uiteindelijk de be-scherming en nam de Dollard aan het einde van de 16e en het begin van de 16e eeuw het gebied in bezit. Vanaf de tweede helft van de 16e eeuw konden delen van het verloren land langzamerhand weer worden ingepolderd. Het oorspronkelijke veenlandschap was nu echter be-dekt met een pakket klei.

3. ECONOMIE

We weten nog niet goed waarom het veengebied voor de middeleeuwse mens aantrekkelijk was. Dat het van-af de eerste ontginningen (12e eeuw) in ras tempo tot een zeer welvarend gebied werd, meen ik te kunnen afleiden uit het feit dat er in betrekkelijk korte tijd een aantal forse kruiskerken gesticht werd: Midwolda (tweede helft 12e eeuw), Eexta (vierde kwart 13e eeuw), Scheemda (twee kruiskerken, de eerste ca. 1200 en de tweede uit het derde à vierde kwart 13e eeuw), Oost-finsterwolde (geen daterende informatie), West-finsterwolde (tweede helft 13e eeuw), Zuidbroek (vierde kwart 13e eeuw) en Noordbroek (eerste helft 14e eeuw). De bouw van dergelijke grote kerken vereiste toch wel enig vermogen van de stichters.

Opgravingen die tot nu toe in het gebied verricht zijn hebben betrekkelijk weinig organisch materiaal opge-leverd, hetgeen hoofdzakelijk te maken heeft met de omvang van de onderzoeken en in mindere mate met de vraagstelling van de onderzoeker. De onderzoeken die tussen de jaren 1940-1960 in het gebied verricht zijn, waren meer gericht op gebouwsporen dan op materiaal als keramiek, bot en hout. Onderzoek op het oude kerkhof van Scheemda leverde een hoeveelheid bot-materiaal op dat in het onderzoekgebied wat omvang betreft (nog) alleen staat. Conclusies uit het onderzoek van dit materiaal kunnen dan ook niet als algemeen geldend voor het hele gebied worden aangenomen; hiervoor is verder onderzoek noodzakelijk. W. Prummel (1990) heeft het botmateriaal bestudeerd en vergeleken met andere middeleeuwse vondstcomplexen. Haar conclusie is dat in Scheemda paarden oververtegenwoordigd zijn en dat het beest niet of nauwelijks als voedselbron diende (geen slachtafval van paard aangetroffen), maar als werkkracht. Rund is het best vertegenwoordigd, direct gevolgd door paard, terwijl varken en schaap/geit de kleinste groepen vormen.

Pollenanalyse van een monster van het veenoppervlak leverde een enorme hoeveelheid roggepollen op.[1] Dat er op het veen rogge werd verbouwd weten we ook uit de kroniek van Wittewierum: "... Want als de bewoners van de Wolden geen goede roggeoogst hebben, kunnen zij niet terugvallen op vee of op zuivelproducten" (Jansen & Janse, 1991).[2] Voor roggeverbouw is het een vereiste dat het veenoppervlak droog is. Lager gelegen en dus nattere delen van het veengebied waren alleen geschikt als weide en hooiland. Rondom het oude kerkhof van Scheemda en bij een verkenning in het tracé van een gasleiding te Midwolda zijn korte greppels gevuld met allerlei afval gevonden. Vermoedelijk dienden deze greppels als winning voor zand, dat met het veenoppervlak werd vermengd, waardoor een betere bouwvoor ontstond. Een dergelijke bouwvoor kon in Scheemda in de nabijheid van het oude kerkhof over een lengte van meer dan 60 meter worden gevolgd. Borger (1989) vermoedt zelfs dat zonder vorm van bodemverbetering geen akkerbouw op het veen mogelijk was. Het afval dat we in de kuilen aantreffen is waarschijnlijk enkel gestort om de ontstane gaten op te vullen.

De polder De Wiede, gelegen bij Muntendam in het stroomdal van de Oude Ae, is rijk aan moerasijzererts. Veldverkenningen in deze polder leverden behalve kogelpotaardewerk een groot aantal ijzerslakken op, duidend op middeleeuwse exploitatie van het ijzererts. Ook bij de opgraving van het oude kerkhof te Scheemda en de bovengenoemde verkenning te Midwolda werd ijzerslak gevonden. In Vriescheloo werd bij de opgraving van een steenhuis een grote ijzerslak en twee pijpjes gevonden die de opgraver H.A. Groenendijk bestempelt als luchtpijpjes van een ertsoven (Groenendijk, 1989: p. 293). Winning en verwerking van het in het woldgebied aanwezige moerasijzererts is een economische activiteit geweest, maar op welke schaal dit gebeurde moet uit komende opgravingen blijken.

4. MEEDEN EN EEXTA

Stratingh en Venema suggereren dat het dorp Meeden vanuit Eexta gesticht is. Zij baseren dit op de vermelding van Meeden in een dekanaatsregister van het bisdom Münster (ca. 1450[3]) als Extengamedum (Stratingh & Venema, 1855: p. 48). Een tweede dekanaatsregister (1559) geeft eveneens een dorpsnaam die een relatie met Eexta zou kunnen doen vermoeden, namelijk Exterigamedum (Schmitz-Kallenberg, 1917; Siemens, 1962: pp. 23 e.v.). De naam zou 'de hooilanden van Eexta' betekenen. In navolging van Stratingh & Venema (1855), verklaren ook De Vries (1946: p. 145), Achterop (1969: p. 58) en ik zelf (Molema, 1990: p. 269) de naam op dezelfde wijze en ook Von Richthofen (1882: p. 867) ziet een verband tussen de twee dorpen, maar hij weidt er niet over uit. Mijn nieuwe opvatting is dat de relatie tussen Meeden en Eexta ten onrechte opgelegd is, doordat men de woorden 'extenga' en 'exteriga' onjuist verklaarde (zie hoofdstuk 5).

Meeden wordt in de late middeleeuwen een aantal keren vermeld. In een waterstaatkundige overeenkomst uit 1391 (Feith, 1853-1858: nr. 1391/3; Stratingh & Venema, 1855: pp. 304-308; Blok et al., 1899: nr. 814) wordt het dorp vermeld als 'Medum'. Een zoenbrief uit 1411 – de afspraken uit 1391 waren geschonden – noemt 'Meeden' (Feith, 1853-1858: nr. 1411/10).[4] In 1420, bij een bekrachtiging van de vorige overeenkomst, lezen we 'Meden' (Stratingh & Venema, 1855: pp. 309-312). In 1435 wordt een verbond gesloten tussen de kerspelen in het Oldambt en de Stad. Er is dan voor het eerst sprake van zowel 'Mieden' als 'Avermieden' (Feith, 1853-1858: nr. 1435/17).[5] In de Dijkbrief van het Oldambt uit 1441 heet het dorp 'Medenn' en ook 'Medum' (Feith, 1853-1858: nr. 1441/6; Stratingh & Venema, 1855: pp. 324-326).[6] In 1451 vindt een transactie plaats waarbij een stuk grond liggend tussen 'Broeck' en 'Meden' wordt verkocht (Feith, 1853-1858: nr. 1451/9).[7] Het bovengenoemde dekanaatsregister uit ca. 1475 geeft 'Extengamedum' en een wijdingstekst uit 1488 in de kerk van Zuidbroek noemt 'de Meden'. Op deze tekst kom ik nog terug.

In oorkonden uit 1435, 1464, 1469 en 1470 wordt Eexta aangeduid als Eghiste en Egeste. Dit leidt tot de volgende verklaring van de plaatsnaam: Geest, garst, ofwel gaast aan de rivier de Ae. Zo komen we tot de naam Aegeest ofwel Eghiste, Egeste.[8] Geest is in het algemeen de benaming voor zand- en leemopduikingen. De geomorfologische situering van Eexta, namelijk de ligging op een keileemwelving in de nabijheid van de Ae, voldoet aan deze omschrijving. De plaatsnamen die we in het onderzoekgebied aantreffen zijn afgeleiden van fysisch markante situaties binnen de parochiegrenzen. Meeden betekent madelanden; de weidelanden. Westerlee en Oosterlee (Heiligerlee) wijzen op de bebossing op de pleistocene rug waarop ze liggen. Noord- en Zuidbroek duiden op de aanwezigheid van laaggelegen (natte) landen.

5. MIEDEN, AVERMIEDEN EN EXTENGAMIEDEN

We spitsen ons nu toe op Meeden en wel de vermeldingen van het dorp in 1435, wanneer sprake is van Mieden en Avermieden, en ca. 1475, wanneer Extengamedum in het Münsterse dekanaatsregister wordt genoemd.

Het bovengeschetste proces van daling van het maaiveld door bewoning en gebruik van het veen was

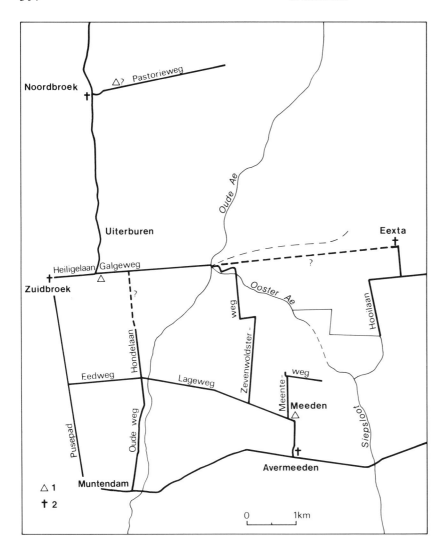

Fig. 1. De situatie voor 1970 van het zuidwestelijk deel van het Wold-Oldambt. Het middeleeuwse weg- en waterpatroon is nog grotendeels intact. 1. Verdwenen kerken; 2. Huidige kerken. Het bestaan van een oud kerkhof bij Noordbroek is niet zeker. De namen 'Galgeweg' en 'Oude weg' zijn niet oorspronkelijk. Het is niet uit te sluiten dat de Galgeweg destijds doorliep naar Eexta. Ten westen van Eexta treffen we de sporen aan van een oude bedding van de Oude Ae (onderbroken lijn). De Siepsloot sloot oudtijds aan op de Ooster Ae. Tussen de Hooilaan en de Ooster Ae loopt een gegraven watering, een deel van de in de overeenkomst van 1391 genoemde Zijtwending (tekening J.H. Zwier, B.A.I.).

een relatief snelle ontwikkeling. Binnen een periode van 300 jaar voltrok zich de catastrofe. Ondanks de maatregelen die blijkens de overeenkomsten van 1391 en 1420 waren getroffen, was het proces onomkeerbaar. Dat men zich hiervan bewust was, of in ieder geval het zekere voor het onzekere wilde nemen, blijkt in het geval van Meeden, waar een proces van dorpsverplaatsing op gang kwam. Dit is af te leiden uit de naam Avermieden. Avermieden ofwel Overmieden geeft aan dat er sprake is van een buurschap dat over, of beter gezegd buiten het oorspronkelijke dorp lag (fig. 1). De nieuwe buurschap was niet op het veen gelegen maar op de langgerekte pleistocene rug waarop het huidige dorp nog steeds ligt. Hier had men weinig last van veen- en grondwater. Het groeien van de buurschap wordt geïllustreerd door twee gebeurtenissen. In 1435 sluit behalve Mieden ook Avermieden een overeenkomst met de stad Groningen en in de tweede helft van de 15e eeuw wordt zelfs de kerk uit Mieden naar Avermieden verplaatst (fig. 2), waardoor ze niet langer een buurschap is. Door

het volledig verdwijnen van het oude dorp, aanvankelijk een langzaam proces met een acuut einde bij het doorbreken van de Dollard, werd de toevoeging 'Aver'-voor het nieuwe dorp overbodig.

De dorpsnaam of buurschapnaam Avermieden geeft reden de naam Extengamedum nader te bekijken. De Vries (1946: p. 145) suggereert dat het een verschrijving is voor Exteramedum en trekt dan, in navolging van Strathingh en Venema en Von Richthofen, de conclusie dat Meeden van Eexta uit gesticht is. Hij gaat voorbij aan de mogelijkheid Extera, of zoals het in het kerkregister staat Extenga, kortweg te zien als 'extra' hetgeen 'buiten' betekent. Exteramedum of Extengamedum is dan niets anders dan een verlatiniseerde naam voor Avermieden of Overmieden en heeft niets met Eexta te maken. Deze verwarring is vermoedelijk ontstaan doordat het kerkregister door een latijn-schrijvende bisschoppelijke ambtenaar is opgesteld die de naam Avermieden 'fraai' vertaalde met Extengamedum.[9]

Er is nog een argument om de veronderstelde relatie

Fig. 2. Kerk en toren te Meeden (foto's
J. Hovinga, Zuidlaren).

tussen Eexta en Meeden te ontkrachten, namelijk het relatief laat voorkomen van de naam Extengamedum, zo rond 1475. Zou men die naam niet vroeger in de tijd willen tegenkomen, bijvoorbeeld in de oorkonden van 1391 of desnoods 1420? Deze tijdstippen staan immers dichter bij de stichtingsdatum van het dorp. En zou men niet juist bij de oudste naamgeving van het dorp de relatie met Eexta benadrukt willen zien? De afsplitsing is dan immers nog vers. Dit is echter niet het geval. We komen de naam Extengamedum juist op een moment tegen waarop ook sprake is van Avermieden en zoals hierboven beargumenteerd is is dit niet toevallig, want de namen zijn identiek. Het naamkundige argument om een relatie tussen de dorpen Eexta en Meeden aan te nemen is hierdoor weggevallen.

Stratingh & Venema (1855: kaartbijlage) en, in navolging, Achterop (1969: p. 23) gebruiken nog een topografisch element om hun veronderstelling dat Meeden van Eexta is afgesplitst te onderbouwen. Zij suggereren dat tussen Meeden en Eexta een weg heeft gelopen welke "de olde weg van de kapel te Meeden" zou zijn geweest. Vanuit Eexta zou de Hooilaan op deze weg hebben aangesloten en zo zou er een verbinding zijn geweest tussen de kerk van Eexta en de kapel van Meeden (Van der Aa, 1846: p. 773).[10] De weg die Stratingh en Venema bedoelen is de Meenteweg (fig. 1). Deze weg liep dood tegen de Siepsloot en heeft voor zover de gegevens bekend zijn nooit aansluiting op de Hooilaan gehad (fig. 1). Het bestaan van een 'weg van de kapel van Meeden' moet dan vooralsnog worden uitgesloten. Een zekere verbinding Meeden-Eexta liep via Westerlee en Heiligerlee. Een mogelijke verbinding liep via de Zevenwoldsterweg-Galgeweg, maar het is niet duidelijk of de Galgeweg tot Eexta doorliep (fig. 1). Het is duidelijk dat het wegennet geen enkele aanleiding geeft om een moeder-dochter-relatie tussen Eexta en Meeden te veronderstellen. Indien we toch in deze trant willen argumenteren, dan is er eerder reden naar Zuidbroek te kijken, een dorp waarmee Meeden goede verbindingen had door middel van de Zevenwoldsterweg (verbastering van Soevenfollt (Achterop, 1969: p. 59), zeven bochten?) en de Lageweg (fig. 1). De Soeven-folltsterweg sloot aan op de Heiligelaan in Zuidbroek. De Lageweg had via de Eedweg (later de Eide- of Eedeweg genaamd) en het langs het huidige Muntendammerdiep (Tusschenklappen) lopende Pusepad (Stratingh & Venema, 1855: p. 103) aansluiting op Zuidbroek. Een derde verbinding via de Hondelaan is niet zeker, omdat de laatste mogelijk een 16e-eeuwse Dollarddijk is.[11]

Los van deze bespiegelingen over de infrastructuur zijn er duidelijke aanknopingspunten voor een relatie tussen Meeden en Zuidbroek. Een eerste aanwijzing vinden we in de grenzen tussen de verschillende kerspelen. Bij voorkeur werden fysische elementen zoals waterlopen als grens genomen. Zo wordt de grens tussen Meeden en Eexta gevormd door de Ooster Ae en een oude tak van de Oude Ae. De grens tussen Meeden en Westerlee is de Siepsloot (een rivier), die tevens de grens tussen de bisdommen Münster en Osnabrück vormde. Meeden lag in het bisdom Münster en Westerlee in het bisdom Osnabrück. Men zou verwachten dat de grens tussen de kerspelen Meeden en Zuidbroek (waaronder ook de buurschap Muntendam viel) door de Oude Ae gevormd zou worden. Dit is echter voor een zeer klein deel het geval. De grens loopt voor het grootste deel evenwijdig met een vrijwel noord-zuid-opstrekkende kavel. Dit wekt de indruk dat deze grens niet behoort tot de primaire kerspelgrenzen in het gebied. Wat ik zou willen suggereren is dat Meeden oorspronkelijk tot het kerspel Zuidbroek behoorde en pas in de loop van de late middeleeuwen zelfstandig is geworden. Dit zou dan ook verklaren waarom de grens door het noord-zuid-opstrekkende kavelpatroon snijdt, terwijl juist deze noord-zuid-georiënteerde kavels zijn vanuit Meeden ontgonnen. Mogelijk zijn individuele bezits-verhoudingen de reden geweest voor deze zo op het oog willekeurige kerspelgrens. Zuidbroek heeft oost-west georiënteerde kavels.

Een tweede, meer concrete aanwijzing voor een relatie tussen de dorpen vinden we terug in de juridische indeling van het onderzoekgebied. Meeden en Zuidbroek vormden met Noordbroek namelijk één rechtstoel. Oorspronkelijk viel de omvang van een rechtstoel samen met die van de kernparochie (de primaire nederzetting). Buurschappen met een groeiend aantal inwoners, zoals Meeden, werden op den duur van de kernparochie afgescheiden en vormden zelfstandige parochies. De rechtstoel behield echter zijn oorspronkelijke omvang en ging verschillende parochies omvatten (Cleveringa, 1927; De Blécourt, 1935).

Een derde aanwijzing voor de veronderstelde relatie vinden we op het westelijke gewelf in de kerk van Zuidbroek. Daar is bij een restauratie een wijdingstekst te voorschijn gekomen (Achterop et al., 1969; Pathuis, 1977: nr. 515). Uit de tekst is een kerkelijke relatie tussen Zuidbroek en Meeden af te leiden.[12] Een van de in de tekst genoemde personen is van belang, want deze (zijn naam is onleesbaar) blijkt vicarius (hulppastor) 'van de Meden' te zijn. Deze vicarius moet als basis wel Zuidbroek hebben gehad, want anders is de reden tot zijn vermelding in de wijdingstekst wel erg duister. De geestelijke verzorging van de Medemer parochie werd kennelijk uit Zuidbroek ondersteund. Het ligt dan voor de hand te suggereren dat deze relatie is ontstaan als een vervolg op de (veronderstelde) afsplitsing van Meeden van Zuidbroek. Mogelijk heeft de buurschap Meeden, dus voorafgaand aan de afscheiding, al over een kapel beschikt die vanuit Zuidbroek werd bediend.

6. BROEK, SUTHABROKE, NORTBROKE, UITERBUREN

De aanwezigheid van een oud kerkhof bij Zuidbroek is diverse keren in de literatuur vermeld. Voor het eerst gebeurde dat in 1573 door de Stadssyndicus Doede Tyarcks (Antonides, 1973: p. 80). Ook de kaart van Nicolaas Visscher 'De Provincie van Stadt en Lande' situeert een oud kerkhof aan de Heiligelaan, bij de molen. Stratingh & Venema (1855: kaartbijlage) plaatsen zowel in Zuidbroek als Noordbroek een oud kerkhof. Dit doet ook Craandijk (1890: pp. 203 en 204) in zijn verslag van wandelingen door Nederland. In 1984 werd de aanwezigheid van het kerkhof in Zuidbroek archeologisch bevestigd: in een tuin aan de Galgeweg (hoek Uiterburen-Heiligelaan, coördinaten 254.450-577.050; fig. 1 en 3) werd een aantal skeletten gevonden die op christelijke wijze waren begraven.[13]

Het door Stratingh en Venema en Craandijk genoemde oude Noordbroekster kerkhof zou enkele honderden meters ten oosten van de huidige kerk gelegen hebben. De beweerde aanwezigheid berust waarschijnlijk op overlevering. Concrete bewijzen voor het bestaan van een oud kerkhof zijn mij niet bekend, maar daarom is de overlevering nog niet direct onjuist. In de toren van de huidige kerk van Noordbroek is gebruikte tufsteen verwerkt als muurvulling (De Olde, 1975: p. 61). Dit is geen bewijs voor het bestaan van een (tufstenen) voorganger, maar het geeft wel aan dat een uitgebreide veldverkenning op de door Stratingh en Venema aangeduide locatie van het oude kerkhof op zijn plaats is.

In de kroniek van Wittewierum wordt in 1273 de naam Broke (Broek) vermeld (Jansen & Janse, 1991: pp. 452 en 453). In 1283 is sprake van de parochie Suthabroke (Jansen & Janse, 1991: p. 460). De naam Nortbroke komen we voor het eerst tegen in de waterstaatkundige overeenkomst uit 1391. Ook de namen Suidbroeck en Broke komen in dit stuk nog een keer voor. Uit de tekst blijkt dat met Broke het gebied rond Noord- en Zuidbroek wordt bedoeld "... alle watertochten tusschen Muntendam ende Broke ....".

Ik stel me de volgende ontwikkeling van de nederzettingen voor, ervan uitgaande dat Noordbroek inderdaad een oud kerkhof heeft. Ten oosten van het huidige Noordbroek wordt in de 12e eeuw een nederzetting gesticht met de naam Broek. Dit is de primaire nederzetting en haar totale grondgebied is de kernparochie (fig. 3). De bewoning breidt zich naar het zuiden uit en in de tweede helft van de 12e of eerste helft van de 13e eeuw wordt van de kernparochie een deel afgescheiden met als nederzetting Zuidbroek (fig. 3). De primaire nederzetting Broek gaat vanaf dat moment Noordbroek heten. De naam Broke blijft echter tot ver in de 15e eeuw in zwang voor Noordbroek.[14]

Binnen het grondgebied van de parochie Zuidbroek volgen nieuwe ontwikkelingen. Ten zuiden en zuidoosten van de nederzetting Zuidbroek ontstaan twee buurschappen namelijk Muntendam (als zodanig vermeld in 1435) en Meeden. In de waterstaatkundige overeenkomst van 1391 is reeds sprake van een dam in de Oude Ae (Munter Ae). Deze dam bestaat nog steeds, en heet Brede weg. Muntendam groeit in de middeleeuwen niet dusdanig dat het een parochie kan gaan vormen.[15] Dit is wel het geval met Meeden dat in ieder geval in de overeenkomst van 1391 de indruk geeft boven de status van buurschap uitgegroeid te zijn en eigen karspelgrenzen en een kerkhof heeft (fig. 3). Hoewel Meeden een zelfstandige parochie is geworden, blijft het, zoals gebruikelijk, onder de oude rechtstoel vallen (Cleveringa, 1927: p. 100; De Blécourt, 1935: pp. 223-229).

De oostelijk gelegen gronden van Noord- en Zuidbroek waren laag en drassig. Dit geeft de term 'broek' reeds aan en ook de kroniek van Wittewierum maakt hiervan in 1273 melding: "Tijdens deze periode van schaarste werden vooral de bewoners van de Wolden in het nauw gedreven, met name in Astawalda (Oostwold bij Siddeburen) en in Broke, vanwege de lage ligging van de akkers ...." (Jansen & Janse, 1991). Het zou kunnen zijn dat door het boven beschreven proces van daling van het maaiveld dat in het gehele woldgebied optrad, de broeklanden reeds in een eerder stadium onbruikbaar waren dan de overige veengronden. De bewoning schoof op naar de bruikbare, hoger gelegen gronden. Er was echter geen sprake van gedwongen verplaatsing door vernatting, maar verplaatsing uit bedrijfstechnische overweging. Men wilde niet te ver van de bruikbare gronden vandaan wonen. In het geval van Zuidbroek schoof niet de hele oude bewoning op, want de streek Uiterburen, die ik als een deel van de

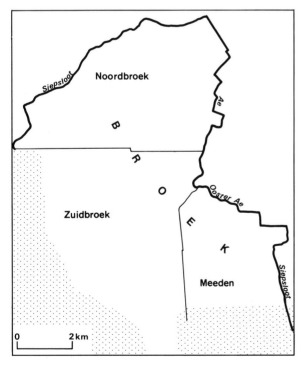

Fig. 3. De kernparochie Broek met daarin de in de loop der middeleeuwen ontstane opdeling in de parochies Noorbroek, Zuidbroek en Meeden. Raster: onontgonnen veen.

oude bewoning beschouw, is continu bewoond gebleven.

In Zuidbroek werd door het opschuiven van de grootste bewoningsconcentratie ook de kerk verplaatst. De ligging van het oude kerkhof in Uiterburen hoog op een keileemwelving, de Garst genaamd, maakt duidelijk dat het nooit vanwege directe waterdreiging (veenwater/Dollardwater) verlaten kan zijn. Los hiervan kunnen we stellen dat het water in de 13e eeuw in het hele woldgebied nog niet bedreigend was. Omstreeks deze tijd werd immers te Scheemda nog een kerk op het veen gebouwd (Molema, 1989: pp. 130-134). Concrete signalen van chronische wateroverlast zijn de genoemde overeenkomsten uit 1391 en 1420 en de gedwongen geleidelijke dorpsverplaatsing van Meeden in de eerste helft van de 15e eeuw. Maar zoals gezegd kan de specifieke lage ligging van het oostelijk deel van de Zuid- en Noordbroekster gronden de reden zijn geweest voor een vroegere dorpsverplaatsing dan we van omringende nederzettingen gewend zijn. Bovendien lagen deze broeklanden in het stroomgebied van de Oude Ae, waardoor ze gevoelig waren voor het buiten de oevers treden van deze rivier.

De huidige Zuidbroekster kerk en de opgegraven sporen van een kerkhof in Uiterburen liggen op dezelfde opstrekkende kavel, de pastorieheerd. De huidige kerk dateert uit het laatste kwart van de 13e eeuw (Ozinga, 1940: pp. 251-253). Dit geeft aan dat we een voorgangerkerk in ieder geval vroeger moeten dateren. De aanwezigheid van baksteen spreekt voor een datering na ca. 1150.

De naam Uiterburen is geen primaire naam, want zoals reeds werd betoogd vereenzelvig ik het oude Zuidbroek met het huidige Uiterburen. De naam Uiterburen is ontstaan toen een groot deel van de bewoning opschoof naar het huidige Zuidbroek. De oude bewoningsconcentratie verviel hiermee tot een buurschap. Er zijn echter nog meer middeleeuwse sporen in Uiterburen gevonden die de veronderstelling van het bestaan van de parochie Broek enige kracht bijzetten.

Er hebben in ieder geval drie steenhuizen gestaan (fig. 4): het Gockinga-steenhuis (de latere Drostenborg; nu nog bestaand als renteniersbehuizing), een steenhuis waarvan de funderingen in het voorhuis van boerderij Uiterburen nr. 49 zijn verwerkt[16], en een steenhuis waarvan de grondslagen zijn opgegraven (Antonides & Van Zeist, 1957).[17] Steenhuizen van dit type, met een eenvoudig rechthoekig of vierkant grondplan, komen in ieder geval al sinds de 13e eeuw voor. De weg door Uiterburen en Noordbroek heeft een opvallend kronkelend karakter in vergelijking met andere middeleeuwse wegen in het gebied (fig. 1 en 4). Mogelijk wijst dit op een hogere ouderdom van de weg.

De huidige Noordbroekster kerk wordt in de eerste helft van de 14e eeuw gedateerd (Ozinga, 1940: pp. 98-100). Dit geeft aan dat het proces van dorpsverplaatsing, die we veronderstellen, later op gang kwam dan in Zuidbroek. Indien de in de toren van de huidige kerk aangetroffen tufsteen afkomstig is van een voorgangerkerk, dan is deze kerk te plaatsen vóór ca. 1150. Hiermee zou deze vooralsnog hypothetische kerk tot de oudste kerken in het Reiderlander en Oldambster woldgebied behoren.

Een kanttekening is op haar plaats bij wat hierboven over de nederzettingen Noordbroek en Zuidbroek gezegd is. Er is al gesteld dat voor de aanwezigheid van een oud kerkhof in Noordbroek geen concreet bewijs voorhanden is. Het is dan ook mogelijk dat er geen oud kerkhof is. Ook bij het bestaan van het oude kerkhof van Zuidbroek is nog een vraagteken te plaatsen. De bij de opgraving gevonden skeletten lagen in zandige grond. In kalkarme zandgrond wordt botmateriaal snel afgebroken. Het is dan ook niet zeker dat het in Uiterburen botmateriaal uit de 12e of 13e eeuw kan stammen. Een ^{14}C-datering van het aangetroffen skeletmateriaal biedt mogelijk meer zicht op het karakter van het kerkhof. Mocht het skeletmateriaal jonger zijn dan verwacht, maar nog wel van middeleeuwse ouderdom, dan zouden we dat moeten verklaren als horend bij een kapelletje dat vanuit Zuidbroek gesticht is. Dat er een kerk of

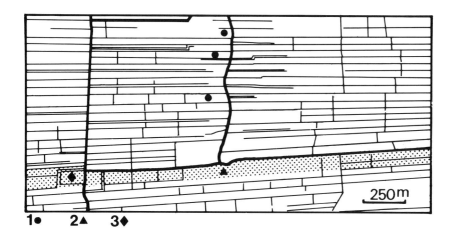

Fig. 4. De topografische kaart van 1911 dient als ondergrond waarop tot nu toe bekende middeleeuwse gebouwsporen in Zuidbroek geprojecteerd zijn. 1. Steenhuizen; 2. Het kerkhof aan de Galgeweg; 3. De huidige kerk van Zuidbroek. De pastorieheerd is gearceerd.

een kapel heeft gestaan lijdt weinig twijfel gezien het puin dat tijdens de opgraving werd aangetroffen.[18] Het toeschrijven van de skeletten aan veroordeelden die aan de galg (Galgeweg) gehangen hebben is vanwege de aanwezigheid van het puin vooralsnog onjuist.

7. KERNPAROCHIES IN HET WOLD-OLDAMBT EN REIDERLAND; EEN HYPOTHESE

Hierboven is het bestaan van een kernparochie met de naam Broek gesuggereerd. Deze suggestie geeft aanleiding het Wold-Oldambt en het Reiderland, gebieden die het onderwerp zijn van een onderzoek van de auteur naar hun middeleeuwse nederzettingsgeschiedenis, eveneens in hypothetische kernparochies in te delen. Deze kernparochies worden in figuur 5 getoond. De indeling is gebaseerd op natuurlijke grenzen, in dit geval rivieren; de grens tussen de bisdommen Münster en Osnabrück, welke gelijk is aan de grens tussen Reiderland en Wold-Oldambt, en als derde element de grenzen van juridische districten, de rechtstoelen, waarvan de omvang samenvalt met die van de kernparochies (Cleveringa, 1927; De Blécourt, 1935). In feite zou de kennis over de omvang van de rechtstoelen voor een indeling in kernparochies voldoende zijn, maar juist over de omvang van deze rechtsgebieden zijn we slecht ingelicht.

De voorgestelde indelingen van het Reiderland en Wold-Oldambt in kernparochies is als volgt. Ik veronderstel dat het Wold-Oldambt twee kernparochies heeft gehad: Broek en Menterwolde. De grens tussen Broek en Menterwolde is de Ooster Ae en de Termunter Ae. Van Broek is bekend dat het een rechtstoel vormde. De grens tussen het Wold-Oldambt en Reiderland wordt grotendeels gevormd door de Siepsloot en de Tjamme. De Pekel A verdeelt op haar beurt Reiderland in twee kernparochies (fig. 5). Het Duitse deel en het door de Dollard opgeslokte deel van Reiderland zijn niet in de hypothese verwerkt. Oorspronkelijk zou het landschap een vijfdeling hebben gehad. Opvallend is dat bij de opsplitsing van de kernparochies in kleinere delen, de parochies, de grenzen geen natuurlijke begrenzingen omvatten. Men maakte altijd gebruik van kunstmatige grenzen, die langs opstrekkende kavels zijn getrokken. Het opdelen van de kernparochies in kleinere parochies is in een snel tempo gebeurd. Dit is af te leiden uit de bouwdata van de kerken. Blijkbaar was er sprake van een snelle bevolkingsgroei.

In het Wold-Oldambt en Reiderland treffen we een aantal lange bewoningslinten aan. Ze passen redelijk binnen de vier voorgestelde kernparochies. We kunnen de hypothese verder uitbreiden door te pogen de primaire nederzettingen binnen de kernparochies aan te wijzen. We doen dit aan de hand van bekende of vermoede bouwdata van kerken. Voor de kernparochie Broek is Noordbroek al als primaire nederzetting voorgesteld. In Menterwolde komt Midwolda in aanmerking met haar

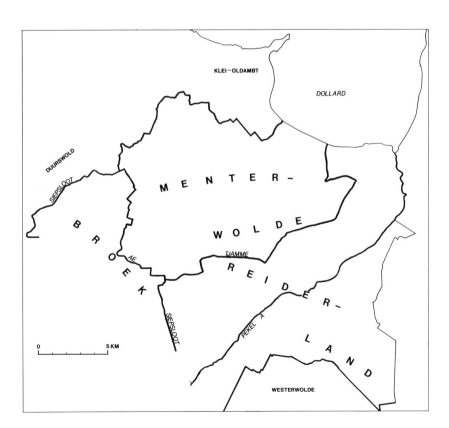

Fig. 5. De hypothetische vierdeling in kernparochies van het Wold-Oldambt en Reiderland. De grenzen tussen de kernparochies (vette lijnen) worden voor het grootste deel gevormd door rivieren.

voormalige 12e-eeuwse kerk. De Reiderlander kernparochie tussen de Tjamme en de Pekel A heeft mogelijk
in Beerta haar primaire nederzetting. De huidige Beerster
kerk heeft nogal wat secundair verwerkte tufsteen, die
vermoedelijk van een voorganger afkomstig is. Deze
voorganger moet gezien de tufsteen van voor ca. 1150
dateren. Voor de ten zuiden van de Pekel A gelegen
kernparochie is het lastiger een mogelijk primaire nederzetting aan te wijzen. Bellingwolde komt voorlopig
het meest in aanmerking. Haar kerk, ten onrechte als
gotisch bestempeld, heeft afgezien van het koor, een
romaanse grondslag.

8. NOTEN

1. Mondelinge mededeling Prof.dr. S. Bottema.
2. Waar de vertalers in de tekst spreken van tarwe moet rogge
   worden gelezen.
3. Enige voorzichtigheid bij het gebruik van de dekanaatsregisters
   is geboden. De datering van de eerste lijst is gebaseerd op
   paleografische gronden (Von Ledebur, 1836) en is dus vaag.
   Bovendien werden lijsten vaak klakkeloos overgeschreven zodat
   ze niet de meest recente stand van zaken in een gebied hoeven te
   weerspiegelen en bovendien fouten kunnen bevatten.
4. Rijksarchief Groningen (R.A.G.), V.V.A. 1411/10.
5. R.A.G, Familiearchief De Sitter nr. 220.
6. Gemeente Archief Groningen, nr. 1381 (rood, vóór de Reductie).
   Stratingh & Venema, 1855: pp. 324-326.
7. R.A.G., inv. 'Losse stukken' nr. 219.
8. Von Richthofen (1880-1882: p. 867) komt tot een halve verklaring en geeft: "Kirchdorf Eexta, d.i. 'Egeste', an der Ee, ...". J.
   Naarding (1958) denkt aan een samenvoeging van 'eek' (van eik)
   en 'sate', doch dit lijkt me onjuist.
9. Dit gebeurde bijvoorbeeld ook met de dorpsnamen Lutjegast,
   vertaald met Minorgast en Tolbert, vertaald met Antiquabercht.
10. Volgens Van der Aa zijn in Meeden de grondslagen van een
    kerkje of kapel (= een niet-parochiale kerk) blootgelegd. Ook
    meldt hij sporen van het oude dorp in de vorm van puin, her en
    der langs de Lageweg.
11. Bestaande wegen werden vaak gebruikt om er dijken op aan te
    leggen en dijken werden op hun beurt vaak gebruikt om er wegen
    op aan te leggen. Oorsprong en datering van de Hondelaan is
    daarom moeilijk in te schatten.
12. De tekst luidt (naar Pathuis, 1977: nr. 515): "Als men scref dusent
    vier hondert jar – was ...s acht und tachtigh – do was dusse kerke
    gesacrert – ...od ...nder ... iel ... – de do ... was bet her hermen –
    her ... vicarius van de meden – her ido syn eerste mysse sanck –
    de kerkvogede sint dus genant – eltio ellyks wat bekent –
    aybbo sibens ock myt siner vlyt – egge fokens god sye benedyt".
13. Opgraving van het Biologisch-Archaeologisch Instituut o.l.v.
    J.W. Boersma met medewerking van G. Delger en K. Klaassens.
    In 1938 zouden op dezelfde plaats ook al een aantal skeletten
    gevonden zijn.
14. Informatie Dr. R.W.M. van Schaïk.
15. Muntendam wordt in 1838 zelfstandig.
16. Melding O.S. Knottnerus, Zuidbroek.
17. De auteurs menen overigens ten onrechte dat ze het Gockinga-
    steenhuis hebben blootgelegd. Deze lag iets verder noordelijk op
    de plaats van de zogenaamde Drostenborg (fig. 4). Na verovering
    door de stad Groningen werd het steenhuis van de Gockinga's
    zetel van de drost.
18. Mijn twijfel aan het bestaan van oude kerkhoven in Zuidbroek en
    Noordbroek (Molema, 1993: p. 269, noot 6) is zeker wat het
    eerste dorp betreft onjuist gebleken.

9. LITERATUUR

AA, A.J. VAN DER, 1946. *Aardrijkskundig woordenboek der Nederlanden*, deel 7. Gorinchem.

ACHTEROP, S.H., P. VAN DER WAL, G.G. WOLTHUIS, 1969. *Meeden; geschiedenis van een Gronings dorp*. Groningen.

ANTONIDES, H., 1973. *Noord- en Zuidbroek in vroeger jaren*. Noordbroek.

ANTONIDES, H. & W. VAN ZEIST, 1957. Een laat-middeleeuwse woontoren bij Zuidbroek. *Groningse Volksalmanak*, pp. 165-170.

BLÉCOURT, A.S. DE, 1935. *Oldambt en de Ommelanden*. Assen.

BLOK, P.J., J.A. FEITH, S. GRATAMA, J. REITSMA & C.P.L. RUTGERS, 1899. *Oorkondenboek van Groningen en Drente*. Groningen.

BORGER, G.J., 1989. Mittelalterliche Kolonisation von Marsch- und Moorgebieten. In: *Wilhelmshavener Tage 2; Ländliche und städtische Küstensiedlung im 1. und 2. Jahrtausend*. Wilhelmshaven, pp. 76-90.

CASPARIE, W.A. & J. MOLEMA, 1990. De ontwikkeling van het middeleeuwse veenontginningslandschap in Oost-Groningen. De opgravingen op het 'Oud Kerkhof' van Scheemda in een veenkundig, hydrologisch en historisch perspectief. *Palaeohistoria* 32, pp. 271-289.

CLEVERINGA, R.P., 1927. *Ontwikkeling van het rechtsbestel der stad Appingedam*. Groningen.

CRAANDIJK, J., 1890. *Wandelingen door Nederland; Friesland, Groningen, Drenthe en Overijssel*. Haarlem.

FEITH, H.O., 1853-1858. *Register van het Archief van Groningen*. Groningen.

GROENENDIJK, H.A., 1989. Dollartflucht oder allmähliche Siedlungsverschiebung? Ein Steinhaus im überschlickten Moor bei Vriescheloo (Gem. Bellingwedde, Prov. Groningen). *Palaeohistoria* 31, pp. 267-305.

JANSEN, H.P.H. & A. JANSE (editie en vertaling), 1991. *Kroniek van het klooster Wittewierum*. Hilversum.

LEDEBUR, L. VON, 1836. *Die fünf münsterschen Gaue und die sieben Seelande Friesland's: ein Beitrag zur Geographie des Mittelalters*. Berlin

MOLEMA, J., 1989. Het 'Ol Kerkhof' te Scheemda. *Groninger Kerken* 4, pp. 130-134.

MOLEMA, J., 1990. De opgravingen op het kerkhof van het verdronken dorp Scheemda. *Palaeohistoria* 32, pp. 247-270.

NAARDING, J., 1958. De plaatsnaam Eekst als oorkonde in de geschiedenis van het Drents. *Driemaandelijkse bladen* 10, pp. 43-49.

OLDE, H.G. DE, 1975. De Ned. Hervormde kerk te Noordbroek. *Publicaties Stichting Oude Groninger kerken* 14, pp. 49-79.

OZINGA, M.D., 1940. *De Nederlandsche monumenten van geschiedenis en kunst*, VI. *Oost-Groningen*. 's-Gravenhage.

PATHUIS, A., 1977. *Groninger gedenkwaardigheden*. Assen, etc.

PRUMMEL, W., 1990. Draught horses and other animals at late-medieval Scheemda. *Palaeohistoria* 32, pp. 299-314.

RAMAER, J.C., 1909. De vorming van den Dollart en de terpen in Nederland in verband met de geografische geschiedenis van ons polderland. *Tijdschrift van het Koninklijk Nederlandsch Aardrijkskundig Genootschap* 26, tweede serie, pp. 1-61.

RICHTHOFEN, K. VON, 1880-1882. *Untersuchungen über friesische Rechtsgeschichte*. Berlin.

SCHMITZ-KALLENBERG, L., 1917. Zur Geschichte des friesischen Offizialats und Archidiakonats der münsterischen Diözese im 16. Jahrhundert. *Zeitschrift für vaterländische Geschichte und Altertumskunde* 75, pp. 281-296.

SIEMENS, B.W., 1962. *Historische atlas van de provincie Groningen*. Groningen.

STRATINGH, G.A. & G.A. VENEMA, 1855. *De Dollard*. Groningen.

STINSEN EN HET ELITE-NETWERK IN DE MIDDELEEUWSE BEWONINGSGESCHIEDENIS VAN SNEEK EN HAAR OMMELAND

A. JAGER

Grote Kerkstraat 224, Leeuwarden, Netherlands

ABSTRACT: This paper deals with fortified stonehouses (*stinsen*) in Sneek and in the adjacent clay area to the north of the town.

In the 10th century settlement on the salt marshes along the southern edge of Middelzee started. In the 11th century those salt marshes were protected by a dike along the Middelzee. Most of the land was in use by then, with the exception of a few badly drained areas. During the 12th century reclamation of the peaty areas south of Sneek started to create problems because large amounts of water had to be drained off towards the Middelzee. By constructing the Hemdijk and the Groene Dijk, and a series of smaller dikes connecting these two with the dike along the edge of the Middelzee, a series of small polders, each with its own water drainage, were formed. From the end of the 12th century onwards the southern part of the Middelzee silted up rapidly. The newly formed salt marshes were endiked, and brought into cultivation as soon as possible.

Most of this new land was owned by private individuals, probably largely belonging to the local nobility. Noblemen in this area were also farmers. They distinguished themselves from the ordinary farmers a.o. by building small motte-castles, in the form of fortified stonehouses on mounds. Several of these mounds are still visible, although the stonehouses have disappeared, and are known as *stinswieren*. The oldest stonehouses date from around 1150 AD, and are supposed to have been used largely as places of refuge. Gradually these were replaced by towerhouses with a well-defined living function. In both cases daily life took place in a farmhouse next to the tower. The fortified stonehouses on *stinswieren* are considered to be an expression of seigniorial rights, in the absence of central authority.

In 1990 the remains of a largely levelled *stinswier* were excavated near Bons. The mound was surrounded by an almost square wooden stockade and a square moat. Of the stonehouse no traces were left. On the basis of pottery found during the excavation the construction of this *stins* can be dated between 1150 and 1250 AD. The building was knocked down in the 15th century.

During the urbanization of Sneek, shortly after the reclamation of land in the surrounding area, stonehouses were also built in the town. These stonehouses differed from the ones in the rural areas in the absence of a mound and a farm. Three *stinsen* are known to have existed in Sneek, the Roedenburg, the Johansmastins and the Gruytersmastins. The Johansmastins is only known from a single reference in a charter of 1442. The other two stonehouses were fortified but had different purposes. The Roedenburg was probably built by its owner on a plot within Sneek bordering on his property north of the town. By building this stonehouse seigniorial rights were claimed on the owner's property inside Sneek. The Gruytersmastins is known from a bird's-eye view of Sneek from 1616. Archaeological excavations in 1984-85 revealed that the construction of the *stins* can be dated to the first half of the 14th century. In 1399 it was owned by the merchant Aylof die Gruyter, whose wife belonged to the Frisian nobility.

The importance of the Roedenburg and the Gruytersmastins as fortifications within the town was demonstrated in 1399-1400. In this period the owners of both stonehouses sided with the Count of Holland in his struggle for the lordship of Friesland. When the Count withdrew from Friesland the Roedenburg was destroyed by the citizens of Sneek, probably because of the claims of its owner. The Gruytersmastins was spared, however, probably because the owner had no seigniorial claims, and because the *stins* was integrated in the economy of the town.

KEYWORDS: Friesland, Sneek, Middle Ages, land reclamation, seigniorial rights, fortified stonehouses, excavation.

1. INLEIDING

In laat-middeleeuws Friesland tekent zich een duidelijke elite af. Onderscheid kan gemaakt worden in elite op het platteland en die in de steden. In het eerste geval hebben we te maken met de traditionele Friese adel, die vooral bezit op het platteland heeft, en daarover een vrijwel autonome macht heeft. In het tweede geval is er sprake van een stadselite. Deze bestaat uit het stadspatriciaat, een toplaag van handelaars en ambachtslieden, maar tevens uit leden van de plattelandsadel.

Van de middeleeuwse weerbare steenhuizen of

stinsen, die de Friese elite op het platteland of in de steden bouwde, is nauwelijks iets overgebleven. Inzicht hierover zal door de archeologie moeten worden verschaft. In de gemeente Sneek zijn voorbeelden van zowel een plattelands- als een stadsstins onderzocht: op het platteland bij Bons een grotendeels geëgaliseerde stinswier, in de stad zelf de resten van de zogenaamde Gruytersmastins. Opvallend is, dat stinsen in Sneek en haar ommeland alleen voorkomen op de kleigronden, en afwezig zijn in het uitgestrekte veengebied ten zuiden van Sneek. Daarom zal in het volgende betoog de aandacht vooral uitgaan naar de kleigronden rond Sneek.

Nagegaan zal worden op welke manier de stinsen functioneerden in de kolonisatie en cultivering van het ommeland, en in de urbanisatie van Sneek. Tevens zal gekeken worden in hoeverre vergelijkingen zijn te trekken tussen stinsen op het platteland en in de stad.

2. LANDSCHAPSVORMING EN OUDSTE BEWO-NINGSGESCHIEDENIS

2.1. Plattelandsstinsen

Belangrijk voor een Friese edelman in de late middeleeuwen was het bezit van een weerbaar stenen huis, of stins, die tot ca. 1250 op een stinswier werd gebouwd. Een stinswier, ook wel hege wier of wier genoemd, was een omgrachte aarden heuvel met een diameter van ca.

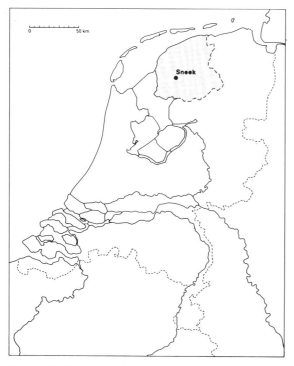

Fig. 1. De ligging van Sneek.

30 m, en een hoogte van 5 à 10 m (Kramer, 1988: p. 214). Omdat in de tweede helft van de 12e eeuw baksteen reeds bekend is als bouwmateriaal, zullen de torens op de wieren hiervan zijn gebouwd. Tot dusver zijn geen stinswieren in Friesland van vóór 1150 bekend. Wanneer er torenheuvels vóór die tijd zijn opgericht, zullen die een bekroning van hout of tufsteen hebben gehad.

De bakstenen toren diende als statussymbool, maar ook als tijdelijk verblijf gedurende conflictsituaties, bijvoorbeeld met andere adellijke families. De oudste stinsen waren stellig niet bestemd voor langdurige bewoning: het waren vluchttorens (Meijer, 1962: p. 262). Pas in de 13e eeuw ontstaan stinstorens die geschikt zijn voor permanente bewoning, zoals de Schierstins te Veenwouden. Dit soort stinsen kan als woontorens beschouwd worden (Temminck Groll, 1963: pp. 11-20). Zowel bij de vlucht- als de woontorens bevond zich buiten de omgrachting een nederhof of boerderij. In het geval van de vluchttoren speelde het dagelijkse leven van de adellijke familie zich in de boerderij af. Bij de woontorens bleef de boerderij gehandhaafd. Boerderijen bij stinswieren hadden een eigen omgrachting. Stenen torens op wieren kunnen het best beschouwd worden als kleine mottekastelen (Janssen, 1990: pp. 228-233).

2.2. Verspreiding der stinswieren

Stinswieren kwamen niet voor in de Friese steden; slechts op het platteland waren ze verspreid. In de omgeving van Sneek bevindt zich een concentratie van stinswieren op het kleidek ten noorden en oosten van Sneek. Dit kleidek is in de vroege middeleeuwen afgezet door de Middelzee (Cnossen, 1971: pp. 62-66). Hierbij werd een groot deel van het veen opgeruimd, dat zich in de late ijzertijd ten noorden van Sneek uitstrekte. De laat-middeleeuwse scheidslijn tussen klei- en veengrond heeft ruwweg een noordoostelijke-zuidwestelijke oriëntatie ten opzichte van Sneek. De overgang tussen klei- en veengebied is niet abrupt; er is namelijk over een brede strook een, tot 3 m dikke kleilaag op veen afgezet (Stiboka, 1974; fig. 2). Zowel in het klei-op-veengebied, als in het noordelijk aangrenzende kleigebied komen stinswieren voor. In het veengebied ten zuiden van Sneek zijn ze afwezig. Stinswieren worden pas opgeworpen als de bewoningsgeschiedenis van het Sneker ommeland al ruim een anderhalve eeuw gaande is.

2.3. Middeleeuwse bewoning

Doordat de Middelzee ten noorden van Sneek in de 10e eeuw dichtslibde, kwam droogvallende kweldergrond beschikbaar voor bewoning. Scherven van reliëfband-amforen met kenmerkende radstempeling uit de 10e eeuw zijn gevonden bij Sate Duinterpen, ten zuiden van Sneek en op de Stadsfenne, ten oosten van Sneek (fig.

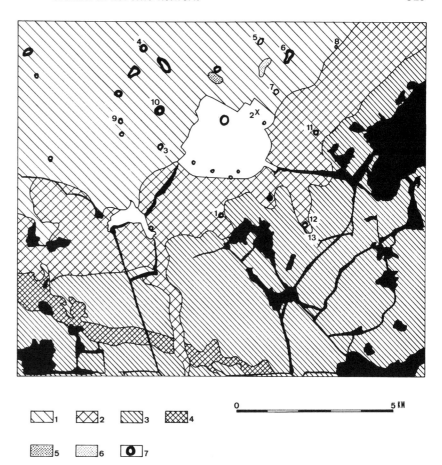

Fig. 2. De bodemkaart van het ommeland van Sneek. De centrale witte vlek is het huidige stadsgebied van Sneek; links daaronder ligt IJlst. Legenda: 1. Klei; 2. Klei-op-veen; 3. Veenmosveen; 4. Zand ondieper dan 120 cm; 5. Zware zavel; 6. Lichte zavel; 7. Terpen. Gehuchten en dorpen: 1. De terp van Sate Duinterpen; 2. Stadsfenne; 3. Bons; 4. Tirns; 5. Scharnegoutum; 6. Goënga; 7. Loënga; 8. Gauw; 9. Folsgare; 10. IJsbrechtum; 11. Offingawier; 12. Oppenhuizen; 13. Uitwellingerga.

2). De buurtschap Bons, ten westen van Sneek, heeft dezelfde ouderdom; 'Bottinge' wordt namelijk vermeld in een 10e-eeuwse lijst als bezit van het klooster Fulda (Halbertsma, 1963: pp. 116-118). Pas in de 11e eeuw zijn nederzettingen over het gehele kweldergebied van Sneek en haar ommeland verspreid, gezien de verspreiding van Pingsdorf- maar ook lokaal aardewerk. Alle dorpen in dit gebied, zoals Tirns, Scharnegoutum, Goënga, Loënga, Gauw, Folsgare, IJsbrechtum, Offingawier, Oppenhuizen, Uitwellingerga, alsook de meeste gehuchten hebben dit materiaal opgeleverd (fig. 2). In deze periode wordt het grootste gedeelte van het gebied gekoloniseerd.[1] De laagst gelegen terreinen konden pas na de bedijking in gebruik worden genomen.

Van de nederzettingen in deze periode veronderstellen we dat het buurtschappen of individuele boerderijen zijn geweest. De enige tot dusver onderzochte buurtschap in dit gebied is Bons, die in de 11e eeuw uit 2 à 4 boerderijen kan hebben bestaan.[2] In verband met de nog niet geregelde waterstaat in dit gebied, manifesteren de nederzettingen uit deze periode zich nog als terpen, die permanent bewoond blijven. Bewoning in het kweldergebied eindigde noordwaarts aan de oever van de Middelzee, die in de 11e eeuw werd bedijkt (Spahr van der Hoek, 1974: pp. 19-20; fig. 3).

In de 12e eeuw worden de gevolgen van de op grote schaal in cultuur gebrachte venen ten zuiden van Sneek merkbaar; overtollig veenwater wordt geloosd op het kweldergebied (de Cock, 1984). Tegen dit water worden binnendijken opgeworpen: de Hemdijk, lopend over ongeveer de noordgrens van het klei-op-veengebied en de Groene Dijk, ten oosten van Sneek.[3] Tussen de zuidelijke Middelzeedijk en de Hemdijk en Groene Dijk worden dijken aangelegd, zodat in de 12e eeuw een reeks polders ontstaat, waarmee getracht wordt de waterstand te beheersen (Rienks & Walther, 1954: pp. 129-134, 138-145 en 150). Lager gelegen, voordien ongeschikte terreinen, worden nu eveneens in gebruik genomen.

Vanaf de late 12e/vroege 13e eeuw heeft men het zuidelijke deel van de Middelzee ingepolderd. Deze inpoldering werd mogelijk doordat de Middelzee versneld dichtslibde. Dit was het gevolg van verschillende zeedoorbraken, vooral door die in 1170, ten westen van Friesland, waardoor de Zuiderzee ontstond. De druk van het zeewater op deze nieuw gevormde zee deed de watervloed in de Middelzee, alsook in de Lauwerszee afnemen (Gottschalk, 1971: p. 194).

Zowel op de kleigrond ten zuiden van de Middelzee, als op de voormalige Middelzeegrond zijn stinswieren opgericht. Van het ommeland van Sneek, het gebied binnen een straal van 12 à 15 km rondom de stad

(Halbertsma, 1963: p. 8), is dit slechts een gedeelte. Wij zullen ons echter beperken tot dit gebied, ruwweg tot aan de noordelijke oever van de Middelzee. De bedoeling is om in dit gebied een eventuele relatie van de inpolderingen met de stinswieren na te gaan.

3. PLATTELANDSELITE EN GRONDBEZIT

3.1. Grootte van de adellijke erven

De uitbreiding van grond door de inpolderingen in het Sneker ommeland, brengt ons bij de kwestie van de grondeigendom: wie bezit hier grond en wie krijgt de nieuw gewonnen grond. De meeste grond is in de late middeleeuwen eigendom van particulieren, waarbij vooral aan de adel moet worden gedacht. Verschillende kloosterorden hebben eveneens veel grondbezit, in het ommeland van Sneek ongeveer 30%.[4]

In de 14e en 15e eeuw kunnen we in de omgeving van Sneek verschillende geslachten aanwijzen die hier uitgestrekt grondbezit en dus invloed hebben, zoals de Bonninga's, de Harinxma's en de Bockema's. Over hun aandeel in de inpoldering en eventuele verwerving van land in de 13e eeuw wordt in de bronnen niets meegedeeld. De meeste adellijke personen zullen zelf boeren zijn geweest, hetgeen in 1511 nog van Tzaling Hottingen te Nijland wordt vermeld.[5]

Over de hoeveel land die gemiddeld behoorde bij de adellijke boerenbedrijven in het ommeland van Sneek in de late middeleeuwen hebben we uit de periode zelf geen gegevens. Aangezien de grote uitbreiding van macht en landbezit van de adel in Friesland zich vooral in de 15e eeuw afspeelt (Vries, 1986: pp. 64-71), krijgen we de indruk dat de adellijke erven voordien qua grootte misschien niet veel van die der pachters verschild hebben. Natuurlijk bewerkte de adel eigen grond en beschikte daarnaast over meer of minder in pacht uitbesteed land. Gegevens over grondbezit in het Sneker ommeland zijn er pas in 1511. Boerenbedrijven waren toen niet groot, gemiddeld 15 à 16 ha.[6] Misschien is dit nog een afspiegeling van de grootte van de laat-middeleeuwse erven.

3.2. Heerlijke rechten en stinswieren

Het verschijnen van stinswieren in het ommeland van Sneek is het gevolg van een machts- of positieverschuiving van de Friese adel in de politieke en maatschappelijke organisatie. Hierin wijkt het studiegebied niet af van de overige gebieden in Friesland met stinswieren. Gedurende de late middeleeuwen vervulden adellijke personen in Friesland bestuursfuncties op zowel lokaal als regionaal niveau. Doordat de adel haar macht uitbreidde, ondermijnde ze het landsheerlijk gezag. Deze ontwikkeling vond waarschijnlijk in de 12e en 13e eeuw plaats (Schuur, 1979: p. 145). De Friese adel realiseerde haar machtsuitbreiding door

heerlijke rechten te claimen op de van oudsher door de eigen familie beheerde grond en door verwerving van meer gezag in bestuursfuncties.

Door vermindering van landsheerlijk gezag zijn de spanningen tussen de edellieden vergroot. Dit blijkt uit het weerbare karakter van de stinswier. De gracht, heuvel en stenen toren kunnen in combinatie een defensief systeem vormen. Voorwaarden hierbij zijn wel dat de gracht voldoende breedte, en de heuvel voldoende hoogte bezit. Drie meter zou de minimum hoogte moeten zijn voor de heuvel om enig strategisch belang te vertegenwoordigen (Besteman, 1981).

4. EEN STINSWIER BIJ BONS

4.1. Inleiding

In de omgeving van Sneek zijn stinswieren bekend van Nijland, Tjalhuizum, Tirns, Goënga, Folsgare en Oppenhuizen, althans volgens de kaart van Schotanus uit 1718, volgens welke er bij Goënga zelfs twee moeten zijn geweest (fig. 3). In het landschap resteert echter nauwelijks iets van deze heuvels. In dit gebied zijn echter meer stinswieren geweest. Min of meer bij toeval werd in 1990 een tot dusver niet bekende stinswier ten westen van Sneek ontdekt. Deze stinswier bevond zich dicht bij de buurtschap Bons en dreigde in 1990, ten gevolge van uitbreiding van het industrieterrein De Hemmen, te verdwijnen (fig. 4). Derhalve werd besloten tot opgraving.[7]

4.2. De stinswier

Om een beeld te kunnen vormen van de heuvel die zich slechts als een lichte welving aftekende, werd een aantal profielsleuven gegraven. Aan de basis van het ophogingspakket tekende zich een gele kleilaag af. Merkwaardig was de verzakking die deze laag naar het centrum van de heuvel toe vertoonde. Buiten de heuvel strekte de gele kleilaag zich 4 à 5 m uit en was hier tamelijk dun. Op de plaats waar deze laag de heuvel bereikte was een neerwaartse knik zichtbaar. Deze knik duidt op een verzakking van de heuvel, veroorzaakt door het gewicht van de heuveltop en bekroning.

De opbouw van de wier, in drie lagen, was tamelijk uniform (fig. 5: A-B, C-D, E-F en G-H). De deklaag bestond uit een sterk gerijpte bouwvoor van ca. 50 cm. Daaronder rustte een grijsblauwe kleilaag op de gele kleilaag. De ondergrond van de wier is bestudeerd in het profiel aan de noordwestelijke zijde (profiel C-D). De heuvel rust op een massief grijsblauw kleipakket, dat Cardiumschelpen bevat, die op een natuurlijk marien sediment wijzen. Onder deze 1,25-1,50 m dikke grijsblauwe kleilaag bevindt zich een rietveenpakket.[8] Gezien de diepte van de veenlaag, ca. 3 m beneden maaiveldniveau, behoort de locatie bodemkundig bij het klei-op-veengebied, al is de locatie volgens de

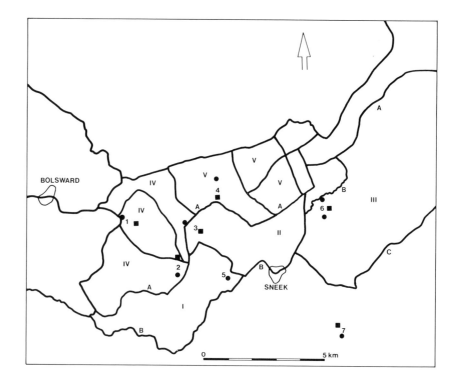

Fig. 3. Overzicht van polders en stinswieren rondom Sneek. A. Zuidelijk Middelzeedijk; B. Hemdijk; C. Groene Dijk; I. Scherwolder Hem; II. Scherhem; III. Hem van Raard; IV. Hem van Exmorra; V. Hem van Scharnegoutum. De stinswieren worden door een rondje en de dorpskernen door een vierkantje aangegeven: 1. Nijland; 2. Folsgare; 3. Tjalhuizum; 4. Tirns; 5. Bons; 6. Goënga; 7. Oppenhuizen.

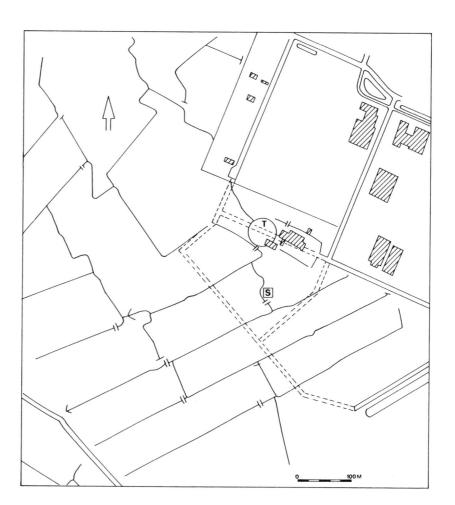

Fig. 4. Het industrieterrein De Hemmen, met de terp van Bons (T) en de stinswier (S).

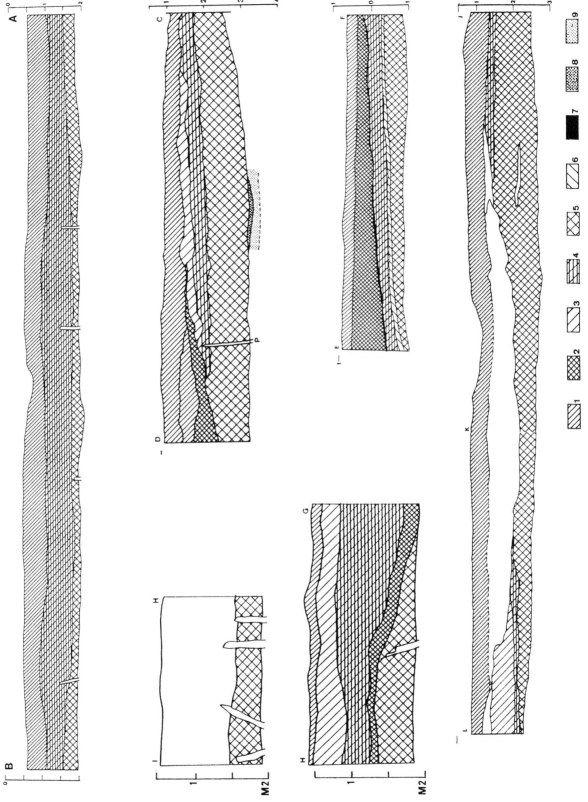

Fig. 5. De profielen van de stinswier van Bons. Legenda: 1. Bouwvoor; 2. Gele klei; 3. Grijsblauwe klei; 4. Blauwe klei; 5. Donkerblauwe klei; 6. Donkerblauwe klei met vlekken; 7. Zavel; 8. Venige klei; 9. Veen.

bodemkaart ten noorden daarvan gelegen (fig. 2).

De bepaling van de heuvelomtrek wordt bemoeilijkt doordat de ophogingslaag (de gele kleilaag) slechts een geleidelijke aanloop bezit. Veiligheidshalve laten we daarom de buitenrand van de heuvel beginnen met de toenemende dikte van de gele kleilaag, juist voor de knik. Aldus zal de heuvel aan de basis afmetingen van ca. 21 bij 18 m hebben, met een uitloper in profiel G-H, die op figuur 5 is aangegeven met een onderbroken lijn. Op maaiveldniveau zijn de afmetingen aan alle kanten ca. 75 à 100 cm minder.

Alles wat nog rest van de wier is een flauwe bult. Het bestaande ophogingspakket heeft een dikte van 1,40-1,50 m. Aanwijzingen voor de oorspronkelijke hoogte van de heuvel werden verkregen bij het graven van een sleuf van 9 bij 7 m in het centrum. Uitgangspunt voor het graven van deze sleuf was het zoeken naar een waterput, die echter niet werd aangetroffen. De gele kleilaag was hier ruim een meter dik. Aan de rand van de heuvel bedroeg de dikte van de gele kleilaag nooit meer dan 25-30 cm.

Uit de waarnemingen kan de oorspronkelijke hoogte van de wier niet zonder meer worden afgeleid. Als de andere ophogingslagen overeenkomstig dikker waren in het centrum, zal de heuvel zeker dan 3 m zijn geweest, en daarnaast nog plaats hebben geboden voor het bouwen van een toren. Zodoende kan de onderhavige heuvel een motte genoemd worden (Besteman, 1981: pp. 40-42).

4.3. Een palissade?

In sommige profielen tekenden zich paaltjes af, waarvan sommige waren aangepunt. Deze palen staken in de onderkant van de grijsblauwe kleilaag. Vlaksgewijs onderzoek van de voet van de wier aan noordoost- en zuidoostzijde gaf een goed beeld van de palenstructuur (fig. 7). Weliswaar is niet de hele structuur onderzocht, maar wel genoeg om vorm en afmetingen te bepalen. De palen tekenen zich af in een bijna vierkante rechthoek van 25,25 bij 24,75 m. Uit de plaatsing van de palen blijkt een zekere systematiek. Zo staan de twee rijen palen aan de binnenzijde in zigzagpatroon. De twee rijen palen aan de buitenzijde zijn paarsgewijs geplaatst. Slechts de zuidwestelijke zijde bezit nog een extra rij individueel geplaatste palen.

Vervolgens doet de vraag zich voor in welke relatie de palenrijen tot de heuvel staan. De palen zijn iets buiten het talud van de heuvel in de grond gedreven. Een functie als versteviging van de heuvelrand lijkt niet waarschijnlijk. Gelet op de zorgvuldige rangschikking van de palen schijnen ze eerder als een soort obstakel te zijn geplaatst. Het staketsel heeft mogelijk dus een defensieve functie gehad. De palen hebben een diameter tussen ca. 6 en 12 cm; de meeste zijn echter ca. 8 cm dik. Slechts weinig palen zijn recht in de grond geplaatst. De scheefheid van het merendeel der palen ondersteunt het denkbeeld van een hindernis. Voor eventuele aanvallers

was het niet zonder meer mogelijk dit obstakel te passeren. Aangezien elzestammetjes zijn gebruikt kan het staketsel bovengronds slechts 10 à 15 jaar houdbaar zijn geweest.[9] Hoever de palen boven de grond staken is niet meer na te gaan.

4.4. De gracht

De wier was omgeven door een gracht, die echter grotendeels gedempt was. In het verlengde van het ZW-NO-profiel (A-B) werd de gracht over de oorspronkelijke breedte aangetroffen. In het profiel tekende deze zich langgerekt, doch ondiep af (fig. 5: J-K en K-L). Aangezien de insteek nu in de bouwvoor ligt is de oorspronkelijke diepte van de gracht niet precies te achterhalen, maar deze moet tenminste 75 cm zijn geweest. De grond tussen de palissade en de gracht, en eveneens buiten de gracht, was ongestoord. Hierop duidde althans gley, ijzerneerslag, in de bodem (Veen, 1986: p. 26).

De vulling van de gracht bevatte verschillende vondsten. Op basis hiervan kan een tijdsbepaling worden verkregen van het moment dat de aarde in de gracht is gestort. Overigens bevatte de vulling nauwelijks humeuze resten. Dat kan betekenen, dat de gracht voornamelijk droog heeft gelegen.

De gracht ligt ca. 6 m buiten de palissade en is ca. 10 m breed. Ten oosten van het opgravingsterrein bevindt zich een laagte in het terrein. Misschien wijst deze op de aanwezigheid van de stinsgracht; boringen toonden eenzelfde grondopbouw aan als die van de gracht. Ten westen van de wier loopt nog een ronde, ca. 1,25 m brede sloot. Dit is een restant van de oude stinsgracht.

Door middel van boringen is getracht de bijbehorende boerderij op te sporen.[10] De boringen hebben sporen van woonlagen opgeleverd, grenzend aan de sloot, aan de westzijde van de wier.

4.5. De ceramiek

De opgraving van de wier heeft niet veel vondsten opgeleverd, en dan ook nog alleen losse vondsten; deze zijn van de stort geraapt of in de bouwvoor aangetroffen. Daarom moet voorzichtigheid worden betracht als geprobeerd wordt met dit materiaal een datering voor de stinswier te verkrijgen.

De oudste import-ceramiek zijn twee scherven van een baksel dat een overgangsvorm is tussen Pingsdorf- en proto-steengoed. Het gaat om gele hardbakken scherven die een grofkorrelig oppervlak hebben en uit de late 12e of vroege 13e eeuw dateren (Bruijn, 1960-1962: pp. 462-508). Het merendeel van de ceramiek zijn scherven van vroeg 13e-eeuws inheems blauwgrijs aardewerk, die veelal van borstelstrepen zijn voorzien (Renaud, 1976: p. 32). Op basis van de ceramiekvondsten kan aangenomen worden dat de heuvel in de periode 1150-1250 is opgeworpen.

De wier heeft ook jonger vondstmateriaal opgeleverd.

Uit de grachtvulling kwam 15e-eeuws gesinterd Rijnlands steengoed met ijzerengobe te voorschijn. Stellig is de gracht volgestort met grond van de heuvel, zoals bij de stinswier van Zweins is geconstateerd (Alders & Kramer, 1982: p. 68). De demping van de stinsgracht te Bons zal in de 15e eeuw hebben plaatsgevonden.

Niet echt talrijk, maar wel nadrukkelijk vertegenwoordigd was baksteenpuin. Fundamentsresten zijn niet aangetroffen, wel puinsporen in de bouwvoor: een waarschijnlijke aanwijzing voor een stenen bekroning van de wier. Het baksteenpuin is namelijk rond de gehele wier en ook in de grachtvulling aangetroffen.

4.6. Het einde van de stinswier

Als de gracht met aarde van de wier zou zijn gedempt, betekent dit dat de stinswier eveneens in de 15e eeuw, of daarvoor moet zijn gesloopt. Dan was deze stinswier dus slechts een kort bestaan beschoren. Door demping van de sloot werd het terrein geëgaliseerd, waarna het

een agrarische bestemming kreeg. Niet de gehele heuvel werd afgegraven, zoals uit de opgraving bleek. In het Register van den Aanbreng uit 1511 wordt een wier ten westen van Sneek vermeld.[11] Of het hier om de torenheuvel van Bons gaat is onzeker.

Waarom is men tot sloop van de stenen toren overgegaan? Waarschijnlijk hebben degenen die de stinswier hebben aangelegd geen rekening gehouden met verzakking van de vers opgeworpen kleilagen. Door de druk van het stenen bouwsel op deze slappe bult is het kleipakket weggedrukt met alle gevolgen van dien.

4.7. Stinswieren en grondbezit

Het beeld van de stinswier van Bons komt overeen met dat van elders in Friesland opgegraven stinswieren. Het gaat om een heuvel die in één keer werd opgeworpen en niet om een meerperioden-heuvel, zoals bijvoorbeeld de Zeeuwse vliedbergen (Vervloet, 1969: pp. 2-28). De wier werd bekroond door een bakstenen toren. Afwijkend in het geval van Bons is alleen de palissade. We

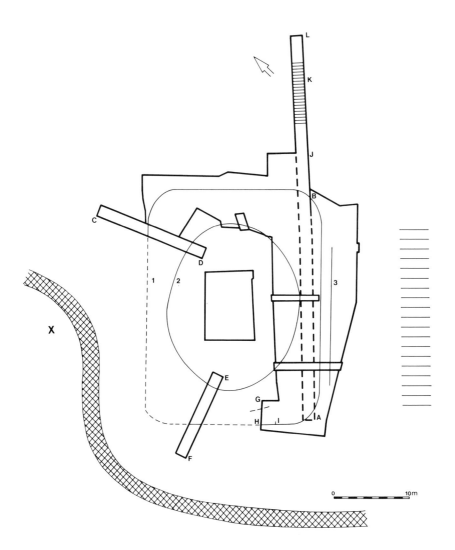

Fig. 6. Overzicht van de gegraven vlakken van de stinswier van Bons. De dikke lijn is de grens van de opgraving; lijn 1 de buitenomtrek van de palissade; lijn 2 de omtrek van de heuvel; lijn 3 de rij enkele palen. De arcering in de profielen J-K-L geeft de gracht aan die in dit profiel is vastgesteld. De kruisarcering, links beneden duidt een sloot aan. De streeparcering rechts geeft een laagte in het terrein aan.

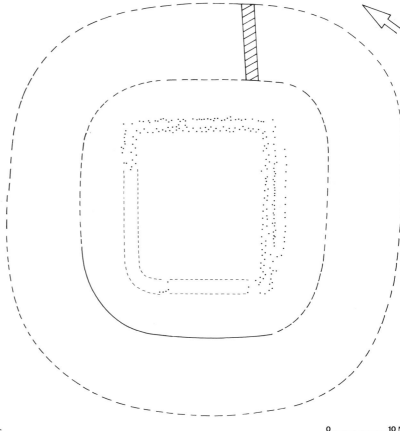

Fig. 7. De palissade van de stinswier van Bons.

0 ▬▬▬▬ 10 M

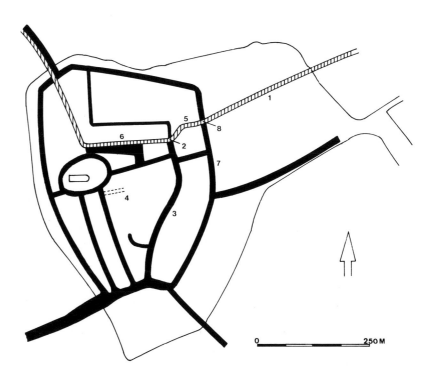

Fig. 8. Reconstructietekening van Sneek rond ca. 1300: 1. Hemdijk; 2. Neltjeszijl; 3. Grootzand; 4. Galigastraat; 5. Roedenburg; 6. Gruytersmastins; 7. Singel; 8. Potters-zijlen.

0 ▬▬▬▬ 250 M

330 A. Jager

gaan er vanuit dat de stinswieren in de omgeving van Sneek niet verschillen van het algemene type, zodat ze omstreeks 1150-1250 zullen zijn opgericht.

Het verspreidingspatroon van de stinswieren in het ommeland van Sneek is tamelijk willekeurig. Bovendien zijn er slechts weinig. Nu zullen zeker niet alle stinswieren bekend zijn, maar waarschijnlijk hebben niet alle adellijke boerderijen een kasteelheuveltje gehad. Er lijkt dus – naast een politiek-maatschappelijke factor – ook een geografische factor mee te spelen. Opvallend is dat de nog bekende stinswieren in dit gebied allemaal gelegen zijn bij een enkele boerderij en niet bij dorpen.[12] Bovendien liggen ze op grond die door inpoldering is verkregen.

Niet alle delen van de kweldergrond zijn in de 10e/11e eeuw geschikt geweest voor cultivering. Pas na de inpoldering in de 12e eeuw kunnen ook lager gelegen terreindelen in gebruik zijn genomen. Op die manier werden de landerijen gewonnen waarop onder andere de boerderij met de stinswier van Bons werd gesticht.[13] Volgens een ander patroon is de voormalige Middelzeegrond in bezit genomen. Door dichtslibbing kwam uitstekende landbouwgrond beschikbaar, waarop boerenbedrijven werden gesticht. Tegelijkertijd, of in een iets later stadium, werd het gewonnen gebied ingedijkt. De inpoldering, uiteraard onder leiding van de adel, moet een collectieve aangelegenheid van de boerengemeenschappen zijn geweest (Spahr van der Hoek, 1969: pp. 22-41). Het gewonnen gebied werd onder de deelnemers verdeeld, met uiteraard een ruimer aandeel voor de adel. In het proces van het zelfstandiger worden van de adel, vanaf ca. 1150, uitten claims op heerlijke rechten op de door inpolderingen gewonnen grond in het ommeland van Sneek zich door stinswieren.

5. ELITE EN STINSEN IN SNEEK

5.1. De ontwikkeling van Sneek tot 1300

In de 11e eeuw moet Sneek een buurtschap of dorp zijn geweest, qua organisatie en voerende economie niet verschillend van de overige nederzettingen in het haar omringende platteland. Vanaf de late 12e eeuw wordt Sneek een centrale plaats voor goederenuitwisseling door een surplusproductie van de agrarische bedrijven in haar ommeland. Op dezelfde manier ontwikkelt zich ook het nabijgelegen IJlst.

Dat Sneek zich sneller tot stad ontwikkelt dan IJlst dankt ze aan een overslaghaven. Deze ontstond in de vroege 13e eeuw, nadat in de late 12e eeuw de Hemdijk was opgeworpen. Sneek ligt op een kruispunt van waterwegen, die afgesloten worden door deze dijk. Om de afwatering te regelen werden in de dijk sluizen aangelegd. Twee daarvan, de Neltjeszijl en de Potterzijlen, blokkeerden respectievelijk het Grootzand en het Singel en alle verkeer hierover, zodat goederen

moesten worden overgeladen (de Cock, 1984; Vernooij, 1987: pp. 2-4). Als gevolg hiervan bloeide handelsactiviteit op rond de Galigastraat, en werd de centrumfunctie van Sneek voor de nederzettingen in het ommeland versterkt (fig. 8).

Het zal de adel in het ommeland niet onwelgevallig zijn geweest dat Sneek zich ontwikkelde als plaats voor goederenuitwisseling en diensten. Hierbij zochten de Snekers bovendien naar een evenwicht met het platteland. Bijzondere rechten die ze proberen te verwerven, zullen vooral op de bevordering van haar economie betrekking hebben gehad. Ze hielden zich wijselijk buiten de bestuurlijke verwikkelingen op het platteland. Een goede relatie tussen de Snekers en de adel in haar ommeland is weliswaar pas aantoonbaar tegen het einde van de middeleeuwen, maar er is geen reden om aan te nemen dat die verhouding in de 13e/14e eeuw anders is geweest.

5.2. Stadsstinsen

In de Friese steden kwamen, als gezegd, geen stinswieren voor. Wel waren hier de stadsstinsen, die veel gelijkenis moeten hebben gehad met de woontorens op het plat-

Fig. 9. De Ilostins te IJlst op de vogelvluchtkaart van N. van Geelkerken uit 1616.

teland. De stadsstins had eveneens een torenvorm, maar geen heuvel, en als woning een huis zonder agrarische bestemming in plaats van een boerderij.

In het urbaniserende Sneek vestigde zich een groeiend aantal ambachtslieden en neringdoenden. De toplaag hiervan: het patriciaat, vormde een elite die mede met de adel het bestuur van de plaats verzorgde. Tot in de late middeleeuwen stond echter een edelman aan het hoofd van het Sneker bestuur. De gemêleerdere elite in Sneek kwam tot uitdrukking in verschillende soorten huizen. De burgers hadden in de 13e en 14e eeuw nog overwegend houten huizen. Alleen de belangrijkste burgers lieten in de 13e en 14e eeuw een stenen huis bouwen. De burgerhuizen waren afgestemd op ambachts- en/of handelsfuncties. Pas in de 15e eeuw werd in Sneek baksteen algemeen in de huisbouw toegepast. Alleen de machtigste edelen bezaten in de stad een weerbaar steenhuis: een stins.

Een goed voorbeeld van een Friese stadsstins is de Ilostins in IJlst. Dit waarschijnlijk in de 14e eeuw gestichte steenhuis van de familie Harinxma lag ten westen van de Wijddraai en beheerste aldus het waterverkeer door IJlst. De Ilostins is in 1710 afgebroken, maar is ons bekend van een afbeelding (fig. 9). Het gaat om een woontoren op een vierkant grondplan, 3 bouwlagen hoog, afgesloten door een zadeldak, met rond de dakrand een weergang met arkeltorens.

5.3. De stinsen in Sneek

In Sneek hebben ooit drie stinsen gestaan: de Roedenburg, de Gruytersmastins en de Johansmastins. Hierbij gaat het om in stedelijke context geïncorporeerde weerbare steenhuizen. Die weerbaarheid kunnen we voor de Johansmastins niet hard maken, omdat hiervan te weinig bekend is. De twee andere steenhuizen nemen beide binnen het urbanisatieproces van Sneek een unieke plaats in.

De Roedenburg was gelegen op de zuidoosthoek van de Nauwe en Wijde Burgstraat, waarvan de namen nog aan de stins herinneren. De sterkte is gesitueerd aan de noordzijde van het Grootzand (fig. 8). Het huis is in 1377 in handen van Rienck Bockema, maar moet al voordien gebouwd zijn. De edelman moet rond 1350 geboren zijn en verblijft van 1377-1385, 1387-1390 en 1391-1396 op de Roedenburg (Napjes, 1772: p. 14). Jammer genoeg wordt in de bronnen over dit huis niets naders meegedeeld. Bockema had voornamelijk grondbezit op het platteland ten noorden van de stad; later kwam dit ten dele in handen van het klooster Thabor (Steensma, 1970: pp. 12-15). Aangezien de Roedenburg nabij Bockema's grond op het platteland

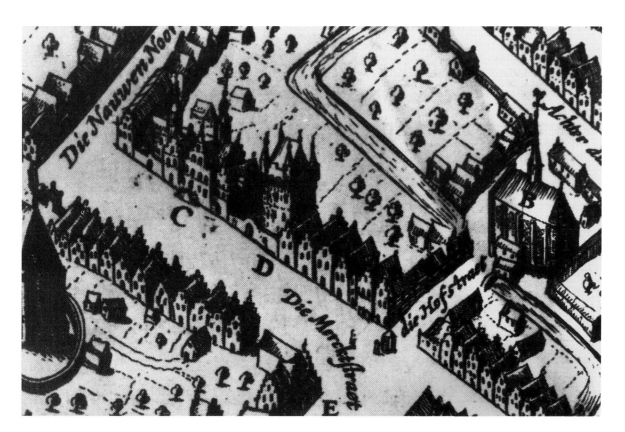

Fig. 10. De Gruytersmastins te Sneek op de vogelvluchtkaart van N. van Geelkerken uit 1616.

was gelegen, zal de edelman een aangrenzend deel van het Sneker grondgebied hebben bezeten. Hierop heeft hij, of een eerdere telg van zijn familie het steenhuis laten oprichten. Met de Roedenburg moet de edelman zijn vermeende heerlijke rechten op zijn grond in de stad hebben willen benadrukken. Hij deed dit door een weerbaar steenhuis op een voor de stad gevoelig punt te bouwen. Met de stins bezat hij de mogelijkheid de overslaghaven en het verkeer op het Grootzand en het Singel te controleren. Ondanks de stedelijke context is de Roedenburg dus vergelijkbaar met de stinswieren in het ommeland van Sneek.

Over het uiterlijk van deze stadsstins weten we niets. Dat geldt ook voor de Johansmastins. Dit steenhuis is alleen bekend van een vermelding in een oud-Friese oorkonde uit 1442. In de oorkonde gaat de edelman Bocka Harinxma een wisselkoop aan met het St. Jans-Hospitaal. Aangezien er van een stins sprake is, en duidelijk onderscheid wordt gemaakt tussen 'huys' en 'stins', zal het om een woontoren gaan (Sipma, 1927: pp. 54-55).

Toch werden de woontorens in Sneek niet altijd als 'stins' aangeduid. In 1399 wordt de Gruytersmastins als 'husinge' aangeduid (Halbertsma & Keikes, 1956: p. 14). Hierbij moet echter worden bedacht dat de betreffende oorkonde door een niet-Fries is opgesteld, en dat deze met het verschil huis-stins waarschijnlijk niet op de hoogte is geweest. Van slechts één der Sneker stinsen kunnen we het uiterlijk controleren, namelijk de Gruytersmastins, waarvan een afbeelding bestaat (fig. 10). De vergelijking met de Ilostins te IJlst is treffend. We gaan er van uit dat zowel de Roedenburg als de Johansmastins soortgelijke woontorens zijn geweest.

6. DE GRUYTERSMASTINS

6.1. Inleiding

Vragen naar bouwdata, grootte en constructie van de Sneker stinsen kunnen met behulp van historische bronnen en topografie niet worden beantwoord. Slechts de archeologie kan ons op deze punten helpen. Dat één van de drie stinsen in Sneek archeologisch is onderzocht, namelijk de Gruytersmastins, lijkt dan ook gerechtvaardigd.

In 1984 en 1985 vonden achter het gemeentehuis te Sneek, aan de Marktstraat, opgravingen plaats.[14] Vooral dankzij de vogelvluchtkaart van Nicolaas van Geelkerken uit 1616, is bekend dat de Gruytersmastins hier heeft gestaan. Enkele historische gegevens van deze stins vullen onze kennis aan.

6.2. Uiterlijk en opzet van de stins

Op de kaart van Van Geelkerken is de Gruytersmastins de enige woontoren van de stad. De stins staat op een bijna vierkant grondplan, bezit drie woonlagen, een zadeldak, en rond de dakrand een weergang met arkeltorens. De ingangspartij is naar de Marktstraat gericht. De Gruytersmastins is aanzienlijk forser weergegeven dan de omringende modale huizen. Bij de beoordeling van deze illustratie is echter voorzichtigheid geboden. Er is wel gedacht dat de Gruytersmastins al in 1611, of kort nadien, zou zijn gesloopt (Ten Hoeve, 1985: p. 40). Toen Van Geelkerken – in 1616 – zijn plattegrond van Sneek vervaardigde bestond de Gruytersmastins dus mogelijk niet meer. Vreemd genoeg komt de stenen toren op vogelvluchtkaarten uit 1649 en 1664 van Sneek ook nog voor. Bovengrondse resten van de Gruytersmastins waren zeker in 1632 verdwenen, aangezien in dat jaar het erf van de stins

Fig. 11. Het Cleyn Stins van de Gruytersmastins. Naar een tekening van J. Stellingwerf uit 1723.

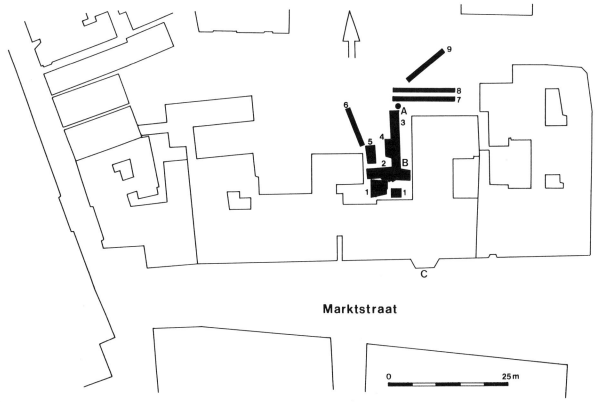

Marktstraat

0   25 m

Fig. 12. Overzicht van het opgravingsterrein achter de Marktstraat ter plaatse van de voormalige Gruytersmastins. A-B geeft de getekende zijde van het profiel uit sleuf 3 aan. C is de erkerachtige uitbouw van het Cleyn Stins. Het rondje tussen de sleuven 3 en 7 is een boom.

werd opgesplitst en verkocht (Ten Hoeve, 1988: p. 15).

Het bijzondere karakter van de Gruytersmastins ten opzichte van de modale huizen in Sneek blijkt, afgezien van het uiterlijk, ook uit de opzet. Het steenhuis wijkt terug van de rooilijn en bestrijkt meerdere kavels, terwijl de modale huizen in Sneek op een enkele kavel staan. Verder was de stins omgracht (Napjes, 1772: p. 72). Tenslotte was aan de linkerzijde van de toren een zijvleugel toegevoegd, het zogenaamde 'Cleyn Stins', het eigenlijke woondeel, waardoor een plein ontstond (fig. 11). Het Cleyn Stins is behouden gebleven en is thans opgenomen in het stadhuiscomplex (Elward & Karstkarel, 1990: p. 156). Dankzij de erkerachtige uitbouw, is het nog in de straat te herkennen (fig. 12).

6.3. De opgravingen

In 1984 werd op de locatie van de Gruytersmastins een vijftal sleuven gegraven (fig. 12: 1-5). In sleuf 1 werd een aantal ophogingslagen van klei vastgesteld. In de onderste laag werden enkele scherven van laat-Pingsdorf en proto-steengoed, daterend uit de late 12e en vroege 13e eeuw aangetroffen. Met de ophoging van de klei-lagen moeten de opgravers de noordflank van de Hem-dijk hebben aangeroerd. Verder werd in deze sleuf een 4,70 m diepe waterput opgegraven die oorspronkelijk

geheel gemetseld moet zijn geweest. Hieruit kwamen 15e-, 16e- en 17e-eeuwse ceramiekscherven. Op de put sloot een afvoer aan (fig. 13). In sleuf 2 vond men een tonput en een vierkante beerput met houten wanden. Uit de tonput kwam een halsfragment van een midden-14e-eeuws Siegburg-kruikje te voorschijn, dat met een fraai wapenapplique was versierd. Verder werd nog een 14e-eeuwse ijzeren kruisboogpijlpunt geborgen.

In sleuf 3 zijn fundamentsresten van de Gruy-tersmastins aangetroffen. Verder is in deze sleuf de bodemopbouw ter plaatse gedocumenteerd, met behulp waarvan de fundamentsresten gedateerd kunnen wor-den. De fundamentsresten, bestaande uit rode roos-winkels (28×13×7 cm), heeft men over een lengte van ca. 6,6 m kunnen volgen. Aan de noordwestzijde, in sleuf 4, tekende zich nog een enkele moppenstructuur af (fig. 13). Opvallend aan het fundament was het onre-gelmatige hoogteniveau, dat in het noorden op 55 cm +NAP lag en zuidwaarts tot 97 cm +NAP steeg. De westwaarts daarvan liggende rij bakstenen bezat in noordzuidwaartse richting eenzelfde hoogteverloop, namelijk van 75-80 tot 105 cm +NAP. De baksteen-structuren moeten gezamenlijk deel uitgemaakt hebben van de westzijde van het stinsfundament. Het betreft hier in beide gevallen een enkele steenlaag, dus de onderste funderingslaag van de stins, welke ca. 1,50-

1,60 m breed moet zijn geweest. Nabij de fundering werd een riooltje van gemetselde kloostermoppen aangetroffen, dat ook weer in noordzuidelijke richting een oplopend niveau vertoonde, van 66-70 tot 110 cm +NAP.

Van de oostwand van sleuf 3 is een profiel getekend (fig. 14). Het uitgediepte vlak van sleuf 3 raakte een wal van schone klei met de vorm van een op z'n kop staande afgeronde V. Onder dit kleirugje bevond zich een opgebrachte laag, waaruit bleek dat dit rugje niet de oudste menselijke aktiviteit ter plaatse vertegenwoordigde. Uit beide lagen werden echter geen dateerbare vondsten gedaan. Aan weerszijden is de kleirug ongeveer tot aan de top aangevuld met bruine, venige grond. Op de min of meer vlakke laag, direct boven het hoogste niveau van de kleirug, werd een zwarte brandlaag aangetroffen. Hieruit zijn Rijnlandse gesinterde grijze steengoedscherven uit de 14e eeuw geborgen (Janssen, 1983: pp. 191-192). Noordwaarts aansluitend op de brandlaag werd een kleilaag met veel bouwpuin aangetroffen, bestaande uit kloostermop- en mortelresten. Boven de brandlaag zette deze puinmantel zich voort. Dit puin is ongetwijfeld afkomstig van de Gruytersmastins. Onder de fundamentsresten in sleuf 3 kwam men de zwarte laag nogmaals tegen, waaruit weer een ongeglazuurde grijze gesinterde steengoedscherf uit het Rijnland te voorschijn kwam. Ditmaal werden ook 13e-eeuwse inheemse blauwgrijze scherven met borstelstreken gevonden. De laatste in de campagne van 1984 gegraven sleuf, nummer 5, leverde niets op.

Gezien het feit dat de woontoren aan de rechterzijde van het Cleyn Stins heeft gestaan (fig. 10 en 11), moeten de opgravers een deel van het westelijk fundament van de toren hebben opgegraven. Belangrijk is de brandlaag onder de funderingen en in het profiel, die mogelijk getuigt van de brand die Sneek in 1294 heeft geteisterd (Napjes, 1772: p. 12). Op de brandlaag is een pakket ter afdekking aangebracht. In de afdeklaag liggen de

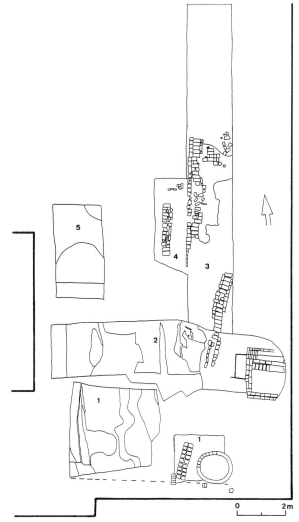

Fig. 13. Vlakgraving van de sleuven 1 t/m 5 van de opgraving van de Gruytermastins uit 1984. Tekening A. Jager naar een opmeting van D.M. Visser.

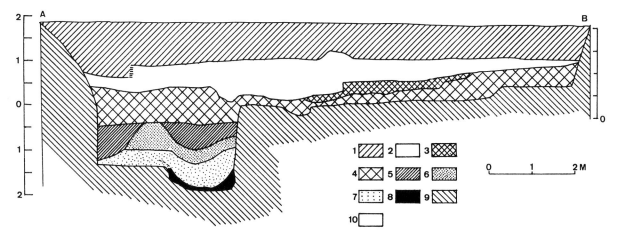

Fig. 14. Profieltekening van de oostzijde van de in 1984 gegraven sleuf 3 op het terrein van de Gruytersmastins. Tekening A. Jager naar een opmeting van D.M. Visser. Legenda: 1. Verstoorde bovengrond; 2. Laag met baksteenpuin en mortel; 3. Opgebrachte gele klei; 4. Zwarte humeuze grond met wat puin; 5. Bruine humeuze grond; 6. Grijsblauwe klei; 7. Zwarte humeuze laag; 8. Donkergrijze klei. De noordzijde van deze sleuf werd begrensd door een boom, vandaar de geronde beëindiging, over de wortels van de boom, van het profiel aan deze zijde.

stinsfunderingen. Deze afdeklaag kan relatief snel na de brand zijn opgebracht, zodat de aanleg van de stins in de eerste helft van de 14e eeuw kan hebben plaatsgevonden.

Met de opgravingsresultaten kan niet worden uitgemaakt of de Gruytersmastins al dan niet een defensieve functie heeft bezeten. Er kan slechts verondersteld worden dat een aanzienlijk persoon een huis heeft laten bouwen op een plek, die vrij is gekomen doordat de brand van 1294 de mogelijk aanwezige bebouwing heeft opgeruimd. Dankzij de opgraving is de reden van afbraak van de stins duidelijk. Uit het opgegraven fundament, met een zeer onregelmatig hoogteverloop, blijkt dat de stins moet zijn verzakt en daardoor in verval is geraakt.

Tijdens de campagne van 1985 is een viertal sleuven (6-9) gegraven, die echter geen gegevens met betrekking tot de Gruytersmastins hebben opgeleverd. In alle negen sleuven is bouwpuin van de Gruytersmastins aangetroffen. Afgezien van sleuven 3 en 4 is hiervan niets in situ. Alle gegevens van de opgravingen op een rijtje gezet, kunnen we slechts concluderen dat een grondige afbraak heeft plaatsgevonden.

6.4. De functie van de Gruytersmastins

Het huis wordt voor het eerst in 1399 vermeld en is dan eigendom van Aylof die Gruyter (Ten Hoeve, 1988: p. 13). Misschien was deze Aylof een koopman. Hij was afkomstig uit Zwolle en trouwde met een adellijke Friezin, zodat we kunnen veronderstellen dat hij een zekere status genoot.

Uit de gebeurtenissen in Sneek rond 1399-1400 blijkt dat de Gruytersmastins een defensieve functie heeft bezeten. In 1399 schaart Aylof de Gruyter zich, evenals Rienck Bockema, aan de zijde van Albrecht van Beieren, de toenmalige graaf van Holland. De graaf tracht in deze periode zijn macht in Friesland uit te breiden en bekleedt daartoe verschillende aanzienlijke personen, zoals Aylof en Rienck met bestuurlijke ambten. Aylof is onder bijzondere bescherming van de graaf geplaatst. In het document dat hiervan rept wordt met een zekere nadruk melding gemaakt van diens 'husinge'. Klaarblijkelijk was niet alleen de persoon Gruytersma van belang, maar tevens diens weerbare huis: "want hi nu ter tijt siin husinge beset heeft binnen der Sneeck mit Sicke sinen zwagher ende anders sinen vrienden up siins selfs cost, dat sy noch houden tot onser behoif jegens onsen vyanden" (Halbertsma, 1968).

De invloed van de graaf van Holland houdt niet lang stand. Al in 1400 trekt hij zich terug uit Friesland en behoudt alleen Staveren. De meeste van zijn aanhangers worden verdreven (Vries, 1986: p. 32). In dat jaar wordt de Roedenburg vernietigd door de burgers van Sneek, maar blijft de Gruytersmastins gespaard. Waarschijnlijk heeft de Gruytersmastins, vanwege haar ligging aan de Marktstraat een mercantiele functie verkregen. Daarnaast was de eigenaar Gruytersma geen edelman,

en oefende deze geen druk uit op de stad met claims op heerlijke rechten, zoals Rienck Bockema met zijn Roedenburg. Tot het einde van de 15e eeuw kan de Gruytersmastins haar defensieve functie hebben behouden. Pas na de omgrachting van Sneek, kort voor 1500, ging deze functie verloren.

8. NOTEN

1. Het verschil in cultiveringspatroon tussen het gebied ten zuiden van Sneek, oftewel de Wolden, en het gebied ten noorden van Sneek, laat zich afleiden uit het verkavelingssysteem. Het veengebied kent lange, maar weinig brede, bijna noord-zuid-verlopende kavels, terwijl blokverkaveling van onregelmatige rechthoeken in het kleigebied overheerst. Verder kenmerken de nederzettingen in het kleigebied zich door permanente bewoning sinds de late middeleeuwen, maar kent het veengebied opschuivende bewoning.

2. Bij een opgraving van de terp van Bons in mei en juni 1990 door de archeologische afdeling van het Fries Museum, is de terp-grootte vastgesteld en zijn een aantal putten gevonden. Op basis van deze gegevens kan een buurtschap van dit formaat worden verondersteld.

3. Over de datering van de Groene Dijk spreken de deskundigen op het gebied van de Friese dijkbouw (Rienks & Walther, 1954: p. 339) zich niet uit. Het lijkt echter waarschijnlijk, gezien het ontstaan van het gehele poldersysteem in dit gebied, dat de Groene Dijk eveneens in de 12e eeuw is ontstaan.

4. De oudste gegevens over het grondbezit van de kloosters in het ommeland van Sneek zijn die uit het Register van den Aanbreng uit 1511:

Oppenhuizen:	7%
Uitwellingerga:	9%
Nijland:	19%
Folsgare:	1%
IJsbrechtum:	57%
Tjalhuizum:	25%
Tirns:	75%
Scharnegoutum:	84%
Goënga:	8%
Gauw:	13%
Loënga:	45%
Offingawier:	25%

5. Van Tzaling Hottingen wordt in het Register van den Aanbreng uit 1511, deel 2, p. 327 gezegd: "eijgen erffd in dessen landen en bruijct die sulven".

6. In het algemeen waren de boerenbedrijven rondom Sneek in 1511, gezien de gegevens van het Register van den Aanbreng niet groot:

Oppenhuizen:	60.04 p. = 16.678 ha
Uitwellingerga:	54.85 p. = 15.236 ha
Nijland:	57.95 p. = 16.097 ha
Folsgare:	48.77 p. = 13.547 ha
IJsbrechtum:	58.06 p. = 16.128 ha
Tjalhuizum:	50.38 p. = 13.877 ha
Tirns:	64.24 p. = 17.844 ha
Scharnegoutum:	62.46 p. = 17.350 ha
Goënga:	58.52 p. = 16.256 ha
Gauw:	49.54 p. = 13.761 ha
Loënga:	56.22 p. = 15.617 ha
Offingawier:	45.58 p. = 12.661 ha

7. Dankzij medewerking van de gemeente Sneek kon het bedreigde object worden onderzocht. De wetenschappelijke leiding berustte bij dr. J.M. Bos en ondergetekende voerde de dagelijkse leiding. Van de archeologische afdeling van het Fries Museum werd door J.K. Boschker en D.M. Visser assistentie verleend. Bij deze ook dank aan A. Bosma uit Sijbrandaburen en O. de Jong uit Beers, die als vrijwilligers meewerkten. Er is gegraven van 27-

31 augustus en 3-8 september. De coördinaten van de op-
gravingslocatie zijn 171.30/560.34 (kaartblad 10 H).
8. Volgens dr. J.M. Bos hebben we hier te maken met een rietveen,
    dat voor ruim 80% uit water kan bestaan.
9. Met dank aan dr. W.A. Casparie van het B.A.I. voor de hout-
    determinatie.
10. Met dank aan J.B. de Voogd uit Groningen voor assistentie bij het
    boorwerk.
11. Register van den Aanbreng, 1511, deel 2, p. 368.
12. Hierbij gaan we ervan uit dat het aantal dorpen in dit gebied sinds
    de late middeleeuwen onveranderd is gebleven. Zodoende is de
    kaart van Schotanus uit 1718, waar ons overzicht der stinswieren
    op gebaseerd is, wat dit betreft een afspiegeling van de laat-
    middeleeuwse toestand.
13. Het maaiveld ter hoogte van de stinswier ligt gemiddeld op 1,3
    m -NAP, terwijl het maaiveld van de landerijen ten noorden
    hiervan gemiddeld een halve meter hoger ligt.
14. De opgraving van de Gruytersmastins vond plaats in de periode
    van 22 okt. tot 12 nov. 1984 en van 4 juli tot 16 sept. 1985, De
    opgraving werd namens het B.A.I. uitgevoerd door de archeo-
    logische afdeling van het Fries Museum. De wetenschappelijke
    leiding berustte bij E. Kramer, de dagelijkse bij D.M. Visser.
    Voorgraver was J.K. Boschker. De coördinaten van de
    opgravingslocatie zijn 173.20/560.68 (kaartblad 10 H).

9. LITERATUUR

ALDERS, G.P. & E. KRAMER, 1982. Onderzoek naar de resten van
   een middeleeuwse 'stinswier' onder Zweins. *De Vrije Fries* 62,
   pp. 65-79.
BESTEMAN, J.C., 1981. Mottes in the Netherlands: a provisional
   survey and inventory. In: T.J. Hoekstra, H.L. Janssen & I.W.L.
   Moerman (eds), *Liber Castellorum*. Zutphen, pp. 40-60.
BRUIJN, A., 1960-1962. Die mittelalterliche keramische Industrie in
   Schinveld. *Berichten van de Rijksdienst voor het Oudheidkundig
   Bodemonderzoek* 12-13, pp. 356-459.
CNOSSEN, J., 1971. *De bodem van Friesland. Toelichting bij blad
   2 van de bodemkaart van Nederland. Schaal 1:200.000. Stichting
   voor Bodemkartering Wageningen*. Meppel.
COCK, J.K. DE, 1984. De veenontginningen rond Sneek en IJlst. *It
   Beaken* 46, pp. 139-149.
ELWARD, R. & P. KARSTKAREL, 1990. *Stinsen en States. Adellijk
   wonen in Friesland*. Meppel.
GOTTSCHALK, M.K.E., 1971. *Stormvloeden en rivier-
   overstromingen in Nederland. Deel 1, de periode voor 1400*.
   Assen.
HALBERTSMA, H., 1963. *Terpen tussen Vlie en Eems. Een geo-
   grafisch-historische benadering*. Groningen.
HALBERTSMA, H., 1968. Leeuwarden. *Bull. K.N.O.B. Archeolo-
   gisch Nieuws* 67, pp. *72-*73.
HALBERTSMA, H. & W.H. KEIKES, 1956. *Sneek. Drie kronen met
   ere*. Sneek.
HOEVE, S. TEN, 1985. Heerlijk wonen. In: J. Greidanus, S. ten

Hoeve & P. Karstkarel, *Sneek, beeld van een stad: vele eeuwen
   stadsleven in woord en beeld*. Drachten, pp. 40-43.
HOEVE, S. TEN, 1988. Het Sneker stadhuis. Meer dan vijf eeuwen
   bouwen. In: *Stadhuis*, pp. 12-16.
JANSSEN, H.L., 1983. Het middeleeuwse aardewerk: ca. 1200- ca.
   1550. In: *Van Bos tot stad*. 's-Hertogenbosch, pp. 188-222.
JANSSEN, H.L., 1990. The archaeology of the medieval castle in the
   Netherlands. Results and prospects for future research. In: J.C.
   Besteman, J.M. Bos & H.A. Heidinga (eds), *Medieval archaeology
   in the Netherlands. Festschrift H.H. Regteren Altena*.
   Assen etc., pp. 219-264.
KRAMER, E., 1988. Onderzoek naar stinswieren in Friesland en
   Groningen. In: M. Bierma, A.T. Clason, E. Kramer & G.J. de
   Langen (red.), *Terpen en wierden in het Fries-Groningse kust-
   gebied*. Groningen, pp. 214-225.
MEIJER, M.W., 1988. De stinsen in Leeuwarden. In: H.M. van den
   Berg et al. (red.), *De stenen Droom*. Zutphen, pp. 161-169.
MOLEN, S.J. VAN DER, 1971. Het Camminghahuis te Franeker en
   zijn geschiedenis. *De Vrije Fries* 51, pp. 33-46.
NAPJES, E., 1772. *Historisch Chronyck of Beschrijvinge van oud en
   nieuw Sneek*. Sneek. Facsimile uitgave 1969.
Register van den Aanbreng van 1511 en verdere stukken tot de
   floreenbelasting betrekkelijk. Friesch Genootschap 1879. Deel 2.
   Leeuwarden.
RENAUD, J.G.N., 1976. *Middeleeuwse ceramiek. Enige hoofdlijnen
   uit de ontwikkeling in Nederland* (= Westerheem monografie no
   3).
RIENKS, K.A. & G.L. WALTHER, 1954. *Binnendiken en slieper-
   diken yn Fryslan*. 2 dln. Leeuwarden.
SCHOTANUS À STERRINGA, B., 1718. *Uitbeelding der
   Heerlijkheit Friesland*. Leeuwarden.
SCHUUR, J.R.G., 1979. *Leeuwarden voor 1435. Een poging tot
   reconstructie van de oudste stadsgeschiedenis*. Zutphen.
SIPMA, P., 1927. *Oudfriese oorkonen, deel I*. 's-Gravenhage.
SPAHR VAN DER HOEK, J.J., 1969. *Samenleven in Friesland*.
   Drachten.
SPAHR VAN DER HOEK, J.J., 1974. Hoe 't lan der hinnelei. In: G.
   Bakker (red.), *Wymbritseradiel*. Boalsert, pp. 9-28.
STEENSMA, R., 1970. *Het klooster Thabor bij Sneek en zijn nage-
   laten geschriften*. Leeuwarden.
Stiboka, 1974; *Bodemkaart van Nederland. Schaal 1: 50.000. Toe-
   lichting bij kaartblad 10 West Sneek blad 10 Oost Sneek*,
   Wageningen.
TEMMINCK GROLL, C.L., 1963. *Middeleeuwse stenen huizen te
   Utrecht en hun relatie met die van andere Noordwesteuropese
   steden*. 's-Gravenhage.
VEEN, A.W.L., 1986. *Het interactie-stelsel bodem/water* (= Rijks-
   universiteit Groningen, Vakgroep Fysische Geografie en
   Bodemkunde. Rapport nr. 21). Groningen.
VERNOOIJ, A.L., 1987. Sneek. *Gemeente Sneek: toelichting bij het
   besluit tot de aanwijzing van Sneek als beschermd stadsgezicht.
   Beschermde stads- en dorpsgezichten*. 's-Gravenhage.
VERVLOET, J.A.J., 1969. *De vliedbergen in het kustgebied van
   Vlaanderen, Zeeland en Zuid-Holland*. Amsterdam, 2e druk.
VRIES, O., 1986. *Het Heilige Roomse Rijk en de Friese Vrijheid*.
   Leeuwarden.